T0311648

MODERN METHODS OF ORGANIC SYNTHESIS

The fourth edition of this well-known textbook discusses the key methods used in organic synthesis, showing the value and scope of these methods and how they are used in the synthesis of complex molecules. All the text from the third edition has been revised, to produce a modern account of traditional methods and an up-to-date description of recent advancements in synthetic chemistry. The textbook maintains a traditional and logical approach in detailing carbon–carbon bond formations, followed by a new chapter on the functionalization of alkenes and concluding with oxidation and reduction reactions. Reference style has been improved to include footnotes, allowing easy and rapid access to the primary literature. In addition, a selection of problems has been added at the end of each chapter, with answers at the end of the book. The book will be of significant interest to chemistry and biochemistry students at advanced undergraduate and graduate level, as well as to researchers in academia and industry who wish to familiarize themselves with modern synthetic methods.

BILL CARRUTHERS was born in Glasgow. He won a bursary to Glasgow University, where he graduated with a first-class honours degree in 1946 and a Ph.D. in 1949. He moved to Exeter in 1956, working first for the Medical Research Council and then, from 1968, as a lecturer then senior lecturer at the Department of Chemistry in the University of Exeter. He died in April 1990, just a few months before he was due to retire.

IAIN COLDHAM was born in Sandbach, Cheshire. He graduated from the University of Cambridge with a first-class honours degree in 1986 and a Ph.D. in 1989. After postdoctoral studies at the University of Texas, Austin, he moved in 1991 to the University of Exeter as a lecturer then senior lecturer. He is currently Reader at the Department of Chemistry in the University of Sheffield and specializes in organic synthesis.

MODERN METHODS OF
ORGANIC SYNTHESIS

W. CARRUTHERS
Formerly of the University of Exeter

IAIN COLDHAM
University of Sheffield

CAMBRIDGE
UNIVERSITY PRESS

CAMBRIDGE UNIVERSITY PRESS
Cambridge, New York, Melbourne, Madrid, Cape Town, Singapore,
São Paulo, Delhi, Dubai, Tokyo

Cambridge University Press
The Edinburgh Building, Cambridge CB2 8RU, UK

Published in the United States of America by Cambridge University Press, New York

www.cambridge.org
Information on this title: www.cambridge.org/9780521778305

First, second and third editions © Cambridge University Press 1971, 1978, 1987
fourth edition © W. Carruthers and I. Coldham 2004

First published 1971
Second edition 1978
Reprinted 1985
Third edition 1986
Reprinted 1987 (with corrections), 1988, 1989, 1990, 1992, 1993, 1996, 1998
Fourth edition 2004
Fifth printing with corrections 2007

A catalogue record for this publication is available from the British Library

Library of Congress Cataloguing in Publication data
Carruthers, W.
Some modern methods of organic synthesis – 4th ed. / W. Carruthers, Iain Coldham.
p. cm.
Includes bibliographical references and index.
ISBN 0 521 77097 1 (hardback) ISBN 0 521 77830 1 (paperback)
1. Organic compounds – Synthesis. I. Coldham, Iain, 1965– II. Title.
QD262.C33 2004
547′.2 – dc22 2004049222

ISBN 978-0-521-77097-2 Hardback
ISBN 978-0-521-77830-5 Paperback

Transferred to digital printing 2010

Contents

Preface to the first edition

This book is addressed principally to advanced undergraduates and to graduates at the beginning of their research careers, and aims to bring to their notice some of the reactions used in modern organic syntheses. Clearly, the whole field of synthesis could not be covered in a book of this size, even in a cursory manner, and a selection has had to be made. This has been governed largely by consideration of the usefulness of the reactions, their versatility and, in some cases, their selectivity.

A large part of the book is concerned with reactions which lead to the formation of carbon–carbon single and double bonds. Some of the reactions discussed, such as the alkylation of ketones and the Diels–Alder reaction, are well established reactions whose scope and usefulness has increased with advancing knowledge. Others, such as those involving phosphorus ylids, organoboranes and new organometallic reagents derived from copper, nickel, and aluminium, have only recently been introduced and add powerfully to the resources available to the synthetic chemist. Other reactions discussed provide methods for the functionalisation of unactivated methyl and methylene groups through intramolecular attack by free radicals at unactivated carbon–hydrogen bonds. The final chapters of the book are concerned with the modification of functional groups by oxidation and reduction, and emphasise the scope and limitations of modern methods, particularly with regard to their selectivity.

Discussion of the various topics is not exhaustive. My object has been to bring out the salient features of each reaction rather than to provide a comprehensive account. In general, reaction mechanisms are not discussed except in so far as is necessary for an understanding of the course or stereochemistry of a reaction. In line with the general policy in the series references have been kept to a minimum. Relevant reviews are noted but, for the most part, references to the original literature are given only for points of outstanding interest and for very recent work. Particular reference is made here to the excellent book by H. O. House, *Modern Synthetic*

Reactions which has been my guide at several points and on which I have tried to build, I feel all too inadequately.

I am indebted to my friend and colleague, Dr K. Schofield, for much helpful comment and careful advice which has greatly assisted me in writing the book.

26 October 1970

Preface to the fourth edition

Some Modern Methods of Organic Synthesis was originally written by Dr W. (Bill) Carruthers, and three popular editions were published that have helped many students of advanced organic chemistry. Unfortunately, Dr Carruthers died in 1990, just prior to his retirement. As his successor at the University of Exeter, it was appropriate that I should take on the task of preparing the fourth edition of this text. In honour of Dr Carruthers, a similar format to previous editions has been taken, although of course the book has been completely re-written and brought up-to-date (through 2003) to take account of the many advances in the subject since the third edition was published. As in previous editions, the text begins with descriptions of some of the most important methods for the formation of carbon–carbon bonds, including the use of enolates and organometallic compounds for carbon–carbon single-bond formation (Chapter 1), methods for carbon–carbon double-bond formation (Chapter 2), pericyclic reactions (Chapter 3), radicals and carbenes (Chapter 4). There has been some re-organization of material and emphasis has been placed on reactions that are useful, high yielding or selective for organic synthesis. For example, Chapter 1 has been expanded to include some of the most popular and contemporary reactions using main-group and transition-metal chemistry (rather than placing reactions of organoboron and silicon compounds into a separate chapter). A new chapter describing the functionalization of alkenes has been devised, covering reactions such as hydroboration, epoxidation and dihydroxylation (Chapter 5). The book concludes with examples of pertinent oxidation (Chapter 6) and reduction (Chapter 7) reactions that are used widely in organic synthesis. The opportunity has been taken to add some problems at the end of each chapter, with answers at the end of the book. References have been compiled as footnotes on each relevant page for ease of use.

In common with the previous editions, the book is addressed principally to advanced undergraduates and to graduates at the beginning of their research careers. My aim has been to bring out the salient features of the reactions and reagents rather than to provide a comprehensive account. Reaction mechanisms are not normally discussed, except where necessary for an understanding of the course or stereochemistry of a reaction. My hope is that the book will find widespread use as a helpful learning and reference aid for synthetic chemists, and that it will be a fitting legacy to Dr Carruthers.

The majority of the text was written at the University of Exeter before my move to the University of Sheffield and I would like to acknowledge the encouragement and help of the staff at Exeter.

Part of one chapter was written while I was a Visiting Professor at the University of Miami, and I am grateful to Professor Bob Gawley for hosting my visit. My thanks extend to various people who have proof-read parts of the text, including Chris Moody, Mike Shipman, Mark Wood, Alison Franklin, Joe Harrity, Steve Pih and Ben Dobson. Finally, I would like to thank my family for their patience during the writing of this book.

I. Coldham
January 2004

1

Formation of carbon–carbon single bonds

The formation of carbon–carbon single bonds is of fundamental importance in organic synthesis. As a result, there is an ever-growing number of methods available for carbon–carbon bond formation. Many of the most useful procedures involve the addition of organometallic species or enolates to electrophiles, as in the Grignard reaction, the aldol reaction, the Michael reaction, alkylation reactions and coupling reactions. Significant advances in both main-group and transition-metal-mediated carbon–carbon bond-forming reactions have been made over the past decade. Such reactions, which have been finding useful application, are discussed in this chapter. The formation of carbon–carbon single bonds by pericyclic or radical reactions are discussed in chapters 3 and 4.

1.1 Main-group chemistry

1.1.1 Alkylation of enolates and enamines

It is well known that carbonyl groups increase the acidity of the proton(s) adjacent (α-) to the carbonyl group. Table 1.1 shows the pK_a values for some unsaturated compounds and for some common solvents and reagents.

The acidity of the C−H bonds in these compounds is caused by a combination of the inductive electron-withdrawing effect of the unsaturated groups and the resonance stabilization of the anion formed by removal of a proton (1.1). Not all groups are equally effective in 'activating' a neighbouring CH; nitro is the most powerful of the common groups, with the series following the approximate order $NO_2 > COR > SO_2R > CO_2R > CN > C_6H_5$. Two activating groups reinforce each other; for example, diethyl malonate has a lower pK_a (≈ 13) than ethyl acetate ($pK_a \approx 24$). Acidity is increased slightly by electronegative substituents

Table 1.1. *Approximate acidities of some activated
compounds and common reagents*

Compound	pKa	Compound	pKa
CH_3CO_2H	5	$C_6H_5COCH_3$	19
$CH_2(CN)CO_2Et$	9	CH_3COCH_3	20
$CH_2(COCH_3)_2$	9	CH_3CO_2Et	24
CH_3NO_2	10	CH_3CN	25
$CH_3COCH_2CO_2Et$	11	$((CH_3)_3Si)_2NH$	26
$CH_2(CO_2Et)_2$	13	$CH_3SO_2CH_3$	31
CH_3OH	16	CH_3SOCH_3	35
$(CH_3)_3COH$	19	$((CH_3)_2CH)_2NH$	36

(e.g. sulfide) and decreased by alkyl groups.

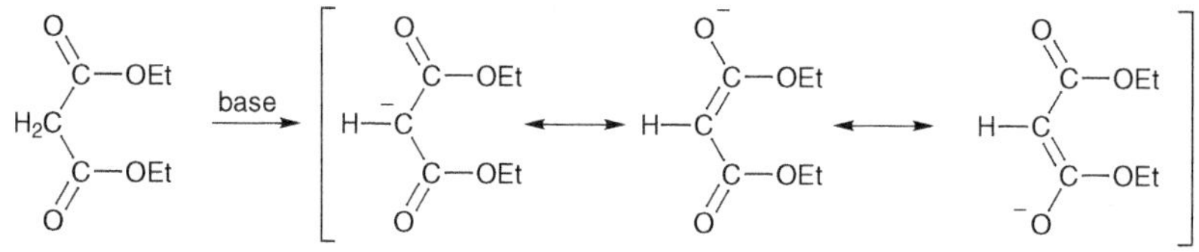

(1.1)

By far the most important activating group in synthesis is the carbonyl group. Removal of a proton from the α-carbon atom of a carbonyl compound with base gives the corresponding enolate anion. It is these enolate anions that are involved in many reactions of carbonyl compounds, such as the aldol condensation, and in bimolecular nucleophilic displacements (alkylations, as depicted in Scheme 1.2).

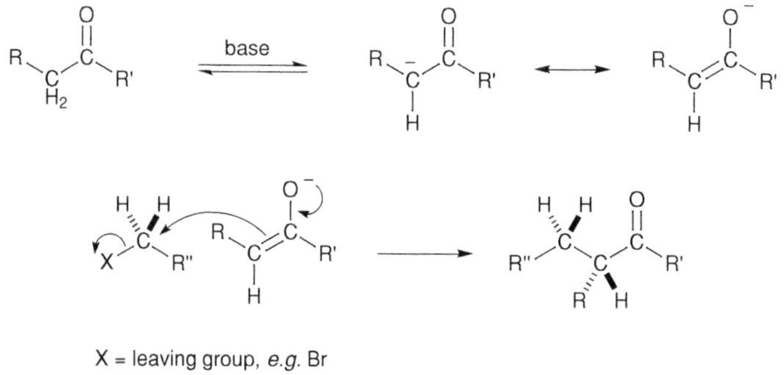

(1.2)

X = leaving group, *e.g.* Br

Enolate anions should be distinguished from enols, which are always present in equilibrium with the carbonyl compound (1.3). Most monoketones and esters contain only small amounts of enol (<1%) at equilibrium, but with 1,2- and 1,3-dicarbonyl compounds much higher amounts of enol (>50%) may be present. In the presence of a protic acid, ketones may be converted largely into the enol form, implicated in many acid-catalysed reactions of carbonyl compounds.

$$R-\underset{H_2}{C}-\overset{O}{\underset{}{C}}-R' \quad \rightleftharpoons \quad R-\underset{H}{C}=\underset{}{\overset{OH}{C}}-R' \qquad (1.3)$$

Table 1.1 illustrates the relatively high acidity of compounds in which a C—H bond is activated by two or more carbonyl (or cyano) groups. It is therefore possible to use a comparatively weak base, such as a solution of sodium ethoxide in ethanol, in order to form the required enolate anion. An equilibrium is set up, as illustrated in Scheme 1.4, in which the conjugate acid of the base (BH) must be a weaker acid than the active methylene compound. Another procedure for preparing the enolate of an active methylene compound is to use sodium hydride (or finely divided sodium or potassium metal) in tetrahydrofuran (THF), diethyl ether (Et$_2$O) or benzene. The metal salt of the enolate is formed irreversibly with evolution of hydrogen gas. β-Diketones can often be converted into their enolates with alkali-metal hydroxides or carbonates in aqueous alcohol or acetone.

$$CH_2(CO_2Et)_2 \quad + \quad B^- \quad \rightleftharpoons \quad {}^-CH(CO_2Et)_2 \quad + \quad BH \qquad (1.4)$$

Much faster alkylation of enolate anions can often be achieved in dimethylformamide (DMF), dimethylsulfoxide (DMSO) or 1,2-dimethoxyethane (DME) than in the usual protic solvents. The presence of hexamethylphosphoramide (HMPA) or a triamine or tetramine can also enhance the rate of alkylation. This is thought to be because of the fact that these solvents or additives solvate the cation, but not the enolate, thereby separating the cation–enolate ion pair. This leaves a relatively free enolate ion, which would be expected to be a more reactive nucleophile than the ion pair.[1] Reactions with aqueous alkali as base are often improved in the presence of a phase-transfer catalyst such as a tetra-alkylammonium salt.[2]

Alkylation of enolate anions is achieved readily with alkyl halides or other alkylating agents.[3] Both primary and secondary alkyl, allyl or benzyl halides may be used successfully, but with tertiary halides poor yields of alkylated product often result because of competing elimination. It is sometimes advantageous to proceed by way of the toluene-*p*-sulfonate, methanesulfonate or trifluoromethane-sulfonate rather than a halide. The sulfonates are excellent alkylating agents and can usually be obtained from the alcohol in a pure condition more readily than

[1] H. E. Zaugg, D. A. Dunnigan, R. J. Michaels, L. R. Swett, T. S. Wang, A. H. Sommers and R. W. DeNet, *J. Org. Chem.*, **26** (1961), 644; A. J. Parker, *Quart. Rev. Chem. Soc. Lond.*, **16** (1962), 163; M. Goto, K. Akimoto, K. Aoki, M. Shindo and K. Koga, *Tetrahedron Lett.*, **40** (1999), 8129.

[2] M. Makosza and A. Jonczyk, *Org. Synth.*, **55** (1976), 91.

[3] D. Caine, in *Comprehensive Organic Synthesis*, ed. B. M. Trost and I. Fleming, vol. 3 (Oxford: Pergamon Press, 1991), p. 1.

the corresponding halides. Primary and secondary alcohols can be used as alkylating agents under Mitsunobu conditions.[4] Epoxides have also been used, generally
reacting at the less substituted carbon atom. Attack of the enolate anion on the
alkylating agent takes place by an S_N2 pathway and thus results in inversion of
configuration at the carbon atom of the alkylating agent (1.5).[5]

$$
\begin{array}{cc}
\underset{\text{CO}_2\text{Et}}{\overset{\text{OSO}_2\text{Me}}{\diagup}} & + \quad \text{CH}_2(\text{CO}_2\text{Et})_2 \quad \xrightarrow[68\%]{\text{CsF}} \quad \underset{\text{CO}_2\text{Et}}{\overset{\text{CH(CO}_2\text{Et})_2}{\diagup}}
\end{array}
\qquad (1.5)
$$

With secondary and tertiary allylic halides or sulfonates, reaction of an enolate
anion may give mixtures of products formed by competing attack at the α- and
γ-positions (1.6). Addition of the enolate anion to a π-allylpalladium complex
provides an alternative method for allylation (see Section 1.2.4).

$$
\underset{\text{Cl}}{\diagup} \quad \xrightarrow[\text{NaOEt, EtOH}]{\text{CH}_2(\text{CO}_2\text{Et})_2} \quad \underset{\text{CH(CO}_2\text{Et})_2}{\diagup} \quad + \quad \underset{\text{CH(CO}_2\text{Et})_2}{\diagup}
\qquad (1.6)
$$

A difficulty sometimes encountered in the alkylation of active methylene compounds is the formation of unwanted dialkylated products. During the alkylation
of the sodium salt of diethylmalonate, the monoalkyl derivative formed initially
is in equilibrium with its anion. In ethanol solution, dialkylation does not take
place to any appreciable extent because ethanol is sufficiently acidic to reduce the
concentration of the anion of the alkyl derivative, but not that of the more acidic
diethylmalonate itself, to a very low value. However, replacement of ethanol by an
inert solvent favours dialkylation. Dialkylation also becomes a more serious problem with the more acidic cyanoacetic esters and in alkylations with very reactive
electrophiles such as allyl or benzyl halides or sulfonates.

Dialkylation may, of course, be effected deliberately if required by carrying out
two successive operations, by using either the same or a different alkylating agent
in the two steps. Alkylation of dihalides provides a useful route to three- to seven-
membered ring compounds (1.7). Non-cyclic products are formed at the same time
by competing intermolecular reactions and conditions have to be chosen carefully
to suppress their formation (for example, by using high dilution).

$$
\text{Br}\underset{n}{\diagup}\text{Br} \quad + \quad \text{CH}_2(\text{CO}_2\text{Et})_2 \quad \xrightarrow[\text{EtOH}]{\text{NaOEt}} \quad \underset{\text{CO}_2\text{Et}}{\overset{\text{CO}_2\text{Et}}{\diagup}}
\qquad (1.7)
$$

$$
n = 0\text{--}4
$$

[4] O. Mitsunobu, *Synthesis* (1981), 1; J. Yu, J.-Y. Lai and J. R. Falck, *Synlett* (1995), 1127; T. Tsunoda, C. Nagino,
   M. Oguri and S. Itô, *Tetrahedron Lett.*, **37** (1996), 2459.
[5] T. Sato and J. Otera, *J. Org. Chem.*, **60** (1995), 2627.

Under ordinary conditions, aryl or alkenyl halides do not react with enolate anions, although reaction can occur with aryl halides bearing strongly electro-negative substituents in the *ortho* and *para* positions. 2,4-Dinitrochlorobenzene, for example, with ethyl cyanoacetate gives ethyl (2,4-dinitrophenyl)cyanoacetate (90%) by an addition–elimination pathway. Unactivated aryl halides may react with enolates under more vigorous conditions, particularly sodium amide in liquid ammonia. Under these conditions, the reaction of bromobenzene with diethyl-malonate, for example, takes place by an elimination–addition sequence in which benzyne is an intermediate (1.8).

Enolate anions with extended conjugation can be formed by proton abstraction of α,β-unsaturated carbonyl compounds (1.9). Kinetically controlled alkylation of the delocalized anion takes place at the α-carbon atom to give the β,γ-unsaturated compound directly. A similar course is followed in the kinetically controlled pro-tonation of such anions.

A wasteful side reaction which sometimes occurs in the alkylation of 1,3-dicarbonyl compounds is the formation of the *O*-alkylated product. For example, reaction of the sodium salt of cyclohexan-1,3-dione with butyl bromide gives the *O*-alkylated product (37%) and only 15% of the *C*-alkylated 2-butylcyclohexan-1,3-dione. In general, however, *O*-alkylation competes significantly with *C*-alkylation only with reactive methylene compounds in which the equilibrium concentration of enol is relatively high (as in 1,3-dicarbonyl compounds). The extent of *C*- versus *O*-alkylation for a particular 1,3-dicarbonyl compound depends on the choice of cation, solvent and electrophile. Cations (such as Li^+) that are more covalently bound to the enolate oxygen atom or soft electrophiles (such as alkyl halides) favour *C*-alkylation, whereas cations such as K^+ or hard electrophiles (such as alkyl sulfonates) favour *O*-alkylation.

Alkylation of malonic esters and other active methylene compounds is useful in synthesis because the alkylated products can be subjected to hydrolysis and decarboxylation (1.10). Direct decarboxylation under neutral conditions with an alkali metal salt (e.g. lithium chloride) in a dipolar aprotic solvent (e.g. DMF) is a popular alternative method.[6]

$$CH_2(CO_2Et)_2 \xrightarrow[\text{R–X}]{\text{NaOEt, EtOH}} RCH(CO_2Et)_2 \xrightarrow[\text{ii, H}_3O^+,\text{ heat}]{\text{i, NaOH}} RCH_2CO_2H \qquad (1.10)$$

$$RCH(CO_2Et)_2 \xrightarrow[\text{DMF}]{\text{LiCl}} RCH_2CO_2Et$$

Proton abstraction from a monofunctional carbonyl compound (aldehyde, ketone, ester, etc.) is more difficult than that from a 1,3-dicarbonyl compound. Table 1.1 illustrates that a methyl or methylene group which is activated by only one carbonyl or cyano group requires a stronger base than ethoxide or methoxide ion to convert it to the enolate anion in high enough concentration to be useful for subsequent alkylation. Alkali-metal salts of tertiary alcohols, such as tert-butanol, in the corresponding alcohol or an inert solvent, have been used with success, but suffer from the disadvantage that they are not sufficiently basic to convert the ketone completely into the enolate anion. This therefore allows the possibility of an aldol reaction between the anion and unchanged carbonyl compound. An alternative procedure is to use a much stronger base that will convert the compound completely into the anion. Traditional bases of this type are sodium and potassium amide or sodium hydride, in solvents such as diethyl ether, benzene, DME or DMF. The alkali-metal amides are often used in solution in liquid ammonia. Although these bases can convert ketones essentially quantitatively into their enolate anions, aldol reaction may again be a difficulty with these bases because of the insolubility of the reagents. Formation of the anion takes place only slowly in the heterogeneous reaction medium and both the ketone and the enolate ion are present at some stage. This difficulty does not arise with the lithium dialkylamides, such as lithium diisopropylamide (LDA) or lithium 2,2,6,6-tetramethylpiperidide (LTMP) or the alkali-metal salts of bis(trimethylsilyl)amine (LHMDS, NaHMDS and KHMDS), which are soluble in non-polar solvents. These bases are now the most commonly used reagents for the generation of enolates.

An example illustrating the intermolecular alkylation of an ester is given in Scheme 1.11. Intramolecular alkylations also take place readily in appropriate cases and reactions of this kind have been used widely in the synthesis of cyclic compounds. In such cases, the electrophilic centre generally approaches the enolate

[6] A. P. Krapcho, *Synthesis* (1982), 805; 893.

from the less-hindered side and in a direction orthogonal to the plane of the enolate anion.

$$
\begin{array}{ccc}
\text{CO}_2\text{Me} & \xrightarrow[\substack{\text{THF, } -78\,°\text{C} \\ \text{ii, } \text{Br}}]{\text{i, LiN}^i\text{Pr}_2\text{ (LDA)}} & \text{CO}_2\text{Me} & \xrightarrow[\substack{\text{THF, } -78\,°\text{C} \\ \text{ii, EtBr}}]{\text{i, LDA}} & \text{CO}_2\text{Me} \quad (1.11) \\
& & & 90\%
\end{array}
$$

A common problem in the direct alkylation of ketones is the formation of di- and polyalkylated products. This difficulty can be avoided to some extent by adding a solution of the enolate in a polar co-ordinating solvent such as DME to a large excess of the alkylating agent. The enolate may therefore be consumed rapidly before equilibration with the alkylated ketone can take place. Nevertheless, formation of polysubstituted products is a serious problem in the direct alkylation of ketones and often results in decreased yields of the desired monoalkyl compound. An explanation for the presence of considerable amounts of polyalkylated product(s) is that enolates of alkylated ketones are less highly aggregated in solution and hence more reactive.[7] Some solutions to this problem use the additive dimethylzinc[8] or the manganese enolate of the ketone.[9] Good yields of the monoalkylated products have been obtained under these conditions (1.12).

$$
\begin{array}{ccccc}
\text{O} & \xrightarrow[\text{MnCl}_2\text{ or MnBr}_2]{\text{LDA or LHMDS}} & \text{OMnX} & \xrightarrow[76\%]{\text{MeI}} & \text{O} \quad \text{Me} \qquad (1.12)
\end{array}
$$

Alkylation of symmetrical ketones or of ketones that can enolize in one direction only can, of course, give just one mono-*C*-alkylated product. With unsymmetrical ketones, however, two different monoalkylated products may be formed by way of the two structurally isomeric enolate anions. If one of the isomeric enolate anions is stabilized by conjugation with another group, such as cyano, nitro or a carbonyl group, then only this stabilized anion is formed and alkylation takes place at the position activated by both groups. Even a phenyl or an alkenyl group provide sufficient stabilization of the resulting anion to direct substitution into the adjacent

[7] A. Streitwieser, Y. J. Kim, and D. Z. R. Wang, *Org. Lett.*, **3** (2001), 2599.
[8] Y. Morita, M. Suzuki and R. Noyori, *J. Org. Chem.*, **54** (1989), 1785.
[9] M. T. Reetz and H. Haning, *Tetrahedron Lett.*, **34** (1993), 7395; G. Cahiez, B. Figadère and P. Cléry, *Tetrahedron Lett.*, **35** (1994), 3065; G. Cahiez, K. Chau and P. Cléry, *Tetrahedron Lett.*, **35** (1994), 3069; G. Cahiez, F. Chau and B. Blanchot, *Org. Synth.*, **76** (1999), 239.

position (1.13).[10]

$$Ph \overset{O}{\underset{}{\bigwedge}} \quad \xrightarrow[\text{MeI}]{\text{KOH, Bu}_4\text{NBr}} \quad Ph \overset{O}{\underset{Me}{\bigwedge}} \quad + \quad Ph \overset{O}{\underset{}{\bigwedge}} Me \qquad (1.13)$$

95%   100 : 0

Sometimes, specific lithium enolates of unsymmetrical carbonyl compounds are formed because of chelation of the lithium atom with a suitably placed substituent. For example, lithiation and alkylation of the mixed ester **1** took place α- to the MEM ester group, presumably as a result of intramolecular chelation of the lithium atom with the ethereal oxygen atom (1.14).[11]

$$\text{MEMO}_2\text{C} \diagdown\diagup \diagdown \text{CO}_2\text{Me} \quad \xrightarrow[\text{ii, PhCH}_2\text{Cl}]{\text{i, LDA, THF}} \quad \text{MEMO}_2\text{C} \diagdown\diagup \diagdown \text{CO}_2\text{Me} \qquad (1.14)$$

1   57%

MEM = CH₃OCH₂CH₂OCH₂–

Alkylation of unsymmetrical ketones bearing α-alkyl substituents generally leads to mixtures containing both α-alkylated products. The relative amount of the two products depends on the structure of the ketone and may also be influenced by experimental factors, such as the nature of the cation and the solvent (see Table 1.2). In the presence of the ketone or a protic solvent, equilibration of the two enolate anions can take place. Therefore, if the enolate is prepared by slow addition of the base to the ketone, or if an excess of the ketone remains after the addition of base is complete, the equilibrium mixture of enolate anions is obtained, containing predominantly the more-substituted enolate. Slow addition of the ketone to an excess of a strong base in an aprotic solvent, on the other hand, leads to the kinetic mixture of enolates; under these conditions the ketone is converted completely into the anion and equilibration does not occur.

The composition of mixtures of enolates formed under kinetic conditions differs from that of mixtures formed under equilibrium conditions. The more-acidic, often less-hindered, α-proton is removed more rapidly by the base (e.g. LDA), resulting in the less-substituted enolate under kinetic conditions. Under thermodynamic conditions, the more-substituted enolate normally predominates. Mixtures of both structurally isomeric enolates are generally obtained and mixtures of products result on alkylation. Di- and trialkylated products may also be formed and it is not always

[10] A. Aranda, A. Díaz, E. Díez-Barra, A. de la Hoz, A. Moreno and P. Sánchez-Verdú, *J. Chem. Soc., Perkin Trans. 1* (1992), 2427.
[11] M. T. Cox, D. W. Heaton and J. Horbury, *J. Chem. Soc., Chem. Commun.* (1980), 799.

Table 1.2. *Composition of enolate anions generated from the ketone and a base*

Ketone	Base (conditions)	Enolate anion composition (%)	

	Base (conditions)		
	LDA, DME, −78 °C (kinetic control)	1	99
	Ph$_3$CLi, DME, −78 °C (kinetic control)	9	91
	Ph$_3$CLi, DME (equilibrium control)	90	10
	t-BuOK, t-BuOH (equilibrium control)	93	7

	LDA, THF, −78 °C (kinetic control)	0	100
	Ph$_3$CLi, DME (equilibrium control)	87	13

easy to isolate the pure monoalkylated compound. This is a serious problem in synthesis as it results in the loss of valuable starting materials.

A number of methods have been used to improve selectivity in the alkylation of unsymmetrical ketones and to reduce the amount of polyalkylation. One procedure is to introduce temporarily an activating group at one of the α-positions to stabilize the corresponding enolate anion; this group is removed after the alkylation. Common activating groups used for this purpose are ester groups. For example, 2-methylcyclohexanone can be prepared from cyclohexanone as shown in Scheme 1.15. The 2-ethoxycarbonyl derivative is obtained from the ketone by reaction with diethyl carbonate (or by reaction with diethyl oxalate followed by decarbonylation). Conversion to the enolate anion with a base such as sodium ethoxide takes place exclusively at the doubly activated position. Methylation with iodomethane and removal of the β-ketoester group with acid gives 2-methylcyclohexanone, free from polyalkylated products.

(1.15)

Another technique is to block one of the α-positions by introduction of a remov-able substituent which *prevents* formation of the corresponding enolate. Selective alkylation can be performed after acylation with ethyl formate and transformation of the resulting formyl (or hydroxymethylene) substituent into a group that is sta-ble to base, such as an enamine, an enol ether or an enol thioether. An example of this procedure is shown in Scheme 1.16, in the preparation of 9-methyl-1-decalone from *trans*-1-decalone. Direct alkylation of this compound gives mainly the 2-alkyl derivative, whereas blocking the 2-position allows the formation of the required 9-alkyl-1-decalone (as a mixture of *cis* and *trans* isomers).

(1.16)

Alkylation of a 1,3-dicarbonyl compound at a 'flanking' methyl or methylene group instead of at the doubly activated C-2 position does not usually take place to any significant extent. It can be accomplished selectively and in good yield, however, by way of the corresponding *dianion*, itself prepared from the dicarbonyl compound and two equivalents of a suitable strong base. For example, 2,4-pentanedione **2** is converted into 2,4-nonanedione by reaction at the more-reactive, less-resonance-stabilized carbanion (1.17).[12]

(1.17)

With unsymmetrical dicarbonyl compounds that could give rise to two different dianions, it is found that in most cases only one is formed and a single product results on alkylation. Thus, with 2,4-hexanedione alkylation at the methyl group greatly predominates over that at the methylene group, and 2-acetylcyclohexanone and 2-acetylcyclopentanone are both alkylated exclusively at the methyl group. In general, the ease of alkylation follows the order $C_6H_5CH_2 > CH_3 > CH_2$.

[12] T. M. Harris and C. M. Harris, *Org. Reactions*, **17** (1969), 155.

Dianion formation can be applied equally well to β-keto esters and provides a useful route to 'mixed' Claisen ester products. The dianions are conveniently pre-pared by reaction with two equivalents of LDA (or one equivalent of sodium hydride followed by one equivalent of butyllithium) and give γ-alkylated products in high yield with a wide range of alkylating agents.[13] This chemistry has been used in the synthesis of a number of natural products. The reaction is used twice in the synthesis of the lactone (±)-diplodialide A **6** (1.18); once to alkylate the dianion generated from ethyl acetoacetate with the bromide **3** and once to introduce the double bond by reaction of the dianion from the β-keto lactone **4** with phenylselenyl bromide to give the selenide **5**. Elimination by way of the selenoxide (see Section 2.2) led to diplodialide A.[14]

$$(1.18)$$

The application of dianion chemistry in synthesis is not confined to γ-alkylation of β-dicarbonyl compounds. Dianions derived from β-keto sulfoxides can be alky-lated at the γ-carbon atom. Nitroalkanes can be deprotonated twice in the α-position to give dianions **7**. In contrast to the monoanions, the dianions **7** give C-alkylated products in good yield (1.19).[15]

$$(1.19)$$

Some solutions to the problem of the formation of a specific enolate from an unsymmetrical ketone were discussed above. Another solution makes use of the structurally specific enol acetates or enol silanes (silyl enol ethers). Treatment of a trimethylsilyl enol ether with one equivalent of methyllithium affords the corre-sponding lithium enolate (along with inert tetramethylsilane). Equilibration of the

[13] S. N. Huckin and L. Weiler, *J. Am. Chem. Soc.*, **96** (1974), 1082.
[14] T. Ishida and K. Wada, *J. Chem. Soc., Perkin Trans. 1* (1979), 323.
[15] D. Seebach, R. Henning, F. Lehr and J. Gonnermann, *Tetrahedron Lett.* (1977), 1161.

enolate does not take place, as long as care is taken to ensure the absence of proton
donors, such as an alcohol or an excess of the ketone. Reaction with an alkyl halide
then gives, predominantly, a specific monoalkylated ketone. It is rarely possible
to obtain completely selective alkylation, because as soon as some monoalkylated
ketone is formed in the reaction mixture it can bring about equilibration of the
original enolate. This difficulty is minimized by using the covalent lithium enolate,
which gives a relatively stabilized enolate whilst maintaining a reasonable rate of
alkylation.

One of the drawbacks of this procedure is that methyllithium is incompatible
with a variety of functional groups. In addition, the lithium enolate may not be sufficiently reactive for alkylation. A solution to these problems has been found in the
use of benzyltrimethylammonium fluoride to generate the enolate anion. The fluoride ion serves well to cleave silyl enol ethers and the ammonium enolates produced
are more reactive than the lithium analogues. Even relatively unreactive alkylating agents such as 1-iodobutane give reasonable yields of specifically alkylated
products.[16]

The success of this approach to specific enolates is dependent on the availability
of the regioisomerically pure silyl enol ethers. The more highly substituted silyl
ethers usually predominate in the mixture produced by reaction of the enolates,
prepared under equilibrium conditions, with trimethylsilyl chloride (1.20).[17] In
some cases this mixture may be purified by distillation or by chromatography. The
less highly substituted silyl ethers are obtained from the enolate prepared from the
ketone under kinetic conditions with lithium diisopropylamide (LDA).

$$(1.20)$$

[16] I. Kuwajima, E. Nakamura and M. Shimizu, *J. Am. Chem. Soc.*, **104** (1982), 1025; I. Kuwajima and E. Nakamura, *Acc. Chem. Res.*, **18** (1985), 181.

[17] M. E. Krafft and R. A. Holton, *Tetrahedron Lett.*, **24** (1983), 1345.

In addition to their use for the preparation of specific lithium enolates, silyl enol ethers are also excellent substrates for *acid*-catalysed alkylation. In the presence of a Lewis acid (e.g. $TiCl_4$, $SnCl_4$, $BF_3 \cdot OEt_2$) they react readily with *tertiary* alkyl halides to give the alkylated product in high yield.[18] This procedure thus complements the more-common base-catalysed alkylation of enolates which fails with tertiary halides. It is supposed that the Lewis acid promotes ionization of the electrophile, RX, to form the cation R^+, which is trapped by the silyl enol ether to give the addition product with cleavage of the silicon–oxygen bond.

Treatment of the thermodynamic silyl enol ether **8** with tert-butyl chloride in the presence of $TiCl_4$ gives the alkylated product **9**, containing two adjacent quaternary carbon atoms, in a remarkable 48% yield (1.21).[19] Alkylation of silyl enol ethers using silver(I) catalysis is also effective.[20]

OSiMe$_3$

i, $^tC_4H_9Cl$, $TiCl_4$
CH_2Cl_2, −23 °C

ii, Na_2CO_3 (aq)

48%

8   **9**

(1.21)

In the presence of a Lewis acid, silyl enol ethers can be alkylated with reactive secondary halides, such as substituted benzyl halides, and with chloromethylphenyl sulfide ($ClCH_2SPh$), an activated primary halide. Thus, reaction of the benzyl chloride **10** in the presence of zinc bromide with the trimethylsilyl enol ether derived from mesityl oxide allowed a short and efficient route to the sesquiterpene ($\pm$)-*ar*-turmerone (1.22).[21] Reaction of $ClCH_2SPh$ with the trimethylsilyl enol ethers of lactones in the presence of zinc bromide, followed by *S*-oxidation and pyrolytic elimination of the resulting sulfoxide (see Section 2.2), provides a good route to the α-methylene lactone unit common in many cytotoxic sesquiterpenes (1.23). Desulfurization with Raney nickel, instead of oxidation and elimination, affords the α-methyl (or α-alkyl starting with RCH(Cl)SPh) derivatives.[22]

i, LDA, THF
−78 °C

ii, Me$_3$SiCl

Me$_3$SiO

10

ZnBr$_2$, CH_2Cl_2

80%

(1.22)

[18] M. T. Reetz, *Angew. Chem. Int. Ed. Engl.*, **21** (1982), 96.
[19] T. H. Chan, I. Paterson and J. Pinsonnault, *Tetrahedron Lett.* (1977), 4183.
[20] K. Takeda, A. Ayabe, H. Kawashima and Y. Harigaya, *Tetrahedron Lett.*, **33** (1992), 951; C. W. Jefford, A. W. Sledeski, P. Lelandais and J. Boukouvalas, *Tetrahedron Lett.*, **33** (1992), 1855; P. Angers and P. Canonne, *Tetrahedron Lett.*, **35** (1994), 367.
[21] I. Paterson, *Tetrahedron Lett.* (1979), 1519.
[22] I. Paterson, *Tetrahedron*, **44** (1988), 4207.

(1.23)

The treatment of an ester (or lactone) with a base and a silyl halide or triflate gives rise to a particular type of silyl enol ether normally referred to as a silyl ketene acetal. The extent of *O*- versus *C*-silylation depends on the structure of the ester and the reaction conditions. The less-bulky methyl or ethyl (or *S*-tert-butyl) esters are normally good substrates for *O*-silylation using LDA as the base. Acyclic esters can give rise to two geometrical isomers of the silyl ketene acetal. Good control of the ratio of these isomers is often possible by careful choice of the conditions. The *E*-isomer is favoured with LDA in THF, whereas the *Z*-isomer is formed exclusively by using THF/HMPA (1.24).[23] Methods to effect stereoselective silyl enol ether formation from acyclic ketones are less well documented.[24]

(1.24)

As an alternative to enolization and addition of a silyl halide or triflate, silyl enol ethers may be prepared by the 1,4-hydrosilylation of an α,β-unsaturated ketone. This can be done by using a silyl hydride reagent in the presence of a metal catalyst. Metal catalysts based on rhodium or platinum are most effective and provide a regiospecific approach to silyl enol ethers (1.25).

(1.25)

Similarly, specific enolates of unsymmetrical ketones can be obtained by reduction of α,β-unsaturated ketones with lithium in liquid ammonia. Alkylation of the intermediate enolate gives an α-alkyl derivative of the corresponding *saturated* ketone which may not be the same as that obtained by base-mediated alkylation of the saturated ketone itself. For example, base-mediated alkylation of 2-decalone generally leads to 3-alkyl derivatives whereas, by proceeding from the enone **11**, the 1-alkyl derivative is obtained (1.26). The success of this procedure depends on

[23] T.-H. Chan, in *Comprehensive Organic Synthesis*, ed. B. M. Trost and I. Fleming, vol. 2 (Oxford: Pergamon Press, 1991), p. 595.
[24] E. Nakamura, K. Hashimoto and I. Kuwajima, *Tetrahedron Lett.* (1978), 2079.

the fact that in liquid ammonia the alkylation step is faster than the equilibration of the initially formed enolate. Lithium enolates must be used since sodium or potassium salts lead to equilibration and therefore mixtures of alkylated products. Alkylations are best with iodomethane, primary halides (or sulfonates) or activated halides such as allyl or benzyl compounds. Reactions with secondary halides are slower, leading to a loss of selectivity.[25]

$$(1.26)$$

The treatment of α,β-unsaturated ketones with organocopper reagents provides another method to access specific enolates of unsymmetrical ketones.[26] Lithium dialkylcuprates (see Section 1.2.1) are used most commonly and the resulting eno- late species can be trapped with different electrophiles to give α,β-dialkylated ketones (1.27). Some problems with this approach include the potential for the inter- mediate enolate to isomerize and the formation of mixtures of stereoisomers of the dialkylated product. The intermediate enolate can be trapped as the silyl enol ether and then regenerated under conditions suitable for the subsequent alkylation. Reac- tion of the enolate with phenylselenyl bromide gives the α-phenylseleno-ketone **12**, from which the β-alkyl-α,β-unsaturated ketone can be obtained by oxidation and selenoxide elimination (1.28).

$$(1.27)$$

$$(1.28)$$

[25] D. Caine, *Org. Reactions*, **23** (1976), 1.
[26] R. J. K. Taylor, *Synthesis* (1985), 364.

If an α,β-unsaturated ketone is treated with a base then proton abstraction can occur on the α'- (1.22) or α-side of the carbonyl group (1.29).[27] The latter regio-selectivity is favoured under equilibrating conditions, for example in the presence of a protic solvent, to give the more-stable dienolate anion **13**. Alkylation of the anion **13** occurs preferentially at the α-carbon atom to give the mono-α-alkyl-β,γ-unsaturated ketone as the initial product. The α-proton in this compound is readily removed by interaction with either the base or the original enolate **13**, since it is activated both by the carbonyl group and the carbon–carbon double bond. In the presence of an excess of the alkylating agent, the resulting anion is again alkylated at the α-position and the α,α-dialkyl-β,γ-unsaturated ketone is pro-duced. If the availability of the alkylating agent is restricted, however, then further alkylation does not occur, and the thermodynamically more stable α-alkyl-α,β-unsaturated ketone gradually accumulates owing to protonation at the γ-position (1.29).

(1.29)

In accordance with this scheme, it is found that dialkylation is diminished by slow addition of the alkylating agent or by use of a less-reactive alkylating agent (for example, an alkyl chloride instead of an alkyl iodide). A disadvantage of this procedure is that it generally gives mixtures of products, particularly in exper-iments aimed at preparing the monoalkylated compound. A solution to this is provided by metalloenamines.[28] Treatment of the unsaturated cyclohexylimine **14** with LDA and iodomethane does not give rise to the dialkylated product because transfer of a proton from the monoalkylated compound to the metalloenamine

[27] K. F. Podraza, *Org. Prep. Proced. Int.*, **23** (1991), 217.
[28] J. K. Whitesell and M. A. Whitesell, *Synthesis* (1983), 517; S. F. Martin, in *Comprehensive Organic Synthesis*, ed. B. M. Trost and I. Fleming, vol. 2 (Oxford: Pergamon Press, 1991), p. 475.

is slow (1.30).

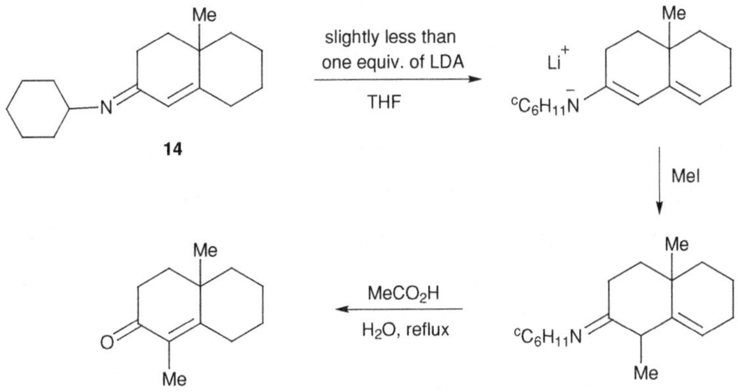

(1.30)

Enamines and metalloenamines provide a valuable alternative to the use of eno-lates for the selective alkylation of aldehydes and ketones.[3,28] Enamines are α,β-unsaturated amines and are obtained simply by reaction of an aldehyde or ketone with a secondary amine in the presence of a dehydrating agent, or by heating in benzene or toluene solution in the presence of toluene-*p*-sulfonic acid (TsOH) as a catalyst, with azeotropic removal of water (1.31). Pyrrolidine and morpholine are common secondary amines useful for forming enamines. All of the steps of the reaction are reversible and enamines are readily hydrolysed by water to reform the carbonyl compound. All reactions of enamines must therefore be conducted under anhydrous conditions, but once the reaction has been effected, the modified car-bonyl compound is liberated easily from the product by addition of dilute aqueous acid to the reaction mixture.

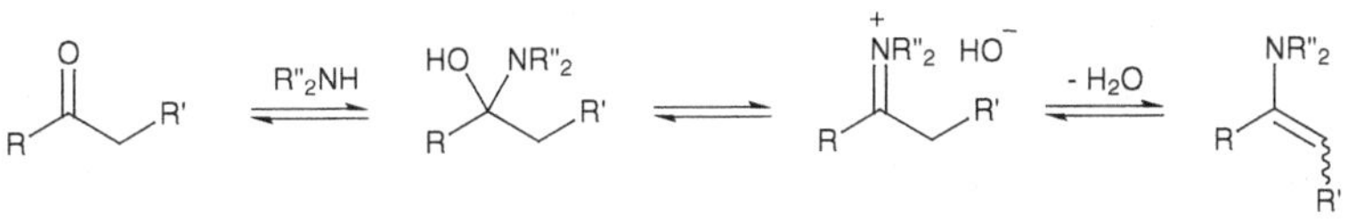

(1.31)

Owing to the spread of electron density, which resides mostly on the nitrogen and β-carbon atoms (1.32), an enamine can act as a nucleophile in reactions with carbon-based electrophiles, leading to the *C*-alkylated and/or *N*-alkylated products. Because no base or other catalyst is required, there is a reduced tendency for wasteful self-condensation reactions of the carbonyl compound and even aldehydes can be alkylated or acylated in good yield.

$$(1.32)$$

A valuable feature of the enamine reaction is that it is regioselective. In the alky-lation of an unsymmetrical ketone, the product of reaction at the *less*-substituted α-carbon atom is formed in greater amount, in contrast to direct base-mediated alkylation of unsymmetrical ketones, which usually gives a mixture of products. For example, reaction of the pyrrolidine enamine of 2-methylcyclohexanone with iodomethane gives 2,6-dimethylcyclohexanone almost exclusively. This selectivity derives from the fact that the enamine from an unsymmetrical ketone consists mainly of the more-reactive isomer in which the double bond is directed toward the less-substituted carbon atom. In the 'more-substituted' enamine, there is decreased inter-action between the nitrogen lone pair and the π-system of the double bond because of steric interference between the α-substituent (the methyl group in Scheme 1.33) and the α-methylene group of the amine.

$$(1.33)$$

Alkylation of enamines with alkyl halides generally proceeds in only poor yield because the main reaction is *N*- rather than *C*-alkylation. Good yields of alkylated products are obtained by using reactive benzyl or allyl halides; it is believed that in these cases there is migration of the substituent group from the nitrogen to the carbon atom. This may take place in some cases by an intramolecular pathway, resulting in rearrangement of allyl substituents or by dissociation of the *N*-alkyl derivative followed by irreversible *C*-alkylation. This difficulty can be circumvented by the use of metalloenamines, which are readily formed from imines and a base.[28] The metal salts so formed give high yields of monoalkylated carbonyl compounds on reaction with primary or secondary alkyl halides. At low temperature, imines derived from methyl ketones are alkylated on the methyl group (1.34); with other dialkyl ketones regioselective alkylation at either α-position can be realised by judicious choice of experimental conditions.

$$
\text{(structure)} \quad \xrightarrow[\substack{\text{ii, CH}_3\text{I} \\ \text{iii, H}_3\text{O}^+}]{\substack{\text{i, LDA} \\ \text{DME, }-60\,°\text{C}}} \quad \text{(structure)} \qquad (1.34)
$$

A useful alternative to the metalloenamine chemistry proceeds not from an imine but from a hydrazone of an aldehyde or ketone.[29] These compounds, on reaction with LDA or *n*-BuLi, are converted into lithium derivatives that can be alkylated with alkyl halides, alkyl sulfonates, epoxides or carbonyl compounds. At the end of the sequence the hydrazone group is cleaved by oxidation, liberating the alkylated aldehyde or ketone. Like metalloenamine chemistry, for the synthetic effort required to prepare and later remove the hydrazone derivative to be worthwhile, the overall benefits of this approach must outweigh the shorter use of the enolate of the carbonyl compound itself. Hydrazones are formed readily by the condensation of a hydrazine and a carbonyl compound. The hydrazone can often be lithiated regioselectively, thereby giving rise, on addition of a carbon-based electrophile, to alkylated products of defined regiochemistry. Stereochemical control can also be afforded, depending on the nature of the substituents. Generally, alkylation takes place at the less-substituted position α- to the original unsymmetrical ketone (unless there is an anion-stabilizing group present). For example, the dimethylhydrazone derived from 2-methylcyclohexanone gave *trans*-2,6-dimethylcyclohexanone (1.35). *Axial* alkylation is favoured with cyclohexanone derivatives. Epoxides give γ-hydroxycarbonyl compounds and hence, by oxidation, 1,4-dicarbonyl compounds. Reaction with aldehydes leads to β-hydroxycarbonyl compounds by a 'directed' aldol reaction (see Section 1.1.3).

$$
\text{(structure)} \quad \xrightarrow[\text{ii, CH}_3\text{I}]{\text{i, LDA}} \quad \text{(structure)} \quad \xrightarrow[\substack{\text{MeOH, H}_2\text{O} \\ 95\%}]{\text{NaIO}_4} \quad \text{(structure)} \qquad (1.35)
$$

97% *trans*

1.1.2 Conjugate addition reactions of enolates and enamines

Section 1.1.1 described the formation of enolates, silyl enol ethers and enamines and their alkylation reactions. An alternative type of alkylation occurs on addition of these nucleophiles to electrophilic alkenes, such as α,β-unsaturated ketones, esters

[29] D. E. Bergbreiter and M. Momongan, in *Comprehensive Organic Synthesis*, ed. B. M. Trost and I. Fleming, vol. 2 (Oxford: Pergamon Press, 1991), p. 503.

or nitriles. High yields of monoalkylated carbonyl compounds can be obtained. The first examples of this chemistry were reported by Michael as early as 1887, and hence this type of reaction is often termed a Michael reaction. The best type of nucleophiles for addition to α,β-unsaturated carbonyl or nitrile compounds are soft in nature, such as organocuprates (see Section 1.2.1) or carbanions stabilized by one, or usually two, electron-withdrawing groups.[30] During conjugate addition, the carbanion adds to the β-carbon of the α,β-unsaturated carbonyl compound. For example, addition of diethyl malonate to the α,β-unsaturated ester **15** under basic conditions gave the product **16** in good yield (1.36). The addition of the stabilized anion to the α,β-unsaturated ester is reversible and leads to the new enolate **17**. Proton transfer (intermolecular) to give the more stable anion **18** can occur. Anion **18**, or the intermediate **17**, is then protonated to give the 1,5-dicarbonyl product **16**.

As the conjugate addition reaction is an equilibrium process, there must be a driving force for the formation of the products, otherwise the starting materials may be recovered. The conjugate addition reaction produces a new anion that can abstract a proton from the original carbonyl compound (diethyl malonate in Scheme 1.36); therefore the base need be present only as a catalyst. Alternatively, the anion **18** may be trapped by addition of an alkylating agent (such as an alkyl halide) in order to generate two carbon–carbon bonds in a single operation. The presence of excess α,β-unsaturated carbonyl compound can lead to a second Michael addition reaction, by reaction of the new anion (e.g. **18**) with the α,β-unsaturated carbonyl compound.

Although the presence of a protic solvent aids these proton-transfer steps, protic solvents are not a necessity for successful Michael addition reactions. Proton abstraction and conjugate addition can be carried out in the presence of a Lewis acid or by using a base in an aprotic solvent. For example, deprotonation of the dicarbonyl compound **19** with sodium hydride in THF and addition of the Michael

---

[30] E. D. Bergmann, D. Ginsburg and R. Pappo, *Org. Reactions*, **10** (1959), 179; M. E. Jung, in *Comprehensive Organic Synthesis*, ed. B. M. Trost and I. Fleming, vol. 4 (Oxford: Pergamon Press, 1991), p. 1; P. Perlmutter, *Conjugate Addition Reactions in Organic Synthesis* (Oxford: Pergamon Press, 1992).

acceptor phenyl vinyl sulfoxide, gave the adduct **20** in reasonable yield (1.37).[31] Heating the product sulfoxide **20** in toluene results in elimination (see Section 2.2) of phenylsulfinic acid to give the vinyl-substituted product **21**.

$$(1.37)$$

A large variety of different Michael acceptors can be used in conjugate addition reactions. The electron-withdrawing group is commonly an ester or ketone, but can be an amide, nitrile, nitro, sulfone, sulfoxide, phosphonate or other suitable group capable of stabilizing the intermediate anion. Likewise, a variety of substituents can be attached at the α- and/or β-position of the Michael acceptor. However, the presence of two substituents at the β-position slows the rate of conjugate addition, owing to increased steric hindrance. Therefore, β,β-disubstituted acceptors are used less commonly in conjugate addition reactions as yields are often poor. This problem can be overcome to some extent by carrying out the reaction under high pressure or by using a β,β-disubstituted acceptor bearing two electron-withdrawing groups to help stabilize the resulting intermediate anion. As an example, the sterically congested Michael adduct **23** has been prepared by conjugate addition of methyl isobutyrate to the doubly activated acceptor **22** under aprotic conditions (1.38).[32]

$$(1.38)$$

With unsymmetrical ketones, a mixture of regioisomeric enolates may be formed, resulting in a mixture of Michael adducts. Deprotonation in a protic solvent is reversible and leads predominantly to the thermodynamically favoured, more-substituted enolate. Reaction with a Michael acceptor then gives the product from reaction at the more-substituted side of the ketone carbonyl group. The 1,5-dicarbonyl compound **24** is the major product from conjugate addition of 2-methylcyclohexanone to methyl acrylate using potassium tert-butoxide in the protic solvent tert-butanol (1.39).[33] In contrast, the major product from Michael addition

[31] G. A. Koppel and M. D. Kinnick, *J. Chem. Soc., Chem. Commun.* (1975), 473.
[32] R. A. Holton, A. D. Williams and R. M. Kennedy, *J. Org. Chem.*, **51** (1986), 5480.
[33] H. O. House, W. L. Roelofs and B. M. Trost, *J. Org. Chem.*, **31** (1966), 646.

using the enamine prepared from 2-methylcyclohexanone is derived from reaction at the less-substituted side of the ketone carbonyl group. Addition of pyrrolidine to 2-methylcyclohexanone and dehydration gives the enamine **25** (see 1.33), which reacts with acrylonitrile to give the product **26** after hydrolysis (1.40). Any *N*-alkylation is reversible and good yields of *C*-alkylated products are normally obtained.

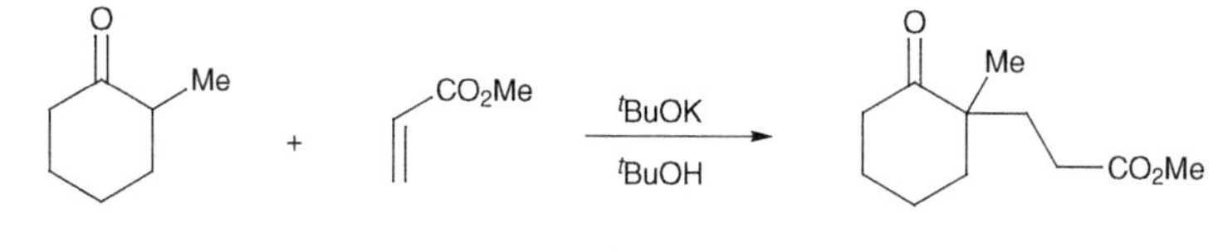

(1.39)

24

(1.40)

25   **26**

Endocyclic enamines, such as pyrrolines and tetrahydropyridines are useful for the synthesis of complex heterocyclic compounds, as found in many alkaloids.[34] Thus, reaction of the enamine **27** with methyl vinyl ketone gave the alkaloid mesembrine (1.41).

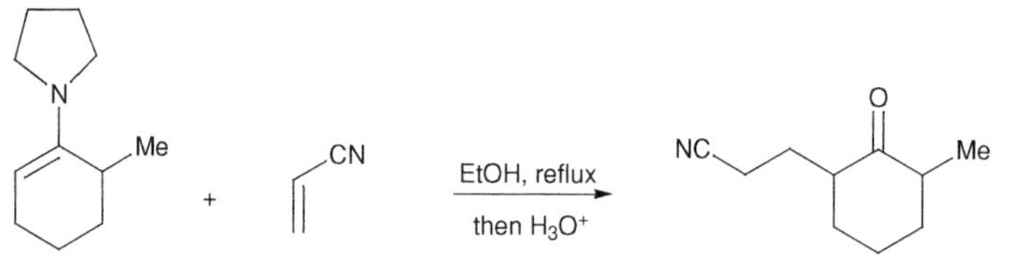

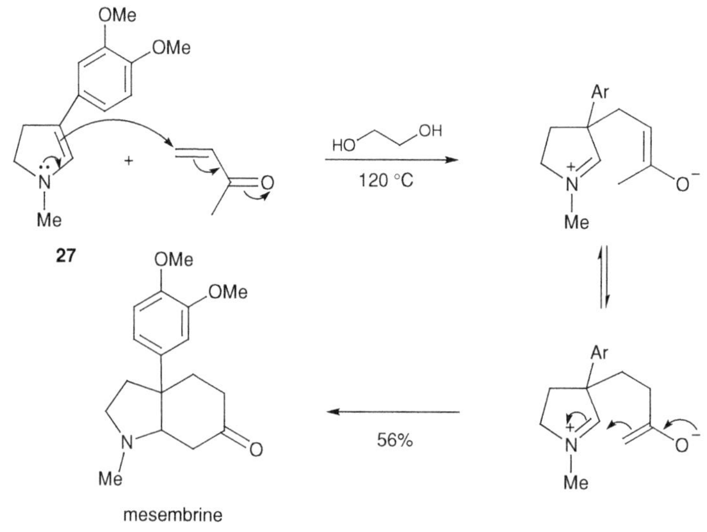

27

(1.41)

56%

mesembrine

[34] R. V. Stevens, *Acc. Chem. Res.*, **10** (1977), 193.

The Michael addition is a useful reaction in organic synthesis as it generates a new carbon–carbon single bond under relatively mild and straightforward conditions. Up to three new chiral centres are generated and recent efforts have focused on stereoselective Michael additions.[35] The enamine **28**, derived from cyclohexanone and morpholine, reacts with 1-nitropropene to give (after hydrolysis) the ketone **29** as the major diastereomer (1.42). The same stereochemical preference for the *syn* stereoisomer has been found in the conjugate addition reaction between the enolate of tert-butyl propionate and the enone **30** (1.43). There are, however, many examples of the formation of approximately equal mixtures of diastereomers or even high selectivity for the *anti* stereoisomer. Careful choice of substituents and conditions may allow the stereocontrolled formation of the desired stereoisomer.

$$(1.42)$$

$$(1.43)$$

The corresponding Michael addition reactions using silyl enol ethers, which require an activator such as a Lewis acid, provide similar stereochemical outcomes.[36] Commonly titanium tetrachloride is used as the Lewis acid, although trityl salts ($Ph_3C^+ X^-$) or other additives have been investigated. For example, conjugate addition of cyclohexanone trimethylsilyl enol ether to the enone **31** gave the 1,5-dicarbonyl product **32** as the major stereoisomer (1.44). The addition of a Michael donor to a cyclic enone leads to the product of attack at the β-carbon from the less-hindered face. This avoids steric hindrance with the substituent on the cyclic enone. However, it has been reported that Lewis acid-assisted conjugate addition, in which chelation to the substituent may take place, can reverse this selectivity. The silyl ketene acetal **33** adds preferentially to the more hindered face of the enone **34** using mercury(II) iodide as the Lewis acid (1.45).

[35] D. A. Oare and C. H. Heathcock, *Top. Stereochem.*, **20** (1991), 87; A. Bernardi, *Gazz. Chim. Ital.*, **125** (1995), 539.

[36] V. J. Lee, in *Comprehensive Organic Synthesis*, ed. B. M. Trost and I. Fleming, vol. 4 (Oxford: Pergamon Press, 1991), p. 139.

$$ (1.44) $$

31   **32**   major

$$ (1.45) $$

33   **34**   *cis : trans* 95 : 5

Michael donors in which the metal attached to the heteroatom is silicon or tin are softer than the corresponding lithium enolates and can enhance 1,4-addition over any undesired 1,2-addition to the carbonyl group. For conjugate addition to occur, tin enolates require an activator such as trimethylsilyl chloride. The tin enolate is prepared from the ketone using a tertiary amine and tin(II) triflate. Addition to an enone in the presence of the Lewis acid gives the 1,5-dicarbonyl product (Scheme 1.46).[37] This procedure requires one equivalent of the Lewis acid, otherwise yields of the product are low. Efforts to avoid a stoichiometric amount of a Lewis acid in Michael addition reactions with tin enolates have uncovered the use of tetrabutyl ammonium bromide (Bu_4NBr) as a catalyst.[38] Treating the tributyltin enolate of acetophenone with methyl acrylate in the presence of 10 mol% Bu_4NBr gave the keto-ester **35** in quantitative yield (1.47).

$$ (1.46) $$

78%

syn : anti 29 : 71

$$ (1.47) $$

>99%   **35**

If the Michael donor and acceptor groups are both located within the same molecule, an intramolecular conjugate addition reaction can take place.[39] This sets up a new ring and up to three new chiral centres. The ease of the reaction depends on a number of factors, including the size of the ring being formed, the geometry of the enolate and Michael acceptor and the *endo* or *exo* nature of the ring closure. Cyclization to give a five- or six-membered ring by the *exo* mode is

[37] T. Mukaiyama and S. Kobayashi, *Org. Reactions*, **46** (1994), 1.
[38] M. Yasuda, N. Ohigashi, I. Shibata and A. Baba, *J. Org. Chem.*, **64** (1999), 2180.
[39] R. D. Little, M. R. Masjedizadeh, O. Wallquist and J. I. McLoughlin, *Org. Reactions*, **47** (1995), 315.

the most facile owing to good orbital overlap between the enolate π-bond and the π-bond of the α,β-unsaturated system. The conditions for intramolecular conjugate addition are in many cases the same as those used for the intermolecular reaction (e.g. catalytic in the metal alkoxide in an alcoholic solvent). Likewise, each step is potentially reversible and the stereoselectivity may be subject to either kinetic or thermodynamic factors.

Intramolecular conjugate addition is most common with a readily enolizable Michael donor, such as a 1,3-dicarbonyl compound. For example, the mild base K_2CO_3 promotes the cyclization of the β-keto-ester **36** by a 5-*exo* ring closure (1.48). The product **37** contains two five-membered rings fused *cis* to each other, as would be expected on the basis of the thermodynamic stability of such bicyclo[3.3.0]octane ring systems.

$$(1.48)$$

Some key features of intramolecular reactions include the need to minimize any intermolecular process (often accomplished by high-dilution conditions) and the requirement that the reagents should react chemoselectively with the desired functional group in the molecule. Cases in which the Michael donor site is not very acidic and therefore requires a strong base may result in proton abstraction in the γ-position of the Michael acceptor, or reaction elsewhere in the molecule. Careful choice of Michael donor and acceptor groups is needed in order to achieve the desired enolate formation. Chemoselective proton abstraction α- to the ester group in the substrate **38** results in the desired enolate and cyclization to give the cyclopentane ring **39** as a single stereoisomer (1.49).[40] Subsequent chemoselective reduction of the ester group in the presence of the carboxylic amide and acid-catalysed cyclization to the lactone gave iridomyrmecin.

$$(1.49)$$

$$(1.50)$$

[40] Y. Yokoyama and K. Tsuchikura, *Tetrahedron Lett.*, **33** (1992), 2823.

The stereochemical preference for the isomer **39** can be rationalized by reaction *via* the conformer shown in Scheme 1.50. Proton abstraction should give predominantly the enolate geometry shown (see Scheme 1.24). The methyl group β- to the ester in the substrate **38** prefers a pseudoequatorial arrangement in the chair-like conformation (1.50) as this avoids 1,3-allylic strain between the methoxy group and the allylic substituents. Therefore, on cyclization, the methyl ester group becomes *trans* to the β-methyl group and *cis* to the Michael acceptor. The stereochemistry of the third new chiral centre (α- to the carboxylic amide group) is determined by the protonation of the enolate resulting from the Michael addition reaction.

An alternative and useful method for intramolecular conjugate addition when the Michael donor is a ketone is the formation of an enamine and its reaction with a Michael acceptor. This can be advantageous as enamine formation occurs under reversible conditions to allow the formation of the product of greatest thermodynamic stability. Treatment of the ketone **40** with pyrrolidine and acetic acid leads to the bicyclic product **41**, formed by reaction of only one of the two possible regioisomeric enamines (1.51).[41] Such reactions can be carried out with less than one equivalent of the secondary amine and have recently been termed 'organo-catalysis' (as opposed to Lewis acid catalysis with a metal salt). The use of chiral secondary amines can promote asymmetric induction (see Section 1.1.4).

$$(1.51)$$

40	R = Me	78%
	R = NHCOMe	85%
	41	

A popular and useful application of the conjugate addition reaction is the combined conjugate addition–intramolecular aldol strategy, commonly known as the Robinson annulation.[30,42] When the Michael donor is a ketone and the Michael acceptor an α,β-unsaturated ketone, the product is a 1,5-diketone which can readily undergo cyclization to a six-membered ring. Typical Michael donor substrates are 2-substituted cyclohexanones, which condense with alkyl vinyl ketones to give the intermediate conjugate addition products **42** (1.52). The subsequent intramolecular

---

[41] A.-C. Guevel and D. J. Hart, *J. Org. Chem.*, **63** (1996), 465; 473.
[42] M. E. Jung, *Tetrahedron*, **32** (1976), 3; R. E. Gawley, *Synthesis* (1976), 777.

aldol reaction to give the cyclohexenone products **43** may take place in the same pot.

$$ (1.52) $$

42   **43**

Good yields of the annulation product are possible, especially with relatively acidic β-dicarbonyl compounds (e.g. Scheme 1.52, R=CO$_2$Et). Typical conditions include the use of a base such as KOH in methanol or ethanol, or sodium hydride in an aprotic solvent such as DMSO. Alternative and effective conditions make use of enamine chemistry, in which reaction at the less-substituted α-carbon of the ketone takes place (1.53).

$$ (1.53) $$

45%

Conjugate addition reactions, including the Robinson annulation, which make use of reactive Michael acceptors such as methyl vinyl ketone, can suffer from low yields of the desired adduct. The basic conditions required for enolate formation can cause polymerization of the vinyl ketone. Further difficulties arise from the fact that the Michael adduct **42** and the original cyclohexanone have similar acidities and reactivities, such that competitive reaction of the product with the vinyl ketone can ensue. These problems can be minimized by the use of acidic conditions. Sulfuric acid is known to promote the conjugate addition and intramolecular aldol reaction of 2-methylcyclohexanone and methyl vinyl ketone in 55% yield. Alternatively, a silyl enol ether can be prepared from the ketone and treated with methyl vinyl ketone in the presence of a Lewis acid such as a lanthanide triflate[43] or boron trifluoride etherate (BF$_3$·OEt$_2$) and a proton source[44] to effect the conjugate addition (followed by base-promoted aldol closure).

1.1.3 The aldol reaction

Sections 1.1.1 and 1.1.2 described the formation of enolates, silyl enol ethers and enamines and their alkylation or conjugate addition reactions. Reaction of these carbon nucleophiles with aldehydes is known generally as the aldol reaction and

[43] S. Kobayashi, *Synlett* (1994), 689.
[44] P. Duhamel, G. Dujardin, L. Hennequin and J.-M. Poirier, *J. Chem. Soc., Perkin Trans. 1* (1992), 387.

is discussed in this section. The aldol reaction is a very good method for making carbon–carbon bonds. The reaction products are β-hydroxycarbonyl compounds, which are common in many natural products. In addition, the hydroxyl and/or carbonyl groups can be converted selectively to other functional groups. Under some circumstances, the aldol product dehydrates to give an α,β-unsaturated carbonyl compound. The general reaction is shown in Scheme 1.54.

$$\text{R—CHO} \quad + \quad \overset{O}{\underset{R'}{\big\Vert}} \quad \xrightarrow{\text{base}} \quad \overset{OH \quad O}{R \underset{R'}{\cdots}} \qquad (1.54)$$

In an aldol reaction, the enolate of one compound reacts with the electrophilic carbonyl carbon of the other carbonyl compound. A problem can arise when the other regioisomeric enolate can form easily or when the electrophilic carbonyl compound is enolizable. In addition, the product is enolizable and the wrong carbonyl compound could act as the electrophile; therefore a mixture of products or predominantly the undesired product may result. An added complication arises when more than one chiral centre is present in the product and therefore two diastereomeric products can be formed. The course of the reaction between unlike components must be 'directed' so that only the product required is obtained, or at least is formed predominantly. In addition, the stereochemical course of the reaction must be controlled. These difficulties have been overcome as a result of intensive study of the aldol reaction,[45] spurred on by the presence of the β-hydroxycarbonyl functional group in the structures of many naturally occurring compounds such as the macrolides and ionophores.

A number of methods have been developed to bring about the 'directed' aldol reaction between two different carbonyl compounds to give a mixed-aldol product. Most of them proceed from the preformed enolate or silyl enol ether of one of the components. With enolates, a number of metal counterions have been used and the best results have been obtained with lithium or boron enolates, although zinc or transition-metal enolates have found widespread use. For example, the aldol reaction of acetone with acetaldehyde under basic aqueous conditions is inefficient

[45] A. T. Nielsen and W. J. Houlihan, *Org. Reactions*, **16** (1968), 1; T. Mukaiyama, *Org. Reactions*, **28** (1982), 203; C. H. Heathcock, in *Comprehensive Organic Synthesis*, ed. B. M. Trost and I. Fleming, vol. 2 (Oxford: Pergamon Press, 1991), pp. 133; 181.

because the acetaldehyde reacts more readily with itself; however, the desired addition can be effected by using the preformed enolate of acetone (1.55).

$$(1.55)$$

Regioselective enolate formation using kinetic deprotonation of an unsymmetrical ketone has been discussed in Section 1.1.1. The specific enolate can react with aldehydes to give the aldol product, initially formed as the metal chelate in aprotic solvents such as THF or Et$_2$O. Thus, 2-pentanone, on deprotonation with lithium diisopropylamide (LDA) and reaction of the enolate with butanal, gave the aldol product **44** in reasonable yield (1.56).

$$(1.56)$$

65%   **44**

Attempts to perform the aldol reaction with the more-substituted (thermodynamic) enolate (formed under equilibrating conditions) from such unsymmetrical ketones normally results in a mixture of aldol products. This is not surprising considering the equilibration with the less-substituted enolate, the possible proton abstraction and self-condensation of the aldehyde and the potential for enolization and further reaction of the product. A solution to the use of the thermodynamic ketone enolate lies in the selective formation and reaction of silyl enol ethers.[46] Treatment of a silyl enol ether with an aldehyde in the presence of a Lewis acid such as titanium tetrachloride results in the formation of the aldol product (Mukaiyama aldol reaction). For example, the β-hydroxyketone **45** was formed by addition of benzaldehyde to the more-substituted silyl enol ether generated from 2-methylcyclohexanone (1.57). A disadvantage of this procedure is that the reaction is not normally stereoselective and so does not allow control of the stereochemistry in reactions that give rise to aldol products containing more than one chiral centre. A recent solution to this problem, providing good levels of *syn* diastereoselection (and which also makes use of water as a solvent), involves addition of 0.1 equivalents of Ph$_2$BOH to catalyse the aldol reaction of the silyl enol ether.[47]

$$(1.57)$$

45

[46] P. Brownbridge, *Synthesis* (1983), 1; ref. 22.
[47] Y. Mori, J. Kobayashi, K. Manabe and S. Kobayashi, *Tetrahedron*, **58** (2002), 8263.

Excellent results have been obtained by using boron enolates (alkenyloxyboranes or enol borinates), in what is commonly known as a boron-mediated aldol reaction.[48] The boron enolates are prepared easily from the corresponding ketone and a dialkylboron trifluoromethanesulfonate (dialkylboron triflate, R_2BOTf) or chloride (R_2BCl) and a tertiary amine base. Boron enolates react readily with aldehydes to give, after oxidative work-up of the resulting borinate species, high yields of the desired aldol product (1.58).

$$(1.58)$$

The aldol reaction is not restricted to the use of ketone enolates and indeed some of the most important examples in this area use carboxylic esters or amides. Proton abstraction with LDA (or other strong base) at low temperature to give the enolate and addition of the aldehyde or ketone gives a β-hydroxyester or β-hydroxyamide product. Likewise the boron or other metal enolates of esters provide alternative methods to effect the aldol reaction.

If the enolate of a carboxylic ester is formed at room temperature then self-condensation of the ester results. This reaction is known as the Claisen condensation and gives a β-keto ester product.[49] A variety of bases including LDA, sodium hydride or sodium alkoxides can be used and the reaction may be driven to completion by the deprotonation of the product, to give the anion of the β-keto ester ($pK_a \approx 11$). The Claisen condensation reaction works best when the two ester groups are the same, to give the self-condensation product (1.59), or when one of the ester groups is non-enolizable. The reaction is less useful in cases where two different enolizable esters are used, as a mixture of up to four β-keto ester products is normally obtained. The product β-keto esters are useful in synthesis as they readily undergo alkylation and decarboxylation reactions (see Section 1.1.1).

$$(1.59)$$

Reaction of an enolate, generated from a β-keto ester or other 1,3-dicarbonyl-type compound with an aldehyde or ketone is known as the Knoevenagel condensation

[48] B. Moon Kim, S. F. Williams and S. Masamune, in *Comprehensive Organic Synthesis*, ed. B. M. Trost and I. Fleming, vol. 2 (Oxford: Pergamon Press, 1991), p. 239.

[49] B. R. Davis and P. J. Garratt, in *Comprehensive Organic Synthesis*, ed. B. M. Trost and I. Fleming, vol. 2 (Oxford: Pergamon Press, 1991), p. 795.

reaction.[50] The reaction conditions are mild and typically employ an amine such as pyridine, sometimes in the presence of the Lewis acid titanium tetrachloride. The product from the initial addition to the aldehyde or ketone dehydrates readily under the reaction conditions to give the α,β-unsaturated dicarbonyl product (1.60). In some cases (particularly with unhindered products) a Michael addition reaction of the product with a second molecule of the original dicarbonyl compound takes place. When the two carbonyl (or other electron-withdrawing) groups in the active methylene compound are different, then condensation with an aldehyde (or unsymmetrical ketone) can give rise to two geometrical isomers. In such cases, the thermodynamically more stable product is normally formed.

$$\text{(1.60)}$$

Enolates, generated by Michael addition reactions of α,β-unsaturated esters or ketones, can add to aldehydes. If the Michael addition is carried out with a tertiary amine (or phosphine) then this is referred to as the Baylis–Hillman reaction.[51] Typically, an amine such as 1,4-diazabicyclo[2.2.2]octane (DABCO) is used. After the aldol reaction, the tertiary amine is eliminated and it can therefore be used as a catalyst (1.61). The reaction is somewhat slow (requiring several days), but rates may be enhanced with other amines such as quinuclidine or quinidine derivatives, the latter effecting asymmetric reaction with high levels of selectivity.[52]

$$\text{(1.61)}$$

1.1.3.1 Stereoselective aldol reactions

The normal product of an aldol reaction between an aldehyde and a monocarbonyl compound is a β-hydroxycarbonyl compound and in many cases a mixture of stereoisomers of the product is formed (1.62). The use of preformed enolates and conditions that are now well established allows control of the stereochemical outcome of aldol reactions. Under appropriate conditions predominantly the *syn* or

[50] G. Jones, *Org. Reactions*, **15** (1967), 204; L. F. Tietze and U. Beifuss, in *Comprehensive Organic Synthesis*, ed. B. M. Trost and I. Fleming, vol. 2 (Oxford: Pergamon Press, 1991), p. 341.
[51] E. Ciganek, *Org. Reactions*, **51** (1997), 201; D. Basavaiah, A. J. Rao and T. Satyanarayana, *Chem. Rev.*, **103** (2003), 811.
[52] Y. Iwabuchi, M. Nakatani, N. Yokoyama and S. Hatakeyama, *J. Am. Chem. Soc.*, **121** (1999), 10 219.

the *anti* (traditionally *erythro* and *threo*) aldol product can be prepared. This is often termed a *diastereoselective* reaction. In addition, with suitable chiral auxiliaries or catalysts, high selectivities for one enantiomer of the *syn* or the *anti* diastereomer can be obtained (see Section 1.1.4 for *enantioselective* reactions).

(1.62)

To achieve good diastereoselection, boron enolates have been most widely used. The stereochemical course of the reactions depends on whether the reaction is run under thermodynamic or kinetic conditions. For the kinetic reaction, enolate geometry is important; it is found in general that *cis*-(or Z-)enolates **46** give mainly the *syn* aldol products, whereas the *trans*-(or E-)enolates **47** (especially for bulky R′ groups) lead to the *anti* aldol products (1.63). The size of the substituent groups affects the diastereomer ratio, with bulkier groups generally enhancing the selectivity. It should be noted that, for consistency, the descriptors for the geometry of the enolate give priority to the OM group over the R′ group (even for ester enolates), despite the conventional Cahn–Ingold–Prelog rules.

(1.63)

The aldol addition reactions are believed to proceed by way of a chair-like six-membered cyclic transition state in which the ligated metal atom is bonded to the oxygen atoms of the aldehyde and the enolate (Zimmerman–Traxler model). For the reaction of a *cis*-enolate **46** with an aldehyde RCHO, the transition state could be represented as **48** (1.64). This places the R group of the aldehyde in a pseudoequatorial position in the chair-like conformation and leads to the *syn* aldol product. Likewise, reaction of the *trans*-enolate proceeds preferentially *via* the transition state **49**, giving the *anti* aldol product.

$$(1.64)$$

As the majority of aldol reactions are carried out under kinetic control, an important issue is whether the proton abstraction generates the *cis*- or the *trans*-enolate. However, the use of geometrically pure enolates does not guarantee the formation of stereoisomerically pure aldol products. In general, the *cis*-enolates are more stereoselective than *trans*-enolates in the aldol reaction. For example, treatment of pentan-3-one with LDA gives a mixture of the *cis*- and *trans*-lithium enolates that react with benzaldehyde to give a mixture of the *syn* and *anti* aldol products (1.65). Using the bulkier 2,2-dimethylpentan-3-one gives almost exclusively the *cis*-enolate and hence the *syn* aldol product.

$$(1.65)$$

R	*cis* : *trans*		*syn* : *anti*	
Et	30 : 70		64 : 36	
^{t}Bu	>98 : <2		>98 : <2	

Better results are normally obtained by using the boron-mediated aldol reaction. This has been ascribed to the shorter boron–oxygen bond length, thereby producing a tighter transition state and enhancing the steric effects compared with that from the lithium enolate (note also the steric effect of ligands attached to the boron atom). Crucially, it is possible to control the enolate geometry by choice of the boron reagent. In general, the use of bulky ligands (such as cyclohexyl) and a relatively poor leaving group (such as Cl) on the boron atom, combined with an unhindered tertiary amine base (such as Et$_3$N) gives rise to predominantly the *trans*-enolate.[53] However, the use of smaller ligands (such as *n*-butyl) and a good leaving group (such as OTf) on the boron atom, combined with a hindered amine (such as

---

[53] K. Ganesan and H. C. Brown, *J. Org. Chem.*, **58** (1993), 7162.

i-Pr$_2$NEt) gives rise to the *cis*-enolate with high selectivity.[54] Thus, the aldol reaction with pentan-3-one occurs with poor selectivity using the lithium enolate (1.65), but excellent selectivity using the *cis*-enolate generated from dibutylboron triflate or the *trans*-enolate generated from dicyclohexylboron chloride (1.66).

(1.66)

Conditions	cis	:	trans		syn	:	anti
Bu$_2$BOTf, ^{i}Pr$_2$NEt Et$_2$O, –78 °C	>97	:	<3		>97	:	<3
(^{c}C$_6$H$_{11}$)$_2$BCl, Et$_3$N pentane, 0 °C	5	:	95		5	:	95

Titanium(IV) enolates, generated by using the ketone, titanium tetrachloride and a tertiary amine base (e.g. Bu$_3$N), give rise, on addition of an aldehyde, to the *syn* aldol products in high yield.[55] This offers an alternative and convenient method for the formation of the *syn* aldol product. Tin(II) enolates, generated using tin(II) triflate and an amine base, also give the *syn* aldol products by a highly selective process.

In a similar way, enolates can be prepared from esters or carboxylic amides. Deprotonation of simple esters (e.g. methyl or tert-butyl propionate with LDA) generates the *trans*-enolate with good selectivity; however, this is not translated into a good selectivity for the *anti* aldol product and approximately an equal mixture of the two diastereomeric products is normally formed. One solution to this is to use a bulky aromatic ester group, which generates predominantly the *anti* aldol product (1.67).

(1.67)

syn : anti
<3 : >97

High selectivity for either the *anti* or the *syn* aldol product can be obtained by using an appropriate thioester and a boron-mediated aldol reaction. The *trans*-enolate and hence the *anti* aldol product can be obtained from the S-tert-butyl ester and dibutylboron triflate (1.68). The *trans*-enolate (note the convention that the O–metal substituent takes a higher priority than the SR substituent) is favoured owing to the steric interaction of the bulky tert-butyl group with the ligands on the boron

[54] D. A. Evans, J. V. Nelson, E. Vogel and T. R. Taber, *J. Am. Chem. Soc.*, **103** (1981), 3099.
[55] Y. Tanabe, N. Matsumoto, T. Higashi, T. Misaki, T. Itoh, M. Yamamoto, K. Mitarai and Y. Nishii, *Tetrahedron*, **58** (2002), 8269.

atom. However, the *cis*-enolate and hence the *syn* aldol product is formed using the S-phenyl ester and 9-borobicyclo[3.3.1]non-9-yl triflate (9-BBNOTf).

$$syn : anti \quad 10 : 90$$

$$syn : anti \quad 97 : 3$$

The situation becomes a little more complex if the aldehyde contains a chiral centre, as in 2-phenylpropanal. In such cases, the aldol product formed contains three chiral centres and there are four possible diastereoisomers – two 2,3-*syn* and two 2,3-*anti* isomers (and their enantiomeric pairs, thereby totalling eight stereoisomers) (1.69). The relative stereochemistry of the substituents at C-2 and C-3 is controlled by the geometry of the enolate. The C-3,C-4 stereochemistry, on the other hand, depends on the direction of approach of the enolate and the carbonyl group to each other. This is illustrated in Scheme 1.70, for the reaction of 2-phenylpropanal with the *trans*-enolate of 2,6-dimethylphenyl propionate.

With a chiral aldehyde, the two faces of the carbonyl group are not equivalent. When a nucleophile, such as an enolate anion, approaches the aldehyde it shows some preference for one face over the other, with the result that unequal amounts of the two diastereomers are formed. This is designated *diastereofacial selectivity*. Thus, in the reaction shown in Scheme 1.70, the 2,3-*anti*, 3,4-*syn* compound **50**, formed by approach of the enolate to the *re* face of the aldehyde (as shown by **51**) is produced in larger amount than the 2,3-*anti*, 3,4-*anti* isomer (both isomers have the 2,3-*anti* configuration because they are derived from the *trans*-enolate).

A good representation of the orientation of addition uses the Newman projection **52**, with the largest group (in this case the phenyl group) perpendicular to the carbonyl group and with approach of the nucleophile 109° to the carbonyl oxygen atom (Bürgi–Dunitz angle) and closer to the smallest group (Felkin–Anh model).[56] However, the Felkin–Anh transition state may not always have the lowest energy and it is common that aldol reactions of *cis*-enolates with α-chiral aldehydes favour the '*anti*-Felkin' product, particularly with substituents larger than phenyl at the α-chiral centre.[57]

$$\text{anti, syn} \qquad 80 : 20 \qquad \text{anti, anti}$$
'Felkin' product

50

51   **52**

Similarly, the reaction of an achiral aldehyde with a chiral enolate leads to some degree of diastereofacial selectivity and the two diastereomeric products are not produced in equal amounts. Further enhancement to the selectivity is possible using a chiral aldehyde with a chiral enolate. When both the nucleophile and the electrophile contain a chiral centre then *double stereodifferentiation* (or *double asymmetric induction*) can occur.[58] Most of the studies in these areas use chiral, non-racemic enolates, which are discussed in the next section.

1.1.4 Asymmetric methodology with enolates and enamines

Excellent levels of asymmetric induction in various carbon–carbon bond-forming reactions, such as alkylation, conjugate addition and aldol reactions, are possible using a suitable chiral enolate and an achiral electrophile under appropriate reaction conditions. A variety of chiral enolates have been investigated, the most common and useful synthetically being those with a chiral auxiliary attached to the carbonyl group. The 2-oxazolidinone group, introduced by Evans, has proved to be an efficient and popular chiral auxiliary.[59] Both enantiomers of the product are

---

[56] For a review on the Cram and Felkin–Anh models, see O. Reiser, *Chem. Rev.*, **99** (1999), 1191.
[57] W. R. Roush, *J. Org. Chem.*, **56** (1991), 4151.
[58] S. Masamune, W. Choy, J. S. Petersen and L. R. Sita, *Angew. Chem. Int. Ed. Engl.*, **24** (1985), 1.
[59] D. A. Evans, in *Asymmetric Synthesis*, ed. J. D. Morrison, vol. 3 (New York: Academic Press, 1984), p. 1.

accessible by choice of either the 2-oxazolidinone **53**, derived from (*S*)-valine, or the 2-oxazolidinone **54** derived from (1*S*,2*R*)-norephedrine (1.71). Other auxiliaries, such as 2-oxazolidinones derived from phenylalanine, are also common.

(1.71)

53   **54**

N-Acylation of the oxazolidinone with the desired carbonyl compound and for-mation of the lithium enolate using LDA gives the respective *cis*-enolate with >100:1 stereoselection. Alkylation with reactive alkylating agents (such as methyl iodide, allyl or benzyl halides) gives alkylated products and occurs with very high levels of diastereoselection. For example, the (2*R*)-imide **55** is formed by using the valine-derived oxazolidinone, whereas the (2*S*)-imide **56** is the major product from reaction with the oxazolidinone derived from norephedrine (1.72). Purifica-tion of the products **55** and **56** gives material with >99:1 diastereomer ratio. The product can be hydrolysed (LiOH, H_2O_2) to give the corresponding carboxylic acid or reduced (LiAlH$_4$ or LiBH$_4$) to give the primary alcohol, essentially as a single enantiomer.

For example, in a synthesis of the immunosuppressant (–)-sanglifehrin A, the alcohol **57** was converted into the chiral alcohol **58** as a single enantiomer (1.73). The procedure involves oxidation of the alcohol **57** with pyridinium chlorochro-mate (PCC) and conversion to the Evans oxazolidinone (via the mixed anhydride), followed by stereoselective enolate formation and alkylation, then reduction to remove the auxiliary.

i, LDA, THF, –78 °C

ii, ⌇Br

71%

55   98:2 (96% d.e.)

(1.72)

i, LDA, THF, –78 °C

ii, ⌇Br

75%

56   98:2 (96% d.e.)

$$(1.73)$$

Alkylation of lithiated hydrazones forms the basis of an efficient method for the asymmetric alkylation of aldehydes and ketones, using the optically active hydrazines (S)-1-amino-2-(methoxymethyl)pyrrolidine (SAMP) **59** and its enantiomer (RAMP) as chiral auxiliaries. Deprotonation of the optically active hydrazones, alkylation and removal of the chiral auxiliary under mild conditions (ozonolysis or acid hydrolysis of the N-methyl salt) gives the alkylated aldehyde or ketone with, generally, greater than 95% optical purity.[60] This procedure has been exploited in the asymmetric synthesis of several natural products. Thus, (S)-4-methyl-3-heptanone, the principal alarm pheromone of the leaf-cutting ant *Atta texana*, was prepared from 3-pentanone in very high optical purity as shown in Scheme 1.74.

$$(1.74)$$

An interesting example of the α-alkylation of α-amino acids without loss of optical activity has been reported by Seebach.[61] Reaction of proline with pivaldehyde gave the single stereoisomer **60** (1.75). Deprotonation with LDA to the chiral non-racemic enolate and addition of an electrophile, such as iodomethane, gives

[60] A. Job, C. F. Janeck, W. Bettray, R. Peters and D. Enders, *Tetrahedron*, **58** (2002), 2253.
[61] D. Seebach, M. Boes, R. Naef and W. B. Schweizer, *J. Am. Chem. Soc.*, **105** (1983), 5390.

the product **61**, in which the alkylation has taken place exclusively on the same side of the bicyclic system as the tert-butyl group (methyl and tert-butyl groups on the *exo* side). Hydrolysis occurs readily to give the optically active α-methylproline. This type of process, in which a chiral centre in the starting material is relayed to another position in order to then functionalize the first position has been termed 'self-regeneration of chirality'.

$$(1.75)$$

60 **61**

An alternative method for the asymmetric α-alkylation of enolates of α-amino-acid derivatives has been reported by Myers.[62] Formation of the crystalline glycine derivative **62**, using the readily available chiral auxiliary pseudoephedrine, followed by treatment with three molar equivalents of lithium hexamethyldisilazide [LiN(SiMe$_3$)$_2$ (LHMDS)] and lithium chloride gives the desired thermodynamic *cis*-enolate **63** (1.76). Addition of an electrophile such as an alkyl halide gives the *C*-alkylated product **64** and this occurs with high levels of diastereoselectivity. For example, when RX is iodoethane (EtI), the product **64** (R=Et) is formed in 83% yield and as a 98:2 ratio of diastereomers. The auxiliary can be cleaved by heating in water to give the α-substituted α-amino acid.

$$(1.76)$$

62

63

64

The presence of a chiral auxiliary attached to the substrate has the advantage that the minor diastereomer formed after the alkylation can be separated easily by chromatography or crystallization. However, the use of a chiral catalyst in substo-ichiometric amounts avoids the need to attach and later remove an auxiliary. To be useful in synthesis, this process needs to be highly enantioselective. One example

[62] A. G. Myers, P. Schnider, S. Kwon and D. W. Kung, *J. Org. Chem.*, **64** (1999), 3322.

of such an asymmetric alkylation is shown in Scheme 1.77. Addition of benzyl bromide to the ester **65** under phase-transfer conditions with the chiral quaternary ammonium salt **66** (based on cinchonidine) as catalyst gave the product **67** with very high selectivity.[63] In this type of reaction, only a low concentration of the enolate associated with the chiral ammonium cation is present in the organic layer and only 0.01 molar equivalents (1 mol%) of the phase-transfer catalyst is needed.

$$(1.77)$$

Asymmetric conjugate addition reactions of carbonyl compounds with α,β-unsaturated systems are known. The simple amine α-methylbenzylamine **68** acts as both the activator (to give the imine and hence the enamine required for alkylation) and as the chiral auxiliary to effect neutral asymmetric conjugate-addition reactions.[64] Thus, condensation of (S)-α-methylbenzylamine **68** with 2-methylcyclohexanone, followed by addition of methyl acrylate (and hydrolysis of the product imine), gave the 2,2-disubstituted cyclohexanone **69** with high enantiomeric purity (1.78).

$$(1.78)$$

Lewis acids promote conjugate addition and the presence of a chiral ligand on the metal can result in high levels of asymmetric induction. A good example in this regard is the addition of the enolate of dicarbonyl compounds (such as dimethylmalonate) with cyclic enones (such as cyclohexenone) in the presence

[63] H. Park, B. Jeong, M.-S. Yoo, J.-H. Lee, M. Park, Y.-J. Lee, M.-J. Kim and S. Jew, *Angew. Chem. Int. Ed.*, **41** (2002), 3036; K. Maruoka and T. Ooi, *Chem. Rev.*, **103** (2003), 3013.

[64] J. d'Angelo, D. Desmaële, F. Dumas and A. Guingant, *Tetrahedron: Asymmetry*, **3** (1992), 459.

of the lithium–aluminium salt of 1,1'-bi-2-naphthol (BINOL) **70** as a catalyst
(1.79). This transformation is extremely enantioselective and was used recently in a
synthesis of the alkaloid (−)-strychnine.[65]

$$(1.79)$$

91%   >99% ee

(*R*)-**70**

Much work has been done on the asymmetric aldol reaction, and high levels of
selectivity have been achieved.[66] The presence of a chiral auxiliary attached to the
carbonyl group can promote a diastereoselective aldol reaction and, after cleavage
of the auxiliary, very highly enantioenriched aldol products. An advantage of the
use of a chiral auxiliary lies in the ease of purification of the product, such that
any unwanted diastereomer can be removed, normally by crystallization or chromatography. Subsequent removal of the auxiliary therefore provides products of
essentially complete optical purity. 2-Oxazolidinones, such as **53** or **54** (1.71), are
effective chiral auxiliaries for aldol (as well as alkylation) reactions. Formation of
the *cis*-boron enolate using, for example, dibutylboron triflate and diisopropylethylamine, followed by addition of the aldehyde, promotes a highly diastereoselective
aldol reaction.[67] Essentially complete selectivity occurs for one of the two possible
syn aldol products, the choice of *syn* product being made by the choice of the chiral
auxiliary group (1.80). The stereoselectivity can be rationalized by a six-membered
(Zimmerman–Traxler) transition state **71**, in which the aldehyde approaches from
the less-hindered face. The product imide can be hydrolysed to give the enantiomerically pure carboxylic acid or derivative, or reduced to give the primary alcohol
product. This procedure therefore allows the formation of either enantiomer of the
desired *syn* aldol product. Interestingly, the use of the titanium enolate of such
imides (prepared by using TiCl$_4$, R$_3$N) can result in the preferential formation of
the opposite *syn* isomer. Of significance is the use of the boron enolate and an

[65] T. Ohshima, Y. Xu, R. Takita, S. Shimizu, D. Zhong and M. Shibasaki, *J. Am. Chem. Soc.*, **124** (2002), 14 546.
[66] C. J. Cowden and I. Paterson, *Org. Reactions*, **51** (1997), 1; P. Arya and H. Qin, *Tetrahedron*, **56** (2000), 917;
T. D. Machajewski and C.-H. Wong, *Angew. Chem. Int. Ed.*, **39** (2000), 1352; C. Palomo, M. Oiarbide and
J. M. García, *Chem. Eur. J.*, **8** (2002), 36.
[67] D. A. Evans, J. Bartroli and T. L. Shih, *J. Am. Chem. Soc.*, **103** (1981), 2127.

aldehyde that has been pre-complexed to a Lewis acid such as diethylaluminium chloride. In such cases one of the two *anti* stereoisomers predominates in the aldol addition reaction. This is thought to be a consequence of an open (acyclic) transition state, rather than the normal chair-like six-membered cyclic transition state. Recently, *anti*-selective asymmetric aldol reactions with the Evans oxazolidinone using Et_3N, Me_3SiCl and 10 mol% $MgCl_2$ have been reported.[68] This provides an aldol reaction which is catalytic in the metal salt (1.81).

i, Bu_2BOTf, $^iPr_2NEt$
ii, PhCHO

88%

ratio of two *syn* stereoisomers >99 : 1   (1.80)

i, Bu_2BOTf, $^iPr_2NEt$
ii, PhCHO

89%

ratio of two *syn* stereoisomers >99 : 1

71

10 mol% $MgCl_2$
Et_3N, TMSCl, PhCHO
then CF_3CO_2H

91%

(1.81)

ratio of major (*anti*) to other stereoisomers 32 : 1

An alternative approach to the asymmetric aldol reaction involves the use of a chiral ligand attached to the metal atom of the enolate species. Such asymmetric aldol reactions give rise to two enantiomeric products and levels of asymmetric induction (arising from diastereomeric transition states) need to be very high. Chiral boron reagents have given very good levels of selectivity using ester or ketone substrates.[48] Masamune introduced the use of the C_2-symmetric dialkyl boron reagent **72**. Formation of the boron enolate using **72** and a bulky thiopropionate ester, followed by addition of an aldehyde, results in the *anti* aldol product with high optical purity (1.82). Another C_2-symmetric boron reagent **73** (or its enantiomer), introduced by Corey, is effective for the asymmetric aldol reaction with propionate esters. Phenylthioesters are good substrates for the formation of *syn* aldol products with

[68] D. A. Evans, J. S. Tedrow, J. T. Shaw and C. W. Downey, *J. Am. Chem. Soc.*, **124** (2002), 392.

high enantioselectivity, whereas the aldol reaction with tert-butyl propionate occurs with high selectivity for one enantiomer of the *anti* aldol product (1.83).

i, ^{i}Pr$_2$NEt

ii, PhCHO, −78 °C

72

71%

anti 99.8% ee (*anti:syn* 97:3)

(1.82)

i, ^{i}Pr$_2$NEt

ii, PhCHO, −78 °C

73

Ar = 3,5-bis(trifluoromethyl)phenyl

90%

syn 97% ee (*anti:syn* 1:99)

(1.83)

i, ^{i}Pr$_2$NEt, **73**

ii, PhCHO, −78 °C

89%

anti 94% ee (*anti:syn* 96:4)

Ketone enolates have also been investigated in the asymmetric boron-mediated aldol reaction. The chiral boron reagents (+)- or (−)-diisopinocampheylboron triflate [(Ipc)$_2$BOTf], derived from α-pinene, allow the formation of the *cis*-enolate and promote enantioselective aldol reactions with aldehydes to give either enantiomer of the *syn* aldol product. For example, the asymmetric aldol reaction between pentan-3-one and 2-methylpropenal takes place in the presence of (−)-(Ipc)$_2$BOTf and diisopropylethylamine to give the *syn* aldol product **74** as the major enantiomer (1.84).

i, (−)-(Ipc)$_2$BOTf, ^{i}Pr$_2$NEt

ii, CHO (Me)

74

78%

syn 91% ee (*anti:syn* 2:98)

(1.84)

When the ketone or the aldehyde contains a chiral centre, then the use of a chiral boron reagent can result in a matched or a mismatched pair. The two chiral groups will either both favour the same stereoisomer of the product, or will work in opposition to one another. Normally, the reaction is carried out first in the absence of the chiral reagent in order to assess the extent of stereoselectivity afforded by the chiral ketone (or aldehyde) alone. One or both enantiomers of the chiral boron reagent can then be used to promote the reaction and to determine the relative influence of the chiral groups. The matched pair enhances the stereoselectivity, whereas the mismatched pair normally gives a lower or even opposite stereoselectivity. The use of a matched pair is known as double asymmetric induction and an example is given in Scheme 1.85. The product **76** is the major diastereomer from the *syn*-selective aldol reaction of the chiral ketone **75** with the achiral dibutylboron triflate reagent. Combining the chiral ketone with the chiral boron triflate derived from α-pinene enhances the selectivity for this *syn* product **76** using (−)-(Ipc)₂BOTf (matched pair). The selectivity is actually reversed using (+)-(Ipc)₂BOTf (an example of reagent control dominating over substrate control).

$$(1.85)$$

L₂BOTf	ratio
ⁿBu₂BOTf	67 : 33
(−)-(Ipc)₂BOTf	92 : 8
(+)-(Ipc)₂BOTf	12 : 88

A disadvantage of the use of chiral ligands attached to the boron atom is that stoichiometric quantities of the chiral ligands are required. The ability to catalyse the aldol reaction with sub-stoichiometric quantities of a chiral ligand has proved possible using the Lewis-acid-mediated Mukaiyama aldol reaction with silyl enol ethers.[69] A variety of Lewis-acid systems have been investigated, such as boranes derived from tartaric acid (or α-amino acids), or tin(II) triflate with a chiral diamine ligand. The latter procedure provides selectively the *syn* aldol product with a >99:1 ratio of enantiomers using the chiral diamine **78** in stoichiometric amount and enantiomeric ratios approaching this value using 20 mol% of the diamine ligand (1.86).

$$(1.86)$$

71%   >98% ee (*anti:syn* 0:100)

Even more efficient catalysis of the Mukaiyama aldol reaction is possible with complexes of transition metals. A number of titanium-based Lewis acids with binaphthyl ligands have been reported to give high enantioselectivities. For example, only 2 mol% of the Lewis acid **80** is required to effect the aldol reaction of

[69] R. Mahrwald, *Chem. Rev.*, **99** (1999), 1095; S. G. Nelson, *Tetrahedron: Asymmetry*, **9** (1998), 357; H. Gröger, E. M. Vogl and M. Shibasaki, *Chem. Eur. J.*, **4** (1998), 1137; C. Gennari, in *Comprehensive Organic Synthesis*, ed. B. M. Trost and I. Fleming, vol. 2 (Oxford: Pergamon Press, 1991), p. 629.

the silyl ketene acetal **79** with aldehydes (1.87). The product β-hydroxy esters are formed in good yield and very high optical purity. Traditionally, aldol reactions with acetate enolates (rather than propionate enolates) occur with low levels of asymmetric induction and this methodology therefore acts as a solution to this problem.

$$\text{(1.87)}$$

72-98% 94-97% ee

79

80

In certain cases, high levels of selectivity in the asymmetric aldol reaction can be achieved in the absence of a metal salt. The amino acid proline catalyses the aldol reaction of aldehydes or ketones (which are enolizable) with aldehydes (preferably non-enolizable or branched to disfavour enolization) to give β-hydroxy-aldehydes or ketones.[70] For example, use of acetone (present in excess) and isobutyraldehyde gave the β-hydroxy-ketone **81** (1.88). The reaction involves an enamine intermediate and is thought to proceed via the usual Zimmerman–Traxler chair-shaped transition state.

$$\text{(1.88)}$$

97% **81** 96% ee

1.1.5 Organolithium reagents

1.1.5.1 Alkyllithium reagents

Organolithium reagents are used extensively in organic synthesis, either as a base or as a nucleophile.[71] They react with a very wide range of electrophiles and the extent of reaction via proton abstraction or nucleophilic attack depends on the structure of the organolithium species, the electrophile and the conditions employed.

[70] K. Sakthivel, W. Notz, T. Bui and C. F. Barbas, *J. Am. Chem. Soc.*, **123** (2001), 5260; B. List, *Tetrahedron*, **58** (2002), 5573; B. Alcaide and P. Almendros, *Eur. J. Org. Chem.* (2002), 1595.
[71] J. Clayden, *Organolithiums: Selectivity for Synthesis* (London: Elsevier, 2002).

Organolithium species are conventionally written as R—Li; however, they are often aggregated structures with significant covalent carbon–lithium bond character. Co-ordinating solvents such as THF, Et$_2$O or *N,N,N′,N′*-tetramethylethylenediamine (TMEDA) can reduce the degree of association of the organolithium species compared with non-polar solvents such as hexane and this can affect the reactivity of the organolithium species. Many simple alkyllithium reagents are available commercially, often as a solution in a non-polar solvent. However, many reactions with organolithium species are done in ethereal solvents, normally at low temperature (−78 °C) in order to avoid problems with abstraction of a proton from the ethereal solvent by the basic alkyllithium species.

Simple, saturated alkyllithium species can be prepared from the corresponding alkyl halide by reaction with lithium metal (1.89). Alternatively, iodine–lithium exchange with tert-butyllithium (Me$_3$CLi) can be used to prepare primary alkyl-lithium species (1.90). Two equivalents of tert-butyllithium are required, since the by-product tert-butyl iodide reacts readily with organolithium species (to give 2-methylpropene). Organolithium species attached to unsaturated carbon centres or bearing a nearby heteroatom may, in some cases, be prepared by proton abstraction, tin–lithium exchange, bromine–lithium exchange or other methods (see Sections 1.1.5.2 and 1.1.5.3).

$$CH_3(CH_2)_3Br \quad + \quad 2 \text{ equiv. Li} \quad \xrightarrow{Et_2O} \quad CH_3(CH_2)_3Li \quad + \quad LiBr \qquad (1.89)$$

$$R\diagup\hspace{-0.3em}\diagdown I \quad \xrightarrow[\text{pentane-Et}_2\text{O, }-78\,°\text{C}]{2 \text{ equiv. }^t\text{BuLi}} \quad R\diagup\hspace{-0.3em}\diagdown Li \qquad (1.90)$$

Alkyllithium species are good bases and a common use is the abstraction of a more-acidic proton. For example, addition of *n*-butyllithium to a solution of diisopropylamine in THF gives lithium diisopropylamide (LDA), a common base. Abstraction of a more-acidic proton attached to a carbon atom can also be effected with *n*-butyllithium, or with stronger bases such as *sec*-butyllithium [EtCH(Me)Li], tert-butyllithium or the complex formed between *n*-butyllithium and potassium tert-butoxide (t-BuOK).

In addition to their basic properties, organolithium species can act as powerful nucleophiles in carbon–carbon bond-forming reactions. The synthetic utility of organolithium reagents can, however, be limited by the ease with which alkyl-lithium species act as a base. Despite this, alkyllithium species are well known to act as nucleophiles with a range of electrophiles, including aldehydes, ketones, carboxylic acids and their derivatives, nitriles, imines, epoxides and alkenes. For example, addition of methyllithium to the aldehyde **82**, derived from alanine, gave the *anti* product **83** as the major diastereomer (1.91). This product is derived from addition to the aldehyde along the direction shown in the Newman projection **85**, as expected from the Felkin–Anh model. This model places the largest group on the α-chiral centre perpendicular to the carbonyl group, with the nucleophile approaching 109° to the carbonyl oxygen atom and closer to the smallest group.[56] Organomagnesium (Grignard) reagents also give similar yields and selectivities. The use of organometallic reagents that allow chelation control (in which the metal coordinates to both the carbonyl oxygen atom and the heteroatom on the α-carbon) can give the corresponding *syn* product **84** as the major isomer (compare with Scheme 1.125).

$$\text{(1.91)}$$

82, Bn = CH$_2$Ph

83
anti   91 : 9   **84**
syn

85

Addition of organolithium species to imines tends to favour deprotonation α-to the imine group. Reducing the basicity of the organometallic species or activating the imine can allow the formation of the carbon–carbon bond by addition to the imine and hence allow a valuable method for the formation of substituted amines.[72] Organocerium reagents, formed by addition of the organolithium species to cerium(III) chloride, are less basic than the organolithium reagents and can add to imines in high yields. Addition of organolithium or organocerium reagents to the *N*-benzyl imine derived from the aldehyde **82** occurs with high selectivity for the *syn* adduct, formed as a result of chelation control (in contrast to the *anti* selectivity in Scheme 1.91). Progress has also been made in the asymmetric synthesis of amines by addition to imines or imine derivatives using chiral auxiliaries or chiral ligands.[73]

[72] R. Bloch, *Chem. Rev.*, **98** (1998), 1407.
[73] S. E. Denmark and D. J.-C. Nicaise, *J. Chem. Soc., Chem. Commun.* (1996), 999; D. Enders and U. Reinhold, *Tetrahedron: Asymmetry*, **8** (1997), 1895; G. Alvaro and D. Savoia, *Synlett* (2002), 651; J. A. Ellman, T. D. Owens and T. P. Tang, *Acc. Chem. Res.*, **35** (2002), 984.

The use of an α,β-unsaturated carbonyl compound as the electrophile tends to lead to the product from 1,2-addition to the carbonyl group, rather than conjugate (1,4-) addition. This is because of the high nucleophilicity and 'hard' nature of organolithium reagents. For example, addition of pentyllithium to the α,β-unsaturated aldehyde **86** gave the 1,2-addition product **87**, used in the synthesis of prostaglandin $F_{2\alpha}$ by Corey (1.92). For 1,4-addition of organometallics to α,β-unsaturated carbonyl compounds, see Section 1.2.1.

i, $^nC_5H_{11}Li$

ii, H_2O

(1.92)

86   **87**

Organolithium reagents add to carboxylic acid derivatives, such as esters, to give tertiary alcohols, as a second equivalent of the alkyllithium species usually adds to the intermediate ketone. If the ketone is the desired product then the carboxylic acid itself can be used as the electrophile. The first equivalent of the organolithium reagent acts as a base to give the lithium carboxylate salt (1.93). A second equivalent can then add to this salt to give the ketone after work-up.[74] Alternatively, addition of alkyllithium reagents to carboxylic amides can give rise to ketone products after hydrolysis of the tetrahedral intermediate.

MeLi

THF, 0 °C

MeLi

then Me_3SiCl
then H_3O^+

(1.93)

85%

Epoxides are opened with organolithium reagents, normally with attack at the less-substituted carbon atom of the epoxide. A potential competing reaction is proton abstraction adjacent to the epoxide, leading to an allylic alcohol.

Organolithium reagents are known to add to alkenes. This generates a carbon–carbon bond and a new organolithium species. Addition to styrene generates a benzyllithium species that normally adds to a second molecule of styene and so on to undergo polymerization, although under suitable conditions the benzyllithium

[74] M. J. Jorgenson, *Org. Reactions*, **18** (1970), 1.

species may be trapped with electrophiles. Intermolecular addition to ethene is also known; however, the addition of an organolithium species to an alkene is most useful in the formation of cyclic systems by intramolecular carbolithiation.[75] For example, the organolithium species **88** can be prepared cleanly from the corresponding iodide using tert-butyllithium (1.94). Cyclization gives the new organolithium species **89**, predominantly as the *cis* isomer, as expected on the basis of a chair-shaped transition state with the methyl substituent in a pseudo-equatorial position. The organolithium species **89** can be trapped with a variety of electrophiles (E$^+$) to give different substituted products.

$$(1.94)$$

1.1.5.2 α-Heteroatom-substituted organolithium reagents

The α-alkylation of amines is a valuable synthetic transformation.[76] The amino group itself is not sufficiently activating to allow conversion of an α-methyl (R_2N-Me) (or methylene) group into an alkali-metal salt (R_2N-CH_2-M), but certain derivatives of secondary amines can be converted into lithium salts with a strong base. The resulting α-amino-organolithium species react readily with alkyl halides, aldehydes, acid chlorides and other electrophiles. Successful results have been obtained with *N*-nitroso derivatives, various sterically hindered amides or formamidines. For example, dimethylamine can be converted into the amines **91** and **92** via the formamidine **90** (1.95).[77]

$$(1.95)$$

[75] M. J. Mealy and W. F. Bailey, *J. Organomet. Chem.*, **646** (2002), 59; W. F. Bailey, A. D. Khanolkar, K. Gavaskar, T. V. Ovaska, K. Rossi, Y. Thiel and K. B. Wiberg, *J. Am. Chem. Soc.*, **113** (1991), 5720.

[76] A. R. Katritzky and M. Qi, *Tetrahedron*, **54** (1998), 2647; S. V. Kessar and P. Singh, *Chem. Rev.*, **97** (1997), 721; R. E. Gawley and K. Rein, in *Comprehensive Organic Synthesis*, ed. B. M. Trost and I. Fleming, vol. 1 (Oxford: Pergamon Press, 1991), p. 459; and vol. 3 (1991), p. 65.

[77] A. I. Meyers, P. D. Edwards, W. F. Rieker and T. R. Bailey, *J. Am. Chem. Soc.*, **106** (1984), 3270.

By using a chiral auxiliary attached to the formamidine, rather than a tert-butyl group, asymmetric α-alkylation of the secondary amine can be achieved. Good results with a valine-derived auxiliary have been reported (1.96). The methodology is particularly effective with cyclic allylic or benzylic amines, an example of the latter being the alkylation in the 1-position of the tetrahydroisoquinoline ring system, which has provided an important entry to isoquinoline alkaloids. For example, proton abstraction α- to the nitrogen atom of the chiral formamidine **93** and alkylation gave, after cleavage of the formamidine with hydrazine, the tetrahydroisoquinoline **94**, used in a synthesis of the alkaloid reticuline.[78]

$$\text{(1.96)}$$

An alternative and increasingly popular method for the alkylation α- to a nitrogen atom is the use of the tert-butyl carbamate of the secondary amine. With cyclic amines or acyclic allylic or benzylic amines, proton abstraction with a strong base and alkylation can take place.[79] The base *sec*-butyllithium and a diamine ligand (such as *N,N,N′,N′*-tetramethylethylenediamine (TMEDA)) is needed to form the organolithium species with cyclic systems such as *N*-tert-butoxycarbonyl (Boc)-pyrrolidine or -piperidine. For example, 2-substituted piperidines **95** (E=Me, CHO, SPh, SnBu$_3$, etc.) can be prepared by using this chemistry (1.97). The combination of *sec*-butyllithium and the chiral diamine ligand (–)-sparteine **96** effects an asymmetric deprotonation of *N*-Boc-pyrrolidine (but not *N*-Boc-piperidine) to give 2-substituted pyrrolidines with high optical purity (1.98). Likewise, this ligand is effective for the synthesis of enantiomerically enriched benzylic derivatives.[80] The benzylamine **97** can be converted to the substituted benzylamine **98** in this way. Removal of the *p*-methoxyphenyl group with ceric(IV) ammonium nitrate (CAN) gives the carbamate **99** with high optical purity (1.99).

$$\text{(1.97)}$$

E$^+$ = MeI, DMF, PhSSPh, Bu$_3$SnCl, etc.

[78] A. I. Meyers and J. Guiles, *Heterocycles*, **28** (1989), 295.
[79] P. Beak, A. Basu, D. J. Gallagher, Y. S. Park and S. Thayumanavan, *Acc. Chem. Res.*, **29** (1996), 552.
[80] D. Hoppe and T. Hense, *Angew. Chem. Int. Ed. Engl.*, **36** (1997), 2282; A. Basu and S. Thayumanavan, *Angew. Chem. Int. Ed.*, **41** (2002), 716.

(1.98)

$$96$$

(1.99)

$$97 \quad\quad 98 \quad\quad 99 \quad 93\text{-}96\% \text{ ee}$$

Ar = p-MeOC$_6$H$_4$

Tf = SO$_2$CF$_3$

R = Me, Et, allyl, benzyl

An alternative method for the formation of α-amino-organolithium species is from the corresponding α-amino-organostannane. Tin–lithium exchange is effective with the N-Boc derivative of the secondary amine or even with unactivated tertiary amines. For example, addition of n-butyllithium to the stannane **100** generates the α-amino-organolithium species **101** that reacts with various electrophiles (1.100).[81] As the process of tin–lithium exchange is known to occur with retention of configuration at the carbon centre, this chemistry allows a study of the stereo-selectivity (retention, inversion or racemization) on reaction with an electrophile. In general, carbonyl-type electrophiles react with retention of configuration at the carbanion centre, whereas alkyl halides tend to react predominantly with inversion of configuration. Racemization, though, can occur with electrophiles capable of promoting single electron transfer (SET).

(1.100)

$$100 \quad\quad 101$$

The use of tin–lithium exchange can allow the formation of various α-heteroatom-substituted organolithium reagents. Alternatively, suitably activated compounds can be deprotonated to form the required organolithium species. Deprotonation at an allylic or benzylic position tends to be easier than at an unactivated alkyl position. The combination of the allylic nature and the O-carbonyl activating group allows the efficient proton abstraction of the substrate **102** with n-butyllithium (1.101).[82] The crystalline (–)-sparteine-complexed organolithium species **103** reacts with various electrophiles, such as tributyltin chloride, to give the

[81] R. E. Gawley and Q. Zhang, *J. Org. Chem.*, **60** (1995), 5763; R. E. Gawley, E. Low, Q. Zhang and R. Harris, *J. Am. Chem. Soc.*, **122** (2000), 3344.

[82] H. Paulsen, C. Graeve and D. Hoppe, *Synthesis* (1996), 141.

allylstannane **104** (for reactions of allylstannanes, see Section 1.1.8). Alternatively, addition of titanium(IV) isopropoxide prior to tributyltin chloride gives the allylically transposed stannane **105**, also with high optical purity.

$$(1.101)$$

Sulfur and selenium can both stabilize a neighbouring carbanion, owing to their electronegativity and the possibility of delocalization of the electron pair in the carbanion. The sulfur atom may be present as a sulfide, a sulfoxide or a sulfone and may be removed from the product after reaction, if desired, by reductive cleavage, by hydrolysis to a carbonyl compound (for an alkenyl sulfide or dithiane) or by elimination to give an alkene. Controlled rearrangement of the alkylated product before removal of the sulfur extends the scope and utility yet further and therefore this chemistry has many applications in organic synthesis.[83] Selenium analogues can often be used in place of the sulfur compounds, but because of the greater expense of the selenium reagents and their toxicity, the sulfur reagents are normally employed, unless some particular advantage is gained by the use of the selenium derivatives.

Alkyl sulfides are not particularly acidic and the presence of another activating group such as a second sulfur atom or a double bond is generally desirable for convenient reaction. Thus, allyl sulfides (allyl phenyl sulfides are usually employed) are readily converted by *n*-butyllithium into organolithium derivatives that can be alkylated with active alkylating agents, mainly at the position α- to the sulfur atom. Reduction of the product to remove the sulfur atom (e.g. with Raney nickel or with lithium and ethylamine) gives the α-alkylated alkene. The sequence provides a useful method for coupling allyl groups; squalene, for example, was synthesized from farnesyl bromide and farnesyl phenyl sulfide. Coupling of two

---

[83] K. Ogura, in *Comprehensive Organic Synthesis*, ed. B. M. Trost and I. Fleming, vol. 1 (Oxford: Pergamon Press, 1991), p. 505; A. Krief, in *Comprehensive Organic Synthesis*, ed. B. M. Trost and I. Fleming, vol. 3 (Oxford: Pergamon Press, 1991), p. 85.

different allyl groups is easily effected (1.102), and intramolecular alkylation can be carried out. A practical advantage is that no dialkylation products are formed because the monoalkylated compound is not sufficiently acidic to form an anion under the reaction conditions.

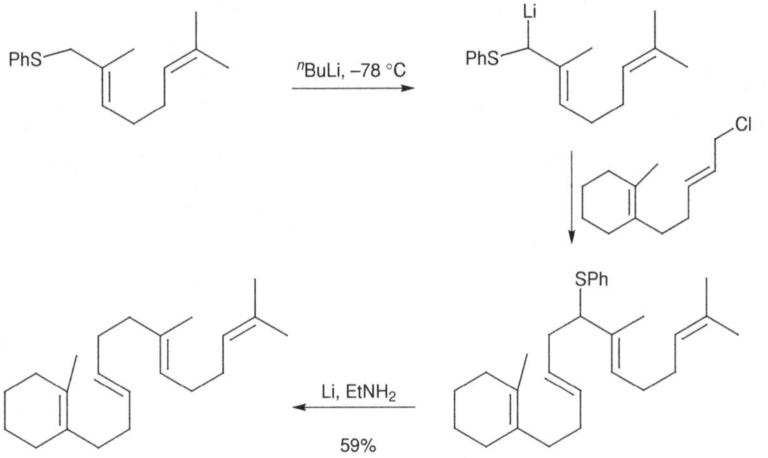

(1.102)

Sulfur ylides are useful reagents in organic synthesis. The ylide is formally a zwitterion in which a carbanion is stabilized by interaction with an adjacent sulfonium centre. They are usually prepared by proton abstraction from a sulfonium salt with a suitable base or by reaction of a sulfide with an alkylating agent such as $Me_3O^+BF_4^-$ or a carbene formed, for example, by metal-catalysed or photolytic decomposition of a diazo compound (1.103).

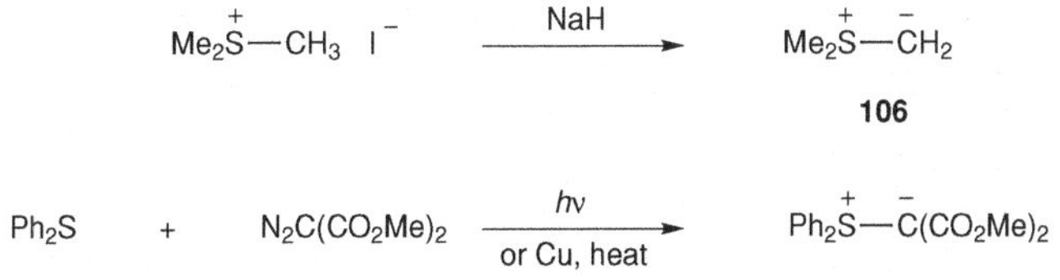

(1.103)

The most useful reaction of a sulfur ylide is with a carbonyl electrophile, in which the major product is an epoxide. Two of the most widely used reagents are dimethylsulfonium methylide **106** and dimethylsulfoxonium methylide **107**.[84] The reaction of the latter ylide with the ketone **108** gave the epoxide **110** *via* the zwitterion **109** (1.104). Unlike the reaction of phosphorus ylides with carbonyl electrophiles (see Section 2.7), which give alkene products, the sulfur ylides lead to the epoxide owing to the lower affinity of sulfur for oxygen, the weak carbon–sulfur

[84] Y. G. Gololobov, A. N. Nesmeyanov, V. P. Lysenko and I. E. Boldeskul, *Tetrahedron*, **43** (1987), 2609.

bond in the zwitterion and the stability of the dimethylsulfide or dimethylsulfoxide and hence its ability to act as a good leaving group.

(1.104)

For other examples of reactions of sulfur ylides with carbonyl compounds to give epoxides, see Schemes 4.104 and 4.105.

Dimethylsulfoxonium methylide **107** is more stable than dimethylsulfonium methylide **106**, although both reagents give epoxides with non-conjugated aldehydes or ketones. However, the two reagents differ slightly in the reactions with, for example, cyclohexanones: in most cases the sulfonium ylide forms an epoxide with a new axial carbon–carbon bond, whereas the sulfoxonium ylide gives an epoxide with an equatorial carbon–carbon bond. This has been ascribed to the fact that the addition of the sulfonium ylide to the carbonyl group to form the intermediate zwitterion is irreversible, whereas addition of the sulfoxonium ylide is reversible, allowing accumulation of the thermodynamically more stable zwitterion.

Dimethylsulfonium methylide and dimethylsulfoxonium methylide also differ in their reactions with α,β-unsaturated carbonyl compounds. The sulfonium ylide reacts at the carbonyl group to form an epoxide, but with the sulfoxonium ylide a cyclopropane derivative is obtained by Michael addition to the carbon–carbon double bond. The difference is again due to the fact that the kinetically favoured reaction of the sulfonium ylide with the carbonyl group is irreversible, whereas the corresponding reaction with the sulfoxonium ylide is reversible, allowing preferential formation of the thermodynamically more stable product from the Michael addition. For example, the cyclopropane **112** is obtained from the reaction of dimethylsulfoxonium methylide with the enone **111** (1.105). Other methods for the formation of cyclopropanes include carbene and Simmons–Smith-type reactions (see Section 4.2).

$$\text{(1.105)}$$

111   **112**

Proton abstraction α- to the sulfur atom of a dialkylsulfoxide is possible with a strong base such as *n*-butyllithium or sodium hydride. The resulting α-sulfinyl carbanion is known to react with carbonyl electrophiles or alkyl halides to give α-alkylated products. With two electron-withdrawing oxygen atoms, sulfones (e.g. $MeSO_2Ph$, pK_a 29) are more acidic than the related sulfoxides (e.g. MeSOMe, pK_a 35) and are therefore deprotonated more readily. Treatment with *n*-butyllithium or a Grignard reagent gives an α-sulfonyl carbanion that undergoes addition to carbonyl or alkyl electrophiles. For example, the sulfone **113** was used in a synthesis of the immunosuppressive agent FK-506. Proton abstraction and addition of the aldehyde **114** gave a mixture of the β-hydroxy-sulfones **115**, which were converted to the ketone **116** (1.106).[85]

$$\text{(1.106)}$$

The removal of the sulfone group can be accomplished under a number of different reductive conditions. Most popular is the concomitant removal of both the sulfone and the derivatized β-hydroxy group to give an alkene and this is commonly termed the Julia olefination reaction (see Section 2.8).

[85] A. B. Jones, A. Villalobos, R. G. Linde and S. J. Danishefsky, *J. Org. Chem.*, **55** (1990), 2786.

When two sulfur atoms are attached to the same carbon atom then proton abstraction at the α-carbon is possible with bases such as *n*-butyllithium. The resulting 2-lithio-1,3-dithianes have proved to be useful reagents in organic synthesis.[86] They react readily with various electrophiles including alkyl halides, epoxides and carbonyl compounds. Subsequent removal of the dithiane group can be accomplished selectively to give a carbonyl compound by hydrolysis or a methylene group by reduction. The ability to hydrolyse the 2-alkylated-1,3-dithiane to an aldehyde or ketone makes the dithiane group a masked carbonyl group and the 2-lithio derivative is classified as an acyl anion equivalent. For example, metallation of 1,3-dithiane and addition of the allyl halide **117** gave the 2-alkylated product **118**, which was hydrolysed to the aldehyde **119** with an aqueous mercury(II) salt (1.107). The 2-lithio-1,3-dithiane acts as an equivalent of the formyl anion $^-$CH=O. This type of process, whereby the normally electrophilic carbon (in this case of the carbonyl group) has been used as a (masked) nucleophilic carbon, and therefore the normal polarization of the functional group has been inverted, is termed 'umpolung'.

$$(1.107)$$

It is possible to abstract a proton from a 2-alkylated-1,3-dithiane and perform a second alkylation reaction. If the 2-alkylated-1,3-dithiane is prepared from the aldehyde with propane-1,3-dithiol, then after alkylation and hydrolysis the sequence provides a method for converting an aldehyde into a ketone. Starting with the commercially available 1,3-dithiane, an alkyl halide or sulfonate can be converted into the homologous aldehyde. Two alkyl groups can actually be introduced successively without isolation of the intermediates and this has been applied to the synthesis of three- to seven-membered cyclic ketones.

Lithiated 1,3-dithianes react readily with epoxides to give thio-acetals of β-hydroxy aldehydes or ketones. Addition of carbonyl electrophiles to 2-lithio-1,3-dithiane is efficient and provides a method for preparing α-hydroxy carbonyl compounds. For example, the ketone **120** can be converted into the hydroxy aldehyde **122** via the alcohol **121** (1.108). The dithiane approaches the ketone from the less-hindered convex face of the fused ring system.

[86] B. T. Gröbel and D. Seebach, *Synthesis* (1977), 357; P. C. B. Page, M. B. van Niel and J. C. Prodger, *Tetrahedron*, **45** (1989), 7643; M. Yus, C. Nájera and F. Foubelo, *Tetrahedron*, **59** (2003), 6147.

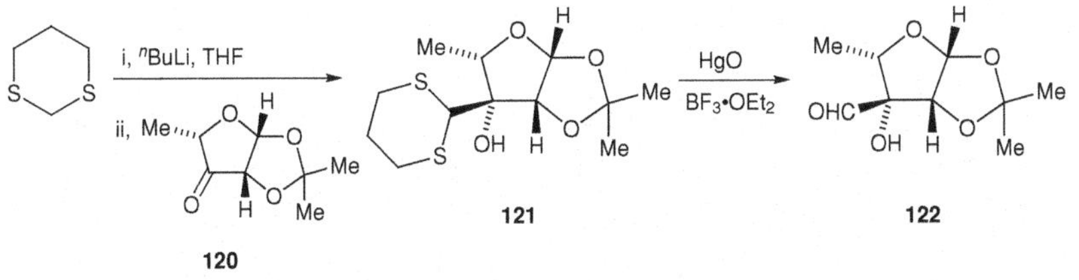

(1.108)

A number of methods for the generation of acyl anion equivalents from aldehydes have been developed. Related to the benzoin condensation, aldehyde cyanohydrins, protected as their ether derivatives, are readily transformed into anions by treatment with lithium diisopropylamide (LDA). Reaction with an alkyl halide gives the protected cyanohydrin of a ketone from which the ketone is liberated easily. Reaction with an aldehyde or ketone leads to the formation of an α-hydroxy ketone (1.109).[87]

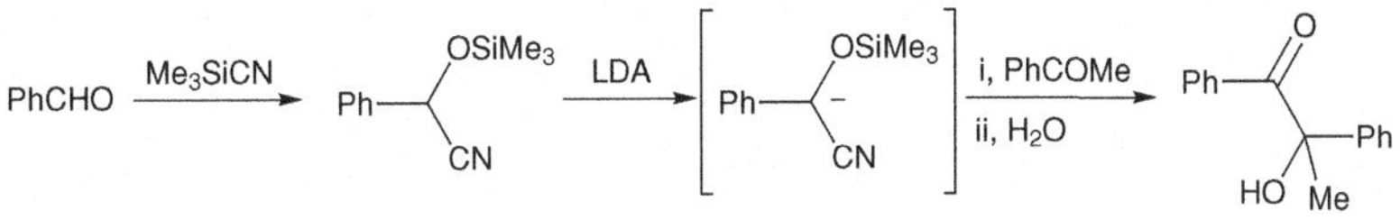

(1.109)

Metallated enol ethers also serve as efficient acyl anion equivalents. The reagents are prepared by action of tert-butyllithium on an enol ether in THF at low temperature. The most useful reagent is α-methoxy-vinyllithium, which reacts with electrophiles to give substituted vinyl ethers, which may be elaborated further or converted, by mild hydrolysis with acid, into the corresponding carbonyl compounds. Reaction with alkyl halides leads to methyl ketones, whereas aldehydes and ketones give α-hydroxy methyl ketones (1.110). Unsaturated carbonyl compounds react at the carbonyl group, although conjugate addition can be achieved by first converting the lithio derivative into the corresponding cuprate (see Section 1.2.1).

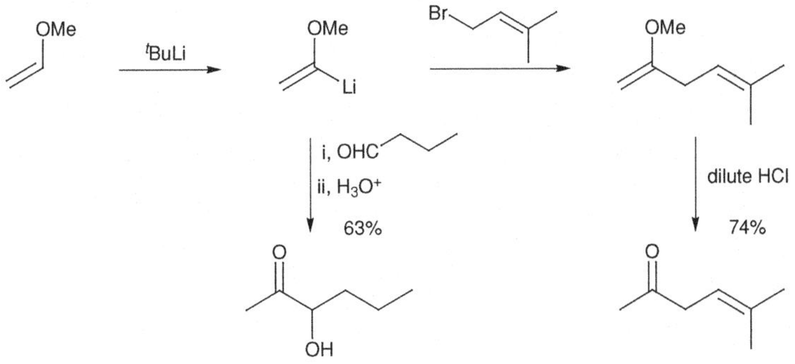

(1.110)

[87] J. D. Albright, *Tetrahedron*, **39** (1983), 3207; A. Hassner and K. M. Lokanatha Rai, in *Comprehensive Organic Synthesis*, ed. B. M. Trost and I. Fleming, vol. 1 (Oxford: Pergamon Press, 1991), p. 541.

Aliphatic nitro compounds serve as good acyl anion equivalents. After electrophilic alkylation at the α-carbon atom, they can be converted into carbonyl compounds by the Nef reaction or by reductive hydrolysis with titanium(III) chloride.
The α-nitro carbanions serve as excellent donors in Michael addition reactions with
α,β-unsaturated systems and therefore the sequence of Michael addition followed
by reductive hydrolysis of the nitro group provides a good route to 1,4-dicarbonyl
compounds. *cis*-Jasmone, for example, was readily obtained by using this strategy
(1.111).

(1.111)

1.1.5.3 Unsaturated organolithium reagents

Proton abstraction of a terminal alkyne gives a metallated alkyne (an acetylide),
suitable for carbon–carbon bond formation with carbon-based electrophiles. The
relatively high acidity of a terminal alkyne ($pK_a \approx 25$) is the result of the high
s-character of the sp-hybridized carbon atom which enhances the stability of the
resulting carbanion. This makes metallated alkynes less basic than sp^3- or sp^2-
hybridized carbanions, but very useful as nucleophiles in addition reactions with
primary alkyl halides or sulfonates, epoxides, aldehydes or ketones.[88] Deprotonation of the alkyne **123** and addition of geranyl bromide has been used in a synthesis of a sesquiterpene (1.112). Addition of metallated alkynes to aldehydes has
proven to be a particularly useful synthetic procedure. For example, formation of
the acetylide from the alkyne **124** and addition of the aldehyde **125** was used in a
synthesis of (+)-allopumiliotoxin 339A, an alkaloid from the South American frog

[88] P. J. Garratt, in *Comprehensive Organic Synthesis*, ed. B. M. Trost and I. Fleming, vol. 3 (Oxford: Pergamon
Press, 1991), p. 271.

of the family *Dendrobatidae* (1.113).

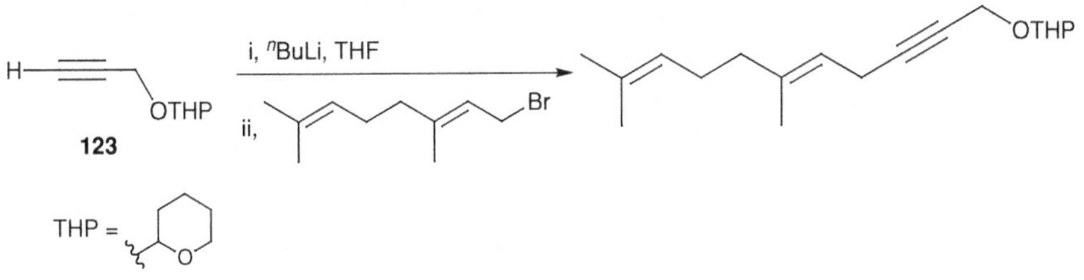

(1.112)

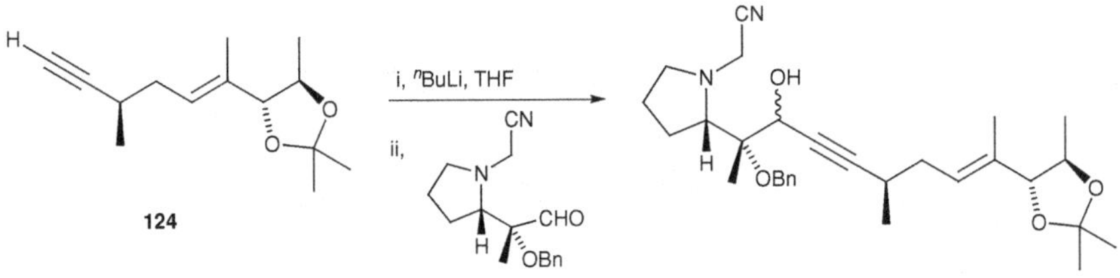

(1.113)

Simple proton abstraction is more difficult in the case of alkenyl substrates ($pK_a \approx 36$, but see Scheme 1.110) and a more-common method for the formation of alkenyllithium (vinyllithium) species is halogen–lithium exchange with *n*- or tert-butyllithium. The exchange reaction is stereospecific, such that the stereochemistry of the starting alkenyl halide determines the stereochemistry of the product alkenyllithium species (1.114). Subsequent reaction with electrophiles also occurs with retention of configuration and this of course is important in order to prepare stereodefined alkenyl products.

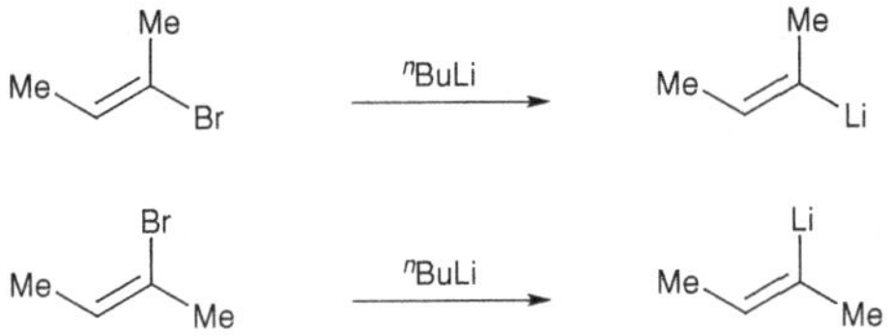

(1.114)

A popular alternative to halogen–lithium exchange is the use of tin–lithium exchange from the corresponding alkenylstannane. Alternatively, conversion of a sulfonyl hydrazone (using the Shapiro reaction) or an alkyne (for example by hydroalumination with diisobutylaluminium hydride) to a metallated alkene are useful procedures (see Sections 2.4 and 2.6).

Metallated alkenes such as alkenyllithium or Grignard species undergo addition reactions with various electrophiles.[89] Reaction with primary alkyl bromides or iodides is possible. Wurtz self-coupled products can be avoided if the alkenyllithium species is generated by tin–lithium exchange, or by insertion of lithium metal into

[89] D. W. Knight, in *Comprehensive Organic Synthesis*, ed. B. M. Trost and I. Fleming, vol. 3 (Oxford: Pergamon Press, 1991), p. 241.

the alkenyl halide. Reaction with carbonyl electrophiles is efficient, leading to allylic alcohol products. For example, Corey prepared the alkenyl organolithium species **126** by tin–lithium exchange with *n*-butyllithium and its reaction with the aldehyde **127** formed part of a synthesis of the fungal metabolite brefeldin A (1.115).[90]

(1.115)

126

127

Direct proton abstraction to form aryllithium species is possible with a suitably substituted aromatic compound. The substituent requires a heteroatom to co-ordinate to the butyllithium base and direct the proton abstraction (sometimes called a complex-induced proximity effect, CIPE).[91] In aromatic compounds, such directed metallation occurs at the position *ortho-* to the substituent and is therefore often referred to as *ortho*-lithiation.[92] A wide variety of directing groups have been used in organic chemistry, the strongest and most common of which, in approximate relative order of directing ability, are shown in Scheme 1.116. The base used for *ortho*-lithiation is normally *n*- or *sec*-butyllithium in THF or Et$_2$O as solvent, often in the presence of the chelating agent TMEDA. An example is the alkylation of diethylbenzamide with iodomethane to give the product **128** (1.117). If X = Br or I, then halogen–lithium exchange occurs in preference to proton abstraction.

X = SO$_2$NR$_2$, OCONR$_2$, CONR, CONR$_2$, OCH$_2$OMe (OMOM), OMe, NCO$_2$R   (1.116)

(1.117)

77%   **128**

The co-operative effect of 1,3-interrelated directing groups is a powerful strategy in synthesis. This allows the otherwise difficult preparation of 1,2,3-trisubstituted aromatic compounds. Hence the lithiation of the aromatic compound **129** is directed by both the methoxy and secondary amide groups (1.118). Two equivalents of the

[90] E. J. Corey and R. H. Wollenberg, *Tetrahedron Lett.* (1976), 4705.
[91] P. Beak and A. I. Meyers, *Acc. Chem. Res.*, **19** (1986), 356.
[92] V. Snieckus, *Chem. Rev.*, **90** (1990), 879.

base are required, the first to deprotonate the amide NH and the second to form the aryllithium species.[93]

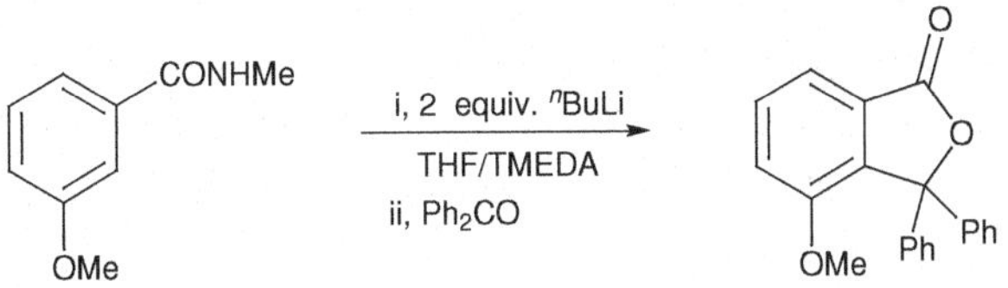

(1.118)

129   65%

Electrophilic quench of aryllithium species with carbonyl electrophiles is particularly efficient. However, alkyl halides (other than iodomethane) are poor electrophiles, probably owing to competing elimination reactions (see Section 2.1). The formation of alkyl-substituted aromatic compounds can be achieved, however, by using epoxide electrophiles or by lithiation and reaction of 2-methylbenzamides, themselves generated by *ortho*-lithiation. For example, the benzamide **128** can be deprotonated at the benzylic position and treated with a variety of electrophiles. Addition of aromatic aldehydes gives, after lactonization, 3-aryl-3,4-dihydroisocoumarins (1.119).

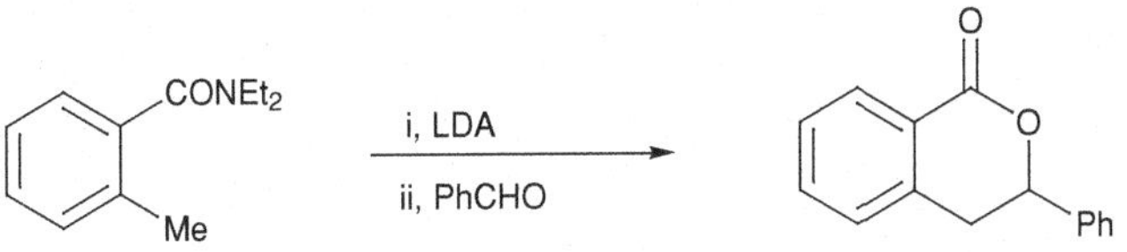

(1.119)

128

Directed lithiation of *O*-aryl carbamates provides a method for the formation of 2-substituted phenols. For example, quenching with dimethylformamide (DMF) gave the aldehyde **130** (1.120). Interestingly, if no electrophile is added but the mixture is allowed to warm to room temperature, then a rearrangement takes place, in which the aryllithium species undergoes attack on the carbamate group (1.121).

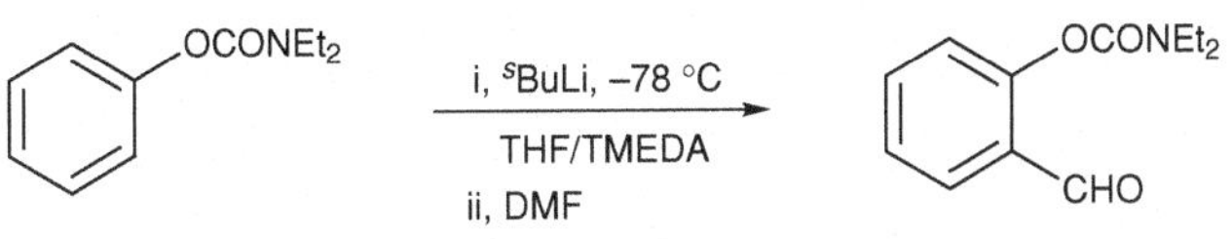

(1.120)

73%   **130**

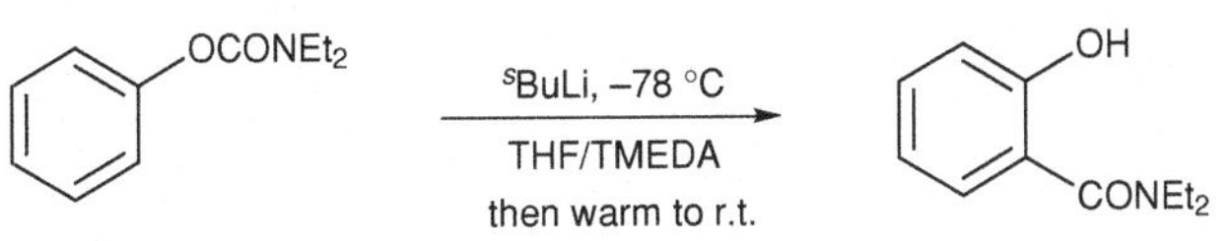

(1.121)

75%

[93] D. W. Slocum and C. A. Jennings, *J. Org. Chem.*, **41** (1976), 3653.

Proton abstraction and electrophilic quench provides a method for the function-alization of heteroaromatic compounds. The five-membered heterocycles furan and thiophen are deprotonated easily with *n*-butyllithium at the 2-position. The resulting 2-lithiofuran or 2-lithiothiophen react with electrophiles such as carbonyl compounds or primary alkyl halides. For example, furan and thiophen have been converted to the 2-substituted derivatives **131** and **132** using such directed lithiation chemistry (1.122).

i, ^{n}BuLi, Et$_2$O, heat

ii, PhCHO

98%   **131**   (1.122)

i, ^{n}BuLi, Et$_2$O, r.t.

ii, ^{n}BuBr

47%   **132**

When the desired aryllithium or heteroaryllithium species is not accessible using directed proton abstraction, then halogen–lithium exchange provides an efficient method for its formation.[94] Bromine–lithium exchange is particularly popular and allows a regiospecific formation of the desired organolithium compound, which can then be used for subsequent carbon–carbon bond formation. For example, 3-lithiofuran can be obtained by treatment of 3-bromofuran with *n*-butyllithium; addition of an aldehyde gives the 3-substituted furan product (1.123).

i, ^{n}BuLi

ii, OHC

60%   (1.123)

An example of the use of both halogen–lithium exchange and directed lithia-tion is shown in Scheme 1.124. Bromine–lithium exchange allows the regiospe-cific formation of the desired 3-lithiophenol derivative from the bromide **133** (an aryllithium species that is not accessible by direct proton abstraction). Addi-tion of the epoxide **134** gave the 3-alkylated aromatic compound **135**. A second

[94] C. Nájera, J. M. Sansano and M. Yus, *Tetrahedron*, **59** (2003), 9255.

aryllithium species is generated by directed lithiation of compound **135**. This occurs *ortho-* to both the OMOM and the $CH_2CH(Ar)OLi$ groups and the resulting aryllithium species was trapped with CO_2. Lactonization and deprotection of the MOM and silyl groups gave the product phyllodulcin.[95]

$$(1.124)$$

1.1.6 Organomagnesium reagents

Organomagnesium reagents are commonly referred to as Grignard reagents. Typically, a Grignard reagent is formed by reaction of an alkyl halide (RX) in ethereal solvent with magnesium to give the species RMgX. An alternative procedure involves the formation of an organolithium species and its conversion to the Grignard reagent with magnesium bromide. The ethereal solvent co-ordinates to the magnesium atom and the Grignard reagents are in equilibrium with the dialkylmagnesium species R_2Mg and MgX_2 (Schlenk equilibrium). Aryl and alkenyl halides can also form Grignard reagents, normally using the more effective co-ordinating solvent THF. A convenient, low-temperature method, which allows the formation of the Grignard reagent in the presence of other (normally reactive) functional groups (such as esters or nitriles) uses iodine–magnesium exchange with i-PrMgBr.[96] Grignard reagents are stable under inert conditions, but reactive towards moisture and air. They have

[95] A. Ramacciotti, R. Fiaschi and E. Napolitano, *J. Org. Chem.*, **61** (1996), 5371.
[96] P. Knochel, W. Dohle, N. Gommermann, F. F. Kneisel, F. Kopp, T. Korn, I. Sapountzis and V. A. Vu, *Angew. Chem. Int. Ed.*, **42** (2003), 4302; M. Rottländer, L. Boymond, L. Bérillon, A. Leprêtre, G. Varchi, S. Avolio, H. Laaziri, G. Quéguiner, A. Ricci, G. Cahiez and P. Knochel, *Chem. Eur. J.*, **6** (2000), 767.

found extensive use in organic synthesis owing to their reactivity towards carbon-based electrophiles, especially carbonyl compounds.[97]

Grignard reagents are less basic than the corresponding organolithium species and they are therefore popular nucleophiles for carbon–carbon bond formation. Addition of Grignard reagents to α-chiral aldehydes or ketones can be highly stereoselective for the Cram or chelation-controlled product (see Section 1.1.5.1). When a heteroatom is present at the α-carbon atom, then chelation often dominates. For example, addition to α-alkoxy ketones gives the cyclic chelate, represented by the Newman projection **137**, and therefore favours the formation of the chelation-controlled product alcohol **136** (1.125).[98]

$$(1.125)$$

>95%   **136**   >99 : 1

137

Asymmetric induction occurs in the addition of a Grignard reagent to an aldehyde or ketone bearing a chiral auxiliary. An example is the use of the 8-phenylmenthol ester **138**, in which Grignard addition to the aldehyde occurs from the front face opposite the bulky substituent on the auxiliary and with the conformation of the carbonyl groups *cis* to one another (due to chelation) as depicted in Scheme 1.126.[99] The product α-hydroxy ester can be reduced to give the corresponding diol with very high optical purity.

$$(1.126)$$

138   86%   99.7 : 0.3

R*OH = 8-phenylmenthol

Despite their reduced basicity in comparison with organolithium species, a potential problem with the use of Grignard reagents is their ability to deprotonate α-to the carbonyl group. This is particularly prevalent with poorly electrophilic or

[97] D. M. Huryn, in *Comprehensive Organic Synthesis*, ed. B. M. Trost and I. Fleming, vol. 1 (Oxford: Pergamon Press, 1991), p. 49.

[98] W. C. Still and J. H. McDonald, *Tetrahedron Lett.*, **21** (1980), 1031.

[99] J. K. Whitesell, A. Bhattacharya and K. Henke, *J. Chem. Soc. Chem. Commun.* (1982), 988.

hindered carbonyl compounds or with bulky Grignard reagents. Bulky Grignard reagents possessing a β-hydrogen atom can act as reducing agents, especially in combination with hindered ketones. Hydride transfer from the Grignard reagent to the carbonyl group gives the alcohol product and an alkene (1.127). For example, addition of isopropylmagnesium bromide to the ketone **139** gave only a low yield of the chelation-controlled addition product **140** (1.128).[100] The major product **141** arises from reduction of the ketone by the Grignard reagent. In addition, a significant amount of enolization occurs with the Grignard reagent acting as a base.

$$(1.127)$$

139   **140** 24%   **141** 43%   **142** 30%

$$(1.128)$$

Grignard addition to the carbonyl group of an ester or lactone results in the formation of a tertiary alcohol product by addition of two equivalents of the organometallic species to the carbonyl carbon. If the ketone is the desired product then a good electrophile is the Weinreb amide RCON(Me)OMe. Addition of the Grignard (or organolithium) species to this amide gives the mono-addition product, which hydrolyses to give the ketone. For example, addition of n-hexylmagnesium bromide to the amide **143** gave the ketone **144** (1.129). The reaction is chemoselective for the amide, leaving the ester group untouched. An alternative method for the formation of ketones is the addition of a Grignard reagent to a nitrile. The intermediate imine is normally hydrolysed easily to the corresponding ketone.

$$(1.129)$$

143   **144**

The addition of a Grignard reagent to a carboxylic ester in the presence of Ti(Oi-Pr)$_4$ and an alkene results in the formation of a hydroxycyclopropane in what is termed the Kulinkovich reaction.[101] The Grignard reagent reacts with Ti(Oi-Pr)$_4$ to give an intermediate titanacyclopropane, which undergoes ligand exchange

[100] M. Tramontini, *Synthesis* (1982), 605.
[101] O. G. Kulinkovich, *Chem. Rev.*, **103** (2003), 2597; O. G. Kulinkovich, S. V. Sviridov and D. A. Vasilevski, *Synthesis* (1991), 234; J. Lee, H. Kim and J. K. Cha, *J. Am. Chem. Soc.*, **118** (1996), 4198.

with the alkene and subsequent reaction with the ester to give, ultimately, the hydroxycyclopropane (1.130).

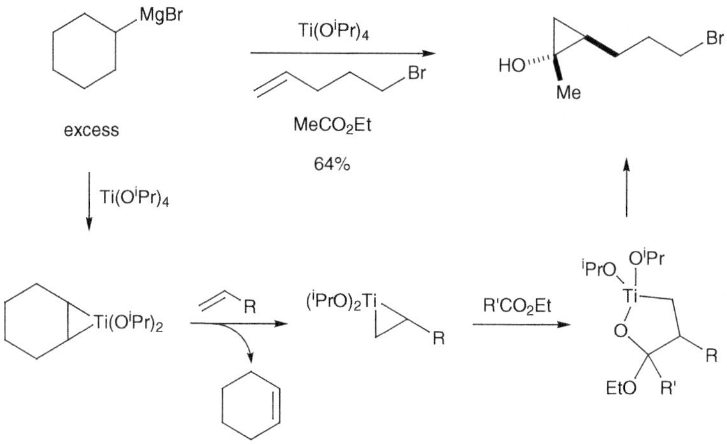

(1.130)

Grignard reagents can give either or both 1,2- and 1,4-addition products on reaction with α,β-unsaturated systems. The extent of conjugate (1,4-) addition depends mostly on the nature of the substituents attached to the unsaturated carbonyl (or other) electrophile. In the absence of significant steric interactions, 1,2-addition takes place using substrates such as α,β-unsaturated aldehydes or unhindered α,β-unsaturated ketones. However, with bulky α,β-unsaturated ketones or esters, in which the carbonyl carbon is more hindered, 1,4-addition predominates. If conjugate addition is the desired pathway, then it is more common to add copper(I) salts to catalyse the reaction. An intermediate organocopper reagent is formed, which adds selectively by 1,4-addition (see Section 1.2.1). Various copper(I) salts are effective, an example being the use of copper(I) iodide in the presence of an equivalent of trimethylsilyl chloride. Under these conditions, conjugate addition generates an intermediate enolate anion, which is trapped as the silyl enol ether in high yield with essentially no 1,2-addition (1.131). The specific lithium enolate can be generated from the silyl enol ether under conditions suitable to allow subsequent alkylation. The toxic additive HMPA may be avoided by using the complex $CuI \cdot 2LiCl$ as the catalyst.[102]

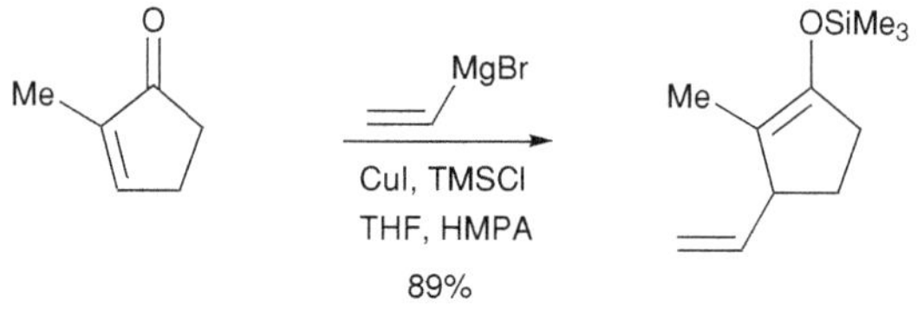

(1.131)

[102] M. T. Reetz and A. Kindler, *J. Organomet. Chem.*, **502** (1995), C5.

Asymmetric conjugate addition with organomagnesium reagents, catalysed by copper(I) salts, has been investigated by using chiral auxiliaries or chiral ligands for the metal atoms. For example, the 2-oxazolidinone-derived auxiliary can lead to high levels of asymmetric induction (1.132).[103] The resulting conjugate addition products can be hydrolysed to give the corresponding carboxylic acid derivatives with high optical purity.

$$(1.132)$$

Epoxides are opened readily by Grignard reagents. Nucleophilic attack normally occurs regioselectively at the less-hindered carbon atom of the epoxide. The ring-opening is assisted by the presence of magnesium halides, present in the Grignard reagent (Schlenk equilibrium), which co-ordinate to the epoxide oxygen atom. In fact the Lewis acidity of the Grignard reagents can cause competing side-reactions, such as rearrangement of the epoxide to a carbonyl compound, which can then react with the Grignard reagent. It can therefore be advantageous in many cases to use copper(I) salts to catalyse the reaction. The ring-opening proceeds with inversion of configuration at the carbon centre being attacked. Alkylation of Grignard reagents or coupling with various types of unsaturated electrophile is possible using copper or other transition-metal catalysis (see Section 1.2).

1.1.7 Organozinc reagents

There has been significant recent growth in the use of organozinc reagents in organic synthesis. Organozinc compounds are less nucleophilic and less basic than the corresponding organolithium or organomagnesium reagents. They can therefore effect chemoselective carbon–carbon bond formation in the presence of otherwise reactive functional groups.[104]

The most common method for the formation of an organozinc reagent involves the insertion of zinc metal into the carbon–iodine bond of an alkyl iodide. For allylic substrates the corresponding bromide or even chloride can be used. The zinc metal normally needs to be 'activated' by washing first with 1,2-dibromoethane and

[103] E. Nicolás, K. C. Russell and V. J. Hruby, *J. Org. Chem.*, **58** (1993), 766; D. R. Williams, W. S. Kissel and J. J. Li, *Tetrahedron Lett.*, **39** (1998), 8593.

[104] P. Knochel and R. D. Singer, *Chem. Rev.*, **93** (1993), 2117; P. Knochel, J. J. Almena Perea and P. Jones, *Tetrahedron*, **54** (1998), 8275; *Organozinc Reagents, A Practical Approach*, ed. P. Knochel and P. Jones (Oxford: Oxford University Press, 1999); A. Bourdier, L. O. Bromm, M. Lotz and P. Knochel, *Angew. Chem. Int. Ed.*, **39** (2000), 4414; P. Knochel, N. Millot, A. L. Rodriguez and C. E. Tucker, *Org. Reactions*, **58** (2001), 417.

trimethylsilyl chloride. Highly activated zinc, which can insert into alkyl (or aryl or alkenyl) bromides, can be prepared by reduction of zinc chloride with lithium naphthalenide.[105] Iodine–zinc exchange can also be carried out with diethylzinc in the presence of catalytic CuI to give a new dialkylzinc species. Organozinc reagents can tolerate many different functional groups, such as esters, ketones or nitriles, any of which may be present within the reagent or electrophile. This can obviously be beneficial, as such functional groups do not require protection, as would be the case with Grignard or organolithium species.

The reactivity of organozinc reagents depends considerably on their structure. Alkyl and aryl zinc reagents themselves are poor nucleophiles and most of their reactions are carried out after prior transmetallation to organocopper or organopalladium species. Reaction with aldehydes can take place in the presence of a Lewis acid (e.g. $BF_3 \cdot OEt_2$ or a titanium(IV) salt). Allylic zinc halides or other organozinc reagents with adjacent unsaturation are relatively good nucleophiles that react with aldehydes without the need for Lewis-acid activation. This is the basis of the Reformatsky reaction, in which an organozinc species is generated from an α-bromoester.[106] Addition to an aldehyde or ketone gives a β-hydroxyester (1.133).

$$\text{PhCHO} \quad + \quad \underset{\underset{\text{Me}}{|}}{\text{Br}}\diagdown\text{CO}_2\text{Me} \quad \xrightarrow[\text{PhH, heat}]{\text{Zn}} \quad \underset{\underset{\text{Me}}{|}}{\overset{\overset{\text{OH}}{|}}{\text{Ph}}}\diagup\diagdown\text{CO}_2\text{Me} \qquad (1.133)$$

80%   syn : anti 63 : 37

Insertion of zinc into allyl bromides occurs readily, for example to give the allyl zinc reagent **145** (1.134). Addition to an aldehyde occurs by attack through the γ-carbon atom to give a homoallylic alcohol. With substrate **145**, bearing a β-ester group, the product homoallylic alcohol cyclizes spontaneously to give the lactone **146**.

$$\underset{\textbf{145}}{\overset{\text{CO}_2\text{Et}}{\diagup\diagdown\text{Br}}} \quad \xrightarrow{\text{Zn}} \quad \underset{}{\overset{\text{CO}_2\text{Et}}{\diagup\diagdown\text{ZnBr}}} \quad \xrightarrow[25\,°C]{\text{PhCHO}} \quad \underset{\textbf{146}}{\text{Ph}\diagdown\text{lactone}} \qquad (1.134)$$

145   88%   **146**

In the presence of a Lewis acid, alkyl zinc halides react with aromatic aldehydes to give secondary alcohols. However, alkyl zinc reagents are less reactive than their allyl derivatives and reaction with aliphatic aldehydes is very sluggish. A solution to this is the use (in the presence of a Lewis acid) of either the dialkyl zinc reagent or the mixed copper–zinc species RCu(CN)ZnX, formed by transmetallation of the alkyl

[105] R. D. Rieke and M. V. Hanson, *Tetrahedron*, **53** (1997), 1925.
[106] A. Fürstner, *Synthesis* (1989), 571; M. W. Rathke and P. Weipert, in *Comprehensive Organic Synthesis*, ed. B. M. Trost and I. Fleming, vol. 2 (Oxford: Pergamon Press, 1991), p. 277.

zinc halide with the THF-soluble copper complex CuCN·2LiCl. The mixed copper–zinc species is more reactive than the organozinc reagent and condenses with acid chlorides (to give ketones), α,β-unsaturated ketones or aldehydes (normally by 1,4-addition), allyl or other activated halides (or sulfonates) (by S_N2' addition), or aromatic or aliphatic aldehydes (in the presence of BF₃·OEt₂).[104] Examples of each of these reactions are shown in Schemes 1.135–1.138. For example, the reagent **148** was prepared from the iodide **147** and reacts with (*R*)-2-phenylpropanal to give the secondary alcohol **149** (1.138). The major product is the *syn* isomer, resulting from Cram (Felkin–Anh) addition.

The coupling of organozinc reagents with unsaturated halides (or sulfonates) is possible in the presence of nickel or palladium(0) complexes. Insertion of zinc metal into an alkyl iodide, followed by palladium-catalysed coupling (see Section 1.2.4) with aryl or alkenyl halides (or sulfonates) is a useful synthetic method and is commonly referred to as the Negishi cross-coupling reaction.[107] Various palladium salts can be used, the simplest of which is tetrakis(triphenylphosphine) palladium(0). For example, coupling of the organozinc iodide **150** with the trifluoromethanesulfonate **151** (formed from hexan-2-one) or the iodide **153** gave the products **152** or **154** respectively (1.139 and 1.140).

[107] E. Negishi, *Acc. Chem. Res.*, **15** (1982), 340; E. Erdik, *Tetrahedron*, **48** (1992), 9577.

$$\text{(1.140)}$$

150   **153**   77%   **154**

Recently it has proved possible to cross-couple alkylzinc halides with primary *alkyl* halides. This can be achieved under nickel catalysis in the presence of tetra-butylammonium iodide and 4-fluorostyrene (1.141).[108] As expected with this chemistry, the reaction tolerates a range of functional groups such as the presence of ketones and carboxylic esters.

$$\text{(1.141)}$$

73%

Dialkyl zinc reagents are more reactive than monoalkyl zinc halides and their addition to aldehydes or enones in the presence of a chiral ligand can occur with high levels of enantioselectivity.[109] Many different catalysts have been employed and the addition of diethylzinc to benzaldehyde is often used as a test reaction for a new chiral ligand. As a result, this transformation can be accomplished with essentially complete enantiocontrol using one of a number of different ligands, such as those shown in Scheme 1.142. For example, the amino alcohol (–)-DIAB gave the addition product with very high selectivity for the (*S*)-enantiomer (1.143).

$$\text{(1.142)}$$

(–)-DIAB   (+)-DBNE   TADDOL   **155**

PhCHO   +   Et₂Zn $\xrightarrow{\text{2 mol\% (–)-DIAB}}$ Ph   (1.143)

98%   99% ee

The enantioselective addition of functionalized dialkylzinc reagents to aldehydes is possible using a chiral ligand and a Lewis-acid catalyst. The dialkylzinc reagents can be prepared by transmetallation (from the corresponding iodides or boranes) with diethylzinc. Selectivities are often best with aromatic aldehydes or

[108] A. E. Jensen and P. Knochel, *J. Org. Chem.*, **67** (2002), 79. For palladium-catalysed couplings with primary alkyl halides, see J. Zhou and G. C. Fu, *J. Am. Chem. Soc.*, **125** (2003), 12527.
[109] R. Noyori and M. Kitamura, *Angew. Chem. Int. Ed. Engl.*, **30** (1991), 49; K. Soai and S. Niwa, *Chem. Rev.*, **92** (1992), 833; B. L. Feringa, *Acc. Chem. Res.*, **33** (2000), 346; L. Pu and H.-B. Yu, *Chem. Rev.*, **101** (2001), 757.

α-substituted α,β-unsaturated aldehydes, although aliphatic aldehydes can give excellent results. The bistrifluoromethanesulfonamide ligand **155** is a particularly effective chiral ligand for such asymmetric transformations (1.144).

$$(1.144)$$

88%   > 96% ee

Asymmetric addition of alkynylzinc reagents to aldehydes has been developed with *N*-methylephedrine (or other chiral amino alcohol) as the chiral ligand.[110] The alkynylzinc reagent is prepared *in situ* from the terminal alkyne and this allows use of the metal salt (zinc triflate) as a catalyst in substoichiometric amount (1.145).

$$(1.145)$$

88%   90% ee

1.1.8 Allylic organometallics of boron, silicon and tin

A useful reaction in organic synthesis is the addition of an allylic organometallic reagent to a carbonyl group.[111] A number of different metals can be employed, although those of boron, silicon and tin have found the most use. The carbon–carbon bond-forming step is often stereoselective and generates the versatile homoallylic alcohol unit (1.146). Oxidative cleavage of the product alkene to the aldehyde (or other carbonyl derivative) provides the β-hydroxy-carbonyl compound and offers an alternative stereoselective approach to the aldol-type product.

$$(1.146)$$

When the allylic organometallic reagent bears an alkyl group in the γ-position (the R group in Scheme 1.146), then a mixture of the *syn* and *anti* products may result. High levels of stereoselectivity can be achieved; for example, an *E*-allylborane or boronate generates predominantly the *anti* product, whereas the *Z*-isomer gives the *syn* product. The reactions with allylboranes or boronates are thought to proceed through a cyclic transition state, in which the substituent attached

---

[110] D. E. Frantz, R. Fässler, C. S. Tomooka and E. M. Carreira, *Acc. Chem. Res.*, **33** (2000), 373; N. K. Anand and E. M. Carreira, *J. Am. Chem. Soc.*, **123** (2001), 9687; L. Pu, *Tetrahedron*, **59** (2003), 9873.
[111] Y. Yamamoto and N. Asao, *Chem. Rev.*, **93** (1993), 2207; W. R. Roush, in *Comprehensive Organic Synthesis*, ed. B. M. Trost and I. Fleming, vol. 2 (Oxford: Pergamon Press, 1991), p. 1; J. W. J. Kennedy and D. G. Hall, *Angew. Chem. Int. Ed.*, **42** (2003), 4732.

to the aldehyde carbon atom prefers a pseudo-equatorial position in the chair-shaped transition state (1.147).

$$E:Z\ 93:7 \qquad\qquad\qquad\qquad\qquad\qquad\qquad\qquad\qquad anti:syn\ 94:6$$

$$(1.147)$$

$$E:Z\ 5:95 \qquad\qquad\qquad\qquad\qquad\qquad\qquad\qquad\qquad anti:syn\ 5:95$$

Allylic organostannanes react with aldehydes under thermal conditions with a high degree of stereocontrol. Like the reactions of allylboranes or boronates, a cyclic transition state has been invoked to explain the preferential formation of the *anti* diastereomer from the *E*-allylstannane (1.148) and the *syn* diastereomer from the *Z*-allylstannane. In contrast, the reaction of allylic organosilanes with aldehydes is sluggish under thermal conditions.

$$E:Z\ 92:8 \qquad\qquad\qquad anti \qquad 87\ :\ 13 \qquad syn$$

$$(1.148)$$

Lewis acids catalyse the addition of allylic organostannanes or organosilanes to aldehydes.[112] In contrast to the thermal reactions of allylboranes or allylstannanes, the use of a Lewis acid promotes reaction *via* an acyclic transition state. With a γ-substituted allylsilane, such as crotyltrimethylsilane **156**, the *E*-isomer reacts with excellent selectivity for the *syn* product (1.149). The corresponding *Z*-isomer (of **156**) also favours the *syn* product, although with reduced selectivity (64:36). The transition state is thought to involve the alignment of the two π-bonds 180° to one another (1.150).

$$\mathbf{156} \qquad\qquad 92\% \qquad\qquad syn \qquad 97\ :\ 3 \qquad anti$$

$$(1.149)$$

[112] I. Fleming, J. Dunoguès and R. Smithes, *Org. Reactions*, **37** (1989), 57; I. Fleming, A. Barbero and D. Walter, *Chem. Rev.*, **97** (1997), 2063; Y. Nishigaichi, A. Takuwa, Y. Naruta and K. Maruyama, *Tetrahedron*, **49** (1993), 7395.

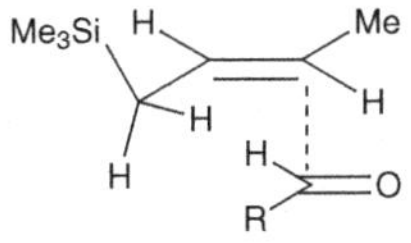

(1.150)

Note that the aldehyde approaches the alkene from the direction *anti* to the silicon atom. Therefore, when a chiral allylsilane or allylstannane with a substituent in the α-position is used, chirality transfer takes place, to generate the homoallylic alcohol with essentially no loss in enantiomeric purity.[113] For example, reaction of the aldehyde **157** with the chiral allylsilane **158**, using boron trifluoride etherate as the catalyst, gave predominantly the *syn* product **159** (1.151). The absolute stereochemistry can be determined by using a model in which the hydrogen atom on the α-carbon of the allylsilane eclipses the alkene (the so-called 'inside hydrogen effect') in order to minimize steric interactions (1.152).

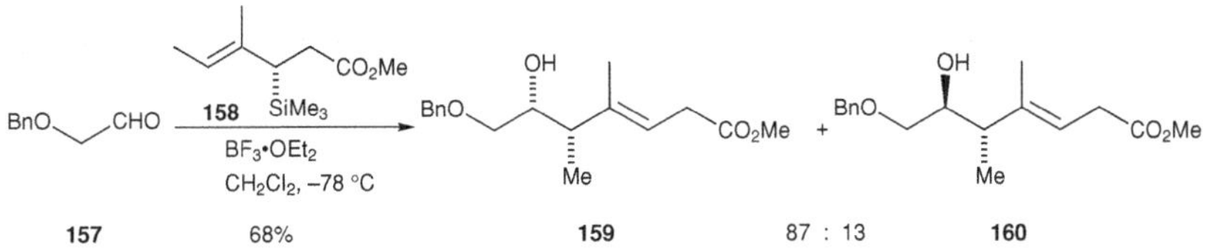

(1.151)

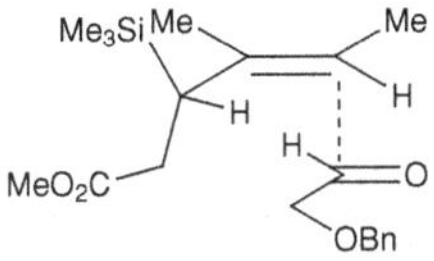

(1.152)

A variety of Lewis acids have been used to promote the addition of an allylsilane or allylstannane to an aldehyde (or ketone or imine). The Lewis acid $BF_3 \cdot OEt_2$ is effective and promotes Cram (Felkin–Anh)-type addition (see Section 1.1.5.1). However, Lewis acids such as $TiCl_4$, $SnCl_4$ or $MgBr_2$ can co-ordinate to a neighbouring (normally α-) heteroatom and promote chelation-controlled addition. For example, allylation of the aldehyde **161** gave either the Cram-type product or the chelation-controlled product depending upon the nature of the Lewis acid (1.153).

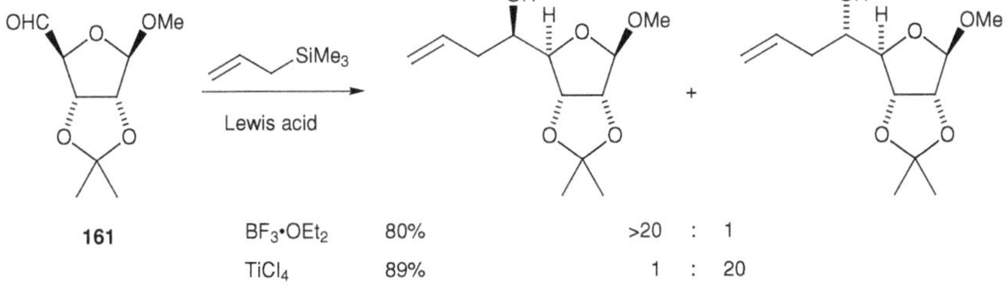

(1.153)

| **161** | $BF_3 \cdot OEt_2$ | 80% | >20 : 1 |
| | $TiCl_4$ | 89% | 1 : 20 |

---

[113] C. E. Masse and J. S. Panek, *Chem. Rev.*, **95** (1995), 1293; J. A. Marshall, *Chem. Rev.*, **96** (1996), 31; E. J. Thomas, *J. Chem. Soc. Chem. Commun.* (1997), 411.

Asymmetric allylation of aldehydes or ketones can be accomplished by using either a chiral auxiliary on the boron atom or a chiral ligand on the Lewis acid. A number of different chiral allylboron or boronate reagents have been developed. The allylboron species **162**, referred to as (Ipc)$_2$B-allyl and derived from (+)-α-pinene, gives very good results in the addition to a variety of aldehyde electrophiles.[114] For example, addition to propanal (propionaldehyde) gave the homoallylic alcohol **163** with high optical purity (1.154). The use of (−)-α-pinene allows the preparation of the other enantiomer of the homoallylic alcohol product. Derivatives of (Ipc)$_2$B–allyl **162** with substituents in the γ-position can be prepared easily and, on addition to an aldehyde, give rise to the *anti* product (from the *E*-alkene) or the *syn* product (from the *Z*-alkene) with high enantiomeric purity.

$$\text{162} \qquad \xrightarrow[\text{-78 °C}]{\text{EtCHO}} \qquad \text{163} \qquad \qquad (1.154)$$

    162   71%   **163** 86% ee

An effective chiral catalyst for asymmetric allylation of aldehydes or ketones is the complex formed between the axially chiral 1,1′-bi-2-naphthol (BINOL) **164** and a titanium(IV) salt.[115] Addition of allyltributylstannane to an aldehyde using this Lewis acid gives the product homoallylic alcohol with high optical purity. For example, allylation of *iso*-butyraldehyde gave the alcohol **165** as a 98:2 ratio of enantiomers (1.155). Use of (*S*)-BINOL gave the other enantiomer of the product.

$$\text{(}R\text{)-BINOL } \mathbf{164} \qquad \xrightarrow[\substack{\text{10 mol\% Ti(O}^i\text{Pr)}_4 \\ \text{10 mol\% BINOL } \mathbf{164} \\ -20 \text{ °C}}]{{}^i\text{PrCHO}} \qquad \text{165} \qquad (1.155)$$

   (*R*)-BINOL **164**   89%   **165** 96% ee

[114] H. C. Brown and P. K. Jadhav, *J. Am. Chem. Soc.*, **105** (1983), 2092; U. S. Racherla and H. C. Brown, *J. Org. Chem.*, **56** (1991), 401.

[115] G. E. Keck, K. H. Tarbet and L. S. Geraci, *J. Am. Chem. Soc.*, **115** (1993), 8467; A. L. Costa, M. G. Piazza, E. Tagliavini, C. Trombini and A. Umani-Ronchi, *J. Am. Chem. Soc.*, **115** (1993), 7001; K. M. Waltz, J. Gavenonis and P. J. Walsh, *Angew. Chem. Int. Ed.*, **41** (2003), 3697. For a review of catalytic asymmetric allylation, see S. E. Denmark and J. Fu, *Chem. Rev.*, **103** (2003), 2763. For a different approach to asymmetric allylation of aldehydes which avoids allyl metal species, see J. Nokami, K. Nomiyama, S. Matsuda, N. Imai and K. Kataoka, *Angew. Chem. Int. Ed.*, **41** (2003), 1273.

1.2 Transition-metal chemistry

The use of transition metals to promote the formation of carbon–carbon bonds has grown tremendously in recent years. This section describes some of the important transformations using the metals copper, chromium, cobalt and palladium. Many other useful reactions with other transition metals have been developed,[116] some of which are described in later sections (see Sections 2.6, 2.9 and 2.10).

1.2.1 Organocopper reagents

Some discussion of the use of organocopper reagents for carbon–carbon bond formation has been described in Section 1.1.6 (use of a Grignard reagent in the presence of a copper salt as a catalyst) and Section 1.1.7 (conversion of organozinc reagents to the corresponding organocopper reagents using $CuCN \cdot 2LiCl$).

There are various types of stoichiometric organocopper reagent, the most common being R_2CuLi, $RCu(CN)Li$ or $R_2Cu(CN)Li_2$. These species have different reactivities and careful choice of the reagent is required.[117] They are prepared *in situ* and not isolated. For example, a lithium dialkylcuprate species, R_2CuLi, often referred to as a Gilman reagent, is most conveniently prepared by reaction of two equivalents of an organolithium compound with copper(I) iodide in diethyl ether (1.156). The composition of the reagent solutions and the state of aggregation of the complexes are not well defined. In diethyl ether solution, lithium dimethylcuprate is thought to exist as a dimer. A drawback with the dialkylcuprate reagents is that only one of the alkyl groups is transferred to the electrophile. This is clearly wasteful and solutions to this problem have been developed, using a mixed organocuprate containing, for example, an alkyne or 2-thienyl group as the unreactive ligand.

$$\text{MeLi} \quad + \quad \text{CuI} \quad \xrightarrow{\text{Et}_2\text{O}} \quad \text{MeCu} \quad \xrightarrow{\text{MeLi}} \quad \text{Me}_2\text{CuLi} \qquad (1.156)$$

Perhaps the most widespread use of organocopper reagents is for conjugate-addition reactions.[117,118] Organocopper reagents are soft in nature and, like enolates (see Section 1.1.2), give good yields of the 1,4-addition product on reaction with α,β-unsaturated carbonyl compounds (1.157). Transfer of the alkyl group to the β-carbon of an α,β-unsaturated ketone usually works well. However, not all

---

[116] *Comprehensive Organometallic Chemistry II*, ed. E. W. Abel, F. G. A. Stone and G. Wilkinson, vol. 12 (Oxford: Elsevier, 1995); *Transition Metals in Organic Synthesis, A Practical Approach*, ed. S. E. Gibson (Oxford: Oxford University Press, 1997).

[117] E. Erdik, *Tetrahedron*, **40** (1984), 641; B. H. Lipshutz and S. Sengupta, *Org. Reactions*, **41** (1992), 135; *Organocopper Reagents, A Practical Approach*, ed. R. J. K. Taylor (Oxford: Oxford University Press, 1994); N. Krause and A. Gerold, *Angew. Chem. Int. Ed.*, **36** (1997), 187.

[118] G. H. Posner, *Org. Reactions*, **19** (1972), 1; J. A. Kozlowski, in *Comprehensive Organic Synthesis*, ed. B. M. Trost and I. Fleming, vol. 4 (Oxford: Pergamon Press, 1991), p. 169.

α,β-unsaturated carbonyl compounds are good substrates. Conjugate addition with β,β-disubstituted enones can be less successful owing to steric crowding. With an α,β-unsaturated aldehyde, 1,2-addition often competes with conjugate addition. With unsaturated esters, conjugate addition is much more sluggish. In these cases, much better yields may be obtained by effecting the reaction in the presence of trimethylsilyl chloride or a Lewis acid such as boron trifluoride etherate (1.158).[119]

A variety of organocopper reagents effect conjugate addition, a popular method being the use of a Grignard reagent in the presence of a copper(I) salt, such as CuI·2LiCl. Stoichiometric organocopper reagents are also effective, including lithium dialkylcuprates or so-called 'higher-order' organocuprates, such as $R_2Cu(CN)Li_2$, formed by the addition of two molar equivalents of the organolithium species to CuCN. Good yields of the conjugate addition product can be obtained (1.159). Highly functionalized organocopper reagents, the formation of which may be incompatible with the required starting magnesium or lithium reagent, can be prepared from the organozinc species (see Section 1.1.7). Insertion of zinc into a functionalized alkyl iodide, followed by transmetallation to the organocopper species with CuCN·2LiCl and conjugate addition can be performed (1.136).

$$(1.157)$$

$$(1.158)$$

$$(1.159)$$

The mechanism of the transfer of the alkyl group from the organocuprate to the β-position of conjugated ketones is uncertain. Evidence points to an initial complexation of the organocopper(I) species to the enone (d-π* complex), followed

[119] E. Nakamura and I. Kuwajima, *J. Am. Chem. Soc.*, **106** (1984), 3368; Y. Yamamoto, *Angew. Chem. Int. Ed. Engl.*, **25** (1986), 947; B. H. Lipshutz, S. H. Dimock and B. James, *J. Am. Chem. Soc.*, **115** (1993), 9283.

by formation of a copper(III) intermediate. Reductive elimination then transfers the alkyl group from the metal to the β-carbon atom (1.160).[120]

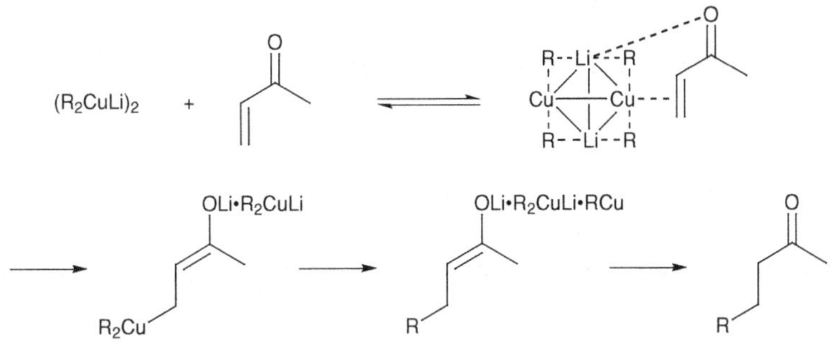

$$(1.160)$$

Whatever the exact mechanism of the conjugate-addition reaction, it seems clear that enolate anions are formed as intermediates and they can be trapped as the silyl enol ether or alkylated with various electrophiles.[121] For example, addition of lithium methylvinyl cuprate (a mixed-cuprate reagent) to cyclopentenone generates the intermediate enolate **166**, that can be alkylated with allyl bromide to give the product **167** (1.161). The *trans* product often predominates, although the *trans:cis* ratio depends on the nature of the substrate, the alkyl groups and the conditions and it is possible to obtain the *cis* isomer as the major product. Examples of intramolecular trapping of the enolate are known, as illustrated in the formation of the *cis*-decalone **168**, an intermediate in the synthesis of the sesquiterpene valerane (1.162).

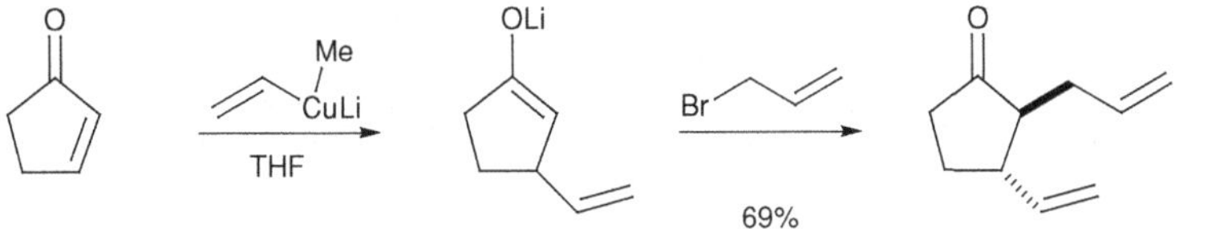

$$(1.161)$$

166　　　　**167**

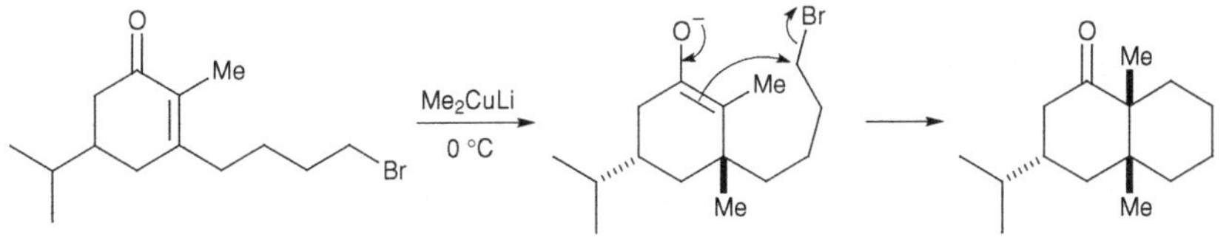

$$(1.162)$$

168

Conjugate addition is also a feature of the reaction of organocuprates with α,β-acetylenic carbonyl compounds. By conducting the reaction at −78 °C, high yields of *cis* addition compounds can be obtained (1.163). This allows the stereocontrolled

[120] S. R. Krauss and S. G. Smith, *J. Am. Chem. Soc.*, **103** (1981), 141; see also H. O. House, *Acc. Chem. Res.*, **9** (1976), 59.
[121] M. J. Chapdelaine and M. Hulce, *Org. Reactions*, **38** (1990), 225.

preparation of trisubstituted alkenes. An alternative approach to α,β-unsaturated carbonyl compounds involves addition–elimination using, for example, a β-iodo-enone. Addition of organometallic species to alkynes is described in Section 2.6.

$$C_7H_{15}\!\!=\!\!=\!\!CO_2Me \xrightarrow[-78\ °C]{Me_2CuLi} \begin{matrix} C_7H_{15} & CO_2Me \\ \diagdown & \diagup \\ Me & Cu \end{matrix} \xrightarrow{H_3O^+} \begin{matrix} C_7H_{15} & CO_2Me \\ \diagdown & \diagup \\ Me \end{matrix} \qquad (1.163)$$

Asymmetric conjugate addition with organocopper reagents has been investigated using chiral auxiliaries or chiral ligands.[122] For example, the 2-oxazolidinone-derived auxiliary can lead to high levels of asymmetric induction (1.132). Other auxiliaries, based on carboxylic esters or amides, have been reported and lead, after cleavage of the auxiliary, to conjugate addition products with high optical purity. Recent developments have focused on the use of chiral ligands to effect asymmetric conjugate addition.[123] Chiral amines or phosphines are good ligands for copper and an example of this approach is outlined in Scheme 1.164. The ligand **169** and a copper salt can be used as catalysts to promote the asymmetric conjugate addition to acyclic or cyclic enones with good to excellent levels of enantioselectivity.

$$\text{(structure of ligand 169)} \qquad \text{(cyclohexenone)} \xrightarrow[\text{5 mol% 169}]{Et_2Zn,\ 2.5\ mol\%\ Cu(OTf)_2} \text{(3-ethylcyclohexanone)} \qquad (1.164)$$

169   >98% ee

A useful carbon–carbon bond-forming reaction involving organocopper reagents is the coupling of the alkyl group on the organocuprate with an alkyl halide.[124] Reactions take place readily at or below room temperature to give high yields of substitution products. Primary alkyl tosylates also react well. Many different functional groups can be tolerated. For example, ketones react only slowly, so selective reaction in the presence of an unprotected ketone carbonyl group is possible. However, aldehydes undergo the normal carbonyl addition unless the temperature is kept below about −90 °C. Some representative reactions are shown in Schemes 1.165–1.167.

$$\text{(hexyl OTs)} \xrightarrow[Et_2O,\ -75\ °C]{Bu_2CuLi} C_{10}H_{22} \qquad (1.165)$$

98%

[122] B. E. Rossiter and N. M. Swingle, *Chem. Rev.*, **92** (1992), 771.
[123] A. Alexakis and C. Benhaim, *Eur. J. Org. Chem.* (2002), 3221; N. Krause and A. Hoffmann-Röder, *Synthesis* (2001), 171; L. A. Arnold, R. Imbos, A. Mandoli, A. H. M. de Vries, R. Naasz and B. L. Feringa, *Tetrahedron*, **56** (2000), 2865; S. J. Degrado, H. Mizutani and A. H. Hoveyda, *J. Am. Chem. Soc.*, **123** (2001), 755.
[124] G. H. Posner, *Org. Reactions*, **22** (1975), 253.

$$(1.166)$$

$$(1.167)$$

Secondary alkyl bromides and iodides generally do not give good yields of product on reaction with lithium dialkylcuprates, R_2CuLi, but this difficulty may be overcome by reaction with the higher-order cuprates, $R_2Cu(CN)Li_2$. These reagents also react very readily with primary alkyl bromides, including some that give only poor yields with the Gilman cuprates. It should be noted, however, that many of these displacement reactions can be effected just as well, or even better, with the appropriate Grignard reagent and a copper(I) salt as a catalyst. This avoids the use of stoichiometric amounts of the copper reagent. Good yields can be obtained by using catalytic lithium tetrachlorocuprate, Li_2CuCl_4 ($CuCl_2 \cdot 2LiCl$), prepared easily from lithium chloride and copper(II) chloride.[125] The active species is believed to be an organocopper(I) complex produced from the copper(II) halide and the Grignard reagent (1.168).

$$(1.168)$$

Reaction of lithium diphenylcuprate with $(-)$-(R)-2-bromobutane takes place with predominant inversion of configuration (1.169) and this (and other features of the reaction of organocuprates with secondary alkyl halides) suggests that they proceed by S_N2 displacement at carbon. However, the corresponding iodide gives a racemic product on reaction with lithium diphenylcuprate, and there is evidence that reaction of cuprates with iodides takes place by a one-electron transfer process and not by S_N2 displacement.[126] In contrast, alkenyl halides react with organocuprates to give the substituted alkene with retention of configuration of the double bond.

[125] G. Cahiez, C. Chaboche and M. Jézéquel, *Tetrahedron*, **56** (2000), 2733.
[126] E. C. Ashby, R. N. De Priest, A. Tuncay and S. Srivastava, *Tetrahedron*, **23** (1982), 5251.

In the example in Scheme 1.170, the *E*-alkene is formed from the *E*-alkenyl halide. Likewise, the *Z*-alkene is formed selectively from the *Z*-alkenyl halide. Alkenyl triflates are also good substrates for alkylation with organocuprate reagents. Treatment of the alkenyl triflate **170** with methyl magnesium bromide and copper(I) iodide as catalyst, gave the alkylated product **171**, used in a synthesis of paeonilactone A (1.171).[127]

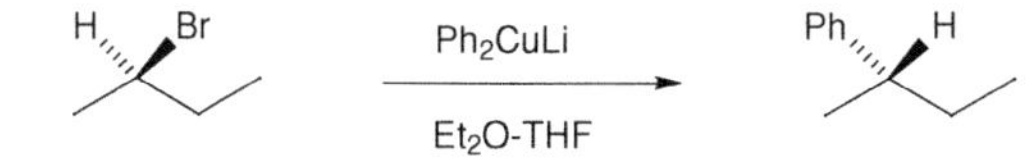

(1.169)

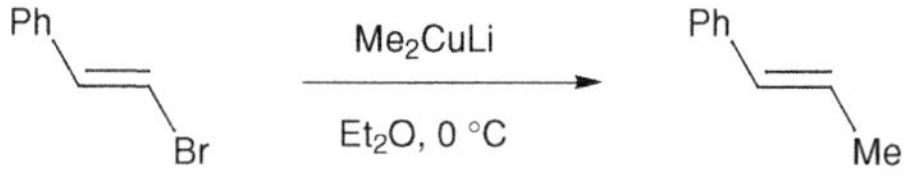

(1.170)

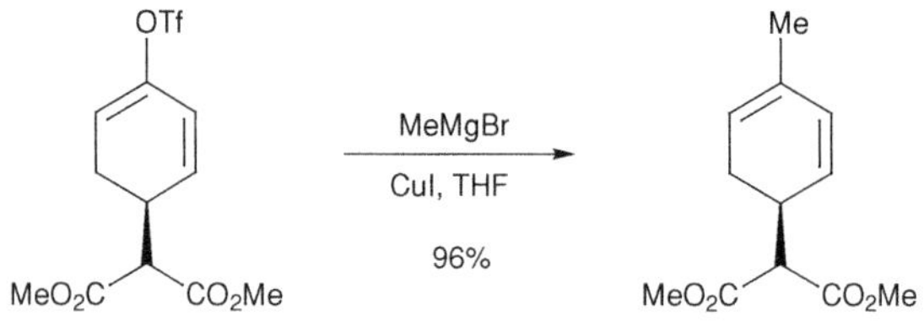

170   **171**

(1.171)

When an alkenyl copper species is used in the displacement reaction, then the alkenyl species reacts with retention of configuration of the double bond (1.172). Alkenyl radicals can interconvert readily and these results suggest, therefore, that alkenyl radicals are not involved in these reactions.

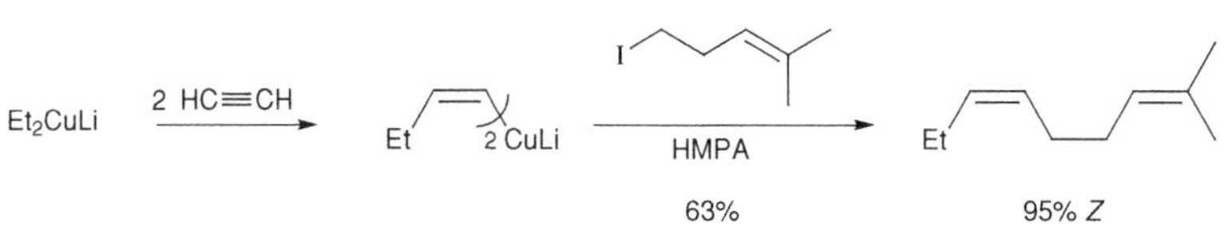

(1.172)

As illustrated in the example in Scheme 1.172, *Z*-alkenylcuprates are obtained readily by *syn* addition of organocuprates to alkynes (see Section 2.6). Alkylation of these cuprates provides a good route to *Z*-1,2-disubstituted alkenes. Reaction with alkenyl halides, best performed in the presence of zinc bromide and Pd(PPh$_3$)$_4$, gives rise to conjugated dienes with high stereoselectivity (1.173). Coupling of alkenyl cuprates with aryl halides gives styrene derivatives with stereochemical control. Allylic halides and acetates also react with organocuprates. Reaction can occur at either end of the allylic system to give the unrearranged (S$_N$2) or rearranged

[127] C. Jonasson, M. Rönn and J.-E. Bäckvall, *J. Org. Chem.*, **65** (2000), 2122.

(S$_N$2′) products.

$$(1.173)$$

Acid chlorides react readily with organocuprates to give ketones.[128] The reaction is chemoselective for the acid chloride and no addition to functional groups such as ketones, esters or nitriles takes place. For example, addition of lithium dibutylcuprate to the acid chloride **172** gave the ketone **173** (1.174).

$$(1.174)$$

The ring-opening of epoxides is often best carried out with an organocopper reagent. Gilman reagents and higher-order cuprates in particular work well, although the use of a Grignard reagent in the presence of copper(I) iodide as catalyst is also very effective. Nucleophilic attack occurs with inversion of configuration at the carbon atom being attacked and takes place, with an unsymmetrical epoxide, at the less sterically hindered carbon atom. Thus, ring-opening of the epoxide **174** gave the alcohol **175** (1.175).[129]

$$(1.175)$$

1.2.2 Organochromium chemistry

Organochromium reagents have found varied use in organic synthesis. Aryl-chromium complexes influence significantly the reactivity of the aromatic ring and have been used widely. Other unsaturated chromium complexes, alkyl-chromium species and chromium carbenes promote useful transformations and continue to attract attention.

[128] R. K. Dieter, *Tetrahedron*, **55** (1999), 4177.
[129] B. H. Lipshutz, R. S. Wilhelm, J. A. Kozlowski and D. Parker, *J. Org. Chem.*, **49** (1984), 3928.

Arylchromium complexes can be prepared easily, simply by heating the arene with chromium hexacarbonyl, $Cr(CO)_6$ or by ligand exchange with (naphthalene) chromium tricarbonyl complex. The desired arylchromium complex is generated, bearing the arene (η^6 species) and three carbon monoxide ligands on the chromium(0) atom (18 electron complex) (1.176). The chromium atom exerts an electron-withdrawing effect on the aromatic ring. This allows nucleophilic attack on the aromatic ring,[130] rather than the normal electrophilic attack. In addition, the electron-deficient arene ring can support a negative charge, thereby allowing metallation on the ring[131] or benzylic position.[132] After reaction, the chromium can be released easily, by mild oxidation.

$$\text{(1.176)}$$

Nucleophilic addition to the arylchromium complex occurs from the face opposite the bulky chromium atom and gives an intermediate η^5-cyclohexadienyl anion complex, such as **176** (1.177). If the nucleophile attacks at the same carbon atom as a halide substituent (*ipso* position), then subsequent loss of the halide leads to an overall nucleophilic substitution. This type of reaction is most effective with fairly 'soft' nucleophiles ($pK_a < 20$), in which nucleophilic attack is thought to be reversible.

$$\text{(1.177)}$$

176

Addition of a nucleophile to an arylchromium complex need not take place at the *ipso* position. Using a 'hard' nucleophile ($pK_a > 20$), in which nucleophilic attack is irreversible, a mixture of products from attack at the *ortho-*, *meta-* or *para-* positions can result. Commonly, attack at the *meta-* position predominates,

[130] M. F. Semmelhack, in *Comprehensive Organometallic Chemistry II*, ed. E. W. Abel, F. G. A. Stone and G. Wilkinson, vol. 12 (Oxford: Elsevier, 1995), p. 979.
[131] M. F. Semmelhack, in *Comprehensive Organometallic Chemistry II*, ed. E. W. Abel, F. G. A. Stone and G. Wilkinson, vol. 12 (Oxford: Elsevier, 1995), p. 1017.
[132] S. G. Davies and T. D. McCarthy, in *Comprehensive Organometallic Chemistry II*, ed. E. W. Abel, F. G. A. Stone and G. Wilkinson, vol. 12 (Elsevier, 1995), p. 1039.

as illustrated in Schemes 1.178 and 1.179; however, the regioselectivity depends on the nature and location of the substituents attached to the aromatic ring and on the nucleophile. Protonation of the cyclohexadienyl anion, hydrogen shift and elimination of HX leads to an overall substitution (but not at the carbon atom that bore the halogen atom) (1.178). Alternatively, if a halogen or other leaving group is not present, then the intermediate cyclohexadienyl anion can be oxidized, leading to an overall alkylation (1.179), or trapped with an electrophile to give the disubstituted cyclohexadiene.

$$(1.178)$$

$$(1.179)$$

Addition of reagents such as *n*- or *s*-butyllithium to the arylchromium complex normally results in lithiation of the aromatic ring, rather than nucleophilic attack. The aryllithium species reacts readily with a wide range of electrophiles to give substituted aromatic compounds. For example, lithiation of the chromium complex of fluorobenzene with *n*-butyllithium (which occurs *ortho*- to the fluorine atom), followed by addition of γ-butyrolactone gave the ketone **177** (1.180). The intermediate aryllithium species adds to the carbonyl carbon atom to generate a new arylchromium complex. This complex is activated to nucleophilic attack and the released alkoxide group substitutes with the fluoride to give **177**.

$$(1.180)$$

Lithiation at the benzylic position of arylchromium complexes can occur readily using a suitable base. Addition of an electrophile then occurs on the side opposite the bulky chromium metal (1.181). As expected, addition of a nucleophile to a chromium-complexed benzylic cation[133] or other electrophilic group occurs from

---

[133] S. G. Davies and T. J. Donohoe, *Synlett* (1993), 323.

the less-hindered, uncomplexed face (1.182). It should be realized that 1,2- and 1,3-disubstituted arylchromium complexes in which the substituents are different are chiral. The two mirror images, in which the chromium tricarbonyl fragment is attached above or below the plane of the aromatic ring, can often be resolved. This can then lead, after decomplexation of the metal, to enantiomerically enriched products.

$$(1.181)$$

$$(1.182)$$

Gaining increasing popularity is the use of allyl or alkenyl (or other unsaturated) chromium species to effect carbon–carbon bond formation.[134] Most common is the addition of $CrCl_2$ to an unsaturated halide followed by coupling with an aldehyde (Nozaki–Hiyama–Kishi reaction). Chromium(II) inserts into the unsaturated halide (or sulfonate) to give the corresponding organochromium(III) reagent. The insertion can be catalysed by nickel salts ($NiCl_2$) or manganese powder (Mn(0)). Organochromium reagents have low basicity and tolerate many different functional groups, reacting chemoselectively with aldehydes (1.183–1.185). The procedure therefore provides a mild method for the synthesis of allylic and homoallylic alcohols (even in the presence of ketones or carboxylic esters). There is current interest in the development of such couplings in an asymmetic and catalytic manner. For example, addition of the alkenyl iodide **178** to the aldehyde **179** in the presence of 10 mol% of the chromium salt and the chiral ligand (*S,S*)-**180**, using manganese(0) for recycling, gave the allylic alcohol **181** with good enantioselectivity (1.185).[135]

$$(1.183)$$

[134] A. Fürstner, *Chem. Rev.*, **99** (1999), 991; L. A. Wessjohann and G. Scheid, *Synthesis* (1999), 1; N. A. Saccomano, in *Comprehensive Organic Synthesis*, ed. B. M. Trost and I. Fleming, vol. 1 (Oxford: Pergamon Press, 1991), p. 173.

[135] A. Berkessel, D. Menche, C. A. Sklorz, M. Schröder and I. Paterson, *Angew. Chem. Int. Ed.*, **42** (2003), 1032.

$$(1.184)$$

$$(1.185)$$

In addition to unsaturated halides, *gem*-dihaloalkanes react with chromium(II) salts. The resulting organochromium species reacts with aldehydes to provide a mild method for alkenylation (see Section 2.9).

Fischer carbene complexes [(CO)$_5$M=CRR′], M=Cr, Mo, W, have found widespread application in organic synthesis. Addition of an organolithium species (RLi) to chromium hexacarbonyl [Cr(CO)$_6$], followed by *O*-methylation, generates the chromium carbene complex **182**. These carbene complexes undergo many types of reaction, giving access to substituted cyclic products.[136] Noteworthy is the Dötz reaction, in which unsaturated complexes (**182**, R=aryl or alkenyl) react with alkynes under thermal conditions to give phenols (1.186).[137] The reaction is thought to proceed by insertion of the alkyne to give the new carbene **183**, followed by insertion of carbon monoxide and cyclization.

$$(1.186)$$

1.2.3 Organocobalt chemistry

The most common use of cobalt in organic synthesis is as its alkyne complex. Addition of dicobalt octacarbonyl [Co$_2$(CO)$_8$] to an alkyne generates the stable organocobalt complex **184** that exists as a tetrahedral cluster (1.187). This complex

[136] J. W. Herndon, *Tetrahedron*, **56** (2000), 1257; W. D. Wulff, *Organometallics*, **17** (1998), 3116; L. S. Hegedus, *Tetrahedron*, **53** (1997), 4105; D. F. Harvey and D. M. Sigano, *Chem. Rev.*, **96** (1996), 271.
[137] K. H. Dötz and P. Tomuschat, *Chem. Soc. Rev.*, **28** (1999), 187.

protects the alkyne and therefore reactions can be carried out elsewhere in the molecule without affecting the alkyne group. The alkyne can be released under mild oxidative conditions.

$$R\!\!=\!\!\!=\!\!\!=\!\!R' \quad + \quad [Co_2(CO)_8] \quad \xrightarrow{\text{heat}} \quad \text{(1.187)}$$

184

An important transformation of dicobalt complexes of alkynes is called the Pauson–Khand reaction, named after its discoverers in the early 1970s. Heating the complex with an alkene generates a cyclopentenone product, an important ring system. The reaction combines the alkyne, alkene and carbon monoxide, in what is formally a [2+2+1] cycloaddition.[138] Good levels of regioselectivity can be obtained, with the larger alkyne substituent adopting the position adjacent to the carbonyl group in the product(s) (**185** and **186**) (1.188). The use of an unsymmetrical alkene, however, commonly results in a mixture of regioisomers (**185** and **186**) of the product cyclopentenone. The mechanism for the transformation is thought to involve loss of CO from the alkyne complex, co-ordination of the alkene into the vacant site, insertion of the alkene into a carbon–cobalt bond (at the less-hindered alkyne carbon atom), then insertion of CO and reductive elimination (1.189).[139]

$$R^S\!\!=\!\!\!=\!\!R^L \quad + \quad \overset{}{\underset{R}{\diagup\!\!\!\diagdown}} \quad \xrightarrow[\text{heat}]{[Co_2(CO)_8]} \quad \text{185} \quad + \quad \text{186} \quad \text{(1.188)}$$

185   **186**

$$\text{(1.189)}$$

185

[138] K. M. Brummond and J. L. Kent, *Tetrahedron*, **56** (2000), 3263; A. J. Fletcher and S. D. R. Christie, *J. Chem. Soc., Perkin Trans. 1* (2000), 1657; N. E. Schore, in *Comprehensive Organometallic Chemistry II*, ed. E. W. Abel, F. G. A. Stone and G. Wilkinson, vol. 12 (Oxford: Elsevier, 1995), p. 703; N. E. Schore, *Org. Reactions*, **40** (1991), 1.
[139] P. Magnus and L. M. Principe, *Tetrahedron Lett.*, **26** (1985), 4851.

As both the formation of the dicobalt complex of the alkyne and the Pauson–Khand reaction occur under thermal conditions, both steps can be accomplished simultaneously. The Pauson–Khand reaction is effective for the formation of simple cyclopentenones by an intermolecular process (1.190), but has found more-widespread application for intramolecular cycloaddition (1.191). The major diastereomer of the bicyclic enone from Scheme 1.191 was used in a formal synthesis of the antitumor sesquiterpene coriolin.[140]

$$\text{(1.190)}$$

$$\text{(1.191)}$$

A significant advance in the Pauson–Khand reaction was made by the discovery that various additives, such as tertiary amine *N*-oxides, promote the cycloaddition reaction.[141] For example, treatment of the dicobalt complexed alkyne **187** with trimethylamine *N*-oxide at only 0 °C provides the cyclopentenone **188** in good yield (1.192). More recent advances have been made in catalytic Pauson–Khand reactions.[142] Only 3 mol% of dicobalt octacarbonyl [$Co_2(CO)_8$] under one atmosphere of CO effects the formation of the cyclopentenone **188** from the alkyne **189** in benzene at 70 °C (an improvement in the yield to 90% was achieved in the presence of the additive $Bu_3P=S$) (1.193).[143]

$$\text{(1.192)}$$

$$\text{(1.193)}$$

[140] C. Exon and P. Magnus, *J. Am. Chem. Soc.*, **105** (1983), 2477.

[141] S. Shambayati, W. E. Crowe and S. L. Schreiber, *Tetrahedron Lett.*, **31** (1990), 5289; N. Jeong, Y. K. Chung, B. Y. Lee, S. H. Lee and S.-E. Yoo, *Synlett* (1991), 204.

[142] S. E. Gibson (née Thomas) and A. Stevenazzi, *Angew. Chem. Int. Ed.*, **42** (2003), 1800; Y. K. Chung, *Coord. Chem. Rev.*, **188** (1999), 297.

[143] M. Hayashi, Y. Hashimoto, Y. Yamamoto, J. Usuki and K. Saigo, *Angew. Chem. Int. Ed.*, **39** (2000), 631.

The formation of a cation adjacent to an alkyne is enhanced when the alkyne is complexed to dicobalt hexacarbonyl. The metal stabilizes the propynyl cation, hence promoting its formation and reaction with various nucleophiles (often called the Nicholas reaction) (1.194).[144] The propynyl cation is commonly generated from the alcohol or an ether or epoxide derivative using HBF_4 or a Lewis acid. Alternatively, addition of a Lewis acid to an aldehyde or acetal is effective. For example, formation of the dicobalt complexed alkyne **190**, followed by treatment with the Lewis acid $BF_3 \cdot OEt_2$ gave the intermediate propynyl cation, which was trapped intramolecularly by the allyl silane to give the cyclic product **191** (5:1 *trans:cis*) (1.195).[145] Subsequent Pauson–Khand reaction gave the tricyclic product **192**. Heteroatom or other carbon-based nucleophiles can add to the propynyl cation. Silyl enol ethers or other metal enolates give rise to alkylation products. Using the chiral boron enolate **193**, high selectivity for the *syn* product **194** was obtained, together with essentially complete asymmetric induction (1.196).[146]

CAN = ceric ammonium nitrate

(1.194)

190   191   192

(1.195)

193   194   >98 : 2 *syn : anti*

(1.196)

[144] A. J. M. Caffyn and K. M. Nicholas, in *Comprehensive Organometallic Chemistry II*, ed. E. W. Abel, F. G. A. Stone and G. Wilkinson, vol. 12 (Oxford: Elsevier, 1995), p. 685; C. Mukai and M. Hanaoka, *Synlett* (1996), 11; B. J. Teobald, *Tetrahedron*, **58** (2002), 4133.

[145] S. L. Schreiber, T. Sammakia and W. E. Crowe, *J. Am. Chem. Soc.*, **108** (1986), 3128.

[146] P. A. Jacobi and W. Zheng, *Tetrahedron Lett.*, **34** (1993), 2581.

Another noteworthy transformation that is particularly effective with organo-cobalt complexes is the cyclotrimerization of alkynes.[147] Co-trimerization of three monoalkynes is not normally synthetically useful as it leads to a mixture of regioi-somers of the product arenes. However, by tethering at least two of the alkynes and using a third bulky alkyne that does not undergo self-condensation, then single iso-mers of substituted bicyclic aromatic compounds can be formed in high yield. The alkyne bis-trimethylsilyl acetylene does not self-trimerize owing to its steric bulk, but does condense with the metallacyclopentadiene intermediate **196** formed from the diyne **195** (1.197). The cobalt complex [CpCo(CO)$_2$] allows efficient [2+2+2] cycloaddition with **195**, $n=0$–2.[148] The cyclotrimerization is not restricted to three alkynes, but can be accomplished with one component as a nitrile (leading to a pyridine)[149] or as an alkene (leading to a cyclohexadiene).

$$(1.197)$$

195   **196**

1.2.4 Organopalladium chemistry

Of all the transition metals, palladium has found the most widespread use in organic synthesis. Organopalladium species tolerate many different functional groups and promote a variety of carbon–carbon (and other) bond-forming reactions with extremely high chemo- and regioselectivity.

Oxidative addition of palladium(0) species into unsaturated halides or triflates provides a popular method for the formation of the σ-bound organopalladium(II) species. It is important to use an unsaturated (e.g. aryl or alkenyl) halide or tri-flate, as β-hydride elimination of alkyl palladium species can take place readily. Oxidative addition of palladium(0) into alkenyl halides (or triflates) occurs stere-ospecifically with retention of configuration. The palladium is typically derived from tetrakis(triphenylphosphine)palladium(0), [Pd(PPh$_3$)$_4$], or tris(dibenzylidene-acetone)dipalladium(0), [Pd$_2$(dba)$_3$], or by *in situ* reduction of a palladium(II) species such as [Pd(OAc)$_2$] or [Pd(PPh$_3$)$_2$Cl$_2$].

Organopalladium species generated by oxidative addition react with organ-ometallic species or with compounds containing a π-bond, such as alkynes or alkenes.[150] Various different organometallic species can be used, although

[147] D. B. Grotjahn, in *Comprehensive Organometallic Chemistry II*, ed. E. W. Abel, F. G. A. Stone and G. Wilkinson, vol. 12 (Oxford: Elsevier, 1995), p. 741.

[148] K. P. C. Vollhardt, *Angew. Chem. Int. Ed. Engl.*, **23** (1984), 539.

[149] G. Chelucci, *Tetrahedron: Asymmetry*, **6** (1995), 811.

[150] J. Tsuji, *Palladium Reagents, Catalysts, Innovations in Organic Synthesis* (New York: Wiley, 1995).

particularly popular is to effect an overall cross-coupling reaction using an organo-stannane (Stille reaction).[151] Two examples of this reaction using alkenyl triflates are shown in Schemes 1.198 and 1.199, although alkenyl, aryl and other unsaturated halides are just as effective.

(1.198)

(1.199)

The Stille reaction is one of the most popular for cross-coupling, owing to the ease of preparation and stability of the organostannanes. The reaction has found consid-erable use in organic synthesis, promoting both inter- and intramolecular couplings, even for the formation of large ring systems.[152] Polystyrene-supported palladium(0) catalyst was used for the macrocyclization of the substrate **197** (1.200).[153] Acidic hydrolysis of the two methoxyethoxymethyl (MEM) groups in the product **198** completed a synthesis of zearalenone.

(1.200)

The mechanism for the coupling involves, after oxidative addition, transmetal-lation of the organopalladium and organometallic species (R—M, e.g. M=SnBu$_3$), to generate a new organopalladium species containing two carbon–palladium σ-bonds. This then undergoes reductive elimination to give the coupled product and regenerate the palladium(0) catalyst. The catalytic cycle is often represented as shown in Scheme 1.201 (ligands on palladium omitted for clarity, X = halogen or OTf).[154]

[151] V. Farina, V. Krishnamurthy and W. J. Scott, *Org. Reactions*, **50** (1997), 1; J. K. Stille, *Angew. Chem. Int. Ed. Engl.*, **25** (1986), 508.
[152] M. A. J. Duncton and G. Pattenden, *J. Chem. Soc., Perkin Trans. 1* (1999), 1235.
[153] A. Kalivretenos, J. K. Stille and L. S. Hegedus, *J. Org. Chem.*, **56** (1991), 2883.
[154] C. Amatore and A. Jutand, *Acc. Chem. Res.*, **33** (2000), 314.

$$(1.201)$$

The transmetallation is normally the slow step, such that the organic halide or triflate must be chosen to avoid more-rapid β-hydride elimination of the intermediate organopalladium(II) species **199**. The organometallic species can, however, have a wide variety of structures as reductive elimination occurs faster than β-elimination. In practice, transmetallation is best with unsaturated organometallics, such as alkynyl-, alkenyl- or arylstannanes, although saturated alkyl zinc reagents can promote effective cross-coupling (see Section 1.1.7, Schemes 1.139 and 1.140).[107] The choice of halide or triflate is also important. Unsaturated iodides are normally most reactive. Cross-couplings with unsaturated bromides or triflates normally work well, especially in the presence of an additive such as LiCl. Recently, it has been found that the less-reactive aryl chlorides can be coupled using electron-rich phosphine ligands such as tri-tert-butylphosphine, P(t-Bu)$_3$.[155]

If the coupling reaction is carried out in the presence of carbon monoxide, insertion of CO into the intermediate organopalladium(II) species **199** occurs. This generates an acyl palladium(II) species that undergoes transmetallation with the organometallic species, leading to a ketone product (1.202).

$$(1.202)$$

Organoboron reagents can be used as the organometallic partner in the cross-coupling (Suzuki reaction).[156] This reaction is attractive as it avoids the formation

[155] A. F. Littke and G. C. Fu, *Angew. Chem. Int. Ed.*, **41** (2002), 4176; A. F. Littke, L. Schwarz and G. C. Fu, *J. Am. Chem. Soc.*, **124** (2002), 6343.
[156] N. Miyaura and A. Suzuki, *Chem. Rev.*, **95** (1995), 2457.

of toxic trialkyltin halide by-products. In addition, alkenyl boronate species can be prepared easily by hydroboration of alkynes. Boronate esters or boronic acids can be used in the coupling reaction. Suzuki cross-couplings (1.203) are best performed in the presence of a base, which is thought to react with the organoboron species to form a more-reactive borate complex, thereby enhancing the rate of transmetallation.

$$(1.203)$$

86%

Like the Stille reaction, unsaturated iodides react fastest, although bromides and triflates can be used. Unsaturated chlorides react very slowly and need high temperatures, although electron-rich phosphine ligands promote their coupling.[157] By using tri-tert-butylphosphine as the ligand, cross-coupling with aryl chlorides can take place even at room temperature. The chloride reacts in preference to the triflate (1.204).

$$(1.204)$$

95%

Coupling reactions of *alkyl* boranes, formed by hydroboration of alkenes, with unsaturated halides (or triflates or phosphonates) is possible, and this reaction is finding increasing use in synthesis.[158] For example, coupling of the alkyl borane derived from hydroboration (with 9-borobicyclo[3.3.1]nonane, 9-BBN) of the alkene **200** with the alkenyl iodide **201** gave the substituted cyclopentene **202**, used in a synthesis of prostaglandin E$_1$ (1.205). This type of *B*-alkyl Suzuki coupling reaction is very useful for the synthesis of substituted alkenes.

$$(1.205)$$

[157] V. V. Grushin and H. Alper, *Chem. Rev.*, **94** (1994), 1047; D. W. Old, J. P. Wolfe and S. L. Buchwald, *J. Am. Chem. Soc.*, **120** (1998), 9722; A. F. Littke, C. Dai and G. C. Fu, *J. Am. Chem. Soc.*, **122** (2000), 4020; see also Reference 154.

[158] S. R. Chemler, D. Trauner and S. J. Danishefsky, *Angew. Chem. Int. Ed.*, **40** (2001), 4544.

Remarkably, cross-couplings of alkyl boranes with *alkyl* bromides or even chlorides are possible using the catalyst [Pd$_2$(dba)$_3$] and the ligand tricyclohexylphosphine, PCy$_3$ (Cy $= C_6H_{11}$).[159] For example, the alkyl chloride **203** was coupled to the alkyl borane **204** (prepared by chemoselective hydroboration with 9-BBN; see Section 5.1) to give the product **205** (1.206). The [Pd$_2$(dba)$_3$]/PCy$_3$ catalyst system overcomes the normally slow oxidative addition of the alkyl halide to the palladium and promotes cross-coupling to alkyl boranes in preference to β-hydride elimination. Such *B*-alkyl Suzuki reactions are likely to be used as key carbon–carbon bond-forming reactions in future synthetic sequences.

$$ (1.206) $$

203   **204**   73%   **205**

Organozinc, organomagnesium or other organometallic species can be effective partners in palladium-catalysed coupling reactions. Metals other than palladium(0), such as nickel(0) or copper(I) (see Section 1.2.1), can alternatively be used to promote cross-coupling of unsaturated halides and organometallic species.[160]

A convenient method for coupling alkynes to unsaturated halides or triflates is the Sonogashira reaction.[161] This uses a terminal alkyne with a copper(I) salt as co-catalyst. An intermediate alkynyl copper species is generated that effects the transmetallation process with the organopalladium(II) species **199** (1.201), following the same catalytic cycle as the Stille and Suzuki reactions. Cross-couplings with aryl halides or triflates using copper(I) iodide at room temperature (or on mild heating) give rise to arylalkynes (1.207). An amine is added to act as a base (to aid formation of the alkynyl copper species) and to reduce the palladium(II) pre-catalyst to the required palladium(0) complex. Additives such as Bu$_4$NI or P(t-Bu)$_3$ can promote low-temperature Sonogashira cross-coupling reactions.[162]

$$ (1.207) $$

68%

[159] M. R. Netherton, C. Dai, K. Neuschütz and G. C. Fu, *J. Am. Chem. Soc.*, **123** (2001), 10 099; J. H. Kirchhoff, C. Dai and G. C. Fu, *Angew. Chem. Int. Ed.*, **41** (2002), 1945.
[160] G. D. Allred and L. S. Liebeskind, *J. Am. Chem. Soc.*, **118** (1996), 2748; V. P. W. Böhm, T. Weskamp, C. W. K. Gstöttmayr and W. A. Herrmann, *Angew. Chem. Int. Ed.*, **39** (2000), 1602.
[161] K. Sonogashira, in *Comprehensive Organic Synthesis*, ed. B. M. Trost and I. Fleming, vol. 3 (Oxford: Pergamon Press, 1991), p. 521.
[162] E. Negishi and L. Anastasia, *Chem. Rev.*, **103** (2003), 1979; R. R. Tykwinski, *Angew. Chem. Int. Ed.*, **42** (2003), 1566.

Cross-couplings with alkenyl halides (or triflates) give rise to enynes. This has found particular application in the preparation of enediyne compounds, present in antitumor agents such as neocarzinostatin and calicheamicin. Two consecutive Sonogashira cross-coupling reactions with Z-1,2-dichloroethene using 2.5 mol% palladium(0) and 5 mol% copper(I) gave the enediyne **206**, which was used to prepare the core structures of calicheamicinone and dynemicin A (1.208).[163] Cross-couplings of two alkynes (Glaser reaction) provides diynes.[164]

One of the most important transformations catalysed by palladium is the Heck reaction.[165] Oxidative addition of palladium(0) into an unsaturated halide (or triflate), followed by reaction with an alkene, leads to overall substitution of a vinylic (or allylic) hydrogen atom with the unsaturated group. For example, formation of cinnamic acid derivatives from aromatic halides and acrylic acid or acrylate esters is possible (1.209). Unsaturated iodides react faster than the corresponding bromides and do not require a phosphine ligand. With an aryl bromide, the ligand tri-*o*-tolylphosphine is effective (1.210).[166] The addition of a metal halide or tetra-alkylammonium halide can promote the Heck reaction. Acceleration of the coupling can also be achieved in the presence of silver(I) or thallium(I) salts, or by using electron-rich phosphines such as tri-tert-butylphosphine.[167]

[163] P. Magnus, S. A. Eisenbeis, R. A. Fairhurst, T. Iliadis, N. A. Magnus and D. Parry, *J. Am. Chem. Soc.*, **119** (1997), 5591.
[164] P. Siemsen, R. C. Livingston and F. Diederich, *Angew. Chem. Int. Ed.*, **39** (2000), 2632.
[165] R. F. Heck, *Org. Reactions*, **27** (1982), 345; A. de Meijere and F. E. Meyer, *Angew. Chem. Int. Ed. Engl.*, **33** (1994), 2379; I. P. Beletskaya and A. V. Cheprakov, *Chem. Rev.*, **100** (2000), 3009.
[166] For Heck reactions of aryl bromides and chlorides, see N. J. Whitcombe, K. K. Hii and S. E. Gibson, *Tetrahedron*, **57** (2001), 7449.
[167] A. F. Littke and G. C. Fu, *J. Am. Chem. Soc.*, **123** (2001), 6989; see also Reference 154.

The reaction is thought to proceed by co-ordination of the alkene with the organopalladium(II) species, followed by carbopalladation. Subsequent β-hydride elimination regenerates an alkene and releases palladium(II). This is reduced (reductive elimination) to palladium(0) in the presence of a base, to allow further oxidative addition and continuation of the cycle (1.211). The carbopalladation and β-hydride elimination steps occur *syn* selectively. Excellent regioselectivity, even for intermolecular reactions, is often observed, with the palladium normally adding to the internal position of terminal alkenes (except when the alkene substituent is electron-rich as in enamines or enol derivatives), thereby leading to linear substitution products.

(1.211)

A mixture of alkenes can result when there is a choice of β-hydrogen atoms for the β-hydride elimination step. As the elimination is reversible, there is a preference for the formation of the more stable alkene. For example, Heck reaction between iodobenzene and the allylic alcohol **207** gave the intermediate organopalladium(II) species **208**. β-Hydride elimination to the more stable enol results (after enol–keto tautomerism) in the formation of the ketone **209** (1.212).[168]

(1.212)

Intramolecular Heck reactions are particularly efficient and have been used considerably in organic synthesis.[169] *In situ* reduction of palladium acetate and oxidative addition of the resulting palladium(0) into the aryl iodide **210** gave an

[168] R. C. Larock, E. K. Yum and H. Yang, *Tetrahedron*, **50** (1994), 305.
[169] E. Negishi, C. Copéret, S. Ma, S.-Y. Liou and F. Liu, *Chem. Rev.*, **96** (1996), 365; J. T. Link, *Org. Reactions*, **60** (2002), 157.

intermediate organopalladium(II) species. Cyclization onto the alkene generates a six-membered ring and a new organopalladium(II) species, which undergoes β-hydride elimination to give the alkene **211**, used in a synthesis of tazettine (1.213).[170]

An efficient synthesis of the alkaloid strychnine makes good use of an intramolecular Heck reaction. Oxidative addition of palladium(0) into the alkenyl iodide **212** and cyclization onto the cyclohexene gave, after β-hydride elimination and silyl deprotection, isostrychnine **213** (1.214).[171] Isomerization of the new alkene to give the enone and cyclization of the alcohol onto the enone gives strychnine.

$$\text{(1.213)}$$

$$\text{(1.214)}$$

In the presence of a chiral ligand, asymmetric Heck reactions can be carried out.[172] The axially-chiral bisphosphine ligand (*R*)- or (*S*)-BINAP promotes good to excellent levels of enantioselectivity in intramolecular Heck reactions. For example, insertion of palladium into the aryl iodide **214** followed by cyclization gave the indolinone **215** in high enantiomeric excess, used in a synthesis of physostigmine (1.215).[173]

[170] M. M. Abelman, L. E. Overman and V. D. Tran, *J. Am. Chem. Soc.*, **112** (1990), 6959.
[171] V. H. Rawal and S. Iwasa, *J. Org. Chem.*, **59** (1994), 2685.
[172] A. B. Dounay and L. E. Overman, *Chem. Rev.*, **103** (2003), 2945; M. Shibasaki and E. M. Vogl, *J. Organomet. Chem.*, **576** (1999), 1 (see also other articles in this issue on organopalladium chemistry).
[173] T. Matsuura, L. E. Overman and D. J. Poon, *J. Am. Chem. Soc.*, **120** (1998), 6500.

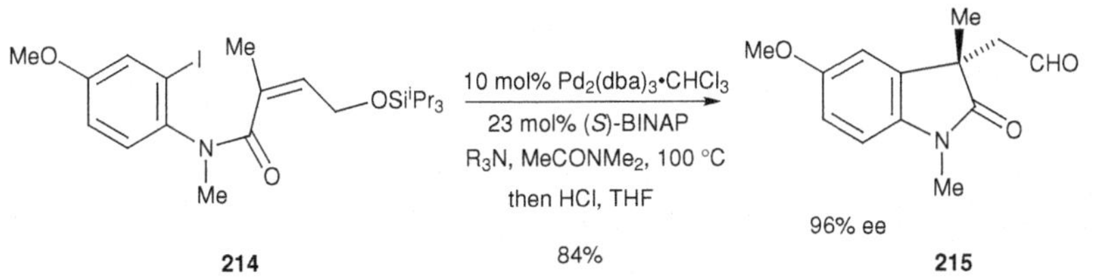

(1.215)

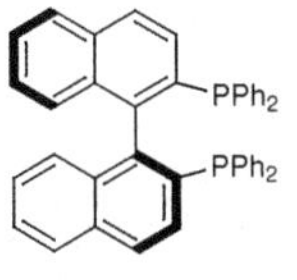

(S)-BINAP

The Heck reaction, in which β-hydride elimination is disfavoured or not possible (e.g. for lack of β-hydrogen atom), is useful since further carbon–carbon bond formation can be promoted using the organopalladium(II) intermediate. The second carbon–carbon bond can be formed intramolecularly, by using a suitably positioned alkene or alkyne (as in an impressive synthesis of the steroid ring system in Scheme 1.216), or intermolecularly, for example by trapping the organopalladium(II) species with carbon monoxide to give an ester (1.217). The conversion of unsaturated halides or triflates to the corresponding carboxylic esters (carbonylation), using CO insertion with an alcoholic co-solvent, is a useful transformation in organic synthesis.

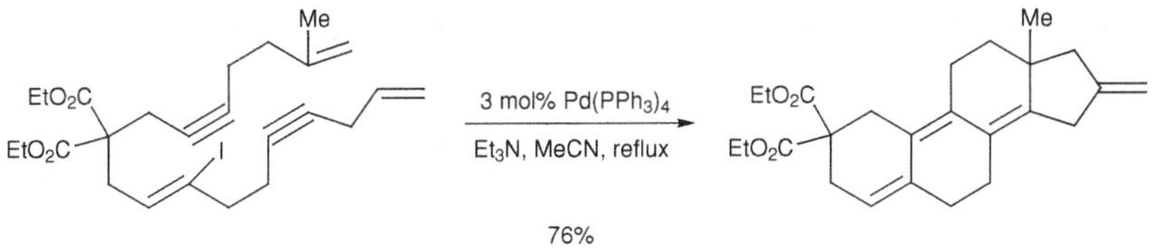

(1.216)

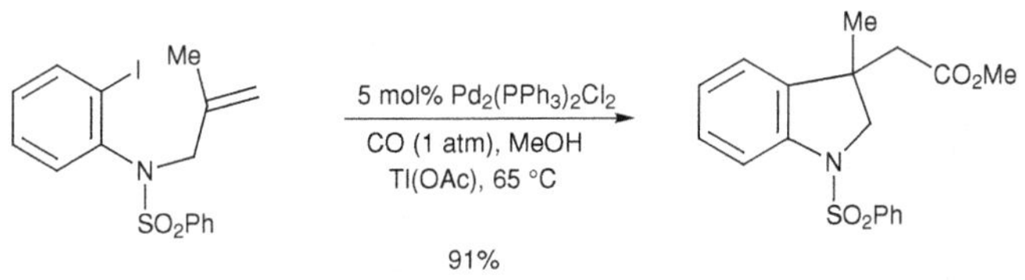

(1.217)

Trapping the organopalladium(II) species, generated by oxidative addition of palladium(0) into an unsaturated halide or triflate, with an organostannane (Stille reaction), an organoborane (Suzuki reaction), an organozinc species (Negishi reaction), an alkynyl metal species (Sonogashira reaction) or an alkene (Heck reaction) have been described above. Enolates can also be used as the nucleophilic component in such reactions. For example, the α-arylation of tert-butyl esters with aryl bromides proceeds in good yield using catalytic Pd(OAc)$_2$ and an electron-rich phosphine ligand.[174] Thus, the non-steroidal anti-inflammatory agent naproxen, as its tert-butyl ester, was prepared using the ligand **216** (1.218).

Although this chapter concentrates on carbon–carbon bond-forming reactions, it is worth pointing out that the use of organopalladium(II) species for the formation of a carbon–heteroatom bond, particularly a C−N bond, using an amine nucleophile is a popular reaction in organic synthesis.[175] Using an aryl halide (or triflate) as the coupling partner, the method allows the preparation of aryl amines and has found considerable use for the formation of biologically active molecules.

$$\text{(1.218)}$$

74%

216

Another important class of organopalladium(II) compounds is based on π-allyl palladium complexes, in which an allyl ligand is co-ordinated η^3 to palladium. The most common method for the formation of these complexes is from the corresponding allylic acetate (or carbonate or other leaving group) and palladium(0). The cationic π-allyl palladium complexes **217** are attacked by a variety of nucleophiles, including amines and soft carbon nucleophiles with pK_a values in the range 10–17, such as the anion of diethyl malonate. After nucleophilic attack, a palladium(0) complex is produced, which interacts with the allylic substrate to release the product, regenerate the π-allyl palladium complex **217** and continue the cycle (1.219).[176]

[174] G. C. Lloyd-Jones, *Angew. Chem. Int. Ed.*, **41** (2002), 953; W. A. Moradi and S. L. Buchwald, *J. Am. Chem. Soc.*, **123** (2001), 7996; M. Jørgensen, S. Lee, X, Liu, J. P. Wolkowski and J. F. Hartwig, *J. Am. Chem. Soc.*, **124** (2002), 12 557; D. A. Culkin and J. F. Hartwig, *Acc. Chem. Res.*, **36** (2003), 234.

[175] D. Prim, J.-M. Campagne, D. Joseph and B. Andrioletti, *Tetrahedron*, **58** (2002), 2041; B. H. Yang and S. L. Buchwald, *J. Organomet. Chem.*, **576** (1999), 125; J. F. Hartwig, *Angew. Chem. Int. Ed.*, **37** (1998), 2046.

[176] A. Heumann and M. Réglier, *Tetrahedron*, **51** (1995), 975; P. J. Harrington, in *Comprehensive Organometallic Chemistry II*, ed. E. W. Abel, F. G. A. Stone and G. Wilkinson, vol. 12 (Oxford: Elsevier, 1995), p. 797.

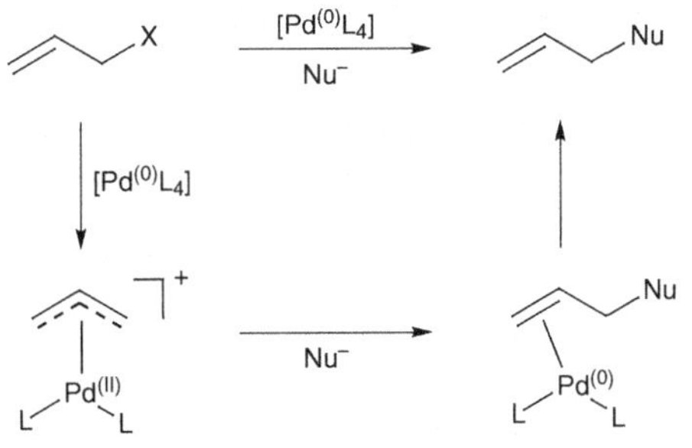

(1.219)

217

Organometallic reagents are generally thought of as nucleophilic, but it should be noted that these π-allyl palladium complexes are electrophilic and react with nucleophiles. Stabilized carbanions or enolates react best, and the use of hard anions, such as those in alkyllithium or Grignard reagents, leads to attack on the palladium rather than the allyl group. With an unsymmetrical π-allyl palladium complex, reaction can take place at either end of the allyl system, although attack at the less-substituted (less sterically hindered) position normally predominates. For example, substitution of the allylic acetate **218** can be effected using this chemistry, in preference to the normally more-reactive bromide (1.220).

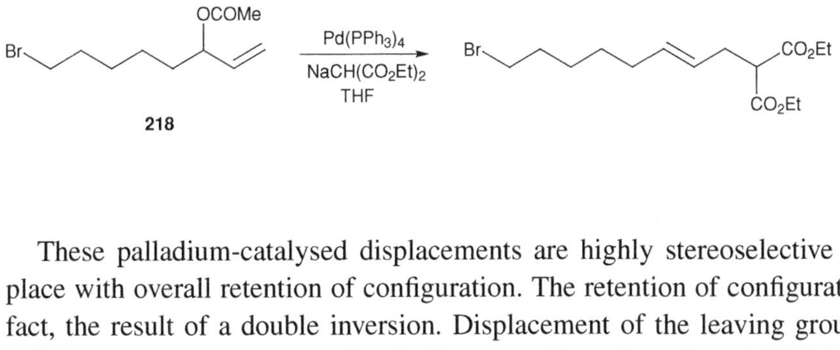

(1.220)

218

These palladium-catalysed displacements are highly stereoselective and take place with overall retention of configuration. The retention of configuration is, in fact, the result of a double inversion. Displacement of the leaving group by the palladium to form the π-allyl palladium complex takes place with inversion of configuration; subsequent attack by the nucleophile on the side opposite the bulky palladium leads again to inversion, with resulting retention of configuration in the product (1.221). Hence, alkylation of the alkenyl epoxide **219** with the malonate **220** occurs to give exclusively the *cis* product **221** (1.222). A key step in a synthesis of the insect moulting hormone ecdysone involved the controlled conversion of the allylic acetate **222** into **223** with retention of configuration (1.223).

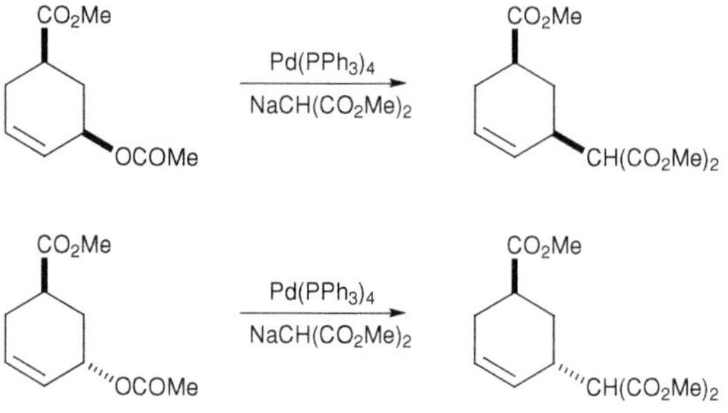

(1.221)

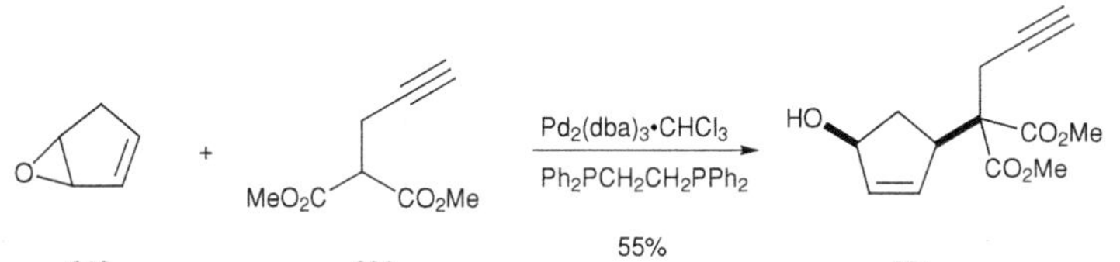

(1.222)

219   **220**   **221**

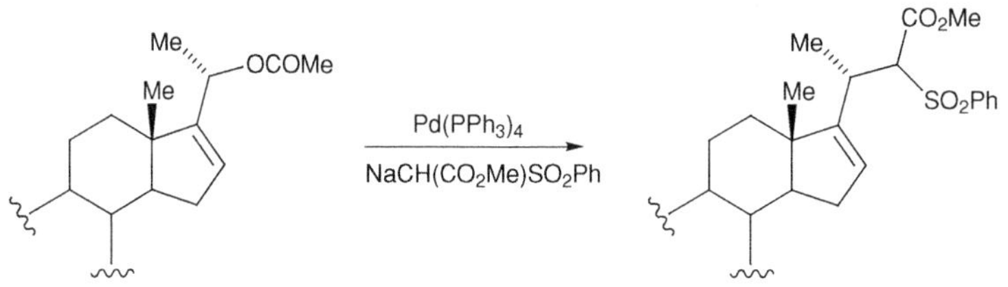

(1.223)

222   **223**

Significant advances have been made in asymmetric nucleophilic additions to π-allyl palladium complexes using chiral ligands.[177] Substitution of allylic acetates in the presence of the chiral phosphine ligand **224** (or other chiral phosphine ligands) can occur with very high levels of enantioselection (1.224).[178] The reaction works best with diaryl-substituted allylic acetates.

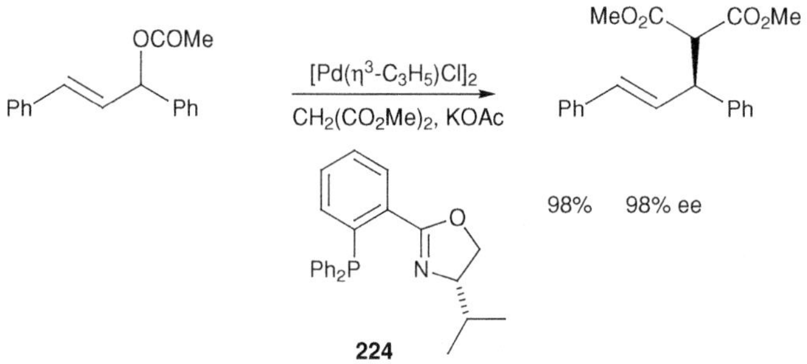

(1.224)

98%   98% ee

224

[177] B. M. Trost and M. L. Crawley, *Chem. Rev.*, **103** (2003), 2921; B. M. Trost and D. L. Van Vranken, *Chem. Rev.*, **96** (1996), 395.

[178] G. Helmchen and A. Pfaltz, *Acc. Chem. Res.*, **33** (2000), 336; J. M. J. Williams, *Synlett* (1996), 705.

Interception of the π-allyl palladium complex by soft nucleophiles, particularly malonates, has been described above. Alkenes, alkynes and carbon monoxide can also insert into the π-allyl palladium complex, generating a σ-alkyl palladium species. When an internal alkene is involved, a useful cyclization reaction takes place (sometimes called a palladium-ene reaction).[179] Addition of palladium(0) to the allylic acetate **225** gave the cyclic product **226** (1.225).[180] The reaction proceeds via the π-allyl palladium complex (formed with inversion of configuration), followed by insertion of the alkene *cis-* to the palladium and β-hydride elimination. In some cases it is possible to trap the σ-alkyl palladium species with, for example, carbon monoxide.

(1.225)

Problems (answers can be found on page 466)

1. Suggest a method for the conversion of pentan-2-one to hexan-3-one and of pentan-2-one to 3-methylpentan-2-one.
2. Suggest a method for the preparation of the following compounds.

[179] W. Oppolzer, *Angew. Chem. Int. Ed. Engl.*, **28** (1989), 38.
[180] W. Oppolzer, T. N. Birkinshaw and G. Bernardinelli, *Tetrahedron Lett.*, **31** (1990), 6995.

3. Explain the formation of the product **1**.

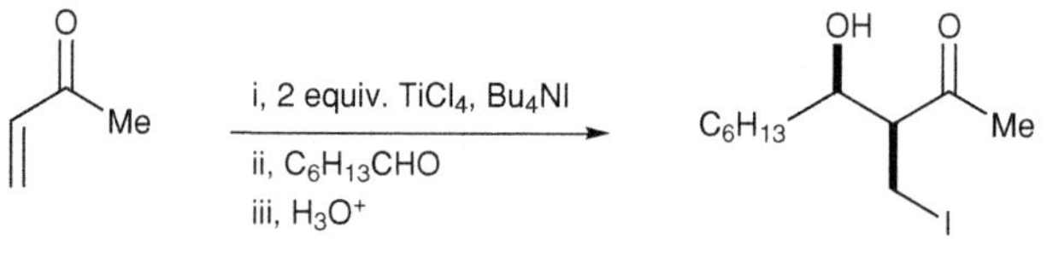

80%

4. The asymmetric aldol reaction using the Evans oxazolidinone auxiliary has been used in a number of syntheses. Suggest reagents for the preparation of the imide **2**, used in a synthesis of cytovaricin and explain the stereoselectivity of this reaction. Suggest how you would obtain the other *syn* aldol product and explain the difference in the stereoselectivity.

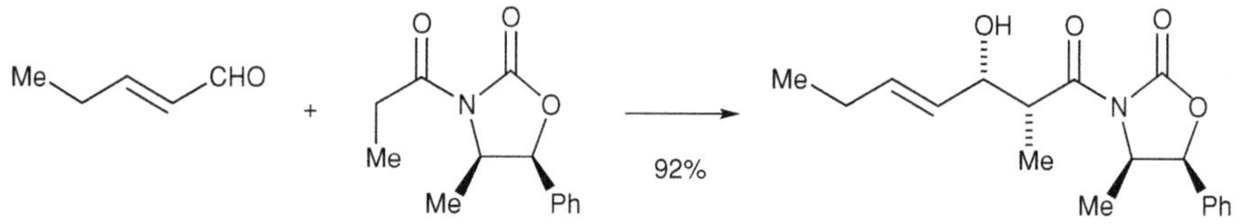

5. Explain the formation of the α-chloro-aldehyde **3**.

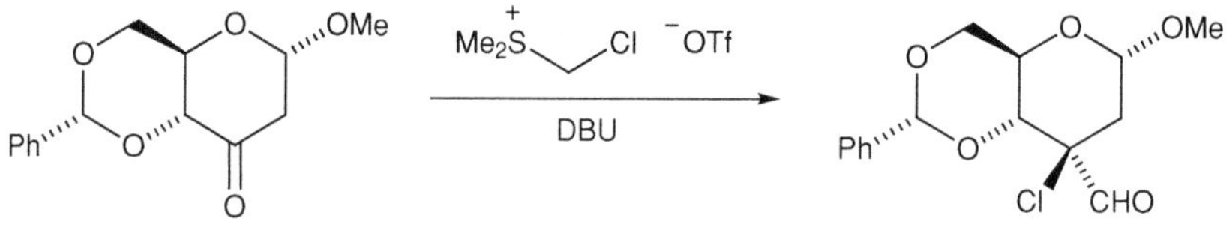

6. Suggest a method for the formation of the organolithium compound $LiCH_2OMOM$, shown below. Draw the structure of the major product from addition of the aldehyde 2-phenylpropanal to this organolithium compound and explain the stereoselectivity.

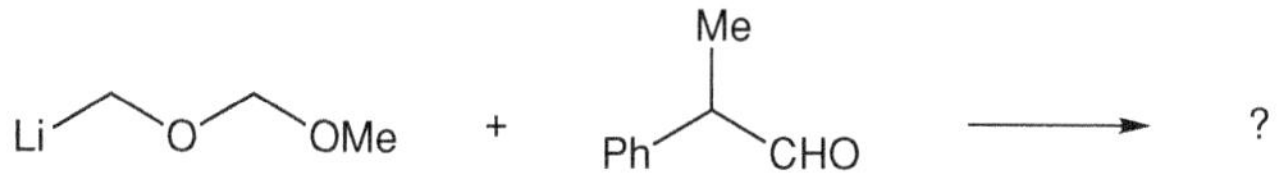

7. Explain why the alcohol **5** is the major diastereomer in the addition of isopropyl magnesium bromide to the ketone **4**. Account for the formation of the alcohol **6** in this reaction and explain the stereochemistry of this product.

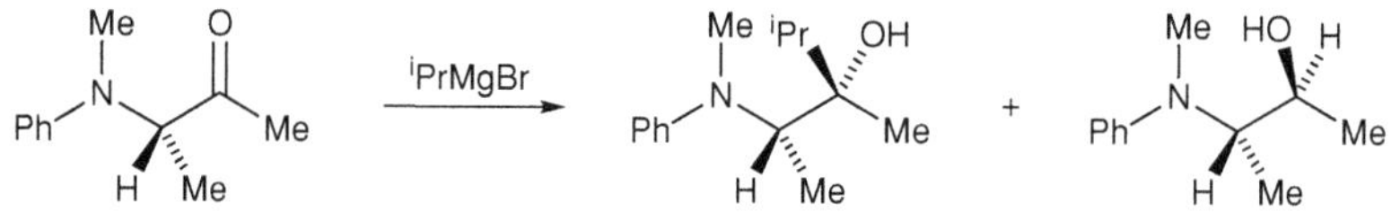

8. Explain the regioselectivity in the formation of the pyrazine **7** and explain the formation of the pyrazine **8**, used in a route to the pesticide septorin.

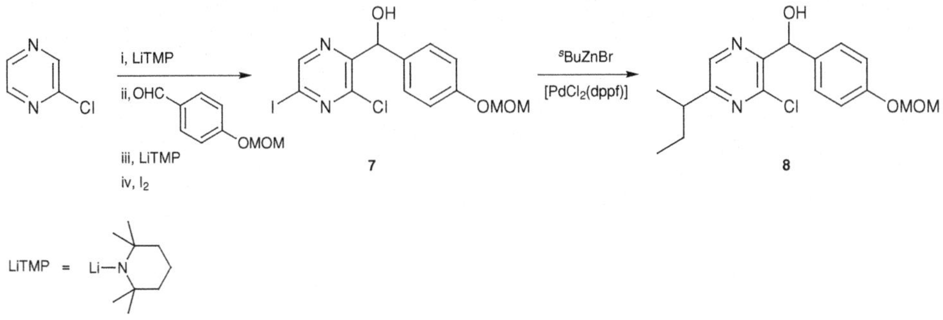

LiTMP = Li—N (structure)

9. Draw the structure of the product from the reaction shown below.

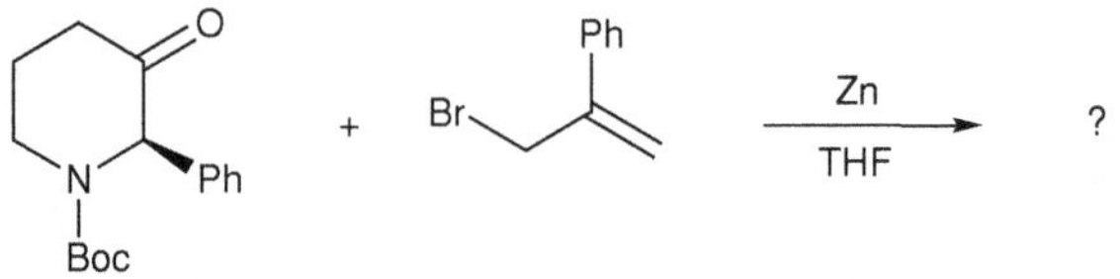

10. Draw the structures of the intermediates in the formation of the disubstituted cyclopentane **9**.

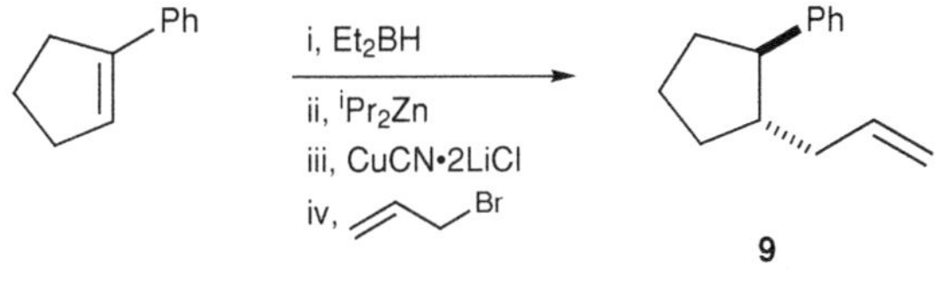

11. Draw the structure of the product **11** from treatment of the alkenyl bromide **10** with chromium(II) chloride.

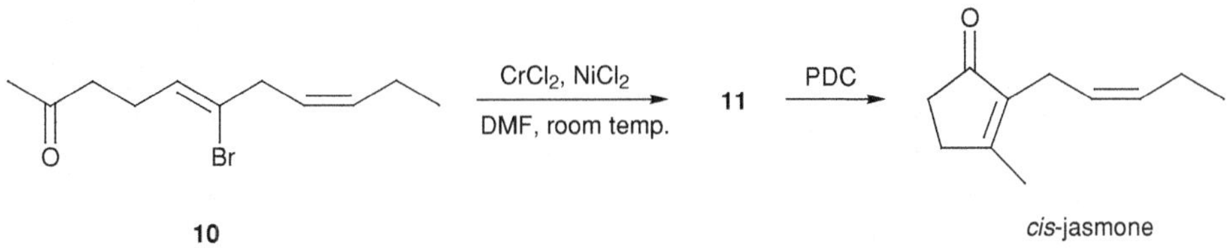

12. Draw the structure of the cobalt complex **12** and explain the formation of the rearranged product **13** on treatment with the Lewis acid TiCl₄.

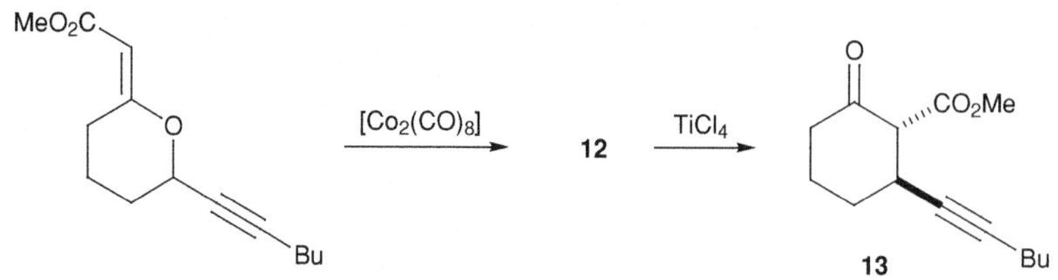

13. Draw the structures of the products of the following reactions.

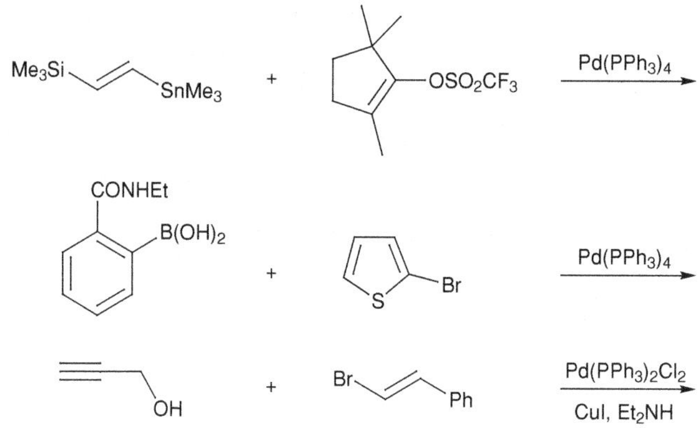

14. Explain the formation of the product **14** by drawing the structures of the intermediates. (Hint: π-allyl palladium and Heck reactions are involved.)

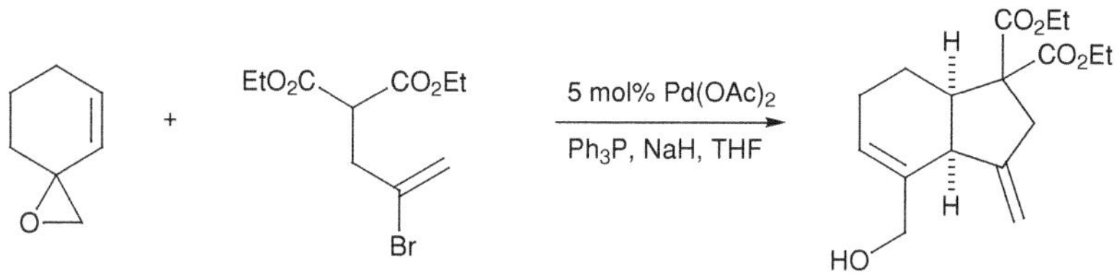

2

Formation of carbon–carbon double bonds

The formation of carbon–carbon double bonds is important in organic synthesis, not only for the obvious reason that the compound being synthesized may contain a double bond, but also because formation of the double bond allows the introduction of a wide variety of functional groups. Methods to construct alkenes[1] are given in this chapter and some reactions that functionalize alkenes are given in Chapter 5. The formation of carbon–carbon double bonds by pericyclic reactions (such as cycloaddition reactions or sigmatropic rearrangements) are discussed in Chapter 3. Methods for the formation of alkenes in which the key step involves preparing the adjacent carbon–carbon single bond are given in Chapter 1 (see for example, Section 1.2.4).

This chapter is divided into reactions that give the alkene π-bond from substrates containing a C–C single bond (typically by elimination) (Sections 2.1–2.5), or from an alkyne substrate (Section 2.6), or from two different precursors in which both the σ- and the π-bonds are formed (Sections 2.7–2.10).

2.1 β-Elimination reactions

One of the most commonly used methods for forming carbon–carbon double bonds is by β-elimination reactions of the types shown in Scheme 2.1, where X = e.g. OH, OCOR, halogen, OSO_2R, $^+NR_3$, etc. Included among these reactions are acid-catalysed dehydrations of alcohols, solvolytic and base-induced eliminations from alkyl halides or sulfonates and the Hofmann elimination from quaternary ammonium salts.[2] They proceed by both E2 (elimination bimolecular) and E1 (elimination

[1] *Preparation of Alkenes, A Practical Approach*, ed. J. M. J. Williams (Oxford: Oxford University Press, 1996); S. E. Kelly, in *Comprehensive Organic Synthesis*, ed. B. M. Trost and I. Fleming, vol. 1 (Oxford: Pergamon Press, 1991), p. 729.

[2] A. Krebs and J. Swienty-Busch, in *Comprehensive Organic Synthesis*, ed. B. M. Trost and I. Fleming, vol. 6 (Oxford: Pergamon Press, 1991), p. 949.

unimolecular) mechanisms. A third mechanism, involving initial proton abstraction as the rate-determining step, followed by loss of X^- is termed E1cB.

β-Elimination reactions, although often used, leave much to be desired as synthetic procedures. One disadvantage is that in many cases elimination can take place in more than one way, so that mixtures of products, including mixtures of geometrical isomers, may be obtained. The direction of elimination in unsymmetrical compounds is governed largely by the nature of the leaving group, but may be influenced to some extent by the experimental conditions. It is found in general that acid-catalysed dehydration of alcohols and other E1 eliminations, as well as eliminations from alkyl halides and sulfonates with base, give the more highly substituted alkene as the major product (the Saytzeff or Zaitsev rule), whereas base-induced eliminations from quaternary ammonium salts and from sulfonium salts give predominantly the less-substituted alkene (the Hofmann rule) (2.2).

Exceptions to these rules are not uncommon, however. If there is a conjugating substituent at one β-carbon atom, then elimination will take place towards that carbon atom to give the conjugated alkene, irrespective of the method used. For example, *trans*-(2-phenylcyclohexyl)trimethylammonium hydroxide on Hofmann elimination gives 1-phenylcyclohexene exclusively. Another exception is found in the elimination of HCl from 2-chloro-2,4,4-trimethylpentane **1**, which gives mainly the terminal alkene product (2.3). In this case it is thought that the transition state leading to the expected Saytzeff product is destabilized by steric interaction between a methyl group and the tert-butyl substituent.

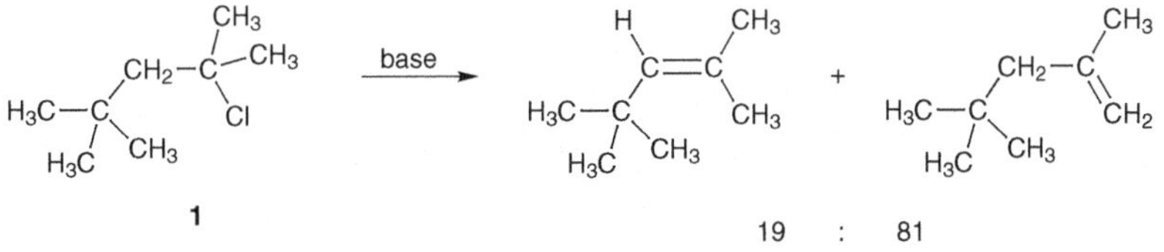

(2.3)

19 : 81

An additional disadvantage of E1 eliminations (such as acid-catalysed dehydration of alcohols) which proceed through an intermediate carbocation, is that elimination is frequently accompanied by rearrangement of the carbon skeleton. Thus, if the alcohol camphenilol **2** is treated with acid, the alkene santene **3** is formed (2.4). Protonation of the alcohol and formation of the secondary carbocation, followed by migration of a methyl group generates a more stable tertiary carbocation; loss of a proton then gives the alkene **3**.

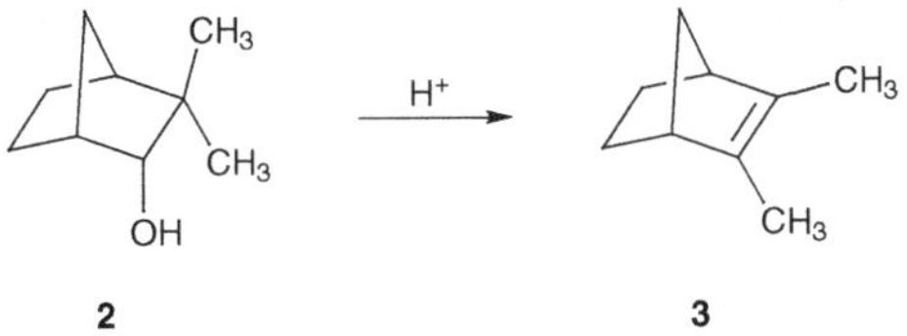

(2.4)

Base-induced eliminations from alkyl halides or sulfonates and the Hofmann reaction with quaternary ammonium salts are generally *anti* elimination processes, such that the hydrogen atom and the leaving group depart from opposite sides of the incipient double bond. This is borne out in the stereochemical course of the reaction, in which one diastereomer eliminates to the *E*-alkene and the other to the *Z*-alkene. For example, the quaternary ammonium salts derived from *syn*- and *anti*-1,2-diphenylpropylamine were found to undergo stereoselective elimination on treatment with sodium ethoxide in ethanol (2.5). In agreement with the relatively rigid requirements of the transition state, the *anti* isomer, in which the phenyl groups become eclipsed in the transition state, reacts more slowly than the *syn* isomer.

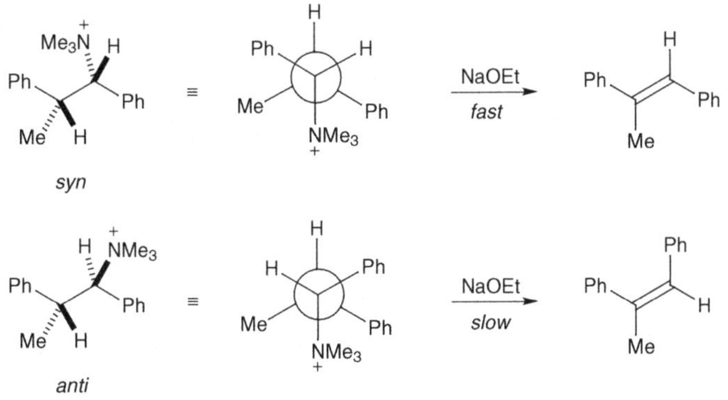

(2.5)

For stereoelectronic reasons, elimination takes place most readily when the hydrogen atom and the leaving group are in an antiperiplanar arrangement, such that orbital overlap is maximized. A number of *syn* eliminations have also been observed, in which the hydrogen atom and the leaving group are eclipsed. In open-chain compounds, the molecule can usually adopt a conformation in which H and X are antiperiplanar, but in cyclic systems this may not be the case, or the lowest energy conformation may not have H and X aligned antiperiplanar. In cyclohexyl derivatives, antiplanarity of the leaving groups requires that they be diaxial, even if this is a less-stable conformation. Menthyl chloride **4**, on treatment with sodium ethoxide in ethanol, gives only 2-menthene, whereas neomenthyl chloride **5** gives a mixture of 2- and 3-menthene, in which the Saytzeff product predominates (2.6). The elimination from **4** is much slower than that from **5** because the molecule has to adopt an unfavourable conformation with axial substituents before elimination can take place.

$$(2.6)$$

Syn eliminations may occur in compounds that cannot adopt a conformation in which the H and X groups are antiperiplanar. This is apparent in some bridged bicyclic compounds, where *anti* elimination is disfavoured by steric or conformational factors (2.7), and in compounds where a strongly electron-attracting substituent on the β-carbon atom favours elimination in that direction, outweighing other effects. Elimination by a *syn* pathway preserves the requirement for coplanarity of the breaking σ-bonds; however, in most cases *anti* elimination is preferred.

$$(2.7)$$

In spite of the disadvantages, acid-catalysed dehydration of alcohols and base-induced eliminations from halides and sulfonates are used widely in the preparation of alkenes. Typical bases include alkali-metal hydroxides and alkoxides, as well as organic bases such as pyridine and triethylamine. Good results have been obtained using the base 1,8-diazabicyclo[5.4.0]undec-7-ene (DBU) **6** and the related 1,5-diazabicyclo[4.3.0]non-5-ene (DBN). For example, elimination of the mesylate **7** gave the unsaturated aldehyde **8** (2.8).

$$\text{(2.8)}$$

Base-induced elimination of epoxides to allylic alcohols is a general and useful reaction that can proceed with good regio- and stereochemical control.[3] A strong base such as lithium diethylamide, or other metal dialkylamide, is often used, as in the regioselective ring-opening of the epoxide **9** (2.9). In this case, the base removes the less-hindered endocyclic C−H to effect the elimination. Using a *meso* epoxide and a chiral amine, asymmetric ring-opening can be promoted with good levels of enantioselectivity.[4] For example, the chiral diamine **10** is deprotonated by LDA and is an effective catalyst that selects one of the two enantiotopic protons of *meso* epoxides such as cyclohexene oxide (2.10).[5]

$$\text{(2.9)}$$

$$\text{(2.10)}$$

A common method for forming alkenes by β-elimination involves the dehydration of an aldol product (see Section 1.1.3). Under appropriate conditions or with suitable substituents, both the aldol reaction and the dehydration steps can be carried out in the same pot. For example, elimination occurs *in situ* to give the conjugated alkene chalcone, on aldol condensation between acetophenone and benzaldehyde (2.11). This reaction works well, as only one component (acetophenone) is enolizable and as benzaldehyde is more electrophilic. Mixtures of products result from

[3] J. K. Crandall and M. Apparu, *Org. Reactions*, **29** (1983), 345.
[4] P. O'Brien, *J. Chem. Soc., Perkin Trans. 1* (1998), 1439; J. Eames, *Eur. J. Org. Chem.* (2002), 393.
[5] M. J. Södergren, S. K. Bertilsson and P. G. Andersson, *J. Am. Chem. Soc.*, **122** (2000), 6610.

aldol condensations between two carbonyl compounds of similar acidity or electrophilicity. Intramolecular aldol reaction followed by dehydration is a useful method for the formation of cyclopentenone and cyclohexenone products (see also the Robinson annulation, Section 1.1.2).

$$(2.11)$$

The product from the addition of an enolate of a 1,3-dicarbonyl-type compound with an aldehyde (or ketone) dehydrates readily to give the α,β-unsaturated dicarbonyl product. This is known as the Knoevenagel condensation reaction, typically carried out under mild conditions with an amine base, and is a useful method for the formation of alkenes bearing electron-withdrawing substituents (see Section 1.1.3).

An alternative method for the formation of an α,β-unsaturated carbonyl compound is the elimination of an initially formed Mannich product. The procedure is particularly effective for the formation of β,β-bis(unsubstituted) α,β-unsaturated carbonyl compounds. The Mannich product **11** can be formed in the presence of a secondary amine and a non-enolizable aldehyde such as formaldehyde (2.12).[6] The Mannich reaction is a useful carbon–carbon bond-forming reaction and the products have found application in the synthesis of, in particular, alkaloid ring systems. The Mannich product may eliminate under the reaction conditions, or can be alkylated to form the quaternary ammonium salt in order to induce elimination. A convenient variation of this method is the use of Eschenmoser's salt, $H_2C=NMe_2{}^+ X^-$. For example, Nicolaou's synthesis of hemibrevetoxin B used this salt in order to introduce the required methylene unit α- to the aldehyde **12** (2.13).[7] The same transformation with the corresponding methyl ester, which is less acidic, requires prior enolization with a strong base (e.g. $NaN(SiMe_3)_2$) and subsequent quaternization of the tertiary amine with iodomethane and elimination using DBU.

$$(2.12)$$

11

[6] M. Arend, B. Westermann and N. Risch, *Angew. Chem. Int. Ed.*, **37** (1998), 1044; E. F. Kleinman, in *Comprehensive Organic Synthesis*, ed. B. M. Trost and I. Fleming, vol. 2 (Oxford: Pergamon Press, 1991), p. 893.

[7] K. C. Nicolaou, K. R. Reddy, G. Skokotas, F. Sato, X.-Y. Xiao and C.-K. Hwang, *J. Am. Chem. Soc.*, **115** (1993), 3558.

12

$$(2.13)$$

2.2 Pyrolytic *syn* eliminations

An important group of alkene-forming reactions, some of which are useful in synthesis, are pyrolytic eliminations.[8] Included in this group are the pyrolyses of carboxylic esters and xanthates, of amine oxides, sulfoxides and selenoxides. These reactions take place in a concerted manner, by way of a cyclic transition state and therefore proceed with *syn* stereochemistry, such that the hydrogen atom and the leaving group depart from the same side of the incipient double bond (in contrast to the eliminations discussed in Section 2.1) (2.14).

$$(2.14)$$

The *syn* character of such eliminations has been demonstrated with all these types of thermal reactions, as illustrated in the examples given in this section. Thus, heating the deuterium-labelled acetate **13** gave *trans*-stilbene **14**, whereas the diastereomeric acetate **15** gave the *trans*-stilbene **16**, in which the deuterium label was still present (2.15). Either the hydrogen or the deuterium atom could be *syn* to the acetoxy group, but the preferred conformations (as shown) are those in which the phenyl groups are as far apart from each other as possible.

Pyrolysis of esters to give an alkene and a carboxylic acid is usually effected at a temperature of about 300–500 °C and may be carried out by heating the ester if its boiling point is high enough, or by passing the vapour through a heated tube. The absence of solvents and other reactants simplifies the isolation of the product. The lack of acidic or basic reagents can be an advantage over conventional

---

[8] P. C. Astles, S. V. Mortlock and E. J. Thomas, in *Comprehensive Organic Synthesis*, ed. B. M. Trost and I. Fleming, vol. 6 (Oxford: Pergamon Press, 1991), p. 1011.

β-eliminations, especially for preparing sensitive or reactive alkenes. For example, 4,5-dimethylenecyclohexene is obtained from the diacetate **17** without extensive rearrangement to *o*-xylene (2.16).

Pyrolytic *syn* eliminations with secondary or tertiary substrates have the disadvantage that elimination can take place in more than one direction, giving mixtures of products. In acyclic compounds, if there is a conjugating substituent in the β-position, elimination takes place to give predominantly the conjugated alkene, but otherwise the composition of the product is determined mainly by the number of hydrogen atoms on each β-carbon. For example, pyrolysis of 2-butyl acetate gives a mixture containing 57% 1-butene and 43% 2-butene, in close agreement with the 3:2 distribution predicted on the basis of the number of β-hydrogen atoms (2.17). Of the 2-butenes, the *E*-isomer is formed in larger amount, because there is less steric interaction between the two methyl groups in the transition state leading to the *E*-alkene.

In cyclic compounds some restrictions are imposed by the conformation of the leaving groups and the necessity to form the cyclic transition state. Thus, the acetate **18**, in which the leaving group is axial, does not form a double bond in the direction of the ethoxycarbonyl group, even though it would be conjugated (2.18). In contrast, the diastereomer **19** can adopt the necessary cyclic transition state that leads to the conjugated alkene.

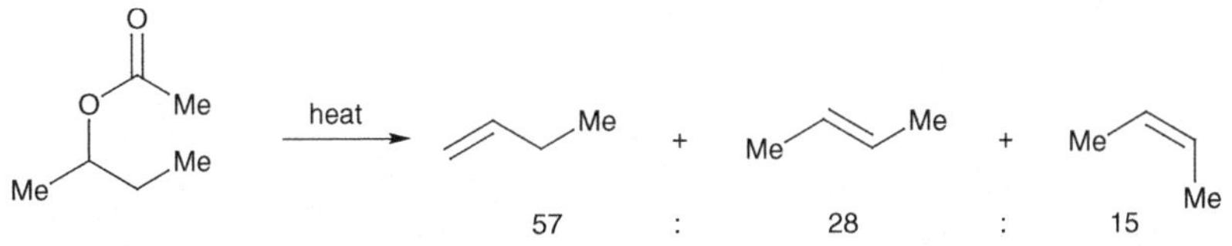

$$(2.17)$$

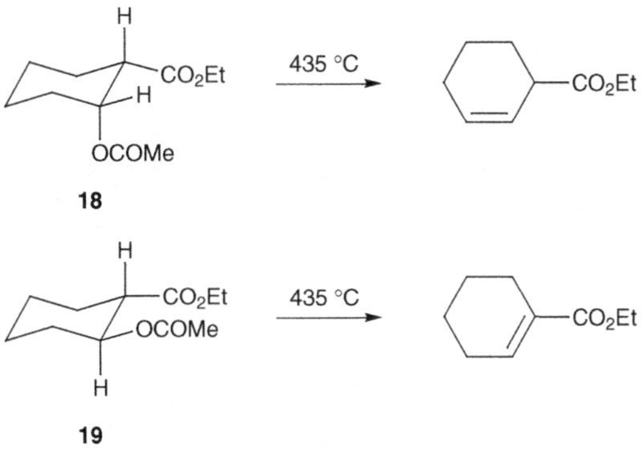

57 : 28 : 15

435 °C

18

$$(2.18)$$

435 °C

19

The high temperatures required for pyrolyses of acetates has limited their synthetic usefulness. The related thiocarbonates or xanthates, however, eliminate at temperatures in the region of 150–250 °C, with the result that further decomposition of the alkene product(s) can often be avoided (Chugaev reaction).[9] On the other hand, separation of the alkene from sulfur-containing by-products can sometimes be troublesome. In the same way as the corresponding acetates, the pyrolysis of xanthates promotes a *syn* elimination. Thus, the xanthate **20**, prepared from its corresponding alcohol with NaH, CS_2 then MeI (91% yield), eliminates regioselectively on heating to give the alkene **21** (2.19). Eliminations using acyclic xanthates give predominantly the *E*-alkene product, although significant amounts of the *Z*-isomer are also obtained and the regioselectivity is often poor.

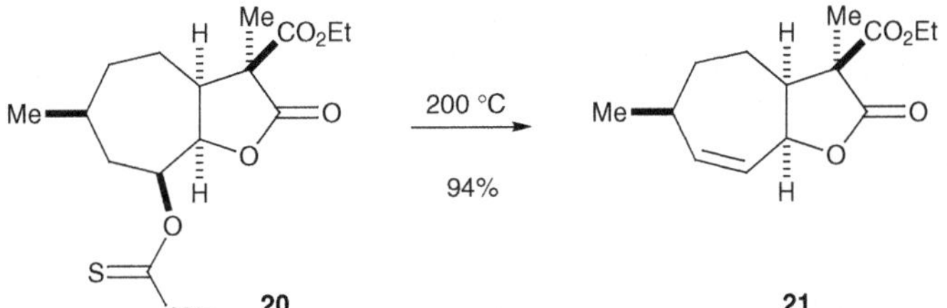

200 °C

94%

20   **21**

$$(2.19)$$

Pyrolysis of tertiary amine oxides (the Cope elimination reaction) also offers relatively mild reaction conditions (100–200 °C).[10] Oxidation of the tertiary amine

[9] H. R. Nace, *Org. Reactions*, **12** (1962), 57.
[10] A. C. Cope and E. R. Trumbull, *Org. Reactions*, **11** (1960), 317.

to the N-oxide (e.g. with H_2O_2 or mCPBA), followed by heating, provides an alternative to the Hofmann elimination of the quaternary ammonium salt (see Section 2.1). The mild conditions of the Cope elimination reaction allow the generation of a new carbon–carbon double bond, without subsequent migration into conjugation with other unsaturated systems in the molecule, as in the synthesis of 1,4-pentadiene (2.20). If an allyl or benzyl group is attached to the nitrogen atom, however, Meisenheimer rearrangement (see Section 3.7) to give an O-substituted hydroxylamine may compete with elimination.

(2.20)

The stereochemistry of the product alkene for acyclic 1,2-disubstituted or trisubstituted alkenes is determined by the configuration of the tertiary amine oxide. A stereospecific *syn* elimination pathway is followed. With cyclic substrates, the ring conformation is important, such that heating dimethylmenthylamine oxide **22** gave a mixture of 2- and 3-menthene, whereas the isomeric neomenthylamine oxide **23** gave only 2-menthene (2.21). This result contrasts with the antiperiplanar β-elimination (2.6).

(2.21)

A notable difference between the pyrolysis of acetates or xanthates and the Cope elimination is found by using 1-methylcyclohexyl derivatives. The acetates and xanthates give mixtures containing 1-methylcyclohexene and methylenecyclohexane in a ratio of about 3:1, whereas pyrolysis of the oxide of 1-dimethylamino-1-methylcyclohexane gives methylenecyclohexane almost exclusively. The reason for this is thought to be that, in the oxide, the five-membered cyclic transition state allows preferential abstraction of a hydrogen atom from the methyl group, whereas with the more flexible six-membered transition state of the ester pyrolyses, hydrogen atom abstraction is also possible from the ring. With larger, more-flexible rings, the cycloalkene is the major product from the amine oxides or the esters.

1-methylcyclohexene   methylenecyclohexane

The Cope elimination is reversible and the intramolecular reverse Cope elimination, involving the addition of a tethered hydroxylamine to an alkene, has found recent application for the stereocontrolled preparation of cyclic amines.[11]

Sulfoxides with a β-hydrogen atom readily undergo *syn* elimination on pyrolysis to form alkenes. These reactions take place by way of a concerted cyclic pathway and are therefore highly stereoselective. The sulfoxide *anti*-**24**, for example, gives predominantly *trans*-methylstilbene, whereas the corresponding *syn*-isomer gives mainly *cis*-methylstilbene (2.22). Because sulfoxides are readily obtained by oxidation of sulfides, the reaction provides another useful method for making carbon–carbon double bonds.

(2.22)

24

Pyrolysis of sulfoxides provides a convenient method for introducing unsaturation at the position α- to carbonyl compounds. Formation of the enolate and reaction with dimethyl (or diphenyl) disulfide gives the α-methylthio (or phenylthio) derivative. Oxidation with a suitable oxidant, such as mCPBA or NaIO$_4$, gives the sulfoxide, which eliminates sulfenic acid on heating to give the α,β-unsaturated carbonyl compound. For example, the methyl ester of a pheromone of queen honey bees was synthesized from methyl 9-oxodecanoate after initial protection of the ketone as the acetal (2.24).[12] The *E*-isomer usually predominates in reactions

[11] E. Ciganek, *J. Org. Chem.*, **60** (1995), 5803.
[12] B. M. Trost and T. N. Salzmann, *J. Org. Chem.*, **40** (1975), 148.

leading to 1,2-disubstituted alkenes.

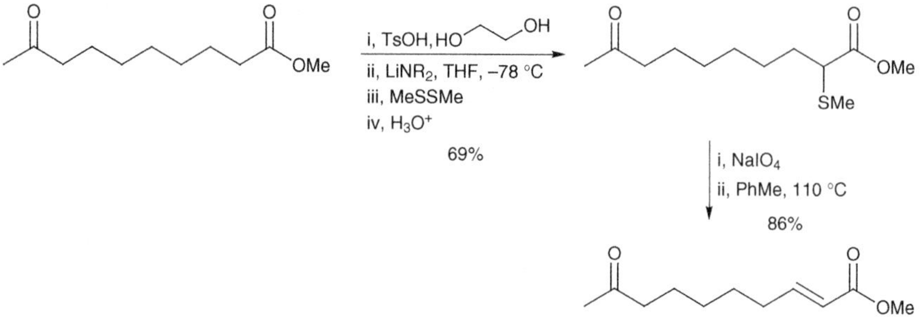

(2.23)

Even better results are obtained by using selenoxides. Alkyl phenyl selenoxides with a β-hydrogen atom undergo *syn* elimination to form alkenes under milder conditions than sulfoxides, owing to the longer, weaker carbon–selenium bond. Elimination occurs at room temperature or below, and this reaction has been exploited for the preparation of a variety of different kinds of unsaturated compounds. The selenoxides are readily obtained from the corresponding selenides by oxidation, and they generally undergo elimination under the reaction conditions to give the alkene directly. Selenides can be prepared by a number of different methods, such as the Mitsunobu reaction of alcohols with *N*-phenylselenophthalimide[13] or the selenylation of an enolate with phenylselenyl bromide.

Sulfides and selenides both stabilize an α-carbanion (see Section 1.1.5.2) and alkylation followed by elimination provides a route to substituted alkenes. Some examples of selenoxide eliminations are given in Schemes 2.24–2.26. Like the other *syn* eliminations described in this section, the regioselectivity of selenoxide elimination can be poor. Elimination normally takes place preferentially towards a conjugating β-substituent or away from an electronegative β-substituent. This latter facet allows a good method for converting epoxides into allylic alcohols.

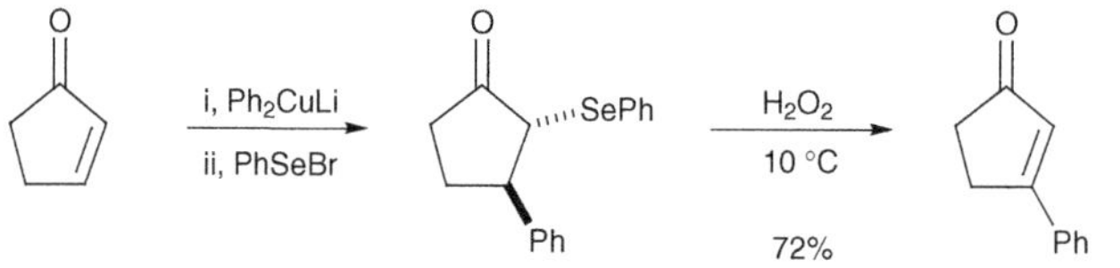

(2.24)

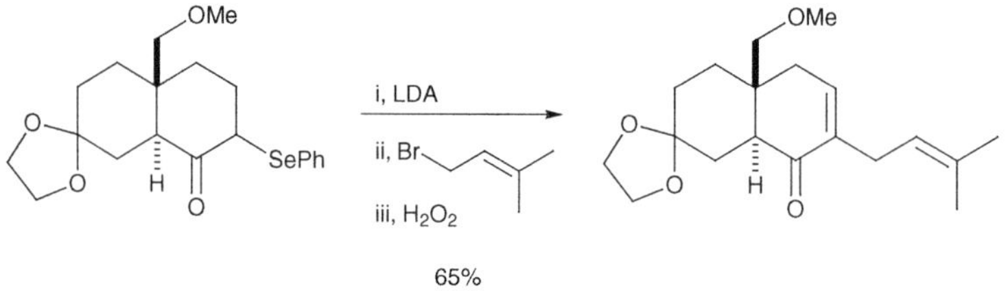

(2.25)

13 P. A. Grieco, J. Y. Jaw, D. A. Claremon and K. C. Nicolaou, *J. Org. Chem.*, **46** (1981), 1215.

$$(2.26)$$

98%

The transformation of carbonyl compounds to α,β-unsaturated carbonyl compounds can be achieved by selenoxide elimination.[14] In fact, this method is superior to the sulfoxide elimination, because of the milder conditions employed and the direct formation of the unsaturated product, without isolation of the selenoxide. Thus, oxidation of the selenide **25** at 0 °C gave the α-methylene lactone **26**, a structural unit found in cytotoxic sesquiterpenes (2.27). The requirement for a *syn* elimination pathway forces the reaction to proceed to give only the product **26** and none of the regioisomeric α,β-unsaturated lactone **28**. However, the lactone **28** is the major product from oxidation of the selenide **27**, illustrating the importance of the stereochemistry of the selenide, derived from the order of addition of the phenylselenyl and methyl groups, in determining the regiochemical outcome.

$$(2.27)$$

Alkenes can also be obtained by elimination of β-hydroxy selenides, under the action of methanesulfonyl chloride, thionyl chloride or other acidic conditions.[15] The reaction is stereospecific, proceeding by *anti* elimination, by way of the episelenonium ion, e.g. **29** (2.28). High yields of di-, tri- or tetrasubstituted alkenes can be obtained and the reaction provides an alternative to the Wittig reaction when the phosphonium salt cannot be readily obtained. The β-hydroxy selenides can be prepared by a number of methods, for example from α-seleno aldehydes or ketones by reduction or reaction with a Grignard reagent, or from α-lithio selenides by

[14] H. J. Reich and S. Wollowitz, *Org. Reactions*, **44** (1993), 1.
[15] H. J. Reich, F. Chow and S. K. Shah, *J. Am. Chem. Soc.*, **101** (1979), 6638; A. Krief, *Tetrahedron*, **36** (1980), 2531.

reaction with an aldehyde.

$$(2.28)$$

29

2.3 Fragmentation reactions

Fragmentation reactions, which are similar to β-elimination reactions (Section 2.1), can be useful for the formation of alkenes, particularly from carbocyclic compounds.[16] Fragmentation reactions occur most easily from conformationally locked 1,3-difunctionalized compounds, in which the breaking C—X and C—C bonds (highlighted) are aligned antiperiplanar (2.29).

$$(2.29)$$

X = leaving group, *e.g.* OSO$_2$Me

    The reaction is referred to as the Grob fragmentation and proceeds by a concerted mechanism, to give an alkene in which the stereochemistry is governed by the relative orientation of the groups in the cyclic precursor. For example, the decalin derivative **30**, in which the tosyloxy group and the adjacent ring junction hydrogen atom are *cis*, gave *E*-5-cyclodecenone in high yield, whereas the isomer **31**, in which the tosyloxy group and the hydrogen atom are *trans*, gave the *Z*-isomer (i.e. in each case the relative orientation of the hydrogen atoms in the precursor is retained in the alkene) (2.30). In these derivatives, there is an antiperiplanar arrangement of the breaking bonds, but this is not so in the isomer **32** and this compound, on treatment with base, gave a mixture of products containing only a very small amount of the *E*-cyclodecenone.

    Fragmentation reactions may be used to prepare cyclic or acyclic alkenes from cyclic precursors. The stereochemistry of the alkene can be set up by controlling the relative stereochemistry of the cyclic substrate, a process that is normally relatively easy. The ketone **35**, for example, an intermediate in a synthesis of juvenile hormone, was obtained stereospecifically from the bicyclic compound **33** using two successive

---

[16] P. Weyerstahl and H. Marschall, in *Comprehensive Organic Synthesis*, ed. B. M. Trost and I. Fleming, vol. 6 (Oxford: Pergamon Press, 1991), p. 1041; C. A. Grob, *Angew. Chem. Int. Ed. Engl.*, **8** (1969), 535.

fragmentation steps (2.31).[17] The geometry of the intermediates **33** and **34** allows easy fragmentation at each stage.

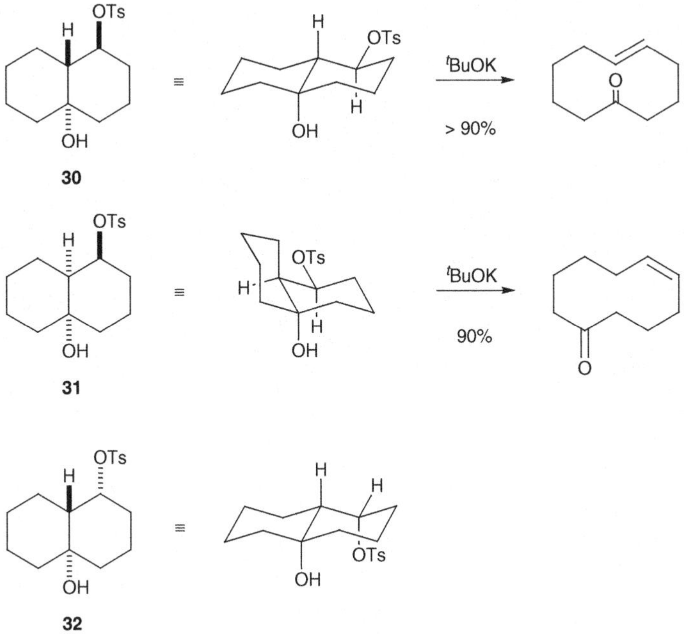

30

(2.30)

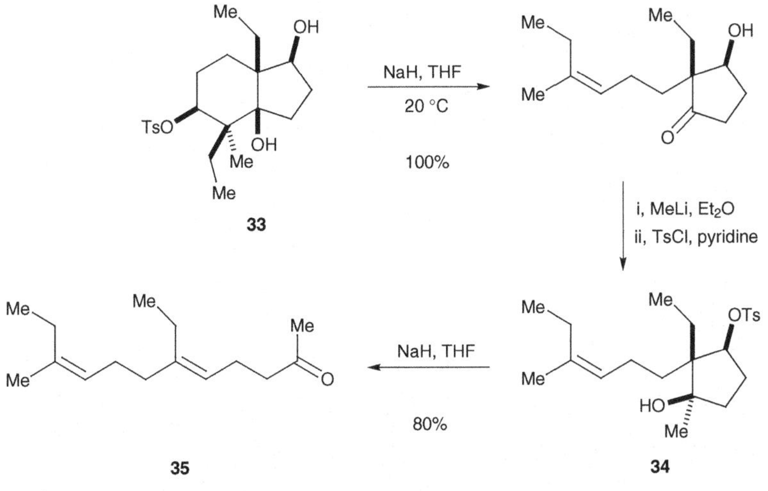

31

32

(2.31)

33

35

34

The fragmentation reaction is not restricted to monosulfonates of 1,3-diols, and various leaving groups together with various electron-releasing groups can be used. The borate species formed by hydroboration (on the less-hindered face) of the

[17] R. Zurflüh, E. N. Wall, J. B. Siddall and J. A. Edwards, *J. Am. Chem. Soc.*, **90** (1968), 6224.

alkene **36**, followed by nucleophilic addition of hydroxide, fragments to the diene **37** (2.32).[18]

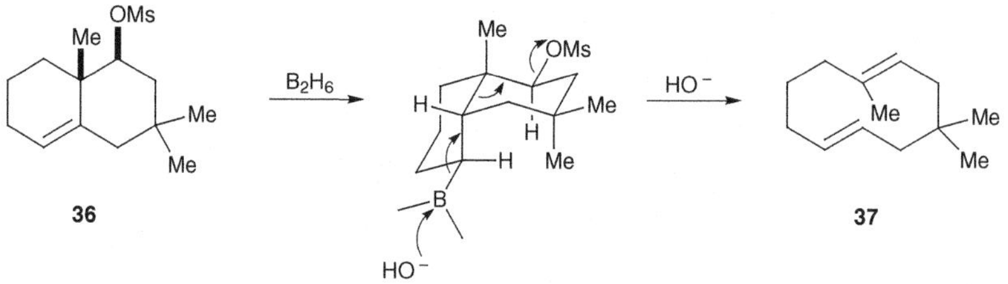

(2.32)

Fragmentation of the hydrazones of α,β-epoxy ketones is known as the Eschenmoser fragmentation. Deprotonation of the hydrazone promotes ring-opening of the epoxide to give an alkoxy species. This alkoxy species then fragments, displacing nitrogen gas and the sulfinate, to give an alkynone. For example, this reaction was made use of in a synthesis of *exo*-brevicomin, starting with the epoxy ketone **38** (2.33).[19]

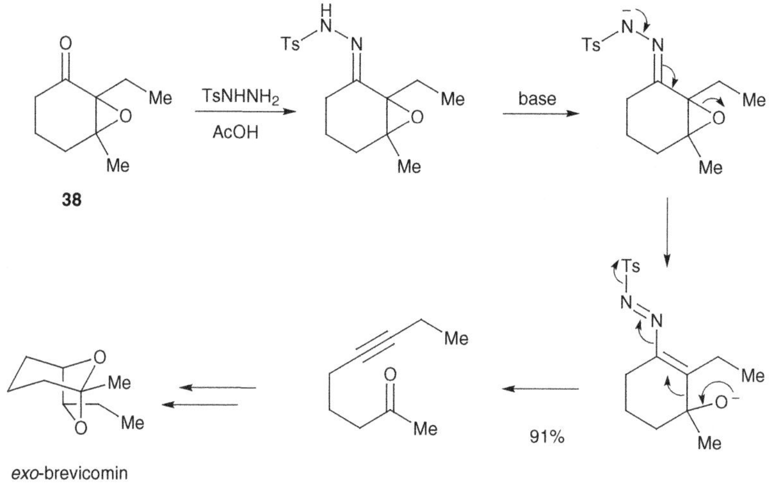

(2.33)

2.4 Alkenes from hydrazones

Hydrazones can be readily prepared by the addition of a hydrazine to an aldehyde or ketone. Treatment of tosyl hydrazones (or other arylsulfonyl hydrazones) with a base has been used for the preparation of alkenes. In the Bamford–Stevens reaction, a mild base, such as NaOMe or KH, is employed and promotes deprotonation of the acidic N—H proton (compare with the Eschenmoser fragmentation, Scheme 2.33).

---

[18] J. A. Marshall, *Synthesis* (1971), 229.
[19] P. J. Kocienski and R. W. Ostrow, *J. Org. Chem.*, **41** (1976), 398.

The resulting salt can then be heated and loses $ArSO_2^-$ to give an intermediate diazo compound **39** (2.34). This compound is not normally isolated, and decomposes on further heating to the carbene species (see Section 4.2), which can undergo a number of different reactions, including rearrangement to the alkene.[20] Typically a mixture of alkenes and C—H insertion products are obtained, with preference for the more-substituted alkene. In some cases, such as with five- or six-membered ring hydrazones, good yields of a single alkene product can be obtained (2.35).

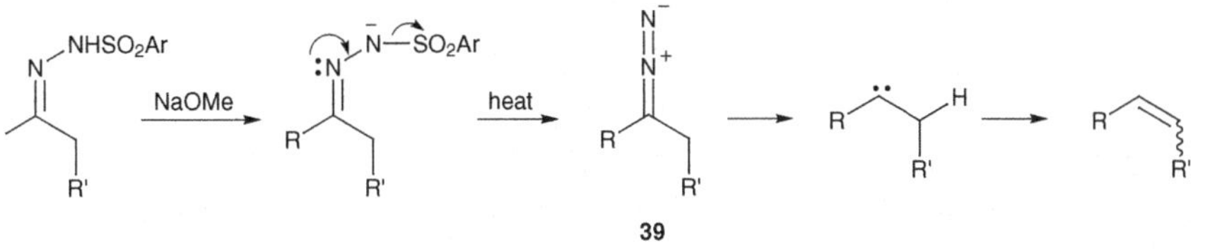

(2.34)

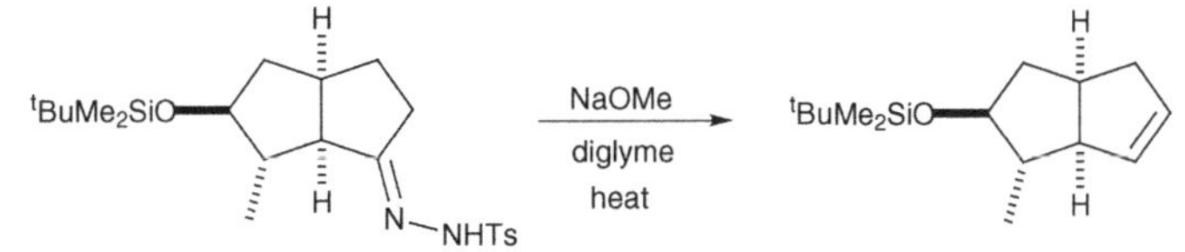

39

(2.35)

Ts = $SO_2C_6H_4$-*p*-Me

In contrast, treatment of the arylsulfonyl hydrazone with two equivalents of a strong base, such as BuLi or lithium diisopropylamide (LDA), effects the Shapiro reaction for the formation of alkenes.[21] Double deprotonation of the hydrazone gives the dianion **40**, which fragments to the alkenyllithium species **41** (2.36). Addition of an electrophile leads to the alkene product. The second deprotonation takes place *syn* to the $ArSO_2N^-$ (aryl sulfonamido anion) group and the less-substituted alkene is typically formed. Thus phenylacetone provides 3-phenylpropene and not the isomeric styrene (2.37). Protonation of the alkenyllithium species **41** to give a 1,2-disubstituted alkene generally leads to predominantly the Z-isomer.

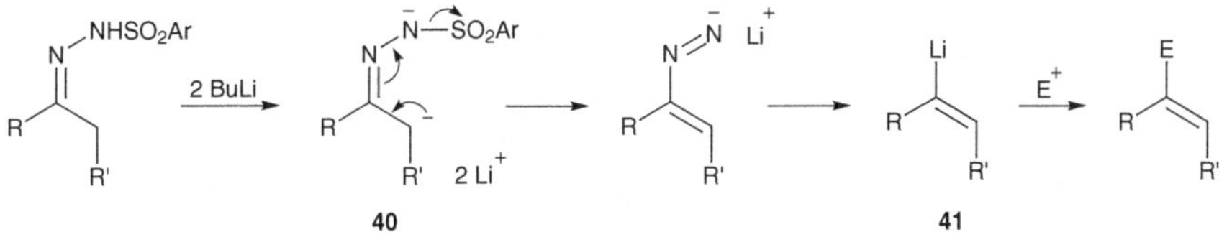

(2.36)

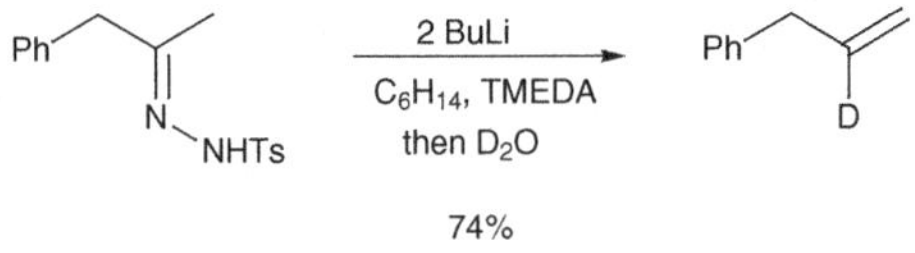

40   **41**

(2.37)

74%

[20] W. R. Bamford and T. S. Stevens, *J. Chem. Soc.* (1952), 4735; R. H. Shapiro, *Org. Reactions*, **23** (1976), 405.
[21] A. R. Chamberlin and S. H. Bloom, *Org. Reactions*, **39** (1990), 1.

Quenching the reaction mixture with deuterium oxide provides an excellent method for the preparation of deuterated alkenes as single regioisomers. In addition to protonation or deuteration, the alkenyllithium species **41** may be halogenated or alkylated by reaction with a suitable electrophile. In this way, di- or trisubstituted alkenes can be formed. It is often advantageous to use 2,4,6-triisopropylbenzenesulfonyl (trisyl) hydrazones as the source of the alkenyllithiums, as this avoids *ortho*-lithiation of the benzenesulfonyl group. If the reaction mixture is kept at low temperature, then the dianion **40** can be trapped with an electrophile and subsequently treated with a further equivalent of BuLi to give the substituted alkenyllithium species, which itself can be trapped with various electrophiles (2.38).[22] If the hydrazone bears a suitably positioned terminal alkene, the intermediate alkenyllithium species may be trapped intramolecularly to give a cyclopentane ring (2.39).[23]

(2.38)

(2.39)

An alternative to tosyl or trisyl hydrazones is the use of phenylaziridinyl hydrazones (e.g. **42**). If the protonated alkene product is desired, then the base LDA can be used as a catalyst. LDA is regenerated by reaction of the alkenyllithium species with the by-product, diisopropylamine (2.40).[24]

(2.40)

With an α,β-unsaturated hydrazone, addition of a nucleophile can lead to a new alkene, in which the double bond has migrated to the position formerly occupied

[22] E. J. Corey and B. E. Roberts, *Tetrahedron Lett.*, **38** (1997), 8919.
[23] A. R. Chamberlin, S. H. Bloom, L. A. Cervini and C. H. Fotsch, *J. Am. Chem. Soc.*, **110** (1988), 4788.
[24] K. Maruoka, M. Oishi and H. Yamamoto, *J. Am. Chem. Soc.*, **118** (1996), 2289.

by the hydrazone group. For example, reduction of the hydrazone with sodium borohydride in acetic acid, in which the reducing agent is probably NaBH(OAc)$_3$, gave an intermediate diazene **43**, which undergoes 1,5-migration of hydride from nitrogen to carbon to generate the reduced compound **44**, in which the alkene has been transposed (2.41).[25]

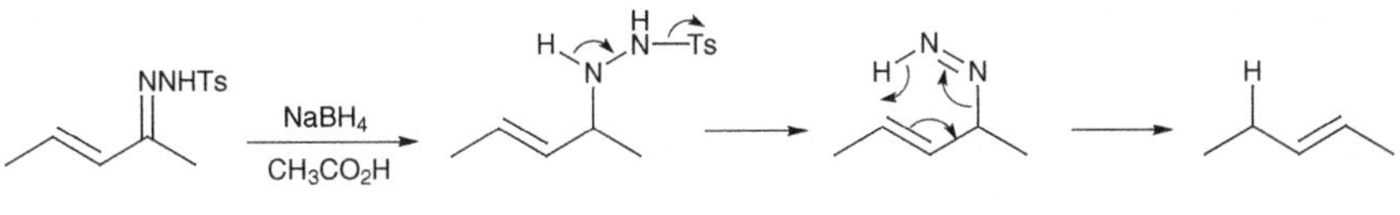

(2.41)

43   **44**

In agreement with this mechanism, reduction with sodium borodeuteride in AcOH or AcOD results in the regioselective introduction of one or two deuterium atoms, respectively. In deuterated acetic acid, exchange of the N−H proton must be faster than reduction and hydride transfer to carbon (2.42).

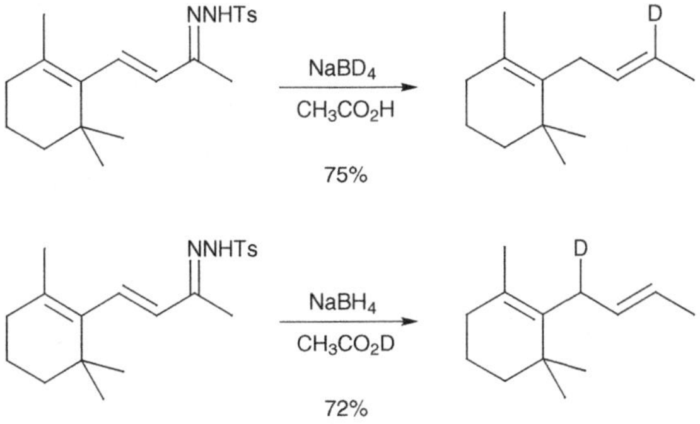

(2.42)

2.5 Alkenes from 1,2-diols

The McMurry reaction (see Section 2.9) can allow the formation of alkenes from dicarbonyl compounds. This reaction generates an intermediate 1,2-diol (pinacol), which is converted on the surface of the titanium to the alkene. The two carbon–oxygen bonds do not break simultaneously and the reaction is not stereospecific. Thus, both *anti* and *syn* acyclic 1,2-diols give mixtures of Z- and E-alkenes. With cyclic 1,2-diols, the two oxygen atoms must be able to bond to a common titanium surface. Thus, the *cis*-diol **45** eliminates to the alkene **46**, whereas the *trans*-diol **47** is inert under these reaction conditions (2.43).[26]

[25] R. O. Hutchins and N. R. Natale, *J. Org. Chem.*, **43** (1978), 2299.
[26] J. E. McMurry, M. P. Fleming, K. L. Kees and L. R. Krepski, *J. Org. Chem.*, **43** (1978), 3555.

(2.43)

Ti(0) can be
generated
from TiCl₃
and K metal

Several approaches for the regio- and stereospecific generation of double bonds
from 1,2-diols have been devised. One of the best methods uses the cyclic thionocarbonates (Corey–Winter reaction), which are readily obtained from the diol with
thiophosgene. As originally conceived, decomposition of the thionocarbonates to
the alkenes required heating with triethyl phosphite, although a milder method
using 1,3-dimethyl-2-phenyl-1,3,2-diazophospholidine $[(CH_2NMe)_2PPh]$ allows
reaction at 25–40 °C, possibly by way of a concerted process or by an intermediate carbene.[27] Under these mild conditions the reaction is applicable to complex
and sensitive molecules containing a variety of functional groups. For example, the
1,2-diol **48** gave the alkene **50** via the thionocarbonate **49** (2.44).

(2.44)

The Corey–Winter reaction proceeds with complete stereospecificity by a *syn*
elimination pathway, allowing the stereospecific synthesis of alkenes. Thus, *anti*-
1,2-diphenylethane-1,2-diol was converted into *cis*-stilbene (2.45), whereas the corresponding *syn*-diol gave *trans*-stilbene. The strained *E*-cyclooctene was prepared
from the *Z*-isomer using this procedure (2.46). An alternative stereospecific route to
alkenes proceeds from the diol with ethyl orthoformate or *N,N*-dimethylformamide
dimethyl acetal,[28] or by conversion of the diol to a 2-phenyl-1,3-dioxolane and
treatment with an organolithium reagent to promote proton abstraction at C-2 and

[27] E. J. Corey and P. B. Hopkins, *Tetrahedron Lett.*, **23** (1982), 1979.
[28] T. Hiyama and H. Nozaki, *Bull. Chem. Soc. Jpn*, **46** (1973), 2248; S. Hanessian, A. Bargiotti and M. La Rue,
Tetrahedron Lett. (1978), 737.

fragmentation,[29] an example of which is provided in Scheme 2.47.

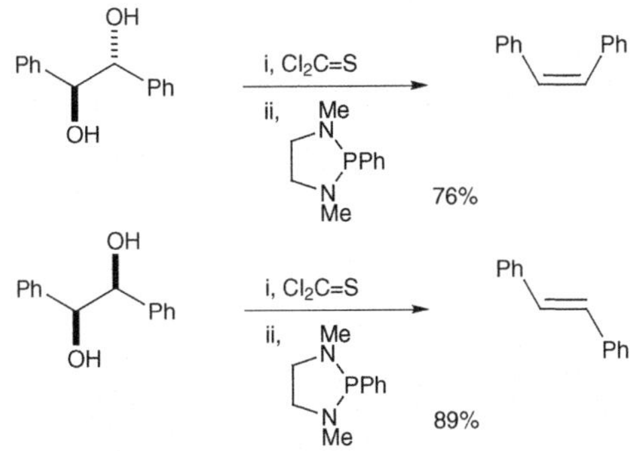

(2.45)

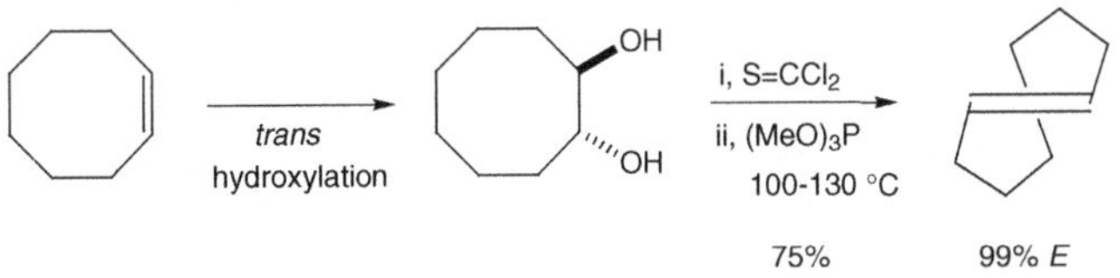

(2.46)

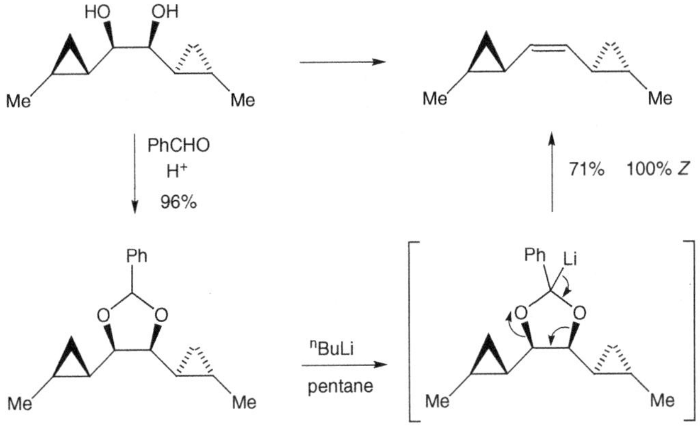

(2.47)

2.6 Alkenes from alkynes

The most obvious method for the formation of alkenes from alkynes is by partial reduction. This reaction can be effected in high yield with a palladium–calcium carbonate catalyst that has been partially deactivated by addition of lead(II) acetate or quinoline (Lindlar's catalyst).[30] It is aided by the fact that the more electrophilic alkynes are adsorbed on the electron-rich catalyst surface more strongly than the corresponding alkenes. An important feature of these reductions is their high stereoselectivity. In most cases the product consists very largely of the thermodynamically

[29] J. N. Hines, M. J. Peagram, E. J. Thomas and G. H. Whitham, *J. Chem. Soc., Perkin Trans. 1* (1973), 2332.
[30] C. A. Henrick, *Tetrahedron*, **33** (1977), 1845.

less stable Z-alkene and partial catalytic hydrogenation of alkynes provides one of the most convenient routes to Z-1,2-disubstituted alkenes. For example, reduction of stearolic acid over Lindlar's catalyst gave 95% of the alkene oleic acid (2.48). Partial reduction of alkynes with Lindlar's catalyst has been invaluable in the synthesis of carotenoids and many other natural products with Z-1,2-disubstituted alkenes.

$$CH_3(CH_2)_7-C\equiv C-(CH_2)_7CO_2H \xrightarrow[\substack{\text{Lindlar's catalyst}\\\text{EtOAc}}]{H_2} \begin{array}{c} CH_3(CH_2)_7 \quad (CH_2)_7CO_2H \\ C=C \\ H \qquad\quad H \end{array} \qquad (2.48)$$

In contrast, reduction of alkynes to E-1,2-disubstituted alkenes is possible using sodium metal in liquid ammonia (see Section 7.2) (2.49). This method therefore complements the formation of Z-alkenes by catalytic hydrogenation. Carbon–carbon double bonds are not normally reduced by metal–ammonia reducing agents and the reduction of the triple bond is therefore selective, such that none of the saturated product is formed. It is thought that the reaction takes place by stepwise addition of two electrons, the first electron adding to the triple bond to give an intermediate radical anion which is protonated by the ammonia to give a vinyl radical. The second electron adds to give a vinyl anion which adopts the more-stable E-configuration and is protonated to give the E-alkene.

$$C_3H_7-C\equiv C-(CH_2)_7OH \xrightarrow[\text{ii, NH}_4Cl]{i, \text{Na, NH}_3} \begin{array}{c} C_3H_7 \qquad H \\ C=C \\ H \qquad (CH_2)_7OH \end{array} \qquad (2.49)$$

Attempts to partially reduce terminal alkynes by this method normally fail as the alkyne reacts to give the alkynyl sodium species (sodium acetylide), which resists reduction because of the negative charge on the alkynyl carbon atom. In the presence of ammonium sulfate, however, the terminal alkyne is preserved and reduction gives the terminal alkene. This method can be preferable to catalytic hydrogenation, which sometimes gives small amounts of the saturated hydrocarbons that may be difficult to separate from the alkene. Reduction of a terminal alkyne can be suppressed by converting it to its sodium salt by reaction with sodium amide, thereby allowing the selective reduction of an internal triple bond in the same molecule. 1,7-Undecadiyne, for example, was converted to E-7-undecen-1-yne in high yield (2.50).[31]

$$C_3H_7C\equiv C(CH_2)_4C\equiv CH \xrightarrow[\text{NH}_3]{\text{NaNH}_2} C_3H_7C\equiv C(CH_2)_4C\equiv C^- \ Na^+ \xrightarrow[\text{ii, NH}_4Cl]{i, \text{Na, NH}_3} \begin{array}{c} C_3H_7 \qquad H \\ C=C \\ H \qquad (CH_2)_4C\equiv CH \end{array} \qquad (2.50)$$

$$75\%$$

---

[31] N. A. Dobson and R. A. Raphael, *J. Chem. Soc.* (1955), 3558.

Partial reduction of alkynes by using hydrogenation with Lindlar's catalyst or by using sodium in liquid ammonia provides *Z*- or *E*-alkenes respectively, although these conditions are not always ideal and other methods have been developed. Reduction of alkynes to *Z*-alkenes is possible by hydroboration (see Section 5.1) and protonolysis. Monohydroboration of alkynes is possible using dialkylboranes, catecholborane **51**, or other substituted boranes. The product alkenylborane is reactive and is not normally isolated. Protonolysis occurs readily with carboxylic acids and takes place with retention of alkene configuration (2.51). Therefore, since hydroboration occurs by *syn* addition of the hydrogen and boron atom, 1,2-disubstituted alkynes can be converted to *Z*-alkenes with high stereoselectivity (2.52).[32]

$$R\!-\!H \quad + \quad R_2B\!-\!OCOR' \qquad (2.51)$$

51

$$82\% \qquad\qquad 99\% \ Z \qquad (2.52)$$

Treatment of the intermediate alkenylborane, such as **52** or **53**, with iodine in the presence of a base (such as sodium hydroxide or methoxide) forms, stereoselectively, a *Z*-1,2-disubstituted or trisubstituted alkene. Transfer of one alkyl group from boron to the adjacent carbon atom occurs stereospecifically, resulting, after *anti* elimination of boron and iodine, in a new alkene in which the two substituents of the original alkyne become *trans* to each other (2.53, 2.54).[33]

$$(2.53)$$

52

99% *Z*

anti elimination

[32] H. C. Brown and G. A. Molander, *J. Org. Chem.*, **51** (1986), 4512.
[33] H. C. Brown, D. Basavaiah and S. U. Kulkarni, *J. Org. Chem.*, **47** (1982), 171.

(2.54)

53   71%

In comparison, no rearrangement can occur using alkenylboronic acids, derived from hydroboration of alkynes with catecholborane **51**, followed by hydrolysis. Treatment of the alkenylboronic acid with sodium hydroxide and iodine results in the replacement of the boronic acid group by iodine with retention of configuration (2.55). However, treatment with bromine, followed by base, results in substitution with inversion of configuration (2.56). In each case the reaction is highly stereoselective. The inversion of configuration in the bromination can be accounted for by invoking the usual *anti* addition of bromine across the double bond, followed by base-induced *anti* elimination of boron and bromine.

(2.55)

(2.56)

An excellent method for the preparation of alkenyl iodides and other substituted alkenes from alkynes uses the Schwartz reagent, bis(cyclopentadienyl)zirconium hydrochloride, [Cp$_2$Zr(H)Cl].[34] Hydrozirconation of terminal alkynes occurs regio- and stereoselectively to give the *E*-alkenylzirconium species. This species can be reacted with a variety of electrophiles, or transmetallated to other alkenyl metals prior to reaction. As an example, hydrozirconation of the alkynes **54** and **56** was used in a synthesis of the antitumor compound FR-901464 (2.57, 2.58).[35] The intermediate alkenylzirconium species **55** was transmetallated with ZnCl$_2$ to the corresponding alkenylzinc chloride and then coupled (using palladium catalysis) with a derivative of the alkenyl iodide **57** to give a diene, which was elaborated

[34] J. A. Labinger, in *Comprehensive Organic Synthesis*, ed. B. M. Trost and I. Fleming, vol. 8 (Oxford: Pergamon Press, 1991), p. 667.
[35] C. F. Thompson, T. F. Jamison and E. N. Jacobsen, *J. Am. Chem. Soc.*, **122** (2000), 10 482. For a review on some reactions of alkenylzirconocenes, see P. Wipf and C. Kendall, *Chem. Eur. J.*, **8** (2002), 1778.

further to give FR-901464.

$$(2.57)$$

$$(2.58)$$

Other useful hydrometallation methods are available for the preparation of substituted alkenes from alkynes. Alkenylalanes are readily prepared by hydroalumination of alkynes with, for example, diisobutylaluminium hydride.[36] In common with hydroboration and hydrozirconation, the reaction takes place by *syn* addition, giving *E*-alkenylalanes. Reaction of these alkenylalanes with halogens proceeds with retention of configuration to give the corresponding alkenyl halides. Thus, iodination of the alane from the reaction of 1-hexyne with diisobutylaluminium hydride produces the isomerically pure *E*-1-iodo-1-hexene (2.59). High selectivity for the other regioisomeric alkenyl iodide is possible by addition of hydrogen iodide to the alkyne.[37]

$$(2.59)$$

In contrast to the reaction with diisobutylaluminium hydride, hydroalumination of disubstituted alkynes with lithium hydridodiisobutylmethylaluminate, obtained from diisobutylaluminium hydride and methyllithium, results in *anti* addition across the triple bond. Subsequent reaction with aldehydes gives allylic alcohols, with CO_2 gives α,β-unsaturated acids and with iodine gives alkenyl iodides, isomeric with the products obtained in the reaction sequences using diisobutylaluminium hydride.[38]

[36] G. Zweifel and J. A. Miller, *Org. Reactions*, **32** (1984), 375; J. J. Eisch, in *Comprehensive Organic Synthesis*, ed. B. M. Trost and I. Fleming, vol. 8 (Oxford: Pergamon Press, 1991), p. 733.
[37] N. Kamiya, Y. Chikami and Y. Ishii, *Synlett* (1990), 675; P. J. Kropp and S. D. Crawford, *J. Org. Chem.*, **59** (1994), 3102.
[38] G. Zweifel and R. B. Steele, *J. Am. Chem. Soc.*, **89** (1967), 5085.

Thus the isomeric α-methylcrotonic acids are obtained from 2-butyne as illustrated in Schemes 2.60 and 2.61.

$$H_3C-C\equiv C-CH_3 \xrightarrow{Li[^iBu_2Al(H)Me]}$$ $$\xrightarrow[ii, H_3O^+]{i, CO_2}$$ (2.60)

72%

$$H_3C-C\equiv C-CH_3 \xrightarrow[\substack{ii, MeLi \\ Et_2O, -30\ ^\circ C}]{i,\ ^iBu_2AlH}$$ $$\xrightarrow[ii, H_3O^+]{i, CO_2}$$ (2.61)

76%

Anti addition across an alkyne can be accomplished with lithium aluminium hydride. This is particularly popular using propargylic alcohols and occurs with very high stereoselectivity. If the reduction is effected in the presence of sodium methoxide and the crude reduction product is treated with iodine, then the final product is exclusively the 3-iodoallylic alcohol (2.62). In contrast, reduction with lithium aluminium hydride and aluminium chloride, followed by iodination, gives the 2-iodoallylic alcohol (2.63).

$$R-C\equiv C-CH_2\!\!-\!\!OH \xrightarrow[NaOMe, THF]{i, LiAlH_4}$$ $$\xrightarrow{ii, I_2, -78\ ^\circ C}$$ (2.62)

$$R-C\equiv C-CH_2\!\!-\!\!OH \xrightarrow[ii, I_2, -78\ ^\circ C]{\substack{i, LiAlH_4 \\ AlCl_3, THF}}$$ (2.63)

Reaction of the alkenyl iodide with a lithium organocuprate, or with an organometallic species in a palladium-catalysed coupling, gives the corresponding substituted allylic alcohol (in which the substituents originally present in the propargylic alcohol are *trans* to each other). This method is applicable to a variety of synthetic problems in which the stereoselective introduction of a trisubstituted carbon–carbon double bond is involved. For example, it formed a key step in a synthesis of juvenile hormone (2.64).

$$\xrightarrow[\substack{ii, I_2 \\ iii, Me_2CuLi}]{\substack{i, LiAlH_4 \\ NaOMe, THF}}$$ (2.64)

Hydrostannylation of alkynes is less selective and mixtures of regio- and stereoisomers often result. The reaction is substrate dependent and, although mixtures of product alkenyl stannanes are normally formed, appropriate choice of the substituted alkyne can promote highly selective hydrostannylation.[39] Under thermal conditions, the hydrostannylation occurs by addition of the trialkyltin radical to the alkyne. Alternatively, metal-catalysed hydrostannylation, particularly with palladium(0), can be employed (2.65). The product alkenyl stannanes are useful for palladium-catalysed coupling reactions with unsaturated halides or triflates (Stille reaction – see Section 1.2.4).

$$
\begin{array}{c}
\underset{\text{HO}}{\diagdown}\!\!\!\!\equiv\!\!\!\!-\text{Me} \quad \xrightarrow[\text{[Pd(PPh}_3)_4]}{\text{Bu}_3\text{SnH}} \quad \underset{\text{Bu}_3\text{Sn}}{\text{HO}}\!\!\!\diagup\!\!\!\diagdown\!\!\text{Me} \quad + \quad \underset{\text{SnBu}_3}{\text{HO}}\!\!\!\diagup\!\!\!\diagdown\!\!\text{Me}
\end{array} \tag{2.65}
$$

84%   83 : 17

Hydrosilylation requires a transition-metal catalyst, such as [H_2PtCl_6], and results in very good yields of the product alkenyl silanes, but often as mixtures of regio- and stereoisomers (2.66).[40] More recently, it has been found that hydrosilylation of internal alkynes using the catalyst [Cp*Ru(MeCN)$_3$]PF$_6$ occurs by *trans* addition, and this can be followed by protodesilylation to provide a route to *E*-alkenes.[41] For example, ruthenium-catalysed hydrosilylation of methyl 2-octynoate with triethoxysilane, followed by fluoride-promoted desilylation of the intermediate regioisomeric alkenyl silanes, gave (*E*)-methyl 2-octenoate (2.67).

$$
\text{H}\!\!-\!\!\!\equiv\!\!\!-\text{C}_4\text{H}_9 \quad \xrightarrow[\text{[H}_2\text{PtCl}_6]}{\text{Et}_3\text{SiH}} \quad \underset{\text{Et}_3\text{Si}}{}\!\!\!\diagup\!\!\!\diagdown\!\!\text{C}_4\text{H}_9 \quad + \quad \underset{\text{SiEt}_3}{}\!\!\!\diagup\!\!\!\diagdown\!\!\text{C}_4\text{H}_9 \tag{2.66}
$$

100%   82 : 18

$$
\underset{}{\diagup\!\!\!\diagdown\!\!\!\diagup\!\!\!\equiv}\!\!\!-\text{CO}_2\text{Me} \quad \xrightarrow[\substack{\text{1 mol\% [Cp*Ru(MeCN)}_3]\text{PF}_6 \\ \text{then Bu}_4\text{NF, CuI}}]{(\text{EtO})_3\text{SiH, CH}_2\text{Cl}_2} \quad \underset{}{\diagup\!\!\!\diagdown\!\!\!\diagup\!\!\!\diagdown\!\!\!\diagup\!\!\!\diagdown}\text{CO}_2\text{Me} \tag{2.67}
$$

Cp* = pentamethylcyclopentadienyl anion   83%

So far in this section, the hydrometallation of alkynes has been described. Carbometallation of alkynes has also been developed into a practical method for the regio- and stereocontrolled formation of substituted alkenes.[42] Organocopper(I) reagents add readily to alkynes by a process known as carbocupration, developed by Normant and co-workers. For example, alkyl copper(I) reagents prepared from Grignard reagents and copper(I) bromide–dimethylsulfide complex add to terminal alkynes with excellent regio- and stereoselectivities. The carbocupration occurs with the copper atom adding to the terminal carbon atom of the alkyne, with

[39] N. D. Smith, J. Mancuso and M. Lautens, *Chem. Rev.*, **100** (2000), 3257.
[40] T. Hiyama and T. Kusumoto, in *Comprehensive Organic Synthesis*, ed. B. M. Trost and I. Fleming, vol. 8 (Oxford: Pergamon Press, 1991), p. 763.
[41] B. M. Trost, Z. T. Ball and T. Jöge, *J. Am. Chem. Soc.*, **124** (2002), 7922.
[42] J. F. Normant and A. Alexakis, *Synthesis* (1981), 841; P. Knochel, in *Comprehensive Organic Synthesis*, ed. B. M. Trost and I. Fleming, vol. 4 (Oxford: Pergamon Press, 1991), p. 865.

syn addition of the alkyl group. The product alkenyl cuprate species react with a variety of electrophiles, including alkyl halides, α,β-unsaturated ketones and epoxides, giving trisubstituted alkenes with almost complete retention of configuration (2.68). The alkenyl cuprate species also react with alkenyl iodides in the presence of [Pd(PPh$_3$)$_4$] to give conjugated dienes.

$$\text{MeMgBr} \xrightarrow[\text{Et}_2\text{O, }-45\text{ °C}]{\text{CuBr•SMe}_2} \text{MeCuMgBr}_2\text{•SMe}_2 \xrightarrow{\text{C}_3\text{H}_7-\text{C}\equiv\text{C}-\text{H}} \tag{2.68}$$

78%

Carboalumination of alkynes is possible in the presence of the zirconium catalyst Cp$_2$ZrCl$_2$.[43] The method is particularly attractive for methylalumination using Me$_3$Al and can tolerate a wide variety of functional groups. Terminal alkynes react with very good regioselectivity (∼95:5) and excellent stereoselectivity (≥98%) in favour of the alkenyl aluminium species, such as **58** (2.69). Mixtures of regioisomers often result when using internal alkynes with two different substituents.

$$\text{H}-\text{C}\equiv\text{C}-\text{C}_5\text{H}_{11} \xrightarrow[\text{Cp}_2\text{ZrCl}_2]{\text{Me}_3\text{Al}} \quad \mathbf{58} \xrightarrow{\text{I}_2} \quad 83\% \tag{2.69}$$

2.7 The Wittig and related reactions

The reaction between an aldehyde or ketone and a phosphonium ylide to form an alkene and a phosphine oxide is known as the Wittig reaction after the chemist, Georg Wittig, who first showed the value of this procedure in the 1950s for the synthesis of alkenes and who was awarded jointly the Nobel prize in 1979 (2.70).[44] The reaction is easy to carry out and proceeds under mild conditions. It is valued in organic synthesis as a method for carbon–carbon bond formation, in which the position of the double bond is unambiguous. The reaction generally leads to high yields of di- and trisubstituted alkenes from aldehydes and ketones but, because of steric effects, yields of tetrasubstituted alkenes from ketones are often poor. Sluggish reactions can sometimes be forced by addition of hexamethylphosphoramide (HMPA),

[43] E. Negishi, D. E. Van Horn and T. Yoshida, *J. Am. Chem. Soc.*, **107** (1985), 6639.
[44] A. Maercker, *Org. Reactions*, **14** (1965), 270; B. E. Maryanoff and A. B. Reitz, *Chem. Rev.*, **89** (1989), 863.

or by conducting the reaction at a higher temperature.

$$(2.70)$$

Phosphonium ylides are resonance-stabilized structures and can be classified as alkylidene phosphoranes, in which there is some overlap between the carbon p orbital and one of the d orbitals of phosphorus. Reaction with a carbonyl compound takes place with formation of a carbon–carbon bond and generation of a four-membered oxaphosphetane **59** (2.71). This fragments to the products, the driving force being provided by the formation of the very strong phosphorus–oxygen bond. Early mechanisms (and many textbooks) portray the formation of an intermediate betaine **60**. This may be formed in certain cases, although it is currently widely accepted that the initial addition is normally concerted, giving directly the oxaphosphetane **59**.[44,45]

$$(2.71)$$

59

60

The reactivity of the phosphonium ylide depends on the nature of the substituents. In practice, the three R groups on phosphorus are nearly always phenyl. If the substituents on the carbanion carbon are electron withdrawing (e.g. a carbonyl group), then the negative charge of the ylide becomes delocalized into the substituent and the nucleophilic character, and reactivity towards carbonyl groups, is decreased. Reagents of this type are much more stable and less reactive than those in which the substituents on the carbanion carbon are alkyl. However, such stabilized ylides are popular and effective Wittig reagents. They are isolable and simple derivatives are commercially available. The reaction is easy to carry out, typically by heating the carbonyl compound with the stabilized ylide in a solvent such as toluene.

[45] E. Vedejs and M. J. Peterson, *Top. Stereochem.*, **21** (1994), 1.

The rate-determining step is the initial addition to the carbonyl group to form the oxaphosphetane. More-electrophilic carbonyl compounds therefore react more readily. For example, (carboethoxymethylene)triphenylphosphorane reacts fairly readily with aldehydes, but can give poor yields in reactions with the less-reactive carbonyl group of ketones (2.72).

$$\tag{2.72}$$

Many Wittig reagents do not possess electron-withdrawing substituents on the carbanion carbon. Such alkyl-substituted phosphonium ylides are referred to as non-stabilized and react readily with carbonyl and other polar groups. Addition of the ylide to the carbonyl group takes place rapidly with aldehydes or ketones, both of which usually react equally well with these reagents. The number and nature of the alkyl substituents on the carbanion carbon normally has little influence on the extent of nucleophilic character of the phosphonium ylide.

Phosphonium ylides (alkylidene phosphoranes) can be prepared by a number of methods,[46] but in practice they are usually obtained by action of a base on (alkyl)triphenylphosphonium salts, which are themselves readily available from an alkyl halide and triphenylphosphine. The phosphonium salt can usually be isolated and crystallized, but the phosphonium ylide is generally prepared in solution and used without isolation. Formation of the phosphonium ylide is reversible, and the reaction conditions and the strength of the base required depend entirely on the nature of the ylide. A common procedure is to add a stoichiometric amount of a solution of *n*-butyllithium to a solution or suspension of the phosphonium salt in ether or THF, followed, after an appropriate interval, by the carbonyl compound. Other bases, such as sodium hydride or sodium or potassium alkoxides, in solution in the corresponding alcohol or in dimethylformamide, are used commonly.

Reactions involving non-stabilized ylides must be conducted under anhydrous conditions and in an inert atmosphere, because these ylides react both with oxygen

---

[46] H. J. Bestmann and R. Zimmerman, in *Comprehensive Organic Synthesis*, ed. B. M. Trost and I. Fleming, vol. 6 (Oxford: Pergamon Press, 1991), p. 171.

and with water. (Benzylidene)triphenylphosphorane, for example, reacts with water to give triphenylphosphonium oxide and toluene. With oxygen, reaction leads in the first place to triphenylphosphonium oxide and a carbonyl compound, which undergoes a Wittig reaction with unoxidized ylide to form a symmetrical alkene. Passing oxygen through a solution of the phosphonium ylide can therefore be a convenient route to symmetrical alkenes.

In the reaction of a phosphonium ylide with an aldehyde or ketone, a mixture of *E*- and *Z*-alkenes can result. In general, it is found that a resonance-stabilized ylide gives rise predominantly to the *E*-alkene, whereas a non-stabilized ylide usually gives more of the *Z*-alkene. The stereochemistry of the alkene product must arise from the stereochemistry of the oxaphosphetane, as the second step (the break-down of the oxaphosphetane) takes place by way of a concerted *syn* elimination. Therefore, of the two diastereomeric oxaphosphetanes, the *cis* isomer leads to the *Z*-alkene and the *trans* isomer to the *E*-alkene (2.73). With a non-stabilized phosphonium ylide, the formation of the oxaphosphetane is thought to be irreversible. Therefore the *Z*–*E* ratio is a reflection of the stereoselectivity in the first, kinetically controlled step. The preference for the formation of the *cis* oxaphosphetane has been attributed to the minimized steric interactions in the transition state involving orthogonally aligned reactants.

$$(2.73)$$

With a stabilized ylide, in which there is conjugation to an electron-withdrawing group, the formation of the oxaphosphetane is thought to be reversible, owing to the greater stability of the ylide. Therefore the ratio of stereoisomers of the alkene product is a reflection of the thermodynamic ratio of the two diastereomeric oxaphosphetanes. Since there are more steric interactions in the *cis* oxaphosphetane, in which the two alkyl groups are on the same side of the four-membered ring, then this diastereomer is normally less stable and breaks down to the starting materials faster than to the product alkene. Recombination will eventually result in a prefer-ence for the thermodynamically more-stable *trans* oxaphosphetane and hence the *E*-alkene product. An alternative explanation lies in an irreversible first step, even for stabilized ylides, but in which there is a late transition state (product-like) with a more planar structure, thereby favouring the *trans* oxaphosphetane. The presence of an intermediate betaine prior to the oxaphosphetane would have to be transient at most. It is likely that there is no universal explanation for all types of substituted phosphonium ylide, carbonyl compound and reaction conditions.

The Wittig reaction has been used widely in organic synthesis.[47] For example, a number of steps in a synthesis of the neurotoxin brevetoxin B make use of the Wittig reaction with both stabilized and non-stabilized phosphonium ylides, two of which are shown in Scheme 2.74.[48] This synthesis also uses a Wittig reaction in a later, key step to combine two large fragments using a non-stabilized phosphonium ylide to prepare a Z-alkene.

(2.74)

The stereoselectivity of the Wittig reaction depends, not only on the substituents, but also on the conditions under which the reaction is effected. The presence of lithium salts tends to favour the E-alkene, so reactions in which the Z-alkene is desired are often carried out using sodium or potassium bases. It is possible to obtain high yields of the E-alkene from a non-stabilized phosphonium ylide by deprotonation, then reprotonation of the intermediate oxaphosphetane or betaine.[49]

An especially useful application of the Wittig reaction is in the formation of exo-cyclic double bonds. Thus, cyclohexanone and (methylene)triphenylphosphorane give (methylene)cyclohexane, whereas the use of the Grignard reaction followed by dehydration leads to the endocyclic isomer.

A valuable group of Wittig reagents is derived from α-haloethers. They react with aldehydes or ketones to form vinyl ethers, which on acid hydrolysis are con-verted into aldehydes containing one more carbon atom. Thus cyclohexanone is converted into cyclohexane carboxaldehyde (2.75). The addition of an aldehyde

[47] K. C. Nicolaou, M. W. Harter, J. L. Gunzner and A. Nadin, *Liebigs Ann.* (1997), 1283.

[48] K. C. Nicolaou, F. P. J. T. Rutjes, E. A. Theodorakis, J. Tiebes, M. Sato and E. Untersteller, *J. Am. Chem. Soc.*, **117** (1995), 10 252.

[49] Q. Wang, D. Deredas, C. Huynh and M. Schlosser, *Chem. Eur. J.*, **9** (2003), 570; M. Schlosser and K. F. Christmann, *Liebigs Ann. Chem.*, **708** (1967), 1.

to the phosphonium ylide $Ph_3P=CBr_2$, generated from carbon tetrabromide, triphenylphosphine and zinc, is called the Corey–Fuchs reaction.[50] Treatment of the product 1,1-dibromo-alkene with *n*-butyllithium and an electrophile provides a useful method for the preparation of substituted alkynes (2.76).

$$\text{(2.75)}$$

$$\text{(2.76)}$$

Intramolecular Wittig reactions can be used for the preparation of cyclic alkenes.[51] The formation of the phosphonium ylide must be compatible with other functionality in the molecule and thus stabilized ylides are used most commonly. Wittig reactions with carbonyl groups other than aldehydes or ketones, such as carboxylic esters, are known.[52] For example, a route to the indole or penem ring systems uses a carboxylic amide or a thioester respectively as the intramolecular electrophile (2.77).

$$\text{(2.77)}$$

The simplest Wittig reagent, (methylene)triphenylphosphorane **61**, does not react easily with unreactive substrates such as some hindered ketones or epoxides. A useful reactive alternative is the doubly deprotonated lithio derivative **62**, which can be prepared from **61** by reaction with one equivalent of tert-butyllithium (2.78).[53] For example, fenchone, which is unaffected by (methylene)triphenylphosphorane itself at temperatures up to 50 °C, reacts with the new reagent to give the exomethylene

[50] E. J. Corey and P. L. Fuchs, *Tetrahedron Lett.* (1972), 3769.
[51] K. B. Becker, *Tetrahedron*, **36** (1980), 1717.
[52] P. J. Murphy and S. E. Lee, *J. Chem. Soc., Perkin Trans. 1* (1999), 3049.
[53] E. J. Corey, J. Kang and K. Kyler, *Tetrahedron Lett.*, **26** (1985), 555.

derivative **63** in high yield.

$$Ph_3P{=}CH_2 \xrightarrow[\substack{^{t}BuLi}]{\text{1 equiv.}} Ph_3P{=}CHLi \xleftarrow[\substack{^{s}BuLi}]{\text{2 equiv.}} \overset{+}{Ph_3P}{-}CH_3 \; X^- \qquad (2.78)$$

$$\underset{\textbf{61}}{} \qquad\qquad\qquad \underset{\textbf{62}}{}$$

$$\xrightarrow[\text{then } ^{t}BuOH]{Ph_3P{=}CHLi}$$

87%   **63**

Wittig reactions with stabilized phosphonium ylides sometimes proceed only slowly. A valuable alternative makes use of phosphonate esters in what is known as the Horner–Wadsworth–Emmons (or Wadsworth–Emmons) reaction.[54] Phosphonate esters are obtained readily from an alkyl halide and a trialkyl phosphite via an Arbuzov reaction. Proton abstraction with a suitable base gives the corresponding carbanion (e.g. **64**), which is more nucleophilic than the related phosphonium ylide, since the negative charge is no longer attenuated by delocalization into d orbitals of the adjacent positively charged phosphorus atom. Such anions react readily with the carbonyl group of an aldehyde or ketone to form an alkene and a water-soluble phosphonate ester (2.79). For example, the anion **64**, from the phosphonate derived from ethyl bromoacetate and triethyl phosphite, reacts rapidly with cyclohexanone at room temperature to give the alkene **65** in 70% yield, compared with a 25% yield obtained for the reaction with the triphenylphosphorane.

$$(2.79)$$

64   70%   **65**

Where applicable, the Horner–Wadsworth–Emmons reaction is generally superior to the Wittig reaction with resonance-stabilized phosphonium ylides and it is employed widely in the preparation of α,β-unsaturated esters and other conjugated systems. It often gives better yields than the Wittig reaction, the phosphonate esters are readily available and it has the practical advantage that the phosphate

---

[54] W. S. Wadsworth, *Org. Reactions*, **25** (1977), 73; see also Reference 44.

by-product is water soluble and easily removed from the reaction mixture. In contrast, the Wittig reaction gives the by-product triphenylphosphine oxide, which is often difficult to remove from the product. However, the Horner–Wadsworth–Emmons reaction is unsuitable for the preparation of alkenes from non-stabilized reagents.

α,β-Unsaturated ketones can be made from β-keto-phosphonates and carbonyl compounds (2.80). β-Keto-phosphonates are themselves obtained by reaction of the lithium salt of dimethyl methylphosphonate with an ester (or by reaction with an aldehyde followed by oxidation of the initial hydroxyphosphonate).

$$(2.80)$$

A mild base such as diisopropylethylamine or DBU can be used for the Horner–Wadsworth–Emmons reaction in the presence of a complexing agent such as lithium chloride.[55] These conditions are useful for sensitive substrates, such as chiral compounds with an enolizable stereocentre. For example, the enone **66** is formed in high optical purity under these conditions (2.81), whereas use of the base potassium tert-butoxide gives the product as a racemic mixture.

$$(2.81)$$

[55] M. A. Blanchette, W. Choy, J. T. Davis, A. P. Essenfeld, S. Masamune, W. R. Roush and T. Sakai, *Tetrahedron Lett.*, **25** (1984), 2183.

The reaction of phosphonate anions with aldehydes normally proceeds with high selectivity for the *E*-alkene. However, the stereochemistry depends on the substitution pattern of the phosphonate and aldehyde, and on the conditions of the reaction, such that the *Z*-alkene can be the predominant or even exclusive product. In reactions that give a mixture of *E*- and *Z*-alkenes, the *E*-selectivity can sometimes be enhanced by using the bulkier bis(isopropyl) phosphonate ester. Still and Gennari showed that high selectivity for the *Z*-alkene could be obtained by using bis(trifluoroethyl) phosphonate esters (2.82).[56] The corresponding reaction with benzaldehyde and the dimethyl phosphonate ester gave almost exclusive ($>$50:1) formation of the *E*-alkene. An alternative and increasingly popular method for the formation of *Z*-alkenes has been reported by Ando, and makes use of the related diaryl phosphonates.[57] For example, deprotonation of ethyl (diphenyl phosphono)acetate with sodium hydride and addition to octanal resulted in the predominant formation of (*Z*)-ethyl 2-decenoate (2.83).

$$PhCHO \quad + \quad \underset{\underset{CF_3CH_2O}{|}}{\overset{\overset{O}{||}}{CF_3CH_2O-P}}\diagdown CO_2Me \quad \xrightarrow[\substack{THF \\ 18\text{-crown-6}}]{KN(SiMe_3)_2} \quad \underset{Ph \qquad CO_2Me}{\diagup\!\!\!=\!\!\!\diagdown} \qquad (2.82)$$

$$>95\% \qquad\qquad Z{:}E \;>50{:}1$$

$$C_7H_{15}CHO \quad + \quad \underset{\underset{PhO}{|}}{\overset{\overset{O}{||}}{PhO-P}}\diagdown CO_2Et \quad \xrightarrow[THF]{NaH} \quad \underset{C_7H_{15} \qquad CO_2Et}{\diagup\!\!\!=\!\!\!\diagdown} \qquad (2.83)$$

$$100\% \qquad\qquad Z{:}E \;9{:}1$$

A variation of the Wittig reaction that can overcome problems with the stereochemical outcome is the Horner–Wittig reaction with phosphine oxides.[58] The oxides are obtained by quaternization of triphenylphosphine and hydrolysis of the phosphonium salt, or by reaction of lithiodiphenylphosphide with an alkyl halide or sulfonate and oxidation of the resulting phosphine with hydrogen peroxide. The derived lithio species react with aldehydes or ketones to give β-hydroxy phosphine oxides, which eliminate on treatment with a base such as sodium hydride or potassium hydroxide to form the alkene. In common with the Horner–Wadsworth–Emmons reaction, the phosphorus by-product is water soluble and easily removed from the product.

An advantage of the Horner–Wittig reaction is that the two diastereomeric β-hydroxy phosphine oxides are stable, isolable compounds and can be separated. The elimination step is stereospecific, such that one diastereomeric β-hydroxy

[56] W. C. Still and C. Gennari, *Tetrahedron Lett.*, **24** (1983), 4405.
[57] K. Ando, *J. Org. Chem.*, **64** (1999), 8406.
[58] J. Clayden and S. Warren, *Angew. Chem. Int. Ed. Engl.*, **35** (1996), 241.

phosphine oxide gives only the *E*-alkene, with the other providing the *Z*-alkene. The elimination occurs by a *syn* pathway, by way of a four-membered cyclic transition state, similar to that in the Wittig reaction. In practice, the reaction of the lithio derivatives of alkyldiphenylphosphine oxides with aldehydes generally leads predominantly to the *anti* alcohols and, hence, on purification and *syn* elimination, to the *Z*-alkene product (2.84). However, the *E*-alkene can be obtained by reduction of the ketone, formed by acylation of the lithio diphenylphosphine oxide with an ester or by oxidation of the *anti*-β-hydroxy phosphine oxide, followed by elimination from the *syn* alcohol. The alkene **67**, a component of a pheromone of the Mediterranean fruit fly, was made in this way (2.85).

There is a silicon version of the Wittig reaction, known as the Peterson reaction.[59] Reaction of an aldehyde or ketone with an α-silyl carbanion forms a β-hydroxy silane, from which elimination of trialkylsilanol, R_3SiOH, provides

[59] D. J. Ager, *Org. Reactions*, **38** (1990), 1; L. F. van Staden, D. Gravestock and D. J. Ager, *Chem. Soc. Rev.*, **31** (2002), 195.

the alkene product. Most commonly, trimethylsilyl derivatives are used and the by-product hexamethyldisiloxane (formed from Me_3SiOH) is volatile and much easier to remove from the reaction product than triphenylphosphine oxide. If the metal counterion forms a fairly covalent bond (e.g. Li^+ or Mg^{2+}), then the intermediate β-hydroxy silanes can be isolated by protonation. A separate elimination step under basic or acidic conditions then provides the alkene. With counterions that give a more ionic intermediate (e.g. Na^+ or K^+), spontaneous elimination often occurs.

Conveniently, both the *E*- and *Z*-alkene products can be separately obtained from a single diastereomer of the β-hydroxy silane, depending on the conditions used for the elimination reaction (2.86). As ordinarily effected, the Peterson reaction would give a mixture of *E*- and *Z*-alkenes, owing to the fact that the β-hydroxy silane is generally obtained as a mixture of *syn* and *anti* isomers. However, the actual elimination reaction is highly stereoselective and, with a pure diastereomer of the hydroxysilane, elimination can be controlled to give either *E*- or *Z*-alkene. Under basic conditions *syn* elimination takes place, probably by way of a cyclic four-membered transition state like that in the Wittig reaction. Under acidic conditions (H^+ or a Lewis acid) the elimination is *anti*, leading to the other geometrical isomer of the alkene. Therefore, with a mixture of β-hydroxy silanes, separation and elimination of each diastereomer under different conditions can lead to the same alkene isomer.

(2.86)

The Peterson reaction requires access to α-silyl carbanions, typically formed by proton abstraction. However, this method is generally only applicable if an electron-withdrawing group is also present on the α-carbon. The α-silyl carbanion can, alternatively, be formed from the corresponding halide and magnesium, or by addition of an organometallic species to an alkenyl silane. The simple Grignard reagent trimethylsilylmethyl magnesium chloride can be prepared readily and is a useful methylenating agent. For example, this reagent was used in a synthesis of periplanone-B, a pheromone of the American cockroach (2.87).[60]

[60] W. Still, *J. Am. Chem. Soc.*, **101** (1979), 2493.

(2.87)

62%

The corresponding organolithium species Me_3SiCH_2Li, prepared from the chloride and lithium metal, or by bromine– or iodine–lithium exchange with butyllithium or sulfur–lithium exchange with lithium naphthalenide, can also prove effective. Addition of $CeCl_3$ to this organolithium species provides a softer reagent that allows its use with sensitive carbonyl compounds.[61]

The addition of an α-substituted α-silyl carbanion to an aldehyde or ketone normally leads to a mixture of the two diastereomeric β-hydroxy silanes. Although these can often be separated, and both converted to the desired *E*- or *Z*-alkene, the most useful Peterson reactions involve the stereoselective formation of the β-hydroxy silane. This can be achieved in only certain cases by addition of an α-silyl carbanion to a carbonyl compound. Reduction of, or organometallic addition to, an α-silyl ketone, or functionalization of an unsaturated silane, provide other, often stereoselective, alternatives. For example, Cram (Felkin–Anh)-type addition of methyllithium to the α-silyl ketone **68** gives predominantly the β-hydroxy silane **69**, and hence either the *E*- or *Z*-alkene, with high selectivity (2.88). Epoxidation of an alkenyl silane and ring-opening of the product α,β-epoxy silane with an organometallic reagent occurs regioselectively at the carbon atom bearing the silyl group and stereoselectively with inversion of configuration, thereby providing a stereocontrolled route to alkenes (2.89). Dihydroxylation of an allylsilane provides another route to the required β-hydroxy silane, from which elimination to the alkene can be accomplished (2.90).

(2.88)

| | KH | 61% | 95 : 5 |
| | TsOH | 60% | 5 : 95 |

[61] C. R. Johnson and B. D. Tait, *J. Org. Chem.*, **52** (1987), 281.

$$(2.89)$$

$$(2.90)$$

2.8 Alkenes from sulfones

Unlike the corresponding phosphonium salts, addition of sulfonium salts to aldehydes results, not in the alkene products, but in the formation of epoxides (see
Section 1.1.5.2). However, sulfones can be used to prepare alkenes, by way of
the α-metallo derivatives, in what is termed the Julia olefination (alkenylation).
Addition of the organometallic species to an aldehyde or ketone gives a β-hydroxy
sulfone which, in the form of its *O*-acyl or *O*-sulfonyl derivative, undergoes
reductive cleavage with, for example, sodium amalgam in methanol to form the
alkene.[62] The reaction is regioselective and can be used to prepare mono-, di- and
trisubstituted alkenes (2.91).

$$(2.91)$$

Although a mixture of the two diastereomeric β-hydroxy sulfones are formed,
the reductive elimination gives predominantly the *E*-alkene product. It is thought
that the initial reductive cleavage of the sulfonyl group generates an anion which,

[62] P. J. Kocienski, in *Comprehensive Organic Synthesis*, ed. B. M. Trost and I. Fleming, vol. 6 (Oxford: Pergamon
Press, 1991), p. 975; see also Reference 1.

whatever its original configuration, is sufficiently long-lived to permit it to adopt the low-energy conformation **71**, in which the two R groups are as far apart as possible, from which the *E*-alkene is formed by loss of acetate ion (2.92). This sequence provides a useful alternative to the Wittig reaction for the preparation of *E*-1,2-disubstituted alkenes. The sulfones are more easily purified than the phosphonium salts and are readily available, even from secondary halides, by reaction with the nucleophilic PhS$^-$ anion and oxidation of the resulting sulfide. Trisubstituted alkenes can be prepared by the Julia alkenylation, although the more-substituted intermediate β-alkoxy sulfone, generated using a ketone electrophile, is prone to revert to the starting materials. In such cases, reversing the fragments, such that a more-substituted sulfone is condensed with an aldehyde is often successful.

$$(2.92)$$

71

If the *Z*-1,2-disubstituted alkene is the desired product then an alternative procedure can be adopted. *O*-Sulfonation of the intermediate β-hydroxy sulfone and elimination using a base (rather than sodium amalgam), gives the corresponding alkenyl sulfone. The *Z*- and *E*-alkenyl sulfones can be obtained selectively from the appropriate diastereomer of the β-hydroxy sulfone, and give, on reductive cleavage of the sulfone, the alkene with retention of stereochemistry. The *E*-sulfone **72**, for example, on treatment with a Grignard reagent in the presence of a palladium catalyst, gave the *Z*-alkene **73** in good yield (2.93).[63] A small amount of the trisubstituted alkene is also formed and, under appropriate conditions, substitution of the sulfone may predominate.

$$(2.93)$$

72   83%   **73** 98.5% *Z*

Oxidation of the intermediate β-hydroxy sulfone to the ketone and reductive desulfonation provide an alternative transformation that has found use in synthesis (see Section 1.1.5.2). The Julia alkenylation reaction normally proceeds in high

[63] J.-L. Fabre and M. Julia, *Tetrahedron Lett.*, **24** (1983), 4311.

overall yield and is a useful method for linking two fragments of a large target molecule. However, it has the disadvantage that three separate steps (carbon–carbon bond formation, *O*-acylation and reductive elimination) are required. A solution to this lengthy procedure is to replace the phenyl sulfone group with a hetero-cyclic sulfone, that permits a one-pot alkenylation.[64] For example, in a synthesis of hennoxazole A, the alkene **76** was prepared by treating the tetrazolyl sulfone **75** with potassium hexamethyldisilazide (KHMDS) and the aldehyde **74** (2.94). In the modified Julia alkenylation, the first-formed alkoxide adds intramolecularly to the heterocycle to give an intermediate anion, e.g. **77** from a benzothiazolyl sulfone, which fragments with loss of SO_2 to give the desired alkene (2.95). This fragmen-tation is stereospecific (*syn* elimination) and therefore it is important that the initial addition to the aldehyde is stereoselective. In practice, a mixture of diastereomeric β-alkoxy sulfones and hence a mixture of alkene geometrical isomers is formed, although a preference for the *E*-alkene is common.

(2.94)

(2.95)

Formation of an α-metalated sulfone in which the α′ position is substituted with a halide promotes alkene formation in what is known as the Ramberg–Bäcklund reaction (or rearrangement).[65] Intramolecular displacement of the halide gives an

[64] J. B. Baudin, G. Hareau, S. A. Julia, R. Lorne and O. Ruel, *Bull. Soc. Chim. Fr.*, **130** (1993), 856; P. R. Blakemore, *J. Chem. Soc., Perkin Trans. 1* (2002), 2563.

[65] L. A. Paquette, *Org. Reactions*, **25** (1977), 1; J. M. Clough, in *Comprehensive Organic Synthesis*, ed. B. M. Trost and I. Fleming, vol. 3 (Oxford: Pergamon Press, 1991), p. 861; R. J. K. Taylor, *J. Chem. Soc., Chem. Commun.* (1999), 217.

episulfone, from which sulfur dioxide is extruded to give the alkene (2.96).

$$\text{X = Cl, Br, I}$$

The Ramberg–Bäcklund reaction permits the synthesis of many different substituted acyclic and cyclic alkenes, including strained alkenes such as cyclobutenes, formed by ring-contraction. Mixtures of acyclic *E*- and *Z*-alkene geometrical isomers are common, the stereoselectivity depending on the substitution pattern and on the conditions used. The Ramberg–Bäcklund reaction is therefore most useful when only one alkene geometry can be formed. An example of the reaction is the addition of the base t-BuOK to the α-chloro sulfones **78**, which promotes the formation of the cyclic *Z*-enediynes **79**, used in a study of the Bergman cycloaromatization reaction (2.97).[66]

A useful variant of the Ramberg–Bäcklund reaction involves *in situ* halogenation and alkene formation.[67] The reagent combination KOH or t-BuOK in CCl_4 or CF_2Br_2 permits chlorination or bromination followed by direct deprotonation, episulfone and hence alkene formation. A route to *exo*-glycals uses this chemistry, with bromination α- to the sulfone, followed by *in situ* Ramberg–Bäcklund reaction (2.98).

[66] K. C. Nicolaou, G. Zuccarello, C. Riemer, V. A. Estevez and W.-M. Dai, *J. Am. Chem. Soc.*, **114** (1992), 7360.
[67] C. Y. Meyers, A. M. Malte and W. S. Matthews, *J. Am. Chem. Soc.*, **91** (1969), 7510; T.-L. Chan, S. Fong, Y. Li, T.-O. Man and C.-D. Poon, *J. Chem. Soc., Chem. Commun.* (1994), 1771.

2.9 Alkenes using titanium or chromium reagents

Alkenes can be obtained from aldehydes or ketones on reductive dimerization by treatment with a reagent prepared from titanium(III) chloride and zinc–copper couple (or LiAlH$_4$), or with a species of active titanium metal formed by reduction of titanium(III) chloride with potassium or lithium metal.[68] This McMurry coupling reaction is of wide application, but in intermolecular reactions generally affords a mixture of the *E*- and *Z*-alkenes (2.99).

$$\text{CHO} \quad \xrightarrow[\text{THF, reflux}]{\text{TiCl}_3,\ \text{K}} \quad \text{(2.99)}$$

77%   *E:Z* 70:30

The reaction takes place in two steps on the surface of the active titanium particles. The first stage, leading to the formation of a new carbon–carbon bond is simply a pinacol reaction. The titanium reagent donates an electron to the carbonyl compound generating a ketyl radical (see Section 7.2), which dimerizes to give the pinacol (1,2-diol). The intermediacy of pinacols in the reaction is supported by the fact that pinacols are smoothly converted into alkenes on treatment with the titanium reagent. In the second stage, de-oxygenation is effected by way of a species formed by co-ordination of the pinacol to the surface of the titanium. Cleavage of the two carbon–oxygen bonds then occurs, yielding the alkene and an oxidized titanium surface (2.100). Under milder reaction conditions (e.g. room temperature), the pinacol can be formed as the major product (2.101).[69]

$$2\ \overset{O}{\underset{}{\parallel}} \quad \xrightarrow{2e} \quad 2\left[\ \overset{O^-}{\underset{}{\cdot}}\ \right] \longrightarrow \ \overset{O^-\ \ O^-}{\underset{}{}} \longrightarrow \ \rangle\!\!=\!\!\langle \quad \text{(2.100)}$$

$$\begin{array}{c}\text{CHO}\\\text{CHO}\end{array} \quad \xrightarrow[\substack{\text{Zn-Cu}\\\text{DME, r.t.}}]{\text{TiCl}_3(\text{DME})_2} \quad \begin{array}{c}\text{OH}\\\text{OH}\end{array} \quad \text{(2.101)}$$

85%

Mixed coupling reactions, using two different carbonyl compounds, can be effected, but they generally lead to mixtures of products and are of limited use in synthesis. Intramolecular reactions with dicarbonyl compounds, on the other hand, provide a good route to cyclic alkenes. The keto-aldehyde **80**, for example, gave the cyclic diterpene kempene-2, despite the presence of a saturated ketone

[68] E. Block, *Org. Reactions*, **30** (1984), 457; J. E. McMurry, *Chem. Rev.*, **89** (1989), 1513; R. G. Dushin, in *Comprehensive Organometallic Chemistry II*, ed. E. W. Abel, F. G. A. Stone and G. Wilkinson, vol. 12 (Oxford: Elsevier, 1995), p. 1071; A. Fürstner and B. Bogdanovic, *Angew. Chem. Int. Ed. Engl.*, **35** (1996), 2443; M. Ephritikhine, *Chem. Commun.* (1998), 2549.

[69] J. E. McMurry and J. G. Rico, *Tetrahedron Lett.*, **30** (1989), 1169.

and an ester group within the substrate (2.102).[70] Although most carbonyl coupling reactions involve dialdehydes, diketones or keto-aldehydes, substrates such as keto-esters can be cyclized by using low-valent titanium. Isocaryophyllene was synthesized by cyclization of the keto-ester **81**, followed by Wittig methylenation of the product **82** (2.103).[71] The unusual *E* to *Z* isomerization of the double bond in this conversion is believed to be induced by strain in the cyclic intermediate in which the two oxygen atoms are bound to titanium.

$$(2.102)$$

$$(2.103)$$

Various titanium-based reagents for alkenylation are known.[72] The titanium–aluminium complex **83**, known as the Tebbe reagent, can effect the methylenation of carbonyl compounds.[73] Aldehydes or ketones can be methylenated and, unlike the Wittig reaction (with $Ph_3P=CH_2$), ester or amide carbonyl groups are good substrates, thereby leading to enol ethers or enamines (2.104). The methylenation of esters or amides may alternatively be carried out using the Petasis reagent Cp_2TiMe_2.[74] These reagents are thought to give the titanium methylidene $Cp_2Ti=CH_2$ as the active methylenating agent.

$$(2.104)$$

[70] W. G. Dauben, I. Farkas, D. P. Bridon and C.-P. Chuang, *J. Am. Chem. Soc.*, **113** (1991), 5883.
[71] J. E. McMurry and D. D. Miller, *Tetrahedron Lett.*, **24** (1983), 1885.
[72] S. H. Pine, *Org. Reactions*, **43** (1993), 1; R. C. Hartley and G. J. McKiernan, *J. Chem. Soc., Perkin Trans. 1* (2002), 2763.
[73] F. N. Tebbe, G. W. Parshall and G. S. Reddy, *J. Am. Chem. Soc.*, **100** (1978), 3611; S. H. Pine, R. Zahler, D. A. Evans and R. H. Grubbs, *J. Am. Chem. Soc.*, **102** (1980), 3270.
[74] N. A. Petasis and E. I. Bzowej, *J. Am. Chem. Soc.*, **112** (1990), 6392.

Another reagent for the methylenation of aldehydes or ketones is the Oshima–Lombardo reagent $TiCl_4$–Zn–CH_2I_2 (or CH_2Br_2).[75] This reagent is non-basic and can therefore be advantageous for base-sensitive substrates (2.105).

A drawback of the Tebbe and related reagents is that they are generally suitable only for methylenation and do not permit the formation of higher alkyl analogues. However, the alkenylation of esters (or amides) has been found possible using the Oshima–Lombardo conditions in the presence of TMEDA (tetramethylethylene-diamine) (2.106).[76] This chemistry requires the prior formation of the alkyl *gem*-dibromide and a more-convenient method, using a dithioacetal, has been reported (2.107).[77]

$$\text{(2.105)}$$

$$PhCO_2Me \quad + \quad RCHBr_2 \quad \xrightarrow[\text{THF}]{TiCl_4,\ Zn,\ TMEDA} \quad \text{(2.106)}$$

R = Me 86%   Z:E 92:8

$$Ph \diagdown CO_2Et \quad + \quad Ph \diagdown\diagup SPh \quad \xrightarrow[\text{THF}]{Cp_2Ti[P(OEt)_3]_2} \quad \text{(2.107)}$$

75%   Z:E 86:14

A related and important alkenylation of aldehydes with *gem*-dihaloalkanes, mediated by chromium salts, is often referred to as the Takai alkenylation.[78] Organochromium reagents are very tolerant of many functional groups and are non-basic, such that this methodology offers a mild and convenient approach to alkenes.[79] A common use of this reaction is for one-carbon homologation of aldehydes to alkenyl halides, which are typically formed with good *E*-selectivity (2.108). The resulting alkenyl iodides are useful substrates for palladium-catalysed coupling reactions (see Section 1.2.4). The alkenylation of ketones is slower than that of aldehydes and the reaction can therefore be used for the chemoselective

[75] K. Takai, Y. Hotta, K. Oshima and K. Utimoto, *Tetrahedron Lett.* (1978), 2417; L. Lombardo, *Tetrahedron Lett.*, **23** (1982), 4293; J. Hibino, T. Okazoe, K. Takai and H. Nozaki, *Tetrahedron Lett.*, **26** (1985), 5579.
[76] K. Takai, K. Nitta and K. Utimoto, *J. Am. Chem. Soc.*, **108** (1986), 7408; K. Takai, Y. Kataoka, J. Miyai, T. Okazoe, K. Oshima and K. Utimoto, *Org. Synth.*, **73** (1996), 73.
[77] Y. Horikawa, M. Watanabe, T. Fujiwara and T. Takeda, *J. Am. Chem. Soc.*, **119** (1997), 1127.
[78] K. Takai, K. Nitta and K. Utimoto, *J. Am. Chem. Soc.*, **108** (1986), 7408.
[79] A. Fürstner, *Chem. Rev.*, **99** (1999), 991.

functionalisation of keto aldehydes (2.109). Although most popular with iodoform, the reaction can be used for the alkenylation of aldehydes with substituted *gem-*dihalides.

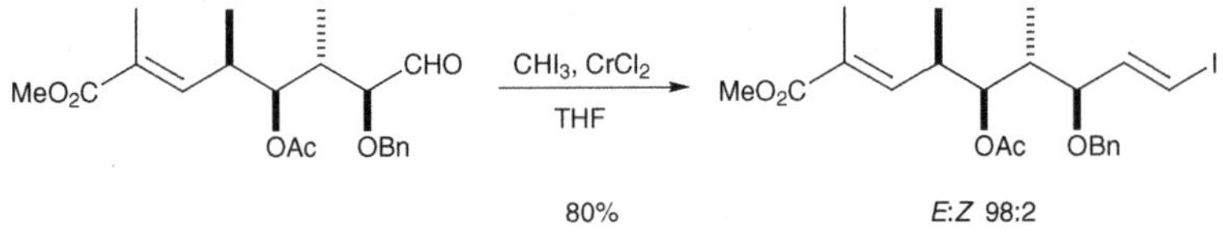

$$2 \; RCH{=}CH_2 \xrightarrow{[L_nM{=}CHR']} RCH{=}CHR \;+\; H_2C{=}CH_2 \quad (2.110)$$

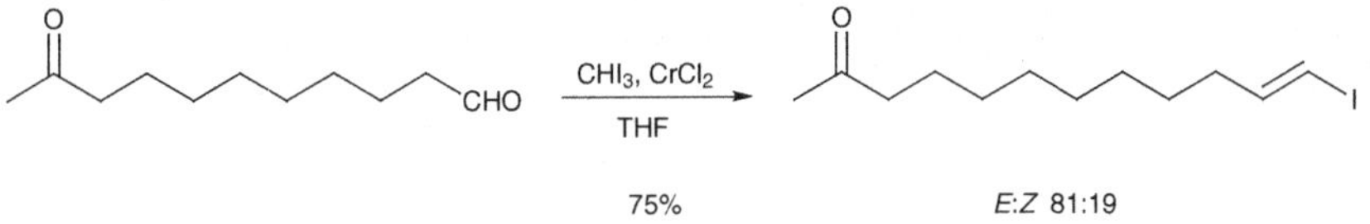

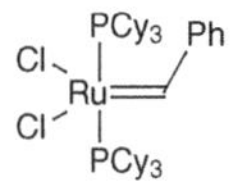

2.10 Alkene metathesis reactions

A significant development for the selective synthesis of alkenes makes use of alkene metathesis.[80] Metathesis, as applied to two alkenes, refers to the transposition of the alkene carbon atoms, such that two new alkenes are formed (2.110). The reaction is catalysed by various transition-metal alkylidene (carbene) complexes, particularly those based on ruthenium or molybdenum. The ruthenium catalyst **84**, developed by Grubbs, is the most popular, being more stable and more tolerant of many functional groups (although less reactive) than the Schrock molybdenum catalyst **85**. More recently, ruthenium complexes such as **86**, which have similar stability and resistance to oxygen and moisture as complex **84**, have been found to be highly active metathesis catalysts.

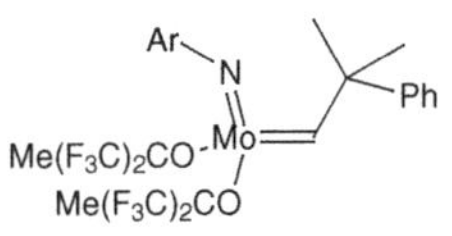

Cy = cyclohexyl
Ar = 2,6-diisopropylphenyl
Mes = 2,4,6-trimethylphenyl

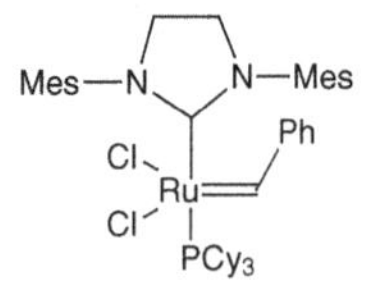

[80] M. Schuster and S. Blechert, *Angew. Chem. Int. Ed. Engl.*, **36** (1997), 2036; R. H. Grubbs and S. Chang, *Tetrahedron*, **54** (1998), 4413; S. K. Armstrong, *J. Chem. Soc., Perkin Trans. 1* (1998), 371; R. R. Schrock, *Tetrahedron*, **55** (1999), 8141; A. Fürstner, *Angew. Chem. Int. Ed.*, **39** (2000), 3012; R. R. Schrock and A. H. Hoveyda, *Angew. Chem. Int. Ed.*, **42** (2003), 4592.

Alkene metathesis occurs by way of an intermediate metallacycle **87**, followed by ring opening to give either the starting materials or one of the new alkenes and a new metallocarbene complex (2.111). Further metallocycle formation using another alkene and ring-opening provides the other product alkene and recovered catalyst to continue the cycle.

$$(2.111)$$

The reaction has shown considerable use in organic synthesis for ring formation (ring-closing metathesis, RCM). The method is not only effective for the preparation of five- and six-membered rings, but can be applied to medium and large ring formation. This has made it popular for the synthesis of many different substituted carbocyclic and heterocyclic ring systems.

Substrates containing two appropriately spaced monosubstituted alkenes often undergo ring-closing metathesis efficiently with the Grubbs catalyst **84**. An example is the cyclization to give the indolizidine ring system from the diene **88** (2.112). Reduction of the product **89** gives the alkaloid (–)-coniceine **90**. Typically, such ring-closing reactions can be accomplished under mild conditions and with only 1–10 mol% of the catalyst. The other newly formed alkene, ethene, is a convenient by-product.

$$(2.112)$$

In cases where metathesis with the catalyst **84** is unsuccessful or very sluggish, then the Schrock catalyst **85** or catalysts such as **86** are often effective. This is particularly the case for the formation of tri- or tetrasubstituted alkenes, which are normally too hindered to be formed using the catalyst **84**. For example, ring-closing metathesis of the diene **91** is unsuccessful with the Grubbs catalyst **84**, but the cyclohexene **92** can be formed in excellent yield with the catalyst **86** (2.113). An efficient synthesis of the medium-ring terpene dactylol **94** was accomplished with the Schrock catalyst **85** (2.114). Attempted metathesis of **93** with the catalyst **84** failed to give any of the medium-ring trisubstituted alkene product.

$$(2.113)$$

$$(2.114)$$

Medium-ring products (containing ring sizes 7–9) are often difficult to prepare by conventional chemistry. Ring-closing metathesis to give medium rings provides a solution to this problem and is particularly successful when a conformational constraint, such as another ring or a stereoelectronic effect, aids the medium-ring formation.[81]

Large-ring products can be accessed readily by ring-closing metathesis. If more than one alkene is present in the substrate then the less-hindered, typically mono-substituted, alkene reacts preferentially. For example, the anticancer epothilone compounds can be prepared by using metathesis as the key ring-forming step. Treatment of the substrate **95** with the catalyst **84** resulted in the formation of both the desired *Z*-alkene **96** and the *E*-alkene **97** (2.115). Control of alkene stereochemistry in macrocycle formation is often difficult unless a conformational constraint

[81] M. E. Maier, *Angew. Chem. Int. Ed.*, **39** (2000), 2073.

promotes one geometrical isomer.

$$\text{95} \xrightarrow[\substack{\text{CH}_2\text{Cl}_2 \\ \text{room temp.}}]{10 \text{ mol\% } \mathbf{84}} \text{96} \quad 46\% \quad + \quad \text{97} \quad 39\% \tag{2.115}$$

SiR$_3$ = SiMe$_2$^{t}Bu

Many other types of heterocyclic and carbocyclic ring systems, with different substitution patterns, can be prepared by using ring-closing metathesis. Metathesis catalysts are also effective for ring-opening metathesis polymerization (ROMP), in which cyclic alkenes can be polymerized.

Cross-metathesis of two different alkenes to give an acyclic alkene is complicated by the possible formation of not only the desired cross-metathesis product, but also self-metathesis products, each as a mixture of alkene isomers. However, some alkenes are amenable to efficient cross-metathesis to give the desired substituted alkene. This is particularly the case with alkenes that are slow to homodimerize, such as α,β-unsaturated carbonyl compounds or alkenes bearing bulky substituents.[82] Hence, cross-metathesis of methyl acrylate with an alkene proceeds efficiently (2.116). The ruthenium catalyst reacts preferentially with the more electron-rich alkene **98**, which then undergoes cross-metathesis with the acrylate or self-metathesis with another molecule of the alkene **98**. The latter reaction is reversible and hence a high yield of the desired substituted acrylate results over time. The use of 1,1-disubstituted alkenes as partners in cross-metathesis provides a route to trisubstituted alkenes. This chemistry is therefore a useful alternative to conventional syntheses of alkenes, such as by the Wittig reaction.

$$\text{98} \quad + \quad \diagup\text{CO}_2\text{Me} \xrightarrow[\substack{\text{CH}_2\text{Cl}_2 \\ \text{room temp.}}]{5 \text{ mol\% } \mathbf{86}} \text{AcO} \diagdown_n \diagup \diagup \text{CO}_2\text{Me} \tag{2.116}$$

94%

[82] S. J. Connon and S. Blechert, *Angew. Chem. Int. Ed.*, **42** (2003), 1900.

Problems (answers can be found on page 469)

1. Explain why the two diastereomeric amine *N*-oxides **1** give, on heating, two different major regioisomeric alkene products.

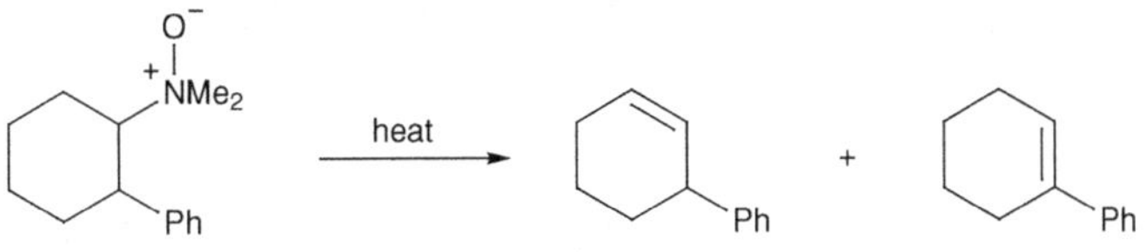

		98	:	2
syn		98	:	2
anti		15	:	85

2. Explain the formation of the Z-α,β-unsaturated ester **2**.

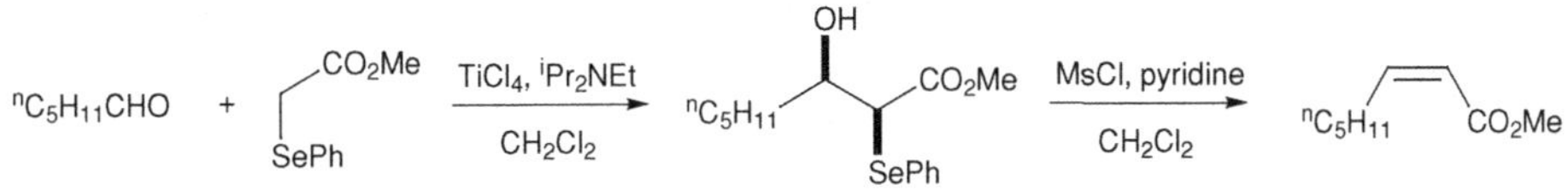

3. Draw a mechanism to account for the formation of methyl chrysanthemate by the transformation shown below.

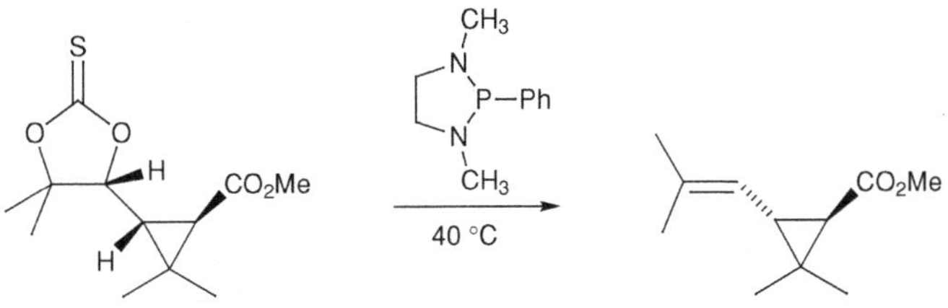

methyl chrysanthemate

4. Suggest a method to prepare the allylic alcohol **3** as a single stereoisomer.

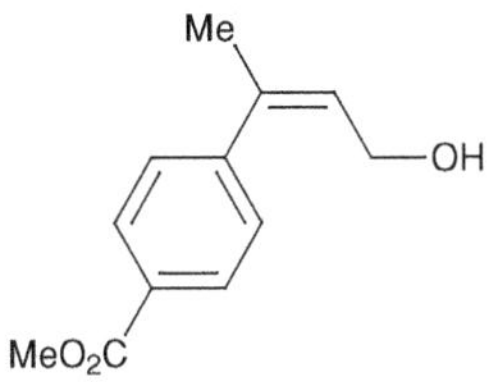

5. Suggest two methods for the conversion of the alkyne **4** into the Z-alkene **5**, which was hydrolysed to the anticancer compound combretastatin A-1.

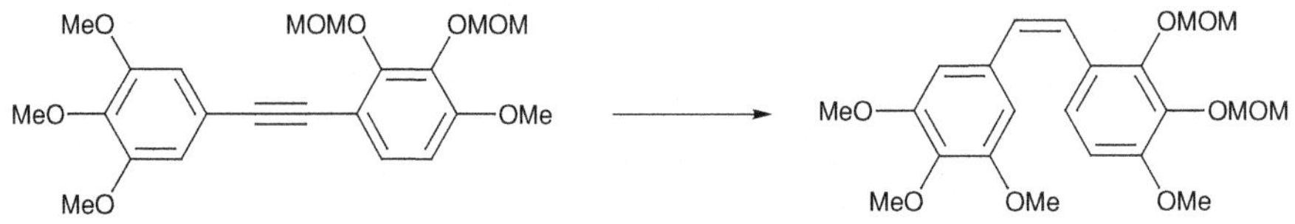

4   **5**

6. Suggest a reagent and conditions to convert the lactol **6** to the alkene **7**.

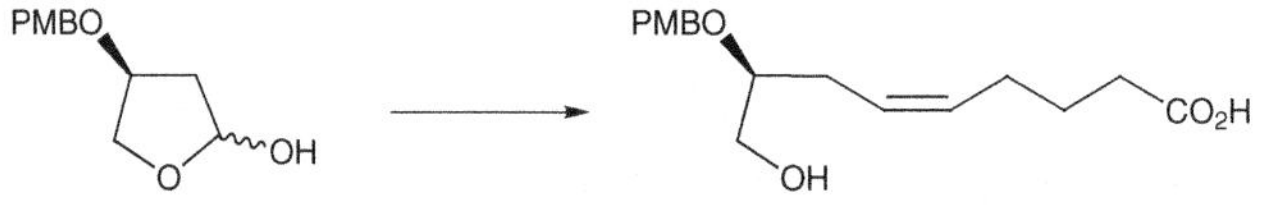

6   **7**

7. Account for the formation of the pyrrolizine **8**.

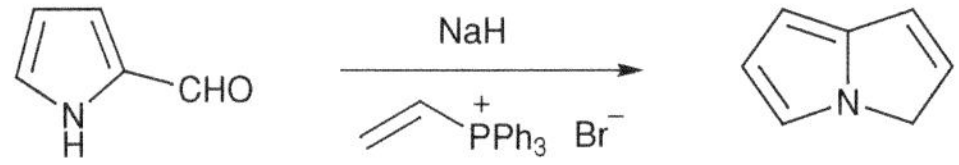

8

8. Explain the formation of the adduct **9**.

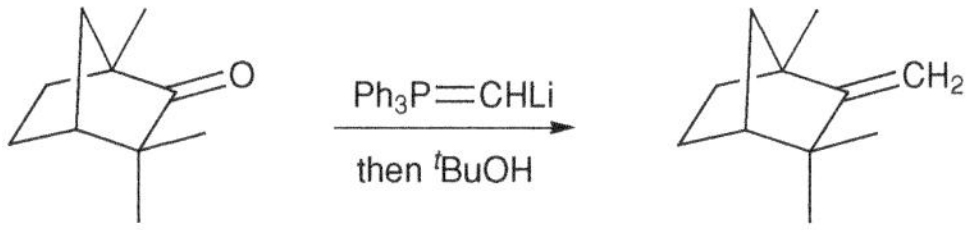

9

9. The diene **11** was used in a synthesis of the anti-inflammatory agent pinnaic acid. Suggest a method to prepare the alkenyl stannane **10** and reagents for the steps from **10** to the diene **11**.

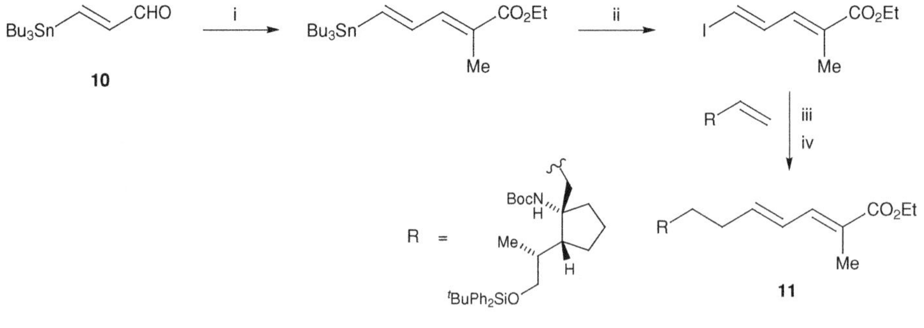

10

11

10. Suggest reagents for the conversion of the silane **12** to the alkene **13** and of the silane **12** to the alkene **14**. Explain the regioselectivity of the elimination in each case.

11. Explain the following chemistry, used in a synthesis of the anti-cancer agent zampanolide.

12. Draw the structure of the product from double ring-closing metathesis of the tetra-ene **15**.

13. Alkene–alkene metathesis reactions are a valuable method to construct cyclic compounds (see Section 2.10). Alkene–alkyne reactions can also be effective. Explain the formation of the bicyclic product **16**.

3

Pericyclic reactions

Pericyclic reactions are concerted processes that occur by way of a cyclic transition state in which more than one bond is formed or broken within the cycle. The classic example of such a process is the Diels–Alder cycloaddition reaction, one of the most common and useful synthetic reactions in organic chemistry. Cycloaddition reactions, sigmatropic rearrangements and electrocyclic reactions all fall into the category of pericyclic processes, representative examples of which are given in Schemes 3.1–3.3. This chapter will discuss these reactions and their use in synthesis.

$$\text{cycloaddition reaction} \qquad (3.1)$$

$$\text{sigmatropic rearrangement} \qquad (3.2)$$

$$\text{electrocyclic reaction} \qquad (3.3)$$

3.1 The Diels–Alder cycloaddition reaction

Of all the pericyclic reactions, the Diels–Alder cycloaddition reaction is the most popular. In the Diels–Alder reaction, a 1,3-diene reacts with a dienophile to form a six-membered ring adduct (3.1). Two new σ-bonds and a new π-bond are formed at the expense of three π-bonds in the starting materials.[1]

---

[1] K. C. Nicolaou, S. A. Snyder, T. Montagnon and G. Vassilikogiannakis, *Angew. Chem. Int. Ed.*, **41** (2002), 1668; W. Oppolzer, in *Comprehensive Organic Synthesis*, ed. B. M. Trost and I. Fleming, vol. 5 (Oxford: Pergamon Press, 1991), p. 315; W. Carruthers, *Cycloaddition Reactions in Organic Synthesis* (Oxford: Pergamon Press, 1990).

In general, the reaction takes place easily, simply by mixing the components at room temperature or by warming in a suitable solvent, although in some cases with unreactive dienes or dienophiles, more vigorous conditions may be necessary. The Diels–Alder reaction is reversible, and many adducts dissociate into their components at quite low temperatures. In these cases, heating is disadvantageous and the forward reaction is facilitated and better yields are obtained by using an excess of one of the components, or a solvent from which the adduct separates readily. Many Diels–Alder reactions are accelerated by Lewis acid catalysts.[2] In a few cases high pressures have been used to facilitate reactions that otherwise take place only slowly or not at all at room temperature.[3]

The usefulness of the Diels–Alder reaction in synthesis arises from its versatility and from its remarkable stereoselectivity. By varying the nature of the diene and the dienophile, many different types of ring structure can be constructed. In the majority of cases all six atoms involved in forming the new ring are carbon atoms, but this is not necessary and ring-closure may take place at atoms other than carbon, giving rise to heterocyclic compounds. It is found, moreover, that although the reaction could give rise to a number of isomeric products, one isomer is very often formed exclusively or at least in predominant amount.

The Diels–Alder reaction and indeed other pericyclic reactions are concerted processes in which there is no intermediate on the reaction pathway. The mechanisms of such processes can be considered in terms of orbital symmetry concepts.[4] A normal Diels–Alder reaction involves an electron-rich diene and an electron-deficient dienophile, and in such cases the main interaction is that between the highest occupied molecular orbital (HOMO) of the diene and the lowest unoccupied molecular orbital (LUMO) of the dienophile (3.4). The smaller the energy difference between these frontier orbitals, the better these orbitals interact and therefore the more readily the reaction occurs.

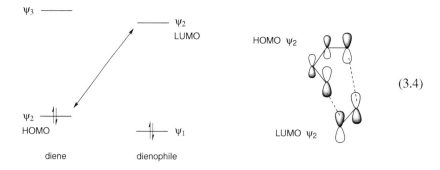

(3.4)

[2] M. Santelli and J.-M. Pons, *Lewis Acids and Selectivity in Organic Synthesis* (New York: CRC Press, 1996).
[3] G. Jenner, *Tetrahedron*, **53** (1997), 2669.
[4] R. B. Woodward and R. Hoffmann, *Angew. Chem. Int. Ed. Engl.*, **8** (1969), 781; K. N. Houk, *Acc. Chem. Res.*, **8** (1975), 361; I. Fleming, *Frontier Orbitals and Organic Chemical Reactions* (London: Wiley, 1976); I. Fleming, *Pericyclic Reactions* (Oxford: Oxford University Press, 1999).

The precise mechanism of the Diels–Alder reaction has been the subject of much debate. There is general agreement that the rate-determining step in adduct formation is bimolecular and that the two components approach each other in parallel planes roughly orthogonal to the direction of the new bonds about to be formed. Formation of the two new σ-bonds takes place by overlap of molecular π-orbitals in a direction corresponding to endwise overlap of atomic p-orbitals. However there is uncertainty about the nature of the transition state and, in particular, about the timing of the changes in covalency that result in the formation of the new bonds.

If the reaction is concerted then there should be a high level of stereoselectivity, as is indeed observed. However, this does not rule out a two-step mechanism should rotation about the bonds in the intermediate be slow compared with the rate of ring-closure.[5] In this connection, it is noteworthy that cycloaddition of *trans*- and *cis*-1,2-dichloroethene to cyclopentadiene is completely stereospecific (3.5). A two-step mechanism via a biradical intermediate might have been expected to be sufficiently long-lived to allow some interconversion, resulting in a mixture of products. Addition of dichlorodifluoroethene to *cis,cis*- and *trans,trans*-2,4-hexadiene is, however, non-stereospecific and is thought to proceed by a two-step mechanism with a biradical intermediate.

(3.5)

Attempts to detect biradical intermediates in the Diels–Alder reaction have been unsuccessful and compounds that catalyse singlet–triplet transitions have no influence on the reaction. Similarly, the kinetic effects of *para* substituents in 1-phenylbutadiene, although large in absolute terms, are considered too small for a rate-determining transition state corresponding to a zwitterion intermediate.

Whether or not both of the new bonds in the concerted mechanism are formed to the same extent in the transition state is an open question. It is likely that in most cases both bonds begin to form at the same time, although this may occur at different rates, such that one bond is formed to a greater extent than the other. There may be a gradation of mechanisms for different Diels–Alder reactions, extending from a completely concerted mechanism with a symmetrical transition state at one

[5] J. Sauer and R. Sustmann, *Angew. Chem. Int. Ed. Engl.*, **19** (1980), 779.

extreme, to something approaching a two-step process at the other. Thus in reactions involving a heterodiene or heterodienophile, the heteroatoms will probably be able to stabilize polar intermediates to a greater extent than carbon so that hetero-Diels–Alder reactions are more likely to be non-concerted or at least to proceed through an unsymmetrical transition state.

3.1.1 The dienophile

Many different types of dienophile can take part in the Diels–Alder cycloaddition reaction. These are normally derivatives of ethene or ethyne, but can also be reagents in which one or both of the reacting atoms is a heteroatom. All dienophiles do not react with equal ease; the reactivity depends on the structure. In general, the greater the number of electron-attracting substituents on the double or triple bond, the more reactive is the dienophile, owing to the lowering of the energy of the LUMO of the dienophile by the substituents. This is because a better interaction between the LUMO of the dienophile and the HOMO of the diene occurs when these orbitals are of similar energy. Thus, whereas maleic anhydride and 1,3-butadiene afford a quantitative yield of adduct in boiling benzene or, more slowly, at room temperature, tetracyanoethene, with four electron-attracting substituents, reacts extremely rapidly even at 0 °C. Similarly, ethyne reacts with electron-rich dienes only under severe conditions, but propynoic acid and derivatives react readily. Table 3.1 gives some values for the rates of addition of a number of dienophiles to cyclopentadiene and 9,10-dimethylanthracene in dioxane at 20 °C.

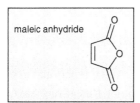

It should be noted, however, that there are a number of Diels–Alder reactions for which the above generalization does not hold, in which reaction takes place between an electron-rich dienophile and an electron-deficient diene. The essential feature is that the two components should have complementary electronic character. These Diels–Alder reactions with inverse electron demand, as they are called, also have their uses in synthesis, but the vast majority of reactions involve an electron-rich diene and an electron-deficient dienophile.

The most commonly encountered activating substituents for the 'normal' Diels–Alder reaction are COR, CO_2R, CN and NO_2. Dienophiles that contain one or more

Table 3.1. *Rates of reaction of dienophiles with cyclopentadiene and 9,10-dimethylanthracene*

Dienophile		Cyclopentadiene $10^5 \ k/\text{mol}^{-1}\text{s}^{-1}$	9,10-Dimethylanthracene $10^5 \ k/\text{mol}^{-1}\text{s}^{-1}$
Tetracyanoethene	$(NC)_2C{=}C(CN)_2$	*c.* 43 000 000	*c.* 13 000 000 00
Tricyanoethene	$(NC)_2C{=}CHCN$	*c.* 480 000	*c.* 590 000
1,1-Dicyanoethene	$H_2C{=}C(CN)_2$	45 500	12 700
Acrylonitrile	$H_2C{=}CHCN$	1.04	0.089
Dimethyl fumarate	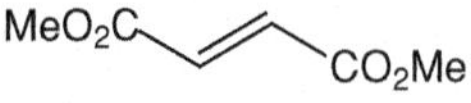 MeO₂C⌒CO₂Me	74	215
Dimethyl acetylene dicarboxylate	MeO₂C≡CO₂Me	31	104

9,10-dimethylanthracene

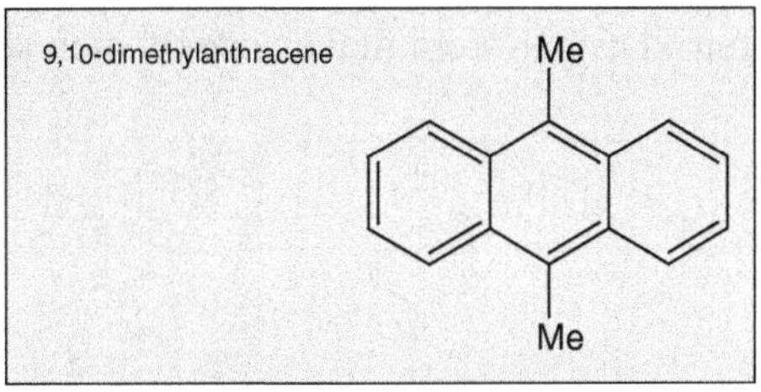

of these groups in conjugation with a double or triple bond react readily with dienes. The most widely used dienophiles are α,β-unsaturated carbonyl compounds and typical examples are acrolein (propenal), acrylic acid and its esters, maleic acid and its anhydride, 2-butyne-1,4-dioic acid (acetylene-dicarboxylic acid) and derivatives of 2-cyclohexenone. Thus acrolein reacts rapidly with butadiene in benzene at 0 °C to give the aldehyde **1** in quantitative yield, and 2-butyne-1,4-dioic acid and butadiene give the diacid **2** (3.6, 3.7).

(3.6)

1

(3.7)

2

Substituents exert a pronounced steric effect on the reactivity of dienophiles. Comparative experiments show that the yields of adducts obtained in the reaction of butadiene (and 2,3-dimethylbutadiene) with derivatives of acrylic acid decrease with the introduction of substituents into the α-position of the dienophile. α,β-Unsaturated ketones with two alkyl substituents in the β-position react very slowly.

Another important group of dienophiles of the α,β-unsaturated carbonyl class are quinones.[6] 1,4-Benzoquinone reacts readily with butadiene at room temperature to give a high yield of the mono-adduct, tetrahydronaphthaquinone (3.8); under more vigorous conditions a bis-adduct is obtained which can be converted into anthraquinone by oxidation of an alkaline solution with atmospheric oxygen. As with other dienophiles, alkyl substitution on the double bond leads to a decrease in activity and cycloaddition of monoalkyl 1,4-benzoquinones with dienes occurs preferentially at the unsubstituted double bond. In addition to steric effects, electronic effects can play a part, such that cycloaddition occurs at the more electron-deficient double bond of the benzoquinone. The first step in an approach to the steroid ring system makes use of such selectivity (3.9).[7]

(3.8)

(3.9)

In contrast to the reactive dienophiles in which the double or triple bond is activated by conjugation with electron-withdrawing groups, ethylenic compounds such as allylic alcohol and its esters and allyl halides are relatively unreactive, although they can sometimes be induced to react with dienes under forcing conditions. Enol ethers or enamines, in which the dienophile bears an electron-donating substituent, take part in Diels–Alder reactions with inverse electron demand. They react with electron-deficient dienes and with α,β-unsaturated carbonyl compounds, the latter to give dihydropyrans. For example, 2-alkoxydihydropyrans are obtained at temperatures between 150 and 200 °C and are useful intermediates for the preparation of glutaraldehydes (3.10). A key step in a synthesis of secologanin, makes use of an inverse electron demand Diels–Alder reaction with the α,β-unsaturated aldehyde **3** and the enol ether **4** (3.11).[8] Reactions with cyclic enamines have been used in the

[6] L. W. Butz and A. W. Rytina, *Org. Reactions*, **5** (1949), 136; V. Nair and S. Kumar, *Synlett* (1996), 1143.
[7] R. B. Woodward, F. Sondheimer, D. Taub, K. Heusler, W. M. McLamore, *J. Am. Chem. Soc.*, **74** (1952), 4223.
[8] L. F. Tietze, K.-H. Glüsenkamp and W. Holla, *Angew. Chem. Int. Ed. Engl.*, **21** (1982), 793.

synthesis of alkaloids.[9] For example, an important step in a synthesis of the alkaloid minovine involved cycloaddition of the enamine **6** with the diene **5** (3.12). It is not certain that such reactions, particularly with the more electron-rich enamines, are concerted cycloadditions; they may well be stepwise, ionic reactions.[10]

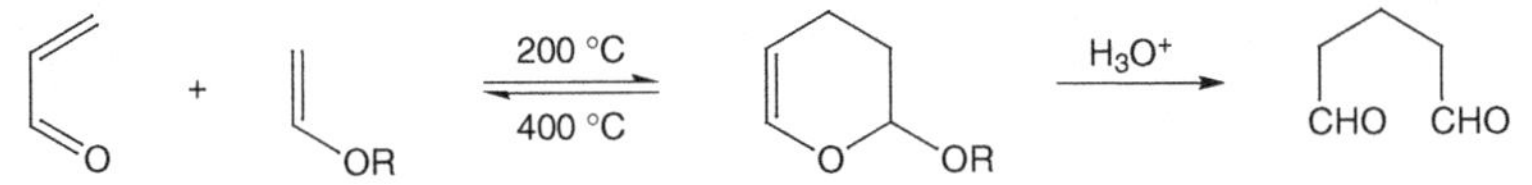

(3.10)

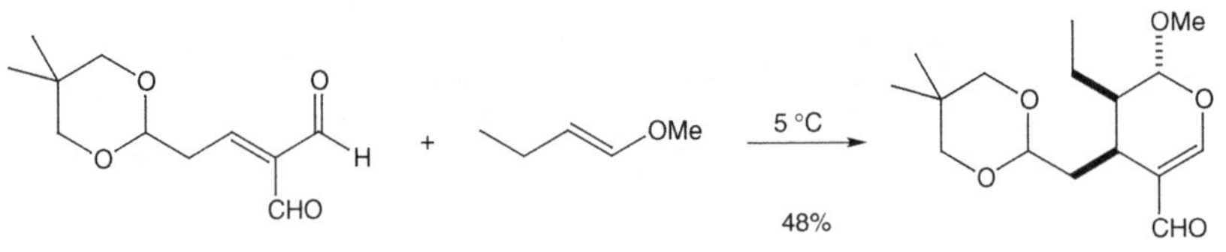

(3.11)

3   **4**

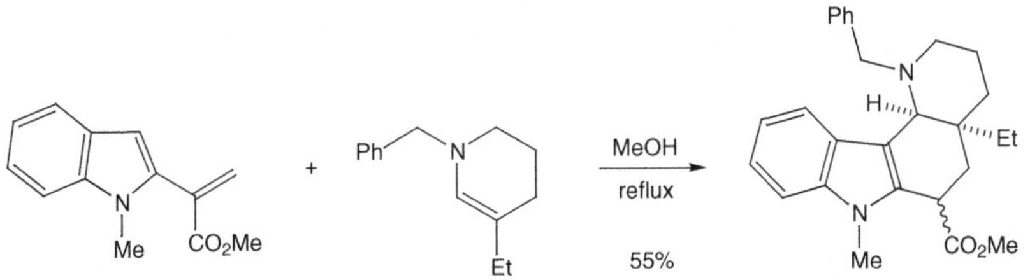

(3.12)

5   **6**

Isolated carbon–carbon double or triple bonds do not usually take part in intermolecular Diels–Alder reactions, but a number of cyclic alkenes and alkynes with pronounced angular strain are reactive dienophiles. The driving force for these reactions is thought to be the reduction in angular strain associated with the transition state for the addition. Thus, cyclopropene reacts rapidly and stereoselectively with cyclopentadiene at 0 °C to form the *endo* adduct **7** in 97% yield, and butadiene gives norcarene **8** in 37% yield (3.13).[11]

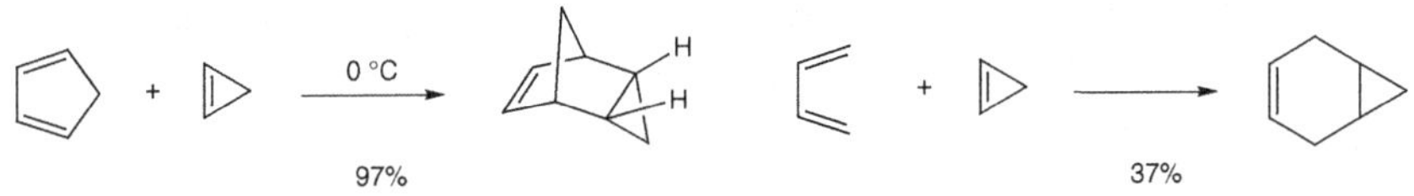

(3.13)

7   **8**

Some cyclic alkynes are also powerful dienophiles. Alkyne-containing ring systems with fewer than nine atoms in the ring are strained, owing to the preferred linear structure of the C–C≡C–C triple bond arrangement. The increasing strain with decreasing ring size in the sequence cyclooctyne to cyclopentyne is shown

[9] F. E. Ziegler and E. B. Spitzner, *J. Am. Chem. Soc.*, **92** (1970), 3492.
[10] I. Fleming and M. H. Kargar, *J. Chem. Soc. (C)* (1967), 226.
[11] M. L. Deem, *Synthesis* (1972), 675.

in an increasing tendency to take part in cycloaddition reactions. Cyclooctyne has been prepared as a stable liquid with significant dienophilic properties. It reacts readily with diphenylisobenzofuran to give the adduct **9** in 91% yield (3.14). The lower cycloalkynes have not been isolated but their existence has been shown by trapping with diphenylisobenzofuran.[12]

$$(3.14)$$

Arynes, such as benzyne (1,2-dehydrobenzene), also undergo Diels–Alder cycloaddition reactions. Benzyne, C_6H_4, is a highly reactive species and can be prepared by elimination of a suitably substituted benzene derivative. It reacts *in situ* with various dienes such as furan, cyclopentadiene, cyclohexadiene and even benzene and naphthalene to give bicyclic or polycyclic cycloadducts (3.15).[12] Analogous addition reactions are shown by dehydroaromatics in the pyridine and thiophene series.

$$(3.15)$$

Indirect methods have been developed for engaging unactivated alkenes as dienophiles in Diels–Alder reactions by temporary introduction of activating groups. Thus, the readily available phenyl vinyl sulfone serves very conveniently as an ethene equivalent (3.16). Reductive cleavage of the sulfone group from the initial adduct with sodium amalgam leads to 1,2-dimethylcyclohexene. Alkylation of the sulfone before reductive cleavage provides access to other derivatives.[13] Likewise, the corresponding ethynyl sulfone (such as ethynyl *p*-tolyl sulfone) serves well in the Diels–Alder reaction with substituted dienes and subsequent reductive cleavage of the adduct with sodium amalgam gives 1,4-cyclohexadiene products. Phenyl vinyl sulfoxide can also be used as an ethyne equivalent; treatment with a reactive diene

[12] G. Wittig, *Angew. Chem. Int. Ed. Engl.*, **1** (1962), 415.
[13] R. V. C. Carr and L. A. Paquette, *J. Am. Chem. Soc.*, **102** (1980), 853.

followed by elimination of benzenesulfenic acid generates the cyclohexadiene (see Section 2.2 for elimination of sulfoxides).[14]

(3.16)

Although many electron-deficient alkenes can function as dienophiles, a notable exception is ketene. The C=C linkage in the ketene $R_2C=C=O$ does react with dienes, but the products are *four-membered* ring compounds, formed by overall [2+2]-cycloaddition (see Section 3.2). Indirect methods are therefore needed to prepare the product corresponding to Diels–Alder addition of a ketene to a 1,3-diene.[15] A good reagent that promotes effective cycloaddition and allows subsequent conversion to the desired ketone is 2-chloroacrylonitrile.[16] Cycloaddition with this dienophile gives an α-chloronitrile adduct that can be converted easily to the ketone (3.17).

(3.17)

Nitroethene and vinyl sulfoxides have also been employed as ketene equivalents. Nitroethene is an excellent dienophile and oxidation of the initial nitro-adduct gives the corresponding ketone.[17] However, the thermal instability of nitroethene limits its application to cycloadditions with reactive dienes. An attractive feature of vinyl sulfoxides as ketene equivalents is that they can be obtained in optically active form because of the chirality of the sulfoxide group, thus allowing enantioselective Diels–Alder reactions. Cycloaddition of *p*-tolyl vinyl sulfoxide with cyclopentadiene requires heat and gives a mixture of all four (two *exo* and two

[14] O. De Lucchi and G. Modena, *Tetrahedron*, **40** (1984), 2585.
[15] V. K. Aggarwal, A. Ali and M. P. Coogan, *Tetrahedron*, **55** (1999), 293.
[16] E. J. Corey, M. M. Weinshenker, T. K. Schaaf and W. Huber, *J. Am. Chem. Soc.*, **91** (1969), 5675; D. A. Evans, W. L. Scott and L. K. Truesdale, *Tetrahedron Lett.* (1972), 121.
[17] P. A. Bartlett, F. R. Green and T. R. Webb, *Tetrahedron Lett.* (1977), 331; D. Ranganathan, C. B. Rao, S. Ranganathan, A. K. Mehrotra and R. Iyengar, *J. Org. Chem.*, **45** (1980), 1185.

endo) isomeric [4+2] cycloadducts, although these are separable. However, prior
O-ethylation of the sulfoxide gives the salt **10**, which reacts at low temperature
and with very high selectivity (3.18).[18] This methodology was used to prepare the
ketone **11**, an intermediate in the synthesis of prostaglandins.

$$(3.18)$$

A number of cycloaddition reactions involving allene derivatives as dienophiles
have been recorded. Allene itself reacts only with electron-deficient dienes but
allene carboxylic acid or esters, in which a double bond is activated by conjugation
with the carboxylic group, react readily with cyclopentadiene to give 1:1 adducts in
excellent yield. For example, the allene **12** gave, with very high yield and selectivity,
the cycloadduct **13**, used in a synthesis of (–)-β-santalene (3.19).[19] An 'allene
equivalent' is vinyl triphenylphosphonium bromide, which is reported to react with
a number of dienes to form cyclic phosphonium salts.[20] These can be converted
into methylene compounds by the usual Wittig reaction procedure (3.20).

$$(3.19)$$

$$(3.20)$$

Cationic dienophiles, in which the alkene is rendered electron deficient, are
good substrates for the Diels–Alder reaction.[21] 2-Vinyl-1,3-dioxolane **13** is very
unreactive towards dienes, however, on protonation, the acetal is in equilibrium

[18] B. Ronan and H. B. Kagan, *Tetrahedron: Asymmetry*, **3** (1992), 115.
[19] W. Oppolzer and C. Chapuis, *Tetrahedron Lett.*, **24** (1983), 4665.
[20] R. Bonjouklian and R. A. Ruden, *J. Org. Chem.*, **42** (1977), 4095.
[21] P. G. Gassman, D. A. Singleton, J. J. Wilwerding and S. P. Chavan, *J. Am. Chem. Soc.*, **109** (1987), 2182; P. A.
Grieco, J. L. Collins and S. T. Handy, *Synlett* (1995), 1155; B. G. Reddy, R. Kumareswaran and Y. D. Vankar,
Tetrahedron Lett., **41** (2000), 10333.

with the oxonium ion **14**, an effective dienophile (3.21). Cycloaddition and re-formation of the acetal gives the Diels–Alder adduct **15**.

$$(3.21)$$

14   65%   **15**

Heterodienophiles

The Diels-Alder reaction is by no means restricted to the all-carbon variant. No significant loss of reactivity is encountered when one or both of the atoms of the dienophile multiple bond is a heteroatom.[22]

Carbonyl groups in aldehydes and ketones add to 1,3-dienes and the reaction has been used to prepare derivatives of 5,6-dihydropyrans.[22,23] Formaldehyde reacts only slowly but reactivity increases with reactive carbonyl compounds bearing electron-withdrawing groups, such as glyoxylate derivatives (3.22).

$$(3.22)$$

In the presence of a Lewis acid such as zinc chloride or boron trifluoride etherate, however, the scope of this reaction is extended greatly. Under these conditions, oxygenated butadiene derivatives react readily with a wide variety of aldehydes to give dihydro-4-pyrones in good yield (3.23).[24] The reactions catalysed by zinc chloride are thought to be true cycloadditions, proceeding through 1:1 cycloadducts **16**, whereas the boron trifluoride reactions appear to be complex, with more than one reaction pathway possible, and under certain conditions proceeding in a stepwise fashion, by way of an open-chain aldol-like intermediate. The stereochemistry of

[22] S. M. Weinreb and R. R. Staib, *Tetrahedron*, **38** (1982), 3087; S. M. Weinreb, in *Comprehensive Organic Synthesis*, ed. B. M. Trost and I. Fleming, vol. 5 (Oxford: Pergamon Press, 1991), p. 401; L. F. Tietze and G. Kettschau, *Top. Curr. Chem.*, **189** (1997), 1; P. Buonora, J.-C. Olsen and T. Oh, *Tetrahedron*, **57** (2001), 6099.

[23] M. D. Bednarski and J. P. Lyssikatos, in *Comprehensive Organic Synthesis*, ed. B. M. Trost and I. Fleming, vol. 2 (Oxford: Pergamon Press, 1991), p. 661.

[24] S. J. Danishefsky, E. Larson, D. Askin and N. Kato, *J. Am. Chem. Soc.*, **107** (1985), 1246.

the dihydropyrone products also depends on a number of factors, but generally favours the *cis*-2,3-disubstituted-4-pyrone **17** by way of an *endo* transition state.

No Lewis acid catalyst is required for the Diels–Alder reaction between an aldehyde and the more-reactive 1-amino-3-siloxy butadienes such as **18** (3.24).[25] Treatment of the cycloadduct **19** with acetyl chloride provides the desired dihydro-4-pyrone.

(3.23)

16   **17**

(3.24)

R = Ph   86%
R = C$_5$H$_{11}$ 73%

18   **19**

Dihydro-4-pyrones are useful intermediates, particularly for the synthesis of carbohydrates.[26] Addition of an activated diene with the carbonyl group of chiral aldehydes is stereoselective, following Cram's rule (see Section 1.1.5.1), and this has been exploited in a number of highly stereoselective syntheses. For example, cycloaddition of the diene **20** (Danishefsky's diene) and the aldehyde **21** gave the dihydro-4-pyrone **22**, which was converted in a number of steps to a derivative of 2,4-dideoxy-D-glucose **23** (3.25).[27]

(3.25)

72%

20   **21**   **22**   **23**

A useful group of heterodienophiles are imines, containing the group RCH=NR. These react with 1,3-dienes to form 1,2,3,6-tetrahydropyridines.[22] Most reactive are imines bearing an electron-withdrawing substituent on one or both of the carbon

[25] Y. Huang and V. H. Rawal, *Org. Lett.*, **2** (2000), 3321.
[26] R. R. Schmidt, *Acc. Chem. Res.*, **19** (1986), 250; T. Kametani and S. Hibino, *Adv. Heterocycl. Chem.*, **42** (1987), 245.
[27] S. J. Danishefsky, S. Kobayashi and J. F. Kerwin, *J. Org. Chem.*, **47** (1982), 1981.

and nitrogen atoms of the C=N group. The *N*-tosyl imine of butyl glyoxylate gives rise, on reaction with cyclopentadiene, to a high yield of the product **24** (3.26).[28] Only the *exo* diastereomer is formed, presumably due to a preference (steric and/or electronic) for the *N*-tosyl group to lie in the *endo* orientation. With *N*-alkyl imines, protonation (e.g. with trifluoracetic acid, TFA) promotes cycloaddition (3.27).[29] Good levels of asymmetric induction can be obtained by using imines bearing a chiral auxiliary (e.g. *N*-α-methylbenzyl imines of glyoxylate esters). Unactivated imines that lack any electron-withdrawing group on the imine can undergo cycloaddition with reactive dienes in the presence of a Lewis acid such as metal halides or triflates.[30]

$$(3.26)$$

$$(3.27)$$

Intramolecular Diels–Alder cycloaddition reactions with imines are useful for the preparation of alkaloid and other polycyclic nitrogen-containing compounds. The intramolecular reactions in many cases are highly stereoselective and can frequently take place in the absence of activating groups on the imine double bond. As an example, heating the acetate **25** gives rise to an intermediate *N*-acyl imine, which undergoes cycloaddition to give the lactams **26** (3.28).[31] The lactams **26** were converted to the fungal neurotoxin slaframine.

$$(3.28)$$

[28] A. Barco, S. Benetti, P. G. Baraldi, F. Moroder, G. P. Pollini and D. Simoni, *Liebigs Ann.* (1982), 960.

[29] P. A. Grieco and S. D. Larsen, *J. Am. Chem. Soc.*, **107** (1985), 1768; P. D. Bailey, G. R. Brown, F. Korber, A. Reed and R. D. Wilson, *Tetrahedron: Asymmetry*, **2** (1991), 1263; H. Abraham and L. Stella, *Tetrahedron*, **48** (1992), 9707.

[30] L. Yu, J. Li, J. Ramirez, D. Chen and P. G. Wang, *J. Org. Chem.*, **62** (1997), 903.

[31] R. A. Gobao, M. L. Bremmer and S. M. Weinreb, *J. Am. Chem. Soc.*, **104** (1982), 7065.

Nitroso compounds react with 1,3-dienes to form oxazine derivatives.[22,32] Aromatic nitroso compounds, Ar—N=O, undergo cycloaddition with most dienes. Thus, butadiene and nitrosobenzene react readily at 0 °C to give *N*-phenyl-3, 6-dihydro-oxazine in high yield (3.29). With unsymmetrical 1,3-dienes, cycloaddition is often highly regioselective.

$$\text{(3.29)}$$

Nitroso compounds bearing an electron-withdrawing group, such as RCON=O derivatives, are particularly effective substrates for hetero-Diels–Alder cycloaddition reactions.[33] Acyl nitroso compounds are prepared *in situ* by oxidation of the corresponding hydroxamic acid, RCONHOH, with periodate. In the presence of the diene, the dihydro-oxazine is produced directly. For example, cycloaddition of the diene **27** (prepared by enzymatic oxidation of bromobenzene) with the nitroso compound derived from the hydroxamic acid **28**, gave the cycloadduct **29** (note that the nitroso dienophile reacts at the less-hindered, convex face of the diene) (3.30).[34] Reductive cleavage of the N–O bond gave the alcohol **30**, used in a synthesis of lycoricidine. The nitroso Diels–Alder reaction, followed by cleavage of the N–O bond, is an effective strategy for the synthesis of 1,4-amino-alcohols.

$$\text{(3.30)}$$

Intramolecular cycloaddition reactions with nitroso dienophiles have been used in a number of syntheses of alkaloids. For example, cycloaddition of the acyl nitroso compound formed from the hydroxamic acid **31** gave the dihydro-oxazine **32**, which was converted to the alkaloid gephyrotoxin 223AB (3.31).[35]

[32] J. Streith and A. Defoin, *Synthesis* (1994), 1107; J. Streith and A. Defoin, *Synlett* (1996), 189.
[33] G. W. Kirby and J. G. Sweeny, *J. Chem. Soc., Perkin Trans. 1* (1981), 3250.
[34] T. Hudlicky, H. F. Olivo and B. McKibben, *J. Am. Chem. Soc.*, **116** (1994), 5108.
[35] Y. Watanabe, H. Iida and C. Kibayashi, *J. Org. Chem.*, **54** (1989), 4088.

(3.31)

31   **32**   gephyrotoxin 223AB

Another type of nitroso dienophile that has found use in organic synthesis bears an α-chloro substituent. Such α-chloronitroso compounds react with dienes to give the usual dihydro-oxazine product, however, in the presence of an alcoholic solvent, this cycloadduct reacts further and the product actually isolated is the *N*-unsubstituted dihydro-oxazine (3.32). Hence the use of α-chloronitroso dienophiles gives the product formed, in effect, by addition of HN=O to the diene. This has been exploited by a number of research groups, an example of which, towards the natural compound conduramine F1, is illustrated in Scheme 3.33.

(3.32)

(3.33)

conduramine F1

Some azo compounds with electron-withdrawing groups attached to the nitrogen atoms, such as diethyl azodicarboxylate (EtO$_2$C−N=N−CO$_2$Et), are reactive dienophiles (owing to their low energy LUMO). Cycloaddition with a 1,3-diene gives a tetrahydropyridazine. The reaction of oxygen with a 1,3-diene to form an endoperoxide is also of interest. The cycloaddition is normally effected under the influence of light, either directly or in the presence of a photosensitizer, and it is thought that singlet oxygen is the reactive species.[36] The photosensitized addition of oxygen to dienes was discovered by Windaus in 1928 in the course of his classical studies of the conversion of ergosterol into vitamin D. Irradiation of a solution of ergosterol in alcohol in the presence of oxygen and a sensitizer led to the formation of a peroxide which was subsequently shown to have been formed

---

[36] W. Adam and M. Prein, *Acc. Chem. Res.*, **29** (1996), 275.

by 1,4-addition of a molecule of oxygen to the conjugated diene system. Numerous other endoperoxides have since been obtained by sensitized photo-oxygenation of other steroids and a variety of other cyclic and open-chain conjugated dienes. Irradiation of cyclohexadiene in the presence of oxygen with chlorophyll as a sensitizer leads to the endoperoxide, which can be converted to *cis*-1,4-dihydroxycyclohexane (3.34).

$$\text{(3.34)}$$

3.1.2 The diene

A wide variety of 1,3-dienes can take part in the Diels–Alder cycloaddition reaction,[37] including open-chain and cyclic dienes, and transiently formed dienes such as *ortho*-quinodimethanes. Heterodienes, in which one or more of the atoms of the diene is a heteroatom, are also known. Acyclic dienes can exist in a *cisoid* or a *transoid* conformation, and an essential condition for reaction is that the diene can adopt the *cisoid* form. If the diene does not have, or cannot adopt, a *cisoid* conformation then no Diels–Alder cycloaddition reaction occurs.

cisoid   transoid

Acyclic dienes

Acyclic conjugated dienes react readily with dienophiles. Butadiene itself reacts quantitatively with maleic anhydride in benzene at 100 °C in 5 h, or more slowly at room temperature, to form *cis*-1,2,3,6-tetrahydrophthalic anhydride (3.35).

$$\text{(3.35)}$$

---

[37] F. Fringuelli and A. Taticchi, *Dienes in the Diels-Alder Reaction* (New York: Wiley, 1990).

Substituents in the butadiene molecule influence the rate of cycloaddition through their electronic nature and by a steric effect on the conformational equilibrium. The rate of the reaction is often increased by electron-donating substituents (e.g. $-NMe_2$, $-OMe$, $-Me$) on the diene, as well as by electron-withdrawing substituents on the dienophile. Reactions with inverse electron demand are favoured by electron-withdrawing substituents on the diene. Bulky substituents that discourage the diene from adopting the *cisoid* conformation hinder the reaction. Thus, whereas 2-methyl-, 2,3-dimethyl- and 2-tert-butylbutadiene react normally with maleic anhydride, the 2,3-diphenyl compound is less reactive and 2,3-di-tert-butylbutadiene is completely unreactive. Apparently 2,3-di-tert-butylbutadiene is prevented from attaining the necessary planar *cisoid* conformation by steric effects of the two bulky *tert*-butyl substituents. In contrast, 1,3-di-*tert*-butylbutadiene, in which the substituents do not interfere with each other even in the *cisoid* form, reacts readily with maleic anhydride.

2,3-di-*tert*-butylbutadiene   1,3-di-*tert*-butylbutadiene   Z-1,3-pentadiene

Z-Alkyl or aryl substituents in the 1-position of the diene reduce its reactivity by sterically hindering formation of the *cisoid* conformation through non-bonded interaction with a hydrogen atom at C-4. Accordingly, an *E*-substituted 1,3-butadiene reacts with dienophiles much more readily than the Z-isomer. Thus Z-1,3-pentadiene gave only a 4% yield of adduct when heated with maleic anhydride at 100 °C, whereas the *E*-isomer formed an adduct in almost quantitative yield in benzene at 0 °C.

Similarly, *E,E*-1,4-dimethylbutadiene reacts readily with many dienophiles, but the *Z,E*-isomer yields an adduct only when the components are heated in benzene at 150 °C. Z,Z-1,4-Disubstituted butadienes are unreactive. 1,1-Disubstituted butadienes also react with difficulty, and with such compounds addition may be preceded by isomerization of the diene to a more reactive species. Thus, in the reaction of 1,1-dimethylbutadiene with acrylonitrile, the diene first isomerizes to 1,3-dimethylbutadiene, which then reacts in the normal way.

Hetero-substituted dienes are excellent substrates for the Diels–Alder reaction. The increased rate of reaction (in comparison with unsubstituted dienes) with electron-deficient dienophiles is ascribed to the higher-energy HOMO of the

hetero-substituted diene, which therefore results in a reduction in the energy difference between the HOMO of the diene and the LUMO of the dienophile. A popular
diene is 1-methoxy-3-trimethylsilyloxybutadiene **20**, sometimes referred to as the
Danishefsky diene.[38] Cycloaddition with this diene, followed by hydrolysis of the
resulting silyl enol ether **33**, gives a cyclohexenone product (3.36).

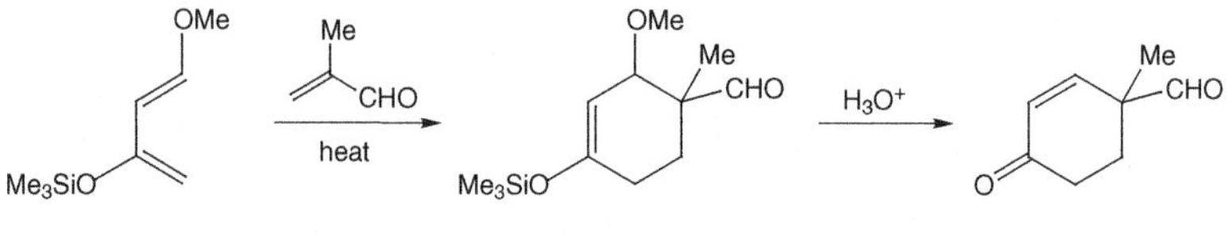

(3.36)

20   **33**

Cycloaddition of the diene **20** with dienophiles bearing a phenyl sulfoxide
substituent leads, after elimination of phenyl sulfenic acid and hydrolysis, to a
4,4-disubstituted cyclohexadienone or a substituted phenol product. For example, an elegant synthesis of disodium prephenate **34** makes use of this chemistry
(3.37).[39]

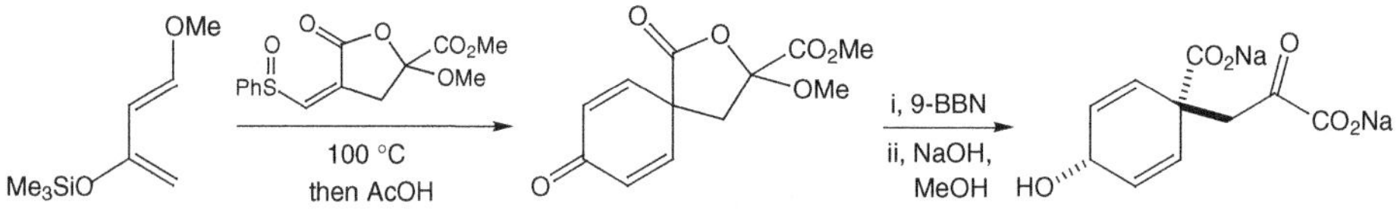

20   **34**

(3.37)

A related group of oxy-substituted butadienes are the vinylketene acetals, such
as 1,1-dimethoxy-3-trimethylsilyloxybutadiene **35**. This type of dienophile reacts
readily with electron-deficient alkynes to provide a convenient synthesis of resor-
cinol derivatives (3.38). The intermediate adduct is not normally isolated and the
aromatic compound is formed directly by heating the components together.

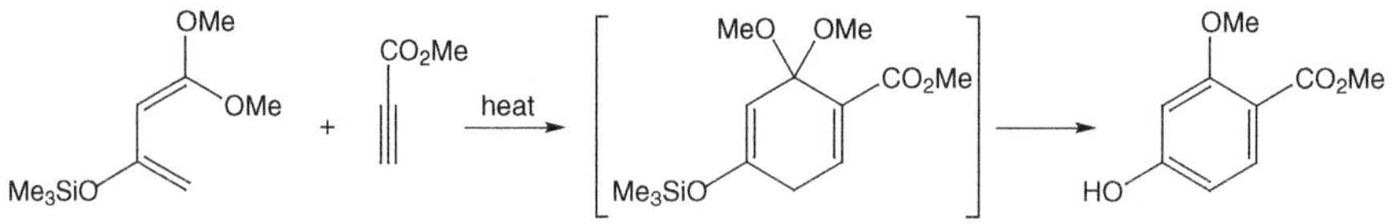

35

(3.38)

[38] S. J. Danishefsky, *Acc. Chem. Res.*, **14** (1981), 400; M. Petrzilka and J. I. Grayson, *Synthesis* (1981), 753.
[39] S. J. Danishefsky, M. Hirama, N. Fritsch and J. C. Clardy, *J. Am. Chem. Soc.*, **101** (1979), 7013.

A valuable group of hetero-substituted dienes are derivatives of 1- and 2-aminobutadiene. The *N,N*-dimethylamino diene **18** (Scheme 3.24) is estimated to react with dienophiles at a rate that is greater than 3000 times that of its methoxy analogue **20**. *N*-Acylaminobutadienes are more stable than their dialkylamino counterparts and can be prepared easily, for example from the α,β-unsaturated aldehyde, primary amine and carboxylic acid chloride. They react readily with dienophiles to give substituted amino-cyclohexenes. The amino group is a powerful directing group and most reactions proceed with high regio- and stereoselectivity. For example, cycloaddition of the diene **36** and crotonaldehyde occurs selectively to give the adduct **37**, used in a synthesis of the poison arrow alkaloid pumiliotoxin C **38** (3.39).[40]

(3.39)

36   **37**   **38**

An alternative synthesis of pumiliotoxin C makes use of a related, but intramolecular Diels–Alder reaction. The diene **39**, prepared in optically active form from (*S*)-norvaline, cyclized to the octahydroquinoline **40** with high selectivity (3.40).[41] In this chemistry, the chiral centre in **39** controls the formation of the three new developing centres in **40**.

(−)-**38**   (3.40)

39   **40**

Other hetero-substituted dienes are also useful in synthesis. 1- and 2-Phenylthiobutadienes form versatile intermediates because the sulfide group can be removed reductively or after oxidation to the sulfoxide followed by elimination to a new alkene or [2,3]-sigmatropic rearrangement to an allylic alcohol. For example, cycloaddition of the 1-phenylthio-substituted diene **41** gave the sulfide **42**, which was oxidized to the intermediate sulfoxide. The sulfoxide undergoes

[40] L. E. Overman and P. J. Jessup, *J. Am. Chem. Soc.*, **100** (1978), 5179.
[41] W. Oppolzer and E. Flaskamp, *Helv. Chim. Acta*, **60** (1977), 204.

[2,3]-sigmatropic rearrangement to give, after treatment with trimethyl phosphite (to cleave the O—S bond), the allylic alcohol **43** (3.41). The product **43** was used in a synthesis of the hypocholesterolemic agent (+)-compactin.

$$(3.41)$$

2-Benzenesulfonyl dienes display an interesting reactivity,[42] being capable of undergoing cycloaddition with both electron-deficient and electron-rich dienophiles. With electron-rich dienophiles the reaction may occur in a stepwise rather than a concerted manner.

Trimethylsilyl dienes can be used in Diels–Alder reactions. The trimethylsilyl group is not a powerful directing group and therefore other substituents tend to have the major influence on the regioselectivity of the cycloaddition reactions. The products are cyclic allylsilanes, which can undergo a range of useful conversions, such as protodesilylation with acid to give cyclohexenes, or epoxidation or dihydroxylation followed by loss of the trimethylsilyl group to give allylic alcohols. Thus, heating 1-acetoxy-4-(trimethylsilyl)butadiene and methyl acrylate gave predominantly the adduct **44** (3.42). Dihydroxylation of the alkene **44** followed by acid-catalysed elimination led to the allylic alcohol **45**, which was converted to shikimic acid.[43]

$$(3.42)$$

[42] J. E. Bäckvall, R. Chinchilla, C. Nájera and M. Yus, *Chem. Rev.*, **98** (1998), 2291.
[43] M. Koreeda and M. A. Ciufolini, *J. Am. Chem. Soc.*, **104** (1982), 2308.

Diels–Alder reactions may be accelerated and the selectivities enhanced if the cycloaddition reactions are conducted in water or under high pressure or in the presence of a Lewis acid (see Section 3.1.3). In water at room temperature, cyclopentadiene reacts with methyl vinyl ketone 700 times faster than in 2,2,4-trimethylpentane and the *endo:exo* selectivity rises from about 4:1 to more than 20:1. This can be ascribed to hydrophobic effects, which promote aggregation of non-polar species. Diels–Alder reactions in water are normally carried out with (at least partially) water-soluble dienes such as sodium salts of dienoic acids. Thus, a key step in a formal synthesis of vitamin D$_3$ involved the cycloaddition of the sodium salt **46** with methacrolein in water, to give the adducts **47** and **48** in high yield and a ratio of 4.7:1 after 16 h (3.43). In contrast, the corresponding methyl ester of the diene in excess neat methacrolein at 55 °C gave only a 10% yield of a 1:1 mixture of isomers after 63 h.[44]

(3.43)

Cyclic dienes

The double bonds of cyclopentadiene are constrained in a planar *cisoid* conformation and this diene therefore reacts easily with a variety of dienophiles. 1,3-Cyclohexadiene is also reactive, but with an increase in the size of the ring, the reactivity of cyclic dienes decreases rapidly. In large rings, the double bonds can no longer easily adopt the necessary coplanar configuration because of non-bonded interaction of methylene groups in the planar molecule. *Cis,cis-* and *cis,trans-*1,3-cyclo-octadienes form only copolymers when treated with maleic anhydride and *cis,cis-* and *cis,trans-*1,3-cyclodecadienes similarly do not form adducts with maleic anhydride. Dienes with 14- and 15-membered rings react with dienophiles but only under relatively severe conditions.

Cyclopentadiene is a very reactive diene and exists as its dimer that needs to be 'cracked' (retro-Diels–Alder reaction) to prepare the diene. Cycloaddition with dienophiles forms bridged compounds of the bicyclo[2.2.1]heptane series. The reaction of cyclopentadiene with mono-and *cis*-disubstituted alkenes could give rise to two stereochemically distinct products, the *endo-* and *exo*-bicyclo[2.2.1]heptene derivatives. It is found in practice, however, that the *endo* isomer predominates,

[44] E. Brandes, P. A. Grieco and P. Garner, *J. Chem. Soc., Chem. Commun.* (1988), 500.

except under conditions where isomerization of the first-formed adduct occurs (see Section 3.1.6).

Bicyclo[2.2.1]heptanes are widely distributed in nature among the bicyclic terpenes, and the Diels–Alder reaction provides a convenient method for their synthesis. Thus, cyclopentadiene and vinyl acetate react smoothly on warming to form the adduct **49**, which is easily transformed into norcamphor **50** and related compounds (3.44).

$$(3.44)$$

49   **50**

Many furan derivatives react with dienophiles to form bicyclic compounds with an oxygen bridge.[45] Most of the adducts obtained from furans are thermally labile and dissociate readily into their components on warming. The adduct from furan and maleic anhydride has been shown to have the *exo* structure **51**, apparently violating the rule that the *endo* isomer predominates (see Section 3.1.4) (3.45). The reason for this is found in the related observation that the normal *endo* adduct formed from maleimide and furan at 20 °C dissociates at temperatures only slightly above room temperature and more rapidly on warming, allowing conversion of the *endo* adduct formed in the kinetically controlled reaction into the thermodynamically more stable *exo* isomer. With the maleic anhydride adduct, equilibration takes place below room temperature so that the *endo* adduct formed under kinetic control is not observed.

$$(3.45)$$

51

Pyrrole derivatives have been used less frequently in the Diels–Alder reaction because this heteroaromatic compound is susceptible to electrophilic substitution. Thiophen and some of its derivatives react with alkynyl dienophiles under vigorous conditions to form benzene derivatives by extrusion of sulfur from the initially formed adduct.

---

[45] C. O. Kappe, S. S. Murphree and A. Padwa, *Tetrahedron*, **53** (1997), 14 179.

Although less reactive than most cyclic conjugated dienes, pyran-2-ones can be used in the Diels–Alder reaction.[46] Cycloaddition with alkynyl dienophiles is followed by loss of carbon dioxide to give benzene derivatives. Thus pyran-2-one itself reacts with dimethyl acetylene-dicarboxylate to give dimethyl phthalate (3.46).

$$(3.46)$$

1,2-Dimethylenecycloalkanes and ortho-quinodimethanes

The *cisoid* conformation of the diene is fixed in 1,2-dimethylenecycloalkanes and this type of diene reacts readily with dienophiles. Cycloaddition provides a convenient route to polycyclic compounds and often occurs in high yield. Thus, 1,2-dimethylenecyclohexane reacts with benzoquinone to give the *bis*-adduct **52** (3.47). This product is readily converted into the corresponding aromatic hydrocarbon pentacene.

$$(3.47)$$

52

Related to the 1,2-dimethylenecycloalkanes are the *ortho*-quinodimethanes (*o*-xylylenes) such as **53**. These are very reactive dienes and form adducts with a variety of dienophiles.[47] The *ortho*-quinodimethanes used in these reactions are generated *in situ* by thermal ring opening of benzocyclobutenes, by photoenolization of *ortho* alkyl aromatic aldehydes or ketones, or by elimination from appropriate derivatives of 1,2-dialkylbenzenes. Thus, 1-hydroxybenzocyclobutene undergoes ring-opening on heating to give the *ortho*-quinodimethane **54** (3.48). The hydroxyl substituent adopts the *E*-configuration, since the thermal conrotatory ring-opening (see Section 3.8) takes place preferentially outward to

[46] K. Afarinkia, V. Vinader, T. D. Nelson and G. H. Posner, *Tetrahedron*, **48** (1992), 9111.
[47] J. L. Segura and N. Martín, *Chem. Rev.*, **99** (1999), 3199; G. Mehta and S. Kotha, *Tetrahedron*, **57** (2001), 625; J. L. Charlton and M. M. Alauddin, *Tetrahedron*, **43** (1987), 2873.

give the sterically less-hindered diene. The same diene is formed by irradiation of *ortho*-methylbenzaldehyde. Cycloaddition with naphthoquinone leads to the adduct **55**.

53

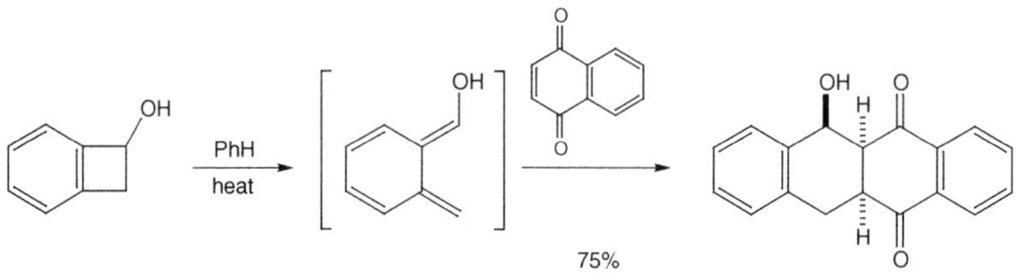

(3.48)

75%

54   **55**

The use of *ortho*-quinodimethanes in the Diels–Alder reaction provides an efficient approach to polycyclic aromatic compounds. A recent example is the cycloaddition of the benzocyclobutene **56** with the quinone **57**, which occurs on heating in toluene (3.49).[48] The adduct **58** was used in a synthesis of rishirilide B.

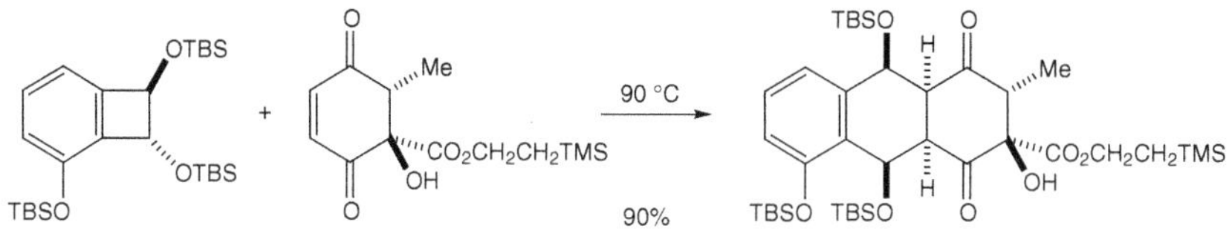

(3.49)

90%

56   **57**   **58**

TBS = SiMe₂Buᵗ

ortho-Quinodimethanes can be obtained by 1,4-elimination from appropriate substituted 1,2-dialkylbenzenes, for example by Hofmann elimination or by debromination of *ortho*-bis-bromoalkyl aromatic compounds. Thus, in a synthesis of 4-demethoxy-daunomycinone, the tetracyclic ketone **61** was obtained by cycloaddition of methyl vinyl ketone to the *ortho*-quinodimethane **60**, itself obtained by iodide induced elimination from the corresponding bis-bromomethyl compound **59** (3.50).[49]

[48] J. G. Allen and S. J. Danishefsky, *J. Am. Chem. Soc.*, **123** (2001), 351.
[49] F. A. J. Kerdesky, R. J. Ardecky, M. V. Lakshmikantham and M. P. Cava, *J. Am. Chem. Soc.*, **103** (1981), 1992.

$$(3.50)$$

59   **60**   **61**

Another mild method for the formation of *ortho*-quinodimethanes proceeds by elimination from *ortho*-(α-trimethylsilylalkyl)benzylammonium salts (such as **62**), triggered by fluoride ion.[50] Addition of tetrabutylammonium fluoride (TBAF) to the silane **62** generates the intermediate *ortho*-quinodimethane **63**, which, in the presence of dimethyl fumarate, gave the tetralin derivative **64** in quantitative yield (3.51).

$$(3.51)$$

62   **63**   **64**

Intramolecular cycloaddition reactions with *ortho*-quinodimethanes are effected readily and have found considerable use in synthesis (see Section 3.1.5).

Heterodienes

Heterodienes, in which one or more of the atoms of the conjugated diene is a heteroatom, can be used in Diels–Alder reactions,[51] although these have not been so extensively employed in synthesis as heterodienophiles.

α,β-Unsaturated carbonyl compounds react as dienes with electron-rich dienophiles such as enol ethers or enamines.[52] With less-reactive dienophiles, dimerization of the α,β-unsaturated carbonyl compound is a competing reaction. A variety of Lewis acids can catalyse the cycloaddition and recent developments have focused on the asymmetric version of the inverse electron demand hetero Diels–Alder reaction. For example, reaction of the α,β-unsaturated ketone **65** and ethyl vinyl ether with the chiral copper(II) catalyst **66** provided the dihydropyran **67** in high yield as essentially a single enantiomer (3.52).[53]

[50] Y. Ito, M. Nakatsuka and T. Saegusa, *J. Am. Chem. Soc.*, **104** (1982), 7609.
[51] D. L. Boger, in *Comprehensive Organic Synthesis*, ed. B. M. Trost and I. Fleming, vol. 5 (Oxford: Pergamon Press, 1991), p. 451.
[52] L. F. Tietze, G. Kettschau, J. A. Gewert and A. Schuffenhauer, *Curr. Org. Chem.*, **2** (1998), 19.
[53] J. Thorhauge, M. Johannsen and K. A. Jørgensen, *Angew. Chem. Int. Ed.*, **37** (1998), 2404; D. A. Evans and J. S. Johnson, *J. Am. Chem. Soc.*, **120** (1998), 4895.

(3.52)

Both 1- and 2-azabutadienes may react with dienophiles to provide access to six-membered nitrogen-containing heterocyclic compounds.[54] Azadienes are less electron-rich than the corresponding all-carbon dienes and therefore typically have diminished reactivity towards electron-deficient dienophiles. α,β-Unsaturated hydrazones, bearing an electron-donating dialkylamino group on the 1-azadiene, react readily with electron-deficient dienophiles to give adducts that can be converted into the corresponding pyridine or dihydropyridine derivatives (3.53).[55] In comparison, α,β-unsaturated imines take part in Diels–Alder reactions preferentially through their enamine tautomers and not as 1-azadienes.

(3.53)

The Diels–Alder reaction of 2-azadienes also benefits from the presence of an electron-donating substituent, to enhance the reactivity with electron-deficient dienophiles.[56] Cycloaddition with alkynyl dienophiles and aromatization leads to substituted pyridines (3.54).[57] Silyloxy-substituted 2-azadienes such as **68** are effective dienes and have been used to prepare substituted 2-pyridones and piperidones after methanolysis (3.55).[58] Asymmetric hetero Diels–Alder reactions with the chiral Lewis acid catalyst **66** provide access to the piperidone products with very high enantioselectivity.[59]

[54] S. Jayakumar, M. P. S. Ishar and M. P. Mahajan, *Tetrahedron*, **58** (2002), 379.
[55] B. Serckx-Poncin, A.-M. Hesbain-Frisque and L. Ghosez, *Tetrahedron Lett.*, **23** (1982), 3261; K. T. Potts, E. B. Walsh and D. Bhattacharjee, *J. Org. Chem.*, **52** (1987), 2285.
[56] M. Behforouz and M. Ahmadian, *Tetrahedron*, **56** (2000), 5259.
[57] A. Demoulin, H. Gorissen, A.-M. Hesbain-Frisque and L. Ghosez, *J. Am. Chem. Soc.*, **97** (1975), 4409.
[58] P. Bayard and L. Ghosez, *Tetrahedron Lett.*, **29** (1988), 6115.
[59] E. Jnoff and L. Ghosez, *J. Am. Chem. Soc.*, **121** (1999), 2617.

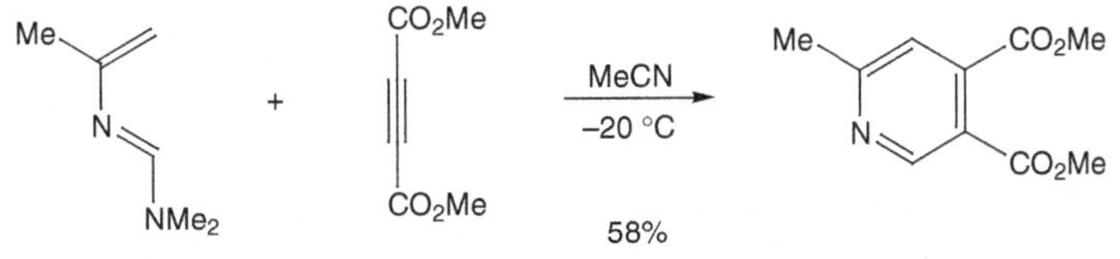

(3.54)

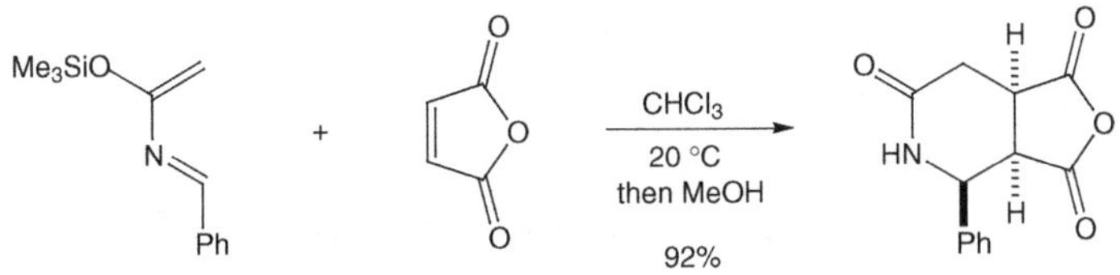

(3.55)

68

Heteroaromatic compounds such as oxazoles, diazines and triazines are useful azadienes. Cycloaddition with the dienophile gives a bridged intermediate that often fragments (e.g. with loss of HCN or N_2) by a retro Diels–Alder reaction to generate a new aromatic compound (see Section 3.1.6).

Nitroalkenes have found some use as dienes in the Diels–Alder reaction. The nitroalkene is electron-deficient and therefore reacts best with electron-rich dienophiles such as enol ethers. Good yields of the cycloadduct can be obtained by using a Lewis acid catalyst such as $SnCl_4$ or $TiCl_2(O^iPr)_2$ at low temperature. For example, cycloaddition with cyclopentene gave the nitronate **69** in high yield (3.56).[60] The nitronate cycloadducts can undergo a variety of different transformations, such as a subsequent 1,3-dipolar cycloaddition with an alkene (see Section 3.4).

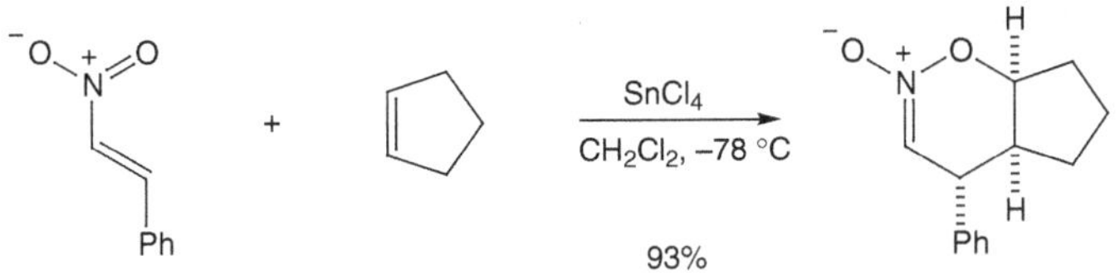

(3.56)

69

3.1.3 Regiochemistry of the Diels–Alder reaction

Addition of an unsymmetrical diene to an unsymmetrical dienophile can take place in two ways to give two structurally isomeric products. It is found in practice, however, that formation of one of the regioisomers is strongly favoured and that this can be predicted prior to reaction. Obviously, this is crucial if the Diels–Alder reaction is to be used successfully in synthesis. Cycloaddition of 1-substituted butadienes with dienophiles such as α,β-unsaturated carbonyl compounds gives rise to

[60] S. E. Denmark, B. S. Kesler and Y.-C. Moon, *J. Org. Chem.*, **57** (1992), 4912.

predominantly the 1,2-disubstituted adduct, whereas cycloaddition of 2-substituted butadienes leads to predominantly the 1,4-adduct. Thus, in the Diels–Alder reaction of a selection of 1-substituted butadienes with acrylic acid derivatives, the 1,2-adduct is favoured in all cases, irrespective of the electronic nature of the substituent (3.57). However, the regioselectivity is lost when the two components are both anionic, presumably because of the Coulombic repulsion of the two charged groups.

(3.57)

R	R'	Temp. (°C)	Ratio of 1,2- : 1,3- adducts		
NEt_2	Et	20	100	:	0
Me	Me	20	95	:	5
CO_2H	H	70	100	:	0
CO_2Na	Na	220	50	:	50

Correspondingly, in the addition of methyl acrylate to 2-substituted butadienes, the 1,4-adduct is formed predominantly, irrespective of the electronic nature of the substituent (3.58). Addition of alkynyl dienophiles to unsymmetrically substituted butadienes also results in preferential formation of 1,2- or 1,4-adducts.

(3.58)

R	Temp. (°C)	Ratio of 1,4- : 1,3- adducts		
OEt	160	100	:	0
Ph	150	82	:	18
CN	95	100	:	0

The regioselectivity can be explained in terms of frontier orbital theory.[4] In a 'normal' Diels–Alder reaction (involving an electron-rich diene and an electron-deficient dienophile), the main interaction in the transition state is between the HOMO of the diene and the LUMO of the dienophile. The orientation of the product obtained from an unsymmetrical diene and an unsymmetrical dienophile is governed largely by the atomic orbital coefficients at the termini of the conjugated systems. The atoms with the larger terminal coefficients bond preferentially, since this leads to better orbital overlap in the transition state. In most cases this leads mainly to the 1,2- or 1,4-adducts. This is represented diagrammatically for butadiene-1-carboxylic acid and acrylic acid, where the size of the circles equates roughly to the size of the orbital coefficients (shaded and unshaded circles represent lobes of opposite sign) (3.59). Similarly, for reaction of 2-phenylbutadiene and methyl acrylate, preferential formation of the 1,4-adduct is predicted (3.60). With 2-methyl-1,3-butadiene, however, where the coefficients of the terminal carbon atoms in the HOMO do not differ from each other so much as they do in the 2-phenyl compound, reaction with methyl acrylate or acrolein gives larger amounts of the 1,3-adduct.

(3.59)

HOMO   LUMO

(3.60)

HOMO   LUMO

The relative amounts of the regioisomeric products formed in the Diels–Alder reaction are strongly influenced by Lewis acid catalysts. In the presence of a Lewis acid, the proportion of the 'expected' isomer is frequently increased and high yields of a single isomer can often be obtained. Thus, in the addition of acrolein to iso-prene, the proportion of the 1,4-adduct was increased in the presence of tin(IV) chloride, so that it became almost the exclusive product of the reaction (3.61). Sim-ilar effects have been noted in many other Diels–Alder reactions. The Lewis acid co-ordinates to the carbonyl oxygen atom, thereby enhancing the electrophilicity of the dienophile (lower LUMO energy) and increasing the size of the orbital coef-ficient at the β-carbon atom relative to that at the α-carbon atom of the dienophile.

(3.61)

| PhMe, 120 °C, no catalyst | 59 | : | 41 |
| PhH, 25 °C, SnCl$_4$•5H$_2$O | 96 | : | 4 |

In substituted dienes, the magnitude of the orienting effect differs, not only with the nature of the substituent but also with its position on the diene. A substituent at C-1 generally has a more pronounced directing effect than that at C-2. With 1,2-disubstituted butadienes, therefore, the substituent at C-1 often controls the regio-chemical outcome. With 2,3-disubstituted butadienes, the structure of the adducts obtained will depend on the nature of the two substituents. For example, reaction of the diene **70** with methyl vinyl ketone gave predominantly the cycloadduct **71**, in which the phenylthio substituent has a stronger directing effect than the methoxy group (3.62).[61] This product can subsequently be converted to the monoterpene carvone. Note that in this cycloaddition reaction the cycloadduct **71** has the methoxy and methyl ketone substituents 1,3-related, and the phenylthio group, which can be removed after cycloaddition, has been used to alter the 'normal' regioselectivity.

$$(3.62)$$

70   **71** ~4:1 in favour of   carvone
                                                       this regioisomer

Regiocontrol in Diels–Alder reactions can be effected by proper choice of activating groups in the dienophile as well as in the diene. In β-nitro-α,β-unsaturated ketones and esters, the nitro group controls the orientation of addition. Thus, reaction of 1,3-pentadiene and 3-nitrocyclohexenone readily affords the cycloadduct **72** (3.63). Reductive removal of the nitro group with tributyltin hydride gives the bicyclic ketone **73**, with orientation opposite to that in the product obtained from reaction of pentadiene with cyclohexenone itself. An alternative method to remove the nitro group is by elimination of nitrous acid using a base such as 1,8-diazabicyclo[5.4.0]undec-7-ene (DBU) or 1,5-diazabicyclo[4.3.0]non-5-ene (DBN).

$$(3.63)$$

72   **73**

3.1.4 Stereochemistry of the Diels–Alder reaction

The great synthetic usefulness of the Diel-Alder reaction depends not only on the fact that it provides easy access to a variety of six-membered ring compounds, but also on its remarkable stereoselectivity. This factor has contributed to its successful application in the synthesis of many complex natural products. It should be noted, however, that the high stereoselectivity applies to the kinetically controlled

[61] B. M. Trost, W. C. Vladuchick and A. J. Bridges, *J. Am. Chem. Soc.*, **102** (1980), 3554.

reaction and may be lost by epimerization of the product or starting materials, or by dissociation of the adduct allowing thermodynamic control of the reaction.

The cis *principle*

The stereochemistry of the adduct obtained in many Diels–Alder reactions can be selected on the basis of two empirical rules formulated by Alder and Stein in 1937. According to the '*cis* principle', the relative stereochemistry of substituents in both the dienophile and the diene is retained in the adduct. That is, a dienophile with *trans* substituents will give an adduct in which the *trans* configuration of the substituents is retained, while a *cis* disubstituted dienophile will form an adduct in which the substituents are *cis* to each other. This aspect is often referred to as the stereospecific nature of the Diels–Alder reaction. For example, in the reaction of cyclopentadiene with dimethyl maleate, the *cis* adducts **74** and **75** are formed, while in the reaction with dimethyl fumarate, the *trans* configuration of the ester groups is retained in the adduct **76** (3.64).

(3.64)

Similarly, with the diene component, the relative configuration of the substituents in the 1- and 4- positions is retained in the adduct; *trans, trans*-1,4-disubstituted dienes give rise to adducts in which the 1- and 4-substituents are *cis* to each other, and *cis, trans*-disubstituted dienes give adducts with *trans* substituents (3.65).

The almost universal application of the *cis* principle provides strong evidence for a mechanism for the Diels–Alder reaction in which both new bonds between the diene and the dienophile are formed at the same time. This includes a mechanism in which the two new σ-bonds are formed simultaneously but at different rates and it does not completely exclude a two-step mechanism, if the rate of formation of the second bond in the (diradical or zwitterionic) intermediate were faster than the rate of rotation about a carbon–carbon bond.

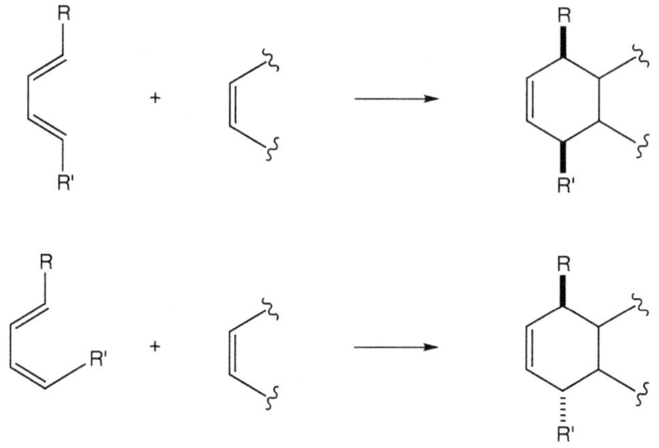

(3.65)

The endo *addition rule*

In the addition of a 1-substituted diene to a dienophile, two different products, the *endo* and the *exo* stereoisomers, may be formed depending on the manner in which the diene and the dienophile are arranged in the transition state. According to the *endo* addition rule, the diene and dienophile arrange themselves in parallel planes, and the most stable transition state arises from the orientation in which there is 'maximum accumulation of double bonds'. The 'double bonds' encompass all the π-bonds in the two components, including those in the activating groups of the dienophile. The rule is by no means always followed and is perhaps best applicable to the addition of cyclic dienes to cyclic dienophiles, but it is a useful guide in many other cycloaddition reactions.

To illustrate this aspect of stereoselectivity, the addition of maleic anhydride to cyclopentadiene gave almost exclusively the *endo* product **77** (3.66). The thermodynamically more stable *exo* compound is formed in yields of less than 2%.

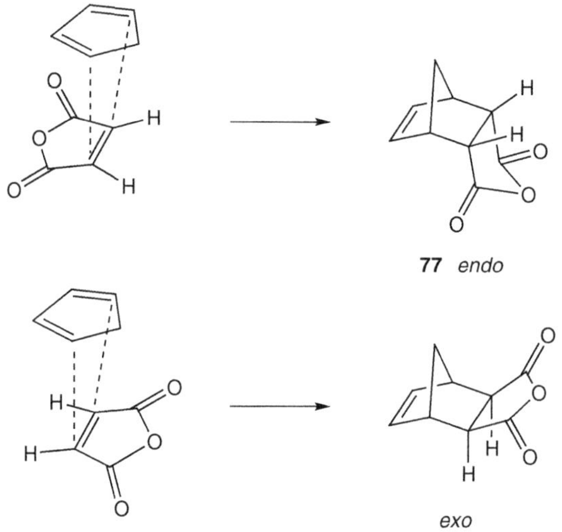

77 *endo*   (3.66)

exo

The products obtained from the cyclic diene furan and maleic anhydride and from diene addition reactions of fulvene do not obey the *endo* rule. The reason for this is that the initial *endo* adducts easily dissociate at moderate temperatures, allowing conversion of the kinetic *endo* adduct into the thermodynamically more stable *exo* isomer. In other cycloadditions, prolonged reaction times may lead to the formation of some *exo* isomer at the expense of the *endo*.

The adducts obtained from acyclic dienes and cyclic dienophiles are frequently formed in accordance with the *endo* rule. A classic example is found in the Woodward synthesis of reserpine, which started with the Diels–Alder reaction of *E*-pentadienoic acid and benzoquinone (3.67).[62] In this cycloaddition reaction, the *endo* adduct **78**, in which the carboxylic acid and the benzoquinone carbonyl groups become *cis* to one another, is obtained as the exclusive product.

(3.67)

78

When the dienophile is acyclic, the *endo* rule is not always obeyed and the composition of the mixtures obtained depends on the structure of the dienophile and diene and on the reaction conditions. For example, in the addition of acrylic acid to cyclopentadiene, the *endo* and *exo* products were obtained in the ratio 75:25; but with α-substituted acrylic acids, the product ratio varies, depending on the nature of the α-substituent (3.68). Variable ratios are also obtained in reactions with β-substituted acrylic acids. With acrylic acid itself, the proportion of the *endo* adduct formed is increased by the presence of a Lewis acid catalyst.

[62] R. B. Woodward, F. E. Bader, H. Bickel, A. J. Frey and R. W. Kierstead, *Tetrahedron*, **2** (1958), 1.

$$(3.68)$$

R	endo		exo
H	75	:	25
Me	35	:	65
Et	0	:	100
Ph	60	:	40
Br	30	:	70

Solvent and temperature may also affect the product ratio. Thus in the kinetically controlled addition of cyclopentadiene to *E*-methyl 2-butenoate, the proportion of *endo* product increases with the polarity of the solvent, and the product ratio was also slightly affected by the temperature of the reaction. A mixture of the *endo* and *exo* products was obtained, with the *exo* isomer predominant in some solvents (e.g. Me_3N at $30\,^{\circ}C$) and the *endo* in others (e.g. EtOH, AcOH).

In the Diels–Alder reaction of acrylic acid and *E*-pentadienoic acid, the temperature has a noticeable effect on the stereoselectivity (3.69). At low or moderate temperatures the *endo* adduct is the major product, but the proportion of the *exo* isomer increases as the temperature of the reaction increases.

$$(3.69)$$

Temp. (°C)	endo		exo
75	100	:	0
90	88	:	12
100	82	:	18
110	67	:	33
130	50	:	50

The factors that determine the steric course of these cycloaddition reactions are still not completely clear. It appears that a number of forces operate in the transition state and the precise composition of the product depends on the balance among these. The preference for the *endo* adduct, in which the dienophile substituents are oriented over the residual unsaturation of the diene in the transition state, has been rationalized by Woodward and Hoffmann in terms of secondary orbital interactions.[4] In this explanation, the atomic orbital at C-2 (and/or C-3) in the HOMO of the diene interacts with the atomic orbital of the activating group in the LUMO of the dienophile. However, there is no evidence for this secondary orbital interaction and the stereoselectivities in the Diels–Alder reaction can be explained in terms of steric interactions, solvent effects, hydrogen-bonding, electrostatic and other forces (3.70).[63]

[63] J. I. García, J. A. Mayoral and L. Salvatella, *Acc. Chem. Res.*, **33** (2000), 658.

In intermolecular Diels–Alder reactions, the combination of these effects, particularly with cyclic dienophiles, often leads to a preference for the *endo* adduct in the kinetically controlled reaction. In intramolecular reactions, the *endo* rule can be a good guide to the stereochemistry of the product, but as described in the next section, geometrical constraints may outweigh other factors and the *exo* product may predominate.

Possible explanations for *endo* selectivity

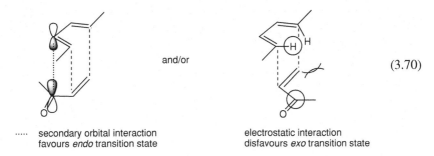

and/or   (3.70)

      secondary orbital interaction   electrostatic interaction
      favours *endo* transition state   disfavours *exo* transition state

3.1.5 Intramolecular Diels–Alder reactions

Intramolecular Diels–Alder reactions have found widespread use in organic synthesis, the cycloaddition providing ready access to polycyclic compounds with excellent levels of regio- and stereoselectivity.[64] Intramolecular reactions often proceed more easily than comparable intermolecular reactions owing to the favourable entropy factor. Heating the *E,E*-dienyl-acrylate **79** gave the *trans-* and *cis*-hydrindanes **80** and **81**, although in the presence of the Lewis acid EtAlCl$_2$ only the *endo* adduct **80** was formed (3.71).[65] Heating the *E,Z*-isomer **82** gave solely the *cis*-hydrindane **81** (3.72). In contrast to the comparatively easy cyclization of **82**, *inter*molecular Diels–Alder reactions with *Z*-1-substituted butadienes generally take place only with difficulty.

79	150 °C	65%	**80**	60 : 40	**81**			(3.71)
	EtAlCl$_2$, 23 °C	60%		100 : 0				

[64] E. Ciganek, *Org. Reactions*, **32** (1984), 1; D. Craig, *Chem. Soc. Rev.*, **16** (1987), 187; W. R. Roush, in *Comprehensive Organic Synthesis*, ed. B. M. Trost and I. Fleming, vol. 5 (Oxford: Pergamon Press, 1991), p. 513; J. D. Winkler, *Chem. Rev.*, **96** (1996), 167; A. G. Fallis, *Acc. Chem. Res.*, **32** (1999), 464; E. Marsault, A. Toró, P. Nowak and P. Deslongchamps, *Tetrahedron*, **57** (2001), 4243.
[65] W. R. Roush, H. R. Gillis and A. I. Ko, *J. Am. Chem. Soc.*, **104** (1982), 2269.

$$(3.72)$$

The intramolecular Diels–Alder reaction is most common with 1-substituted butadienes and favours the fused-ring products (such as **80**), there being very few examples of the opposite regioselectivity leading to bridged-ring compounds. The stereoselectivity across the fused-ring system (*trans* or *cis*) depends on a number of factors, including the length of the tether and the substitution pattern. The *endo* product frequently predominates, although conformational factors due to geometrical constraints or steric factors in the transition state may favour the formation of the *exo* adduct or result in mixtures of stereoisomers. Enhanced selectivity for the *endo* stereoisomer can sometimes be achieved by carrying out the reactions at reduced temperature in the presence of a Lewis acid.

With 2-substituted butadienes, intramolecular cycloaddition necessarily forms bridged-ring compounds and reactions of this kind have been used to make bridge-head double bonds. This substitution pattern allows access to natural product systems such as the taxane ring system (3.73). Conformational factors and the strain energy in forming the bridgehead alkene manifests itself in the vigorous conditions that are frequently required if such Diels–Alder reactions are carried out under thermal conditions; in many cases, however, reactions catalysed by Lewis acids can be effected under much milder conditions.[66]

$$(3.73)$$

Most applications of the intramolecular Diels–Alder reaction use 1-substituted butadienes and these reactions often form a key step in the synthesis of polycyclic natural products. Substituents in the connecting chain may influence the facial selectivity of the cycloaddition reaction, as well as the *endo:exo* selectivity. For example, in a synthesis of the antibiotic indanomycin, the triene **83** gave, on heating,

[66] K. J. Shea and J. W. Gilman, *J. Am. Chem. Soc.*, **107** (1985), 4791; P. A. Brown and P. R. Jenkins, *J. Chem. Soc., Perkin Trans. 1* (1986), 1303; B. R. Bear, S. M. Sparks and K. J. Shea, *Angew. Chem. Int. Ed.*, **40** (2001), 820.

the indane derivative **84** in a reaction in which four new chiral centres are set up selectively in one step (3.74).[67] The high stereoselectivity in this cycloaddition can be ascribed to a preference for the *endo* transition state **85**, which is favoured over the alternative *endo* arrangement **86** owing to the steric interaction between the vinylic proton and the allylic alkyl group in **86** (3.75).

$$ (3.74) $$

R = SiPh₂^{t}Bu or CH₂OCH₂CH₂OMe

R = SiPh$_2$^{t}Bu or CH$_2$OCH$_2$CH$_2$OMe

R' = Me or Et

$$ (3.75) $$

Intramolecular cycloaddition reactions are possible in the absence of electron-withdrawing substituents on the dienophile, although more-vigorous conditions are normally required. Cycloaddition of the triene **87** at 200 °C gave, after hydrolysis of the silyl ether, a mixture of the diastereomeric *trans*-fused ring systems **88**, which were converted into the eudesmane sesquiterpene **89** (3.76).[68]

$$ (3.76) $$

[67] K. C. Nicolaou and R. L. Magolda, *J. Org. Chem.*, **46** (1981), 1506; W. R. Roush and A. G. Myers, *J. Org. Chem.*, **46** (1981), 1509; M. P. Edwards, S. V. Ley, S. G. Lister, B. D. Palmer and D. J. Williams, *J. Org. Chem.*, **49** (1984), 3503.
[68] S. R. Wilson and D. T. Mao, *J. Am. Chem. Soc.*, **100** (1978), 6289.

Hetero-substituted dienes and dienophiles have been used extensively in intramolecular cycloaddition reactions, for example towards the synthesis of alkaloid ring systems (e.g. see Scheme 3.40). A synthesis of the alkaloid manzamine A makes use of the substituted triene **90**, which cyclizes to the tricyclic ring system **91** (3.77).[69] In this example, the one stereocentre in the triene **90** controls the formation of the three new chiral centres in the product **91**.

$$\text{(3.77)}$$

A useful protocol, particularly when the intermolecular Diels–Alder reaction gives a mixture of stereoisomers, or when the desired cycloadduct is not the 1,2- or 1,4-regioisomer, is to use a tethered Diels–Alder reaction.[70] For example, heating the triene **92**, in which the diene and the dienophile are tethered by a silicon atom, gave a single regio- and stereoisomer of the cycloadduct **93** (3.78). Subsequent treatment with acid removes the silyl group and promotes cyclization to the lactone **94**. Attempts to perform the related intermolecular reaction lead to a mixture of regio- and stereoisomers.

$$\text{(3.78)}$$

An interesting variant of the tethered Diels–Alder reaction has been discovered that uses vinyl magnesium species, such as **95** (3.79).[71] In this case, the cycloadduct

[69] J. M. Humphrey, Y. Liao, A. Ali, T. Rein, Y.-L. Wong, H.-J. Chen, A. K. Courtney and S. F. Martin, *J. Am. Chem. Soc.*, **124** (2002), 8584.

[70] P. J. Ainsworth, D. Craig, J. C. Reader, A. M. Z. Slawin, A. J. P. White and D. J. Williams, *Tetrahedron*, **51** (1995), 11601; M. Bols and T. Skrydstrup, *Chem. Rev.*, **95** (1995), 1253; L. Fensterbank, M. Malacria and S. McN. Sieburth, *Synthesis* (1997), 813.

[71] G. Stork and T. Y. Chan, *J. Am. Chem. Soc.*, **117** (1995), 6595.

that is obtained is formally the product of using ethene as the dienophile.

(3.79)

The intramolecular version of the hetero Diels–Alder reaction is particularly valuable in synthesis. Imines as heterodienophiles are useful for the preparation of alkaloid ring systems. For example, addition of aqueous formaldehyde to the amine **96** generated the required intermediate imine (or iminium ion), which undergoes cycloaddition to the quinolizidines **97** and **98** (3.80).[72] Reduction of the alkene **98** and simultaneous hydrogenolysis of the benzyl group using H_2, Pd/C gave the alkaloid lupinine. A related example of a imine in an intramolecular Diels–Alder reaction is given in Scheme 3.28.

(3.80)

Intermolecular cycloaddition reactions of 1-aza- and 2-azadienes are generally sluggish and therefore the intramolecular reaction, in which the entropy factor is more favourable, can lead to a more efficient cycloaddition of these heterodienes. Acylation of *N*-silyl-α,β-unsaturated imines provides a convenient one-pot method for the formation of 1-azadienes required for the Diels–Alder reaction.[73] For example, addition of allyl chloroformate to the *N*-trimethylsilyl imine of acrolein gave, after heating, the cycloadduct **99** (3.81).

Intramolecular cycloaddition reactions with 2-azadienes have found limited application in organic synthesis, however an efficient route to the daphniphyllum alkaloids makes use of such a cycloaddition.[74] The 2-azadiene **101** was formed *in situ* from the diol **100** by Swern oxidation (to the dialdehyde) and addition of ammonia (3.82). Cycloaddition in the presence of acetic acid at room temperature gave the complex ring system **102**. If the reaction mixture is warmed to 70 °C

[72] P. A. Grieco and D. T. Parker, *J. Org. Chem.*, **53** (1988), 3325.
[73] T. Uyehara, I. Suzuki and Y. Yamamoto, *Tetrahedron Lett.*, **31** (1990), 3753.
[74] C. H. Heathcock, M. M. Hansen, R. B. Ruggeri and J. C. Kath, *J. Org. Chem.*, **57** (1992), 2544; C. H. Heathcock and J. A. Stafford, *J. Org. Chem.*, **57** (1992), 2566.

then the cycloaddition product **102** undergoes cyclization (of the alkene onto the protonated imine) to give a more advanced intermediate in high overall yield.

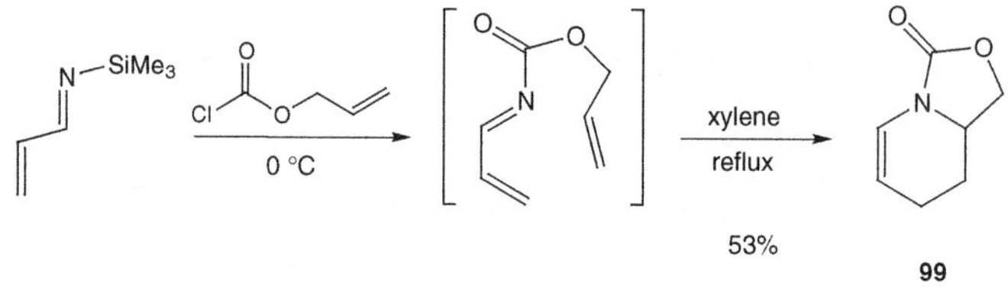

(3.81)

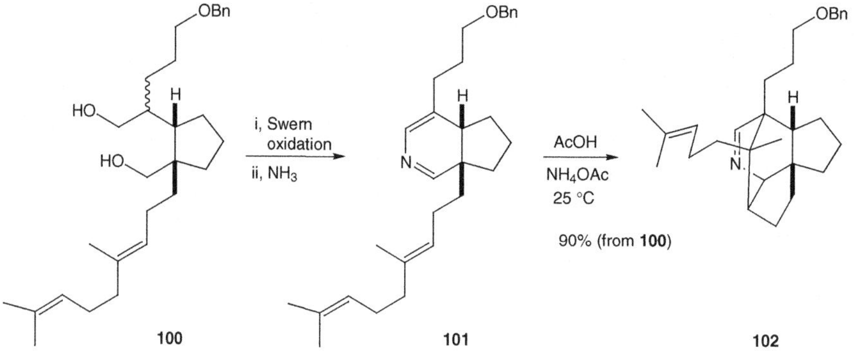

(3.82)

Intramolecular Diels–Alder reactions of *ortho*-quinodimethanes have been used widely in the synthesis of natural products, particularly in the steroid and alkaloid fields.[75] *ortho*-Quinodimethanes are reactive dienes and even unreactive dienophiles can be used to form cycloadducts. The diene is prepared *in situ*, using one of a number of procedures (see Section 3.1.2) such as the thermal ring-opening of benzocyclobutenes. One method makes use of the ready loss of sulfur dioxide from a sulfone such as **103** (3.83).[76] Thus, heating the sulfone **103** gave the intermediate *ortho*-quinodimethane **104**, which cyclized to the tetracycle **105**. This product can be converted readily into the steroid (+)-estradiol.

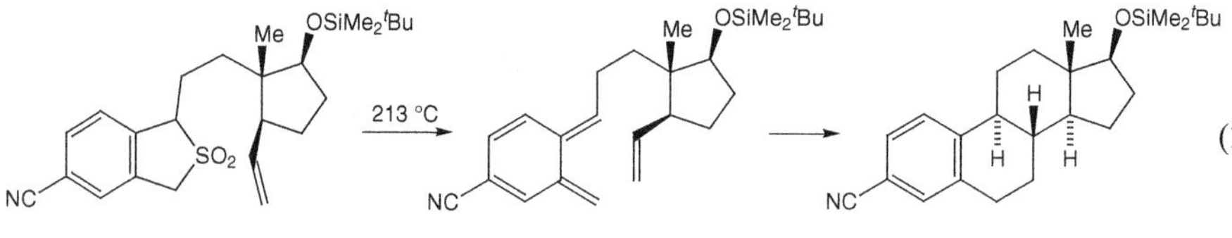

(3.83)

[75] H. Nemoto and K. Fukumoto, *Tetrahedron*, **54** (1998), 5425.

[76] W. Oppolzer and D. A. Roberts, *Helv. Chim. Acta*, **63** (1980), 1703; see also K. C. Nicolaou, W. E. Barnette and P. Ma, *J. Org. Chem.*, **45** (1980), 1463; T. Kametani, H. Matsumoto, H. Nemoto and K. Fukumoto, *J. Am. Chem. Soc.*, **100** (1978), 6218; R. L. Funk and K. P. C. Vollhardt, *J. Am. Chem. Soc.*, **102** (1980), 5253; S. Djuric, T. Sarkar and P. Magnus, *J. Am. Chem. Soc.*, **102** (1980), 6885; Y. Ito, M. Nakatsuka and T. Saegusa, *J. Am. Chem. Soc.*, **103** (1981), 476.

Alkaloids containing the indole ring system are common in nature, many of them having a six-membered ring fused to the indole 2,3-position. This ring system can be ideally set up by using the *ortho*-quinodimethane strategy. For example, *N*-acylation of the indole **106** with the mixed anhydride **107** is thought to give rise to the intermediate 2,3-dimethylene-indole **108**, which undergoes cycloaddition to the adduct **109** with *cis* stereochemistry at the ring junction (3.84).[77] This chemistry has been used to prepare *aspidosperma* alkaloids.

(3.84)

3.1.6 *The retro Diels–Alder reaction*

Diels–Alder reactions are reversible, and many adducts dissociate into their components on heating.[78] This can be made use of in, for example, the separation of anthracene derivatives from mixtures with other hydrocarbons through their adducts with maleic anhydride. More interesting are reactions in which the original adduct is modified chemically and subsequently dissociated to yield a new diene or dienophile. Thus, in a synthesis of sarcomycin methyl ester, the enantiomerically pure adduct **110** was elaborated in a number of steps to the diastereomers **111** and **112** (3.85).[79] Retro Diels–Alder reaction of **111** by flash vacuum pyrolysis gave optically pure (+)-(*S*)-sarcomycin methyl ester **113**, with elimination of cyclopentadiene. The isomer **112** similarly gave (−)-(*R*)-sarcomycin methyl ester. Sarcomycin and other sensitive alkene-containing compounds, such as the prostaglandins, that are difficult to prepare because of their propensity for isomerization or further reaction, are ideally suited to the retro Diels–Alder reaction.

[77] P. Magnus, T. Gallagher, P. Brown and P. Pappalardo, *Acc. Chem. Res.*, **17** (1984), 35; M. Ladlow, P. M. Cairns and P. Magnus, *Chem. Commun.* (1986), 1756.

[78] A. J. H. Klunder, J. Zhu and B. Zwanenburg, *Chem. Rev.*, **99** (1999), 1163; B. Rickborn, *Org. Reactions*, **52** (1998), 1; R. W. Sweger and A. W. Czarnik, in *Comprehensive Organic Synthesis*, ed. B. M. Trost and I. Fleming, vol. 5 (Oxford: Pergamon Press, 1991), p. 551; A. Ichihara, *Synthesis* (1987), 207; M.-C. Lasne and J.-L. Ripoll, *Synthesis* (1985), 121.

[79] G. Helmchen, K. Ihrig and H. Schindler, *Tetrahedron Lett.*, **28** (1987), 183.

$$(3.85)$$

110   **111**   **112**

650 °C, 0.02 mm Hg

113

Cyclopentadiene is a popular moiety to release in a retro Diels–Alder reaction, as the cyclohexene portion of the bicyclo[2.2.1]heptene is locked in a boat conformation, as required in the transition state for the retro (and indeed forward Diels–Alder) reaction. Other bridged bicyclic systems are also more prone to undergo retro Diels–Alder reaction. The bridged adduct from furan and maleic anhydride readily undergoes the retro reaction and, although the *endo* isomer is formed at a much faster rate, the reversible nature of the reaction leads to the accumulation of the more stable *exo* isomer.

High temperatures are normally required for the retro Diels–Alder reaction and this is not always convenient. Flash vacuum pyrolysis is carried out typically at temperatures in the region of 400–600 °C, although the retro reaction can sometimes take place at lower temperatures (180–250 °C) by refluxing in 1,2-dichlorobenzene or diphenyl ether. In contrast, retro Diels–Alder reactions of some anionic intermediates take place under relatively mild conditions.[80] For example, the adduct **114** fragments at only 35 °C after treatment with potassium hydride (3.86). Subsequent deprotection of the alcohol functional groups from the product **115** gave conduritol A. The unusual ease of this reaction can be ascribed to the high ground-state energy of the anion together with the stabilization of the anion in the transition state, thereby leading to a decrease in the activation energy for the pericyclic reaction.

[80] M. E. Bunnage and K. C. Nicolaou, *Chem. Eur. J.*, **3** (1997), 187.

$$\text{(3.86)}$$

114

R = CH₂OCH₂Ph

115

The two σ-bonds that are cleaved in the retro Diels–Alder reaction need not necessarily be the same as those formed in the initial forward reaction. Many valuable retro Diels–Alder reactions involve the cleavage of one carbon–carbon and one carbon–heteroatom bond, or two carbon–heteroatom bonds, that were not set up in a forward Diels–Alder reaction.[81] For example, the otherwise relatively inaccessible 3-substituted and 3,4-disubstituted furans can be prepared by way of a tandem Diels–Alder then retro Diels–Alder reaction of 4-phenyloxazole and substituted alkynes, as illustrated in Scheme 3.87.

$$\text{(3.87)}$$

Fragmentation of an adduct with release of a nitrile, CO_2 or N_2 are most common and the latter provide an irreversible method for the formation of a new diene or aromatic compound. Cycloaddition of a pyran-2-one or a 1,2-diazine (pyridazine) with an alkyne gives an intermediate bridged compound that loses CO_2 or N_2 to generate a benzene derivative (see Scheme 3.46). Many other aromatic and heteroaromatic compounds can be prepared likewise. For example, a synthesis of lavendamycin made use of the inverse electron demand Diels–Alder reaction between the 1,2,4-triazine **116** and the enamine **117**, followed by *in situ* elimination of pyrrolidine and retro Diels–Alder reaction, releasing N_2 and the substituted pyridine **118** (3.88).[82]

$$\text{(3.88)}$$

116　　　　　　**117**　　　　　　**118**

[81] B. Rickborn, *Org. Reactions*, **53** (1998), 223.

[82] D. L. Boger, S. R. Duff, J. S. Panek and M. Yasuda, *J. Org. Chem.*, **50** (1985), 5782 (a small amount of the other regioisomer was also formed).

3.1.7 Asymmetric Diels–Alder reactions

Two main approaches to the preparation of a single enantiomer of a Diels–Alder cycloadduct have been used in synthetic organic chemistry. A chiral auxiliary attached to the diene, or more commonly to the dienophile, is a popular method. Alternatively, the use of an external chiral catalyst can promote the preferential formation of one of the two enantiomeric cycloadducts. Many examples of highly selective Diels–Alder reactions have been documented.[83]

A substituent on the diene or dienophile that bears a chiral centre may promote facially selective cycloaddition, to give a non-statistical mixture of optically active diastereomers. The products of cycloaddition will be diastereomers because the original stereogenic centre is an element of the product. The more facially selective the cycloaddition, the higher the diastereomeric ratio will be, and the ideal scenario is one in which a single diastereomer of the product is formed. In general, the normal thermal reaction gives only poor selectivity and best results have been obtained in reactions catalysed by Lewis acids at low temperatures. Lowering the temperature enhances the difference in the enthalpy of activation for the two diastereomeric transition states. The Lewis acid increases the reactivity (by lowering the energy of the LUMO) of the dienophile and restricts its conformational freedom, thereby reducing the number of possible transition states.

Most studies have centred on the use of optically active dienophiles, particularly with esters or amides of acrylic acid. Having performed their directing function, the optically active auxiliary group (alcohol or amine) is removed from the product and in some cases may be reused. Many optically active alcohols R*OH, have been employed in this sequence. For example, good results with acrylic esters of the neopentyl alcohols **119** or its enantiomer **120**, derived from (*R*)-(+)- or (*S*)-(−)-camphor, have been achieved (3.89). With cyclopentadiene as the diene component, the adduct **121** was formed with almost complete diastereomeric selectivity (3.90).[84] Reduction of the purified product with lithium aluminium hydride regenerated the chiral auxiliary alcohol **119** and gave the optically pure *endo* alcohol **122**. The reaction is thought to take place by addition of the diene to the dienophile in the conformation **123**, in which access to the *re*-face of the double bond (i.e. the front face as drawn) is hindered by the tert-butyl group. Note that the Lewis acid coordinates to the carbonyl oxygen atom and that the acrylate prefers to sit in the *s-trans* orientation.

[83] W. Oppolzer, *Angew. Chem. Int. Ed. Engl.*, **23** (1984), 876; H. B. Kagan and O. Riant, *Chem. Rev.*, **92** (1992), 1007; T. Oh and M. Reilly, *Org. Prep. Proc. Int.*, **26** (1994), 129.
[84] W. Oppolzer, C. Chapuis, G. M. Dao, D. Reichlin and T. Godel, *Tetrahedron Lett.*, **23** (1982), 4781.

(3.89)

119   **120**

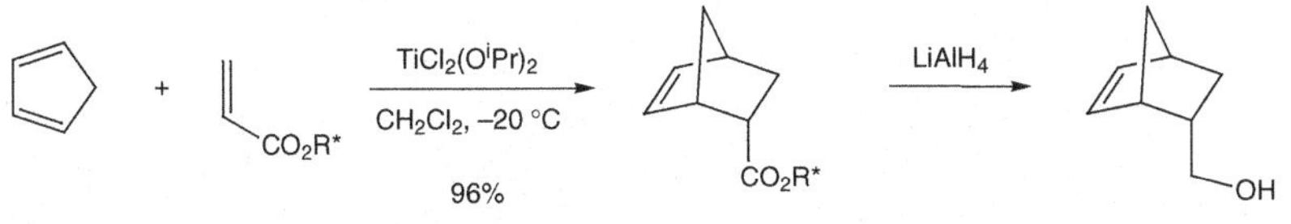

(3.90)

121   96% *endo*
          99.4% d.e.

122

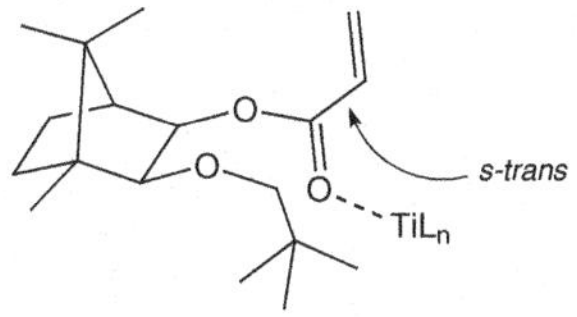

123

It is clear that a restricted orientation of the dienophile is crucial to the success of the asymmetric Diels–Alder reaction. A good method to lock the conformation is to use an auxiliary containing a carbonyl group, such that the two carbonyl groups of the dienophile can chelate to a Lewis acid. Thus, high levels of diastereofacial selectivity can be achieved in Diels–Alder reactions of the acrylates of ethyl lactate or of pantolactone, in the presence of the Lewis acid $TiCl_4$ (3.91).[85] The adduct **124** is formed almost exclusively (93:7 ratio of diastereomers) using butadiene and the acrylate of (*R*)-pantolactone and can be purified easily by crystallization. Simple hydrolysis gives enantiomerically pure carboxylic acid **125**. In such chelated systems, the metal is co-ordinated *anti* to the alkene of the dienophile and the acrylate therefore adopts the *s-cis* conformation **126**.

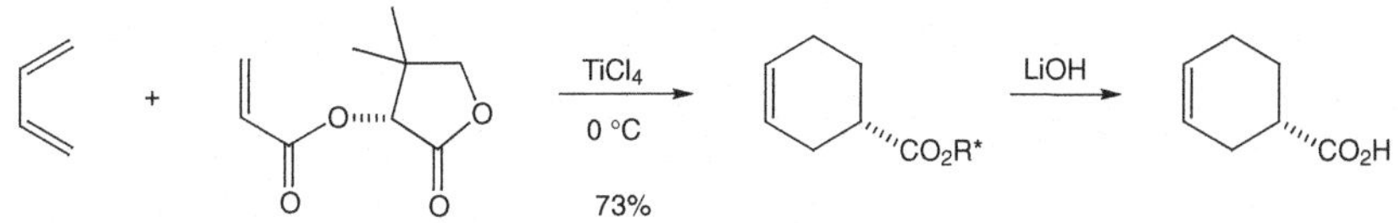

(3.91)

124   86% d.e.

125

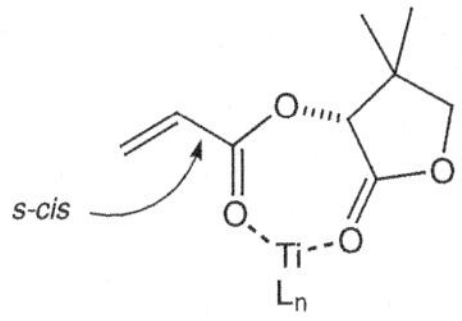

126

[85] T. Poll, A. Sobczak, H. Hartmann and G. Helmchen, *Tetrahedron Lett.*, **26** (1985), 3095.

Highly selective Diels–Alder reactions with other chiral auxiliaries attached to the dienophile have been documented. For example, chiral 2-oxazolidinones or the camphor sultam auxiliaries have proven particularly useful. Such cycloaddition reactions, catalysed by an alkylaluminium chloride, occur with a variety of dienes to give adducts in high yield and with very high diastereoselectivity. In many cases these adducts can be obtained diastereomerically pure by crystallization. The reactions are thought to occur by way of complexed ion pairs (e.g. **129**), in which the substituent on the auxiliary shields one face of the dienophile from attack by the diene. For example, 2-methylbutadiene (isoprene) gave the adduct **127**, which was converted into (*R*)-(+)-α-terpineol **128** (3.92).[86]

$$(3.92)$$

Chelation is also thought to play an important part in directing the facial selectivity of cycloadditions with the camphor sultam auxiliary. A variety of dienes can be used and adducts are obtained with very high diastereomeric excesses.[87] Both inter- and intramolecular cycloaddition reactions are amenable to the use of a chiral auxiliary. An intramolecular example is illustrated in Scheme 3.93, in which the diene and dienophile are tethered and in which cycloaddition leads to predominantly one of the two diastereomeric *trans*-fused bicyclic (*endo*) products.[88] The dienophile is thought to adopt the *s-cis* conformation, with the aluminium atom complexed to the carbonyl and one of the two sulfone oxygen atoms.

$$(3.93)$$

[86] D. A. Evans, K. T. Chapman and J. Bisaha, *J. Am. Chem. Soc.*, **106** (1984), 4261; *J. Am. Chem. Soc.*, **110** (1988), 1238.

[87] W. Oppolzer, C. Chapuis and G. Bernardinelli, *Helv. Chim. Acta*, **67** (1984), 1397; C. Thom, P. Kocienski and K. Jarowicki, *Synthesis* (1993), 475.

[88] W. Oppolzer and D. Dupuis, *Tetrahedron Lett.*, **26** (1985), 5437.

Reactions of dienes containing optically active auxiliary groups have not been so widely studied as those of chiral dienophiles. There are, however, examples of the use of various chiral auxiliaries attached to either C-1 or C-2 of the diene. The 1-substituted diene **130**, derived from mandelic acid, undergoes cycloaddition with dienophiles in the presence of boron trifluoride or boron triacetate (3.94).[89] With the dienophile juglone, the adduct **131** was formed with virtually complete asymmetric induction. The absolute configuration of the product corresponds to reaction of the diene in the conformation in which one face of the diene is shielded by the phenyl substituent.

A number of examples have been reported of the asymmetric hetero Diels–Alder reaction of heterodienes or heterodienophiles such as imines, nitroso or carbonyl compounds.[90] The chiral auxiliary is commonly attached to the nitrogen atom of the imine or nitroso compound, or a chiral ester or amide substituent may provide the necessary asymmetric induction. As an example, the α-chloronitroso dienophile **132**, bearing a sugar-derived auxiliary, has been found to be effective for the formation of cyclic hydroxylamines with high optical purity (3.95). The initial cycloadduct breaks down readily (see Scheme 3.32), via an iminium ion with subsequent methanolysis to release the chiral auxiliary and the product **133**.

The use of a chiral auxiliary attached to the dienophile or the diene has the advantage that the diastereomeric cycloadducts are normally readily separable. Therefore products of essentially complete optical purity can be obtained. The major drawback to this methodology is that it requires the initial preparation of the chiral substrate and, after cycloaddition, the subsequent removal of the auxiliary.

---

[89] B. M. Trost, D. O'Krongly and J. L. Belletire, *J. Am. Chem. Soc.*, **102** (1980), 7595; C. Siegel and E. R. Thornton, *Tetrahedron Lett.*, **29** (1988), 5225.

[90] H. Waldmann, *Synthesis* (1994), 535; P. F. Vogt and M. J. Miller, *Tetrahedron*, **54** (1998), 1317.

A more convenient way to promote an enantioselective Diels–Alder reaction is to perform the desired cycloaddition in the presence of an external chiral catalyst.

Early reports of the use of a chiral catalyst to effect the asymmetric Diels–Alder reaction showed somewhat variable results. Over the past decade or so, many metals and chiral ligands have been tested as chiral Lewis acids, and some very efficient catalysts with good turnover rates have been developed that give rise to high yields of highly enantiomerically enriched products from a range of dienes and dienophiles.

Particularly effective catalysts are the chiral copper(II) bisoxazoline complexes **66** and **134** (3.96).[91] Best results are obtained when the dienophile has two sites for co-ordination to the metal. For example, the catalyst chelates to the two carbonyl groups of acrylimide dienophiles (as in structure **135**) and cycloaddition with a diene leads to the adduct in high yield and with high optical purity (3.97).[92]

$$\text{(3.96)}$$

Me Me

2 OTf⁻

'Bu   Cu   'Bu

66

Me Me

2 SbF₆⁻

'Bu   Cu   'Bu

134

135   *Re* face

$$\text{(3.97)}$$

10 mol% **134**

CH₂Cl₂, 25 °C

89%

83:17 *cis:trans*   *cis*: 94% ee

The SbF₆-derived complex **134** is approximately 20 times more reactive than its triflate analogue **66**. The use of copper as the metal allows a well-defined catalyst with (distorted) square-planar geometry, and analysis of the catalyst–substrate complex **135** allows the prediction that the diene component will approach from the less hindered *Re* face of the dienophile (note that the dienophile adopts the *s-cis* conformation). The bisoxazoline ligand has C₂-symmetry and this is beneficial as it reduces the number of competing diastereomeric transition states.

The bisoxazoline catalysts, sometimes abbreviated to [Cu-(*S*,*S*)-t-Bu-box]X₂, are suitable for inter- and intramolecular asymmetric Diels–Alder reactions with

[91] J. S. Johnson and D. A. Evans, *Acc. Chem. Res.*, **33** (2000), 325; A. K. Ghosh, P. Mathivanan and J. Cappiello, *Tetrahedron: Asymmetry*, **9** (1998), 1.

[92] D. A. Evans, S. J. Miller, T. Lectka and P. von Matt, *J. Am. Chem. Soc.*, **121** (1999), 7559; D. A. Evans, D. M. Barnes, J. S. Johnson, T. Lectka, P. von Matt, S. J. Miller, J. A. Murry, R. D. Norcross, E. A. Shaughnessy and K. R. Campos, *J. Am. Chem. Soc.*, **121** (1999), 7582.

acyclic and cyclic dienes. Using furan as the diene resulted in the adduct **136** with very high enantiomeric purity (3.98). Equilibration of the adduct (retro Diels–Alder reaction) occurs at −20 °C to give predominantly the *exo* diastereomer as a racemic mixture and it is imperative to conduct this cycloaddition reaction at low temperature. Conversion of the adduct **136** to its corresponding methyl ester, followed by enolization and β-elimination gave the cyclohexadiene **137**, which was converted (by dihydroxylation, desilylation and ester hydrolysis) to the enantiomer of natural shikimic acid.

$$\text{(3.98)}$$

136   80% *endo* 97% ee

137   *ent*-shikimic acid

Copper complexes of the bisoxazoline ligands have been shown to be excellent asymmetric catalysts not only for the formation of carbocyclic systems, but also for the hetero-Diels–Alder reaction.[93] Chelation of the two carbonyl groups of a 1,2-dicarbonyl compound to the metal atom of the catalyst sets up the substrate for cycloaddition with a diene. Thus, the activated diene **20** reacts with methyl pyruvate in the presence of only 0.05 mol% of the catalyst **66** to give the adduct **138** with very high enantiomeric excess (3.99).

$$\text{(3.99)}$$

20   **138** 98.4% ee

The use of C_2-symmetric catalysts is popular for promoting enantioselective Diels–Alder reactions. Good results have been obtained using metal complexes of

[93] K. A. Jørgensen, *Angew. Chem. Int. Ed.*, **39** (2000), 3558; K. A. Jørgensen, M. Johannsen, S. Yao, H. Audrain and J. Thorhauge, *Acc. Chem. Res.*, **32** (1999), 605; see also References 53 and 59.

tartaric acid-derived diols (TADDOLs), binaphthol (BINOL) derivatives and 1,2-diaryl-ethylenediamine derivatives. For example, the titanium TADDOL complex **140** is an effective catalyst for the Diels–Alder reaction between cyclopentadiene and the bidentate dienophile **139** (3.100).[94] A cationic trigonal bipyramidal titanium complex **141** is proposed to account for the results, with the *Si* face of the acrylimide blocked by one of the aryl groups.

$$(3.100)$$

139

94%

87% *endo*
88% ee

140, Ar = 2-naphthyl   **141**

Most success in the asymmetric Diels–Alder reaction with an external chiral catalyst has been achieved by using a dienophile such as **139**, in which there are two points of attachment of the dienophile to the metal complex. However, some catalysts are known, and new ones are being developed, that allow highly enantioselective cycloaddition with dienophiles capable of only single-point binding to the metal. To promote asymmetric induction in such systems, a secondary electronic or steric interaction is required in order to favour one diastereomeric transition state. The catalyst BINOL-TiCl$_2$ **143** has been found to be effective in certain cases.[95] Thus, in a synthesis of the alkaloid (−)-ibogamine, the Diels–Alder reaction of the simple dienophile benzoquinone and the diene **142**, catalysed by (*S*)-BINOL-TiCl$_2$ **143** gave the cycloadduct **144** with high optical purity (3.101). The origin of the enantioselectivity must lie with the axially chiral BINOL ligand, but it is not yet

[94] D. Seebach, A. K. Beck and A. Heckel, *Angew. Chem. Int. Ed.*, **40** (2001), 92; D. Seebach, R. Dahinden, R. E. Marti, A. K. Beck, D. A. Plattner and F. N. M. Kühnle, *J. Org. Chem.*, **60** (1995), 1788; K. Narasaka, N. Iwasawa, M. Inoue, T. Yamada, M. Nakashima and J. Sugimori, *J. Am. Chem. Soc.*, **111** (1989), 5340.
[95] K. Mikami, Y. Motoyama and M. Terada, *J. Am. Chem. Soc.*, **116** (1994), 2812; J. D. White and Y. Choi, *Org. Lett.*, **2** (2000), 2373.

clear how this is relayed to the transition state of the cycloaddition reaction.

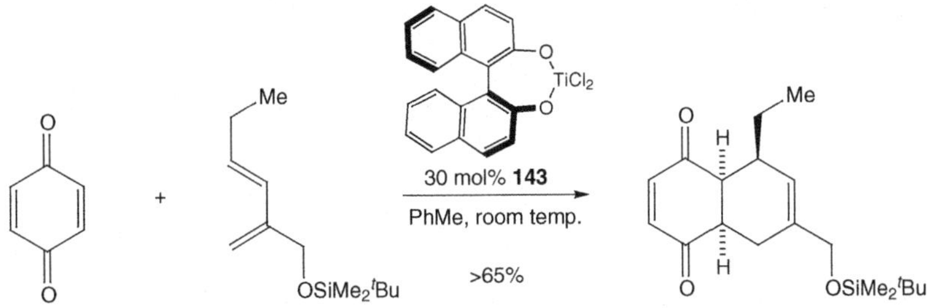

(3.101)

The BINOL ligand complexed to a metal is an effective catalyst for a number of organic reactions. Complexes of derivatives of BINOL, such as the chiral borane **145**, have been found to be excellent catalysts for the asymmetric Diels–Alder reaction.[96] This catalyst has been shown to allow very good enantioselectivity even for cycloadditions with alkynyl dienophiles. Thus, cycloaddition of cyclopentadiene and 3-iodopropynal, in the presence of 20 mol% of **145**, gave the adduct **146** with high optical purity (3.102). The iodine atom in the adduct **146** can be replaced by various other functional groups, thereby allowing the preparation of a variety of enantiomerically enriched bicyclo[2.2.1]heptadienes. Partially reduced derivatives of BINOL, such as octahydro-binaphthol, complexed to titanium(IV) have shown excellent selectivity in the hetero-Diels–Alder reaction of aldehydes with the Danishefsky diene (**20**).[97] For example, cycloaddition of benzaldehyde using 20 mol% of the catalyst 'Ti-(*R*)-H$_8$-BINOL' gave (after treatment with TFA) the cycloadduct **147** in very high yield and enantiopurity (3.103).

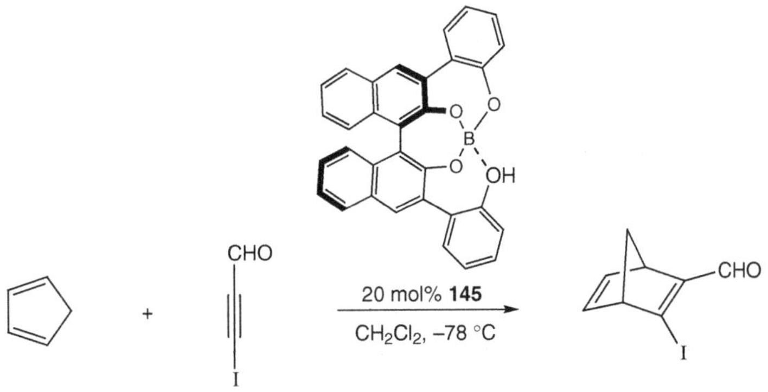

(3.102)

96 K. Ishihara, H. Kurihara, M. Matsumoto and H. Yamamoto, *J. Am. Chem. Soc.*, **120** (1998), 6920; K. Ishihara, S. Kondo, H. Kurihara, H. Yamamoto, S. Ohashi and S. Inagaki, *J. Org. Chem.*, **62** (1997), 3026.
97 B. Wang, X. Feng, Y. Huang, H. Liu, X. Cui and Y. Jiang, *J. Org. Chem.*, **67** (2002), 2175; Y. Yuan, J. Long, J. Sun and K. Ding, *Chem. Eur. J.*, **8** (2002), 5033.

$$\text{PhCHO} \quad + \quad \textbf{20} \xrightarrow[\substack{\text{PhMe, 0 °C} \\ \text{then } CF_3CO_2H}]{20\ \text{mol}\%\ Ti(O^iPr)_4} \quad \textbf{147}$$

20   92%   **147**   97% ee

A ligand having C_2-symmetry is not a necessity and a variety of catalysts are known that promote highly enantioselective Diels–Alder reactions. For example, the oxazaborolidine **148** is a good catalyst for the cycloaddition of cyclopentadiene and 2-bromoacrolein (3.104).[98] The cycloaddition reaction is highly diastereo- and enantioselective in favour of the *exo*-aldehyde. However, the enantioselectivity is poor with this catalyst when the dienophile lacks a substituent (for example bromine) in the 2-position. A solution to this problem is the use of the protonated oxazaborolidine **149**, which promotes highly selective cycloaddition of cyclopentadiene with a range of dienophiles, such as ethyl vinyl ketone or ethyl acrylate (3.105).[99] Asymmetric cycloaddition of α,β-unsaturated aldehydes or ketones with various dienes can alternatively be achieved in the presence of a chiral secondary amine as a catalyst.[100]

95%   96% *exo*   99% ee   (3.104)

94%   97% *endo*   98% ee   (3.105)

[98] E. J. Corey and T.-P. Loh, *J. Am. Chem. Soc.*, **113** (1991), 8966; E. J. Corey, A. Guzman-Perez and T.-P. Loh, *J. Am. Chem. Soc.*, **116** (1994), 3611; for a review, see E. J. Corey, *Angew. Chem. Int. Ed.*, **41** (2002), 1650.
[99] D. H. Ryu, T. W. Lee and E. J. Corey, *J. Am. Chem. Soc.*, **124** (2002), 9992.
[100] A. B. Northrup and D. W. C. MacMillan, *J. Am. Chem. Soc.*, **124** (2002), 2458.

Asymmetric hetero-Diels–Alder cycloaddition of a diene with an aldehyde using the chiral catalyst **150** has been shown to proceed with very high levels of enantioselectivity, as illustrated in a key step of a synthesis of the natural product FR901464 (3.106).[101] This catalyst is also effective for highly enantioselective hetero-Diels–Alder cycloadditions of α,β-unsaturated aldehydes with ethyl vinyl ether.[102]

$$(3.106)$$

3.2 [2+2] Cycloaddition reactions

The [2+2] cycloaddition reaction has found considerable use in synthesis, particularly for the formation of compounds containing a four-membered ring. The combination of two alkenes leads to a cyclobutane ring, although most alkenes do not undergo a thermal [2+2] cycloaddition reaction with another alkene. Tetrafluoroethene is unusual in that it is able to form (tetrafluoro)cyclobutanes with many alkenes under thermal conditions. Ketenes ($R_2C=C=O$) react with alkenes under thermal conditions to give cyclobutanones (*vide infra*). However, many [2+2] cycloaddition reactions are carried out under photochemical conditions. Simple alkenes absorb light in the far ultra-violet and, in the absence of sensitizers, undergo mainly fragmentations and *E–Z* isomerization, but conjugated alkenes which absorb at longer wavelengths form cycloaddition compounds readily. Irradiation of butadiene in dilute solution with light from a high-pressure mercury arc, leads to cyclobutene and bicyclo[1.1.0]butane. Such electrocyclic reactions of conjugated polyenes are discussed in Section 3.8. In the presence of a sensitizer, butadiene dimerizes to form, mainly, *trans*-1,2-divinylcyclobutane.

A common type of photochemical [2+2] cycloaddition reaction involves an α,β-unsaturated carbonyl compound.[103] Since these compounds absorb at sufficiently long wavelengths, sensitizers are not required in these reactions. In general, photocycloaddition is brought about by irradiation with light of wavelength greater than

[101] C. F. Thompson, T. F. Jamison and E. N. Jacobsen, *J. Am. Chem. Soc.*, **123** (2001), 9974.
[102] K. Gademann, D. E. Chavez and E. N. Jacobsen, *Angew. Chem. Int. Ed.*, **41** (2002), 3059.
[103] M. T. Crimmins and T. L. Reinhold, *Org. Reactions*, **44** (1993), 296; T. Bach, *Synthesis* (1998), 683.

300 nm, often by conducting the reaction in a pyrex vessel. In this way, the destructive effect of short wavelength irradiation is avoided. The reaction typically takes place through a triplet excited state of the enone formed by intersystem crossing from the initial $n \to \pi^*$ excited singlet.[104] Combination of the enone in its excited triplet state with an alkene leads to an intermediate 1,4-biradical, from which the cyclobutane is formed.

Both inter- and intramolecular [2+2] cycloaddition reactions have been used in synthesis. In intermolecular reactions, a common problem is that mixtures of regioisomers (sometimes referred to as head-to-tail and head-to-head products) in addition to more than one stereoisomer may be formed. In general, the head-to-tail regioisomer is the major product using an electron-rich alkene, whereas the head-to-head regioisomer is favoured using an electron-deficient alkene. For example, the first step in Corey's synthesis of caryophyllene involved addition of cyclohexenone to isobutene to give predominantly the *trans*-cyclobutane (head-to-tail) derivative **151** (3.107).[105]

$$(3.107)$$

151 26.5%   6.5%   6%

Cycloaddition of an enone with a cyclic alkene, such as **152**, can occur with good stereoselectivity in favour of the thermodynamically more stable *exo* diastereoisomer (**153** in this case) (3.108).[106] Cycloaddition of cyclohexenone and Z-but-2-ene or E-but-2-ene gave the same mixture of addition products in each case, suggesting that the reactions proceed in a stepwise manner through radical intermediates. Alkynes also add to enones on irradiation to form cyclobutenes.

$$(3.108)$$

152   **153**

Intramolecular cycloaddition reactions are a powerful strategy for the formation of bicyclic and polycyclic compounds. For example, two rings and three contiguous

[104] D. I. Schuster, G. Lem and N. A. Kaprinidis, *Chem. Rev.*, **93** (1993), 3.
[105] E. J. Corey, R. B. Mitra and H. Uda, *J. Am. Chem. Soc.*, **86** (1964), 485; E. J. Corey, J. D. Bass, R. LeMahieu and R. B. Mitra, *J. Am. Chem. Soc.*, **86** (1964), 5570.
[106] P. A. Wender and J. C. Lechleiter, *J. Am. Chem. Soc.*, **100** (1978), 4321.

quaternary chiral centres are formed on intramolecular cycloaddition of the enone **154** to give **155**, used in a synthesis of the sesquiterpene isocomene (3.109).[107] The conformational constraints of intramolecular reactions can enhance or reverse the regioselectivity of the cycloaddition in comparison with the intermolecular reaction. Thus, cycloaddition of the enone **156** gave the regioisomer **157** (3.110).[108] In these cases good diastereoselectivity is achieved, with control of the stereochemistry at the three new chiral centres by the existing chiral centre.

(3.109)

(3.110)

The synthetic usefulness of these cycloaddition reactions extends beyond the immediate formation of cyclobutane derivatives. Rearrangements or ring-opening reactions, encouraged by the relief of strain in the cyclobutane ring, can be used to construct complex ring systems. Photocycloaddition of an alkene to the eno-lized form of a 1,3-dicarbonyl compound results in the formation of a β-hydroxy-carbonyl compound, which can undergo retro-aldol reaction, with ring-opening, to give a 1,5-dicarbonyl product.[109] Thus, irradiation of a solution of 2,4-pentanedione (acetylacetone) in cyclohexene gave the 1,5-diketone **159** by spontaneous retro-aldol reaction of the intermediate β-hydroxy-ketone **158** (3.111).

[107] M. C. Pirrung, *J. Am. Chem. Soc.*, **103** (1981), 82.
[108] S. Faure, S. Piva-Le-Blanc, C. Bertrand, J.-P. Pete, R. Faure and O. Piva, *J. Org. Chem.*, **67** (2002), 1061.
[109] P. de Mayo, *Acc. Chem. Res.*, **4** (1971), 41; W. Oppolzer, *Acc. Chem. Res.*, **15** (1982), 135; J. D. Winkler, C. M. Bowes and F. Liotta, *Chem. Rev.*, **95** (1995), 2003.

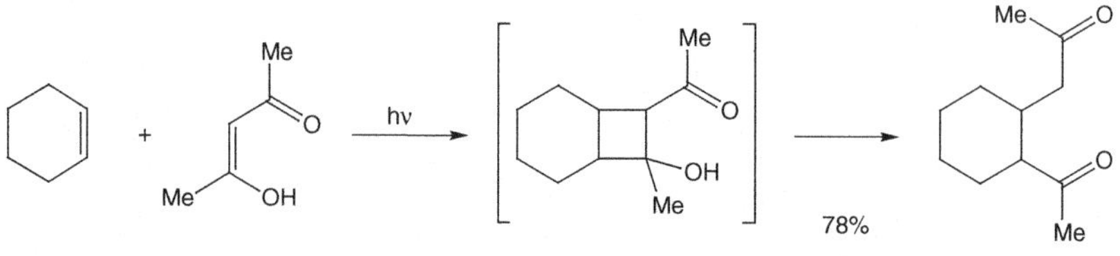

(3.111)

158   **159**

Cycloaddition using an enol derivative of a cyclic 1,3-dicarbonyl compound, followed by retro-aldol reaction, results in ring expansion by two carbon atoms. Reactions of this kind have been applied to the synthesis of a number of polycyclic natural products. For example, irradiation of the enol benzoate **160** gave the tricyclic product **161** in almost quantitative yield (3.112).[110] Dimethylation followed by hydrolysis and retro-aldol reaction gave the eight-membered ring diketone **162**, used in a synthesis of the sesquiterpene *epi*-precapnelladiene.

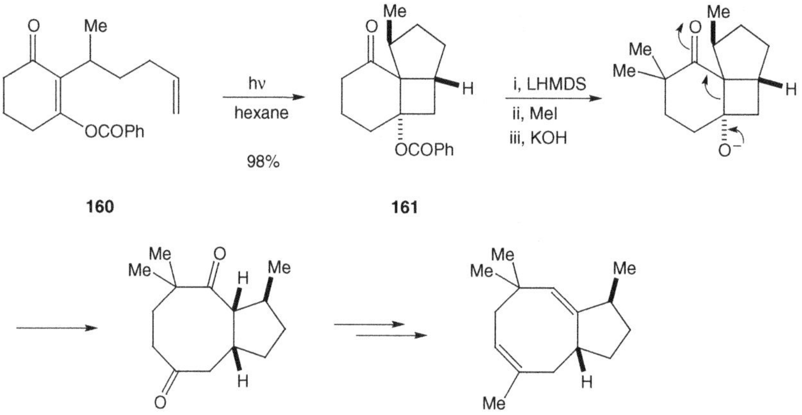

(3.112)

160   **161**

162   *epi*-precapnelladiene

Ring-opening is not restricted to the retro-aldol reaction. Conversion of the ketone product to the epoxide, the alcohol or its sulfonate or other leaving group have been used to promote fragmentation of the four-membered ring to provide novel products. Photocycloaddition of an enamine can be used as an alternative to the enol cycloaddition. Subsequent ring-opening of the cyclobutane occurs by a retro-Mannich fragmentation. Thus, in a synthesis of the alkaloid manzamine A, cycloaddition of the enamine **163** led to the cyclobutane **164**, which fragmented and cyclized to give the aminal **165** (3.113).[111] Isomerization of the aminal **165** gave the ABCE ring system **166** of manzamine A.

[110] A. M. Birch and G. Pattenden, *J. Chem. Soc., Chem. Commun.* (1980), 1195.
[111] J. D. Winkler and J. M. Axten, *J. Am. Chem. Soc.*, **120** (1998), 6425.

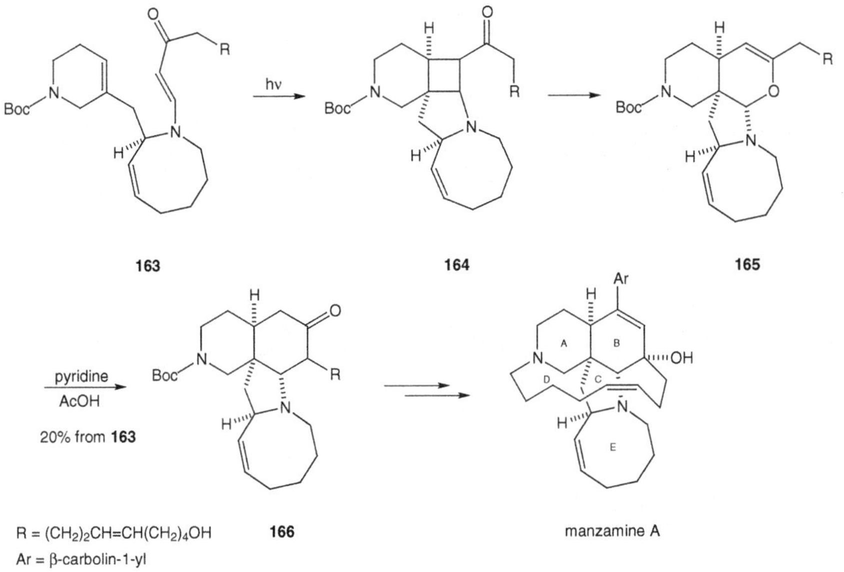

163 **164** **165**

pyridine
AcOH

20% from **163**

166 manzamine A

R = (CH₂)₂CH=CH(CH₂)₄OH
Ar = β-carbolin-1-yl

$R = (CH_2)_2CH=CH(CH_2)_4OH$
$Ar = \beta\text{-carbolin-1-yl}$

If an aldehyde or ketone π-system replaces one of the alkene units, then photo-chemical [2+2] cycloaddition is termed the Paternò–Büchi reaction and an oxetane product is produced.[112] Typically a mixture of stereo- and regioisomers of the oxetane is formed in intermolecular cycloadditions of this type, although the use of small-ring alkenes favours the *cis*-fused ring products. Thus, in a synthesis of the antifungal agent (+)-preussin, cycloaddition of benzaldehyde with the dihydropyr-role **167** led to the *cis*-fused products **168** and **169** (3.114). Hydrogenolysis of the benzylic C–O bond and reduction of the carbamate of the diastereomer **168** gave the target preussin.

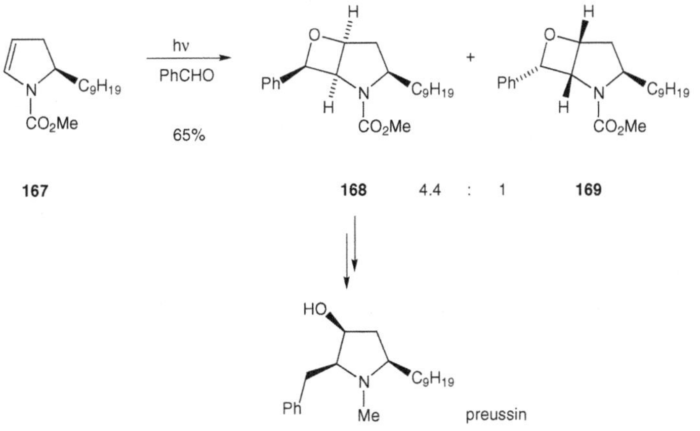

167 **168** 4.4 : 1 **169**

preussin

112 J. A. Porco and S. L. Schreiber, in *Comprehensive Organic Synthesis*, ed. B. M. Trost and I. Fleming, vol. 5 (Oxford: Pergamon Press, 1991), p. 151; T. Bach, *Synlett* (2000), 1699.

High selectivity has been obtained in the Paternò–Büchi reaction of aldehydes with furans and this reaction has formed the basis of a number of total syntheses. Thus, photocycloaddition of furan with nonanal gave the *exo* product **170**, which was converted to the antifungal metabolite avenaciolide (3.115).

(3.115)

In contrast with the photochemical cycloaddition reaction of two alkenes, the [2+2] cycloaddition of a ketene and an alkene occurs under thermal conditions.[113] The ketene is formed typically from an acid chloride and a mild base such as Et_3N, or from an α-halo-acid chloride and zinc. Cycloaddition with an alkene occurs stereospecifically, such that the geometry of the alkene is maintained in the cyclobutanone product. The regioselectivity is governed by the polarization of the alkene, with the more electron-rich end of the alkene forming a bond to the electron-deficient central carbon atom of the ketene. Thus, the product from cycloaddition of dimethylketene with the enol ether *Z*-**171** is the cyclobutanone *cis*-**172**, whereas with *E*-**171**, the isomer *trans*-**172** is formed (3.116).[114]

(3.116)

The high level of stereospecificity in cycloaddition reactions with ketenes points to a concerted mechanism in which both carbon–carbon bonds are formed simultaneously (although not necessarily at the same rate). Orbital symmetry considerations predict that the thermal [2+2] cycloaddition reaction is disallowed, however,

[113] J. A. Hyatt and P. W. Raynolds, *Org. Reactions*, **45** (1994), 159; T. T. Tidwell, *Ketenes* (New York: Wiley, 1995).

[114] R. Huisgen, L. A. Feiler and G. Binsch, *Chem. Ber.*, **102** (1969), 3460.

the presence of two orthogonal π-bonds in the LUMO of the ketene can be invoked to permit the approach of the HOMO of the alkene at an angle to the ketene.

Cycloaddition reactions of ketenes with dienes are rapid and normally lead to the cyclobutanone product, rather than the Diels–Alder adduct. There is evidence that the cyclobutanone forms by a two-step process, involving an initial [4+2] cycloaddition of the diene and the ketene carbonyl group, followed by a [3,3]-Claisen rearrangement.[115]

Dichloroketene is particularly reactive, and reductive dechlorination of the product with zinc and acetic acid allows access to the cyclobutanone from formal addition of ketene itself.[116] Thus, cycloaddition of dichloroketene with cyclopentadiene, followed by dechlorination and Baeyer–Villiger oxidation gave the lactone **173**, a useful precursor to various oxygenated cyclopentane products (3.117). Intramolecular cycloaddition reactions of ketenes can allow the formation of bicyclic and polycyclic products using otherwise unstable ketene intermediates.[117]

The synthesis of cyclobutanones can in some cases be accomplished more efficiently by addition of a ketene-iminium salt or a chromium carbene to an alkene.[118] For example, under photochemical conditions, the chromium carbene **174** gave the cyclobutanone **175** as a single diastereomer (3.118). The product **175** was converted to the antifungal antibiotic (+)-cerulenin by way of the lactone **176**.

R = CH₂CH₂C≡CSi^{i}Pr₃
R' = CH₂CH₂C≡CH

[115] S. Yamabe, T. Dai, T. Minato, T. Machiguchi and T. Hasegawa, *J. Am. Chem. Soc.*, **118** (1996), 6518.
[116] W. T. Brady, *Tetrahedron*, **37** (1981), 2949.
[117] B. B. Snider, *Chem. Rev.*, **88** (1988), 793.
[118] L. S. Hegedus, *Tetrahedron*, **53** (1997), 4105.

Silylketenes are more stable than the parent ketenes and have been used in cycloaddition reactions, particularly with aldehydes to form β-lactones.[119] Cycloaddition of a ketene with an aldehyde normally requires a Lewis acid catalyst, such as $AlCl_3$, $BF_3 \cdot OEt_2$ or $ZnCl_2$ and under such conditions good yields of the β-lactone products can be obtained.[113] For example, cycloaddition of the aldehyde **177** with the ketene **178** using the Lewis acid $EtAlCl_2$ gave the β-lactone **179**, used in a synthesis of lipstatin (3.119). The need for a Lewis acid has prompted cycloadditions in the presence of catalysts bearing a chiral ligand and some high levels of enantioselectivity have been reported.[120] An alternative is to use a chiral tertiary amine to catalyse the reaction, and cinchona alkaloids such as quinidine have given excellent levels of enantioselectivity.[121]

$$ (3.119) $$

177   **178**   **179**   (9:1 1,3-diastereoselectivity)

The addition of an acid chloride to an imine is an important method for the preparation of β-lactams and is often referred to as the Staudinger reaction.[122] The reaction allows a convenient and mild approach to the β-lactam antibiotics and has therefore received considerable attention. Good stereoselectivity in favour of the *cis* 3,4-disubstituted product is common. For example, the β-lactam **182** was formed in reasonable yield by condensation of the acid chloride **180** and the imine **181** (3.120). The reaction is not thought to be a concerted cycloaddition with the ketene, but to take place via a zwitterionic intermediate. Almost complete asymmetric induction in the synthesis of β-lactams by the Staudinger reaction using a chiral auxiliary or a chiral tertiary amine, such as benzoylquinine, has been reported.[123]

[119] A. Pommier, P. Kocienski and J.-M. Pons, *J. Chem. Soc., Perkin Trans. 1* (1998), 2105.

[120] S. G. Nelson, T. J. Peelen and Z. Wan, *J. Am. Chem. Soc.*, **121** (1999), 9742.

[121] H. Wynberg and E. Staring, *J. Org. Chem.*, **50** (1985), 1977.

[122] L. Ghosez and J. Marchand-Brynaert, in *Comprehensive Organic Synthesis*, ed. B. M. Trost and I. Fleming, vol. 5 (Oxford: Pergamon Press, 1991), p. 85; F. H. van der Steen and G. van Koten, *Tetrahedron*, **47** (1991), 7503.

[123] C. Palomo, J. M. Aizpurua, I. Ganboa and M. Oiarbide, *Eur. J. Org. Chem.* (1999), 3223; I. Ojima and F. Delaloge, *Chem. Soc. Rev.*, **26** (1997), 377; A. E. Taggi, A. M. Hafez, H. Wack, B. Young, D. Ferraris and T. Lectka, *J. Am. Chem. Soc.*, **124** (2002), 6626; B. L. Hodous and G. C. Fu, *J. Am. Chem. Soc.*, **124** (2002), 1578.

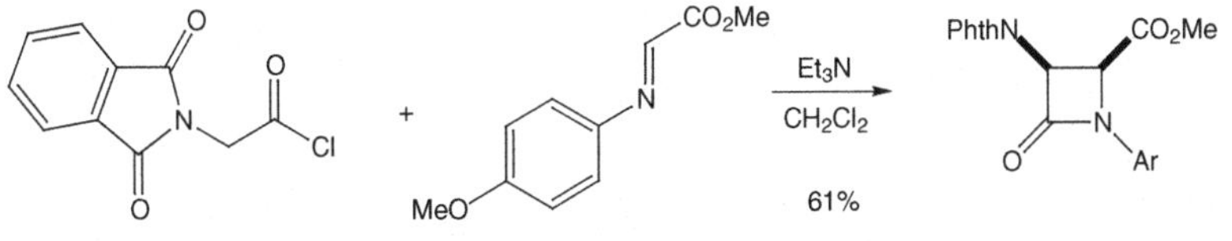

$$(3.120)$$

180 **181** **182**

An alternative approach to the β-lactam ring system uses the cycloaddition of an alkene with an isocyanate such as chlorosulfonyl isocyanate (O=C=N−SO$_2$Cl).[124] For example, reaction of cyclopentadiene with chlorosulfonyl isocyanate gave the β-lactam **183** (3.121). The *N*-unsubstituted β-lactam is formed under these conditions owing to the ease of removal of the SO$_2$Cl group. The regioselectivity can be explained by combination of the more electron-rich end of the alkene with the electron-deficient carbon atom of the isocyanate.

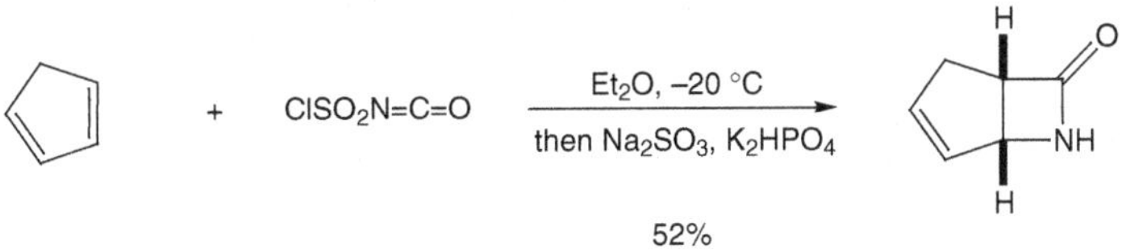

$$(3.121)$$

183

3.3 Cycloaddition reactions with allyl anions and allyl cations

The Diels–Alder reactions discussed in Section 3.1 are concerted [4+2] cycloadditions involving six π-electrons to give six-membered rings. The possibility of analogous six π-electron cycloadditions involving allyl anions and allyl cations to give five- and seven-membered rings respectively is predicted by the Woodward–Hoffmann rules (3.122).[4] Examples of both processes have been observed, although the synthetic scope of the reactions, particularly with allyl anions, is limited.

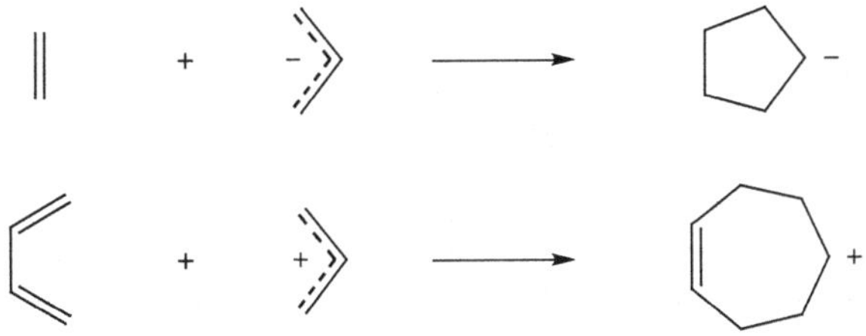

$$(3.122)$$

Allyl anions tend to react as nucleophilic carbanions with various electrophiles, rather than undergo cycloaddition reactions, and the few known cycloaddition

[124] J. K. Rasmussen and A. Hassner, *Chem. Rev.*, **76** (1976), 389. See also G. S. Singh, *Tetrahedron*, **59** (2003), 7631.

reactions with alkenes are confined to allyl anions bearing an electron-withdrawing group at C-2 of the allyl unit. These, and the better-known examples of cycloaddition reactions of 2-aza-allyl anions, may be stepwise rather than concerted processes.[125] The 2-aza-allyl anion adds to a variety of alkenes to form pyrrolidine derivatives.[126] Thus, proton abstraction at the methyl group of the imine **184** gave an intermediate 2-aza-allyl anion, which undergoes cycloaddition to give the 2,3-diaryl-pyrrolidine **185** (3.123). An alternative method, which also allows the formation of unstabilized 2-aza-allyl anions, makes use of tin–lithium exchange and has been used in inter- and intramolecular cycloaddition reactions. For example, the key step in a synthesis of the alkaloid (−)-amabiline involved transmetallation and subsequent intramolecular cycloaddition of the stannane **186** (3.124).[127]

(3.123)

184   85%   **185**   *trans : cis* 95 : 5

(3.124)

186   74%   5 : 1

Cycloaddition of allyl cations to conjugated dienes provides a route to seven-membered carbocycles.[128] One of several methods can be used to generate the allyl cation, such as from an allyl halide and silver trifluoroacetate, or from an allyl alcohol by way of its trifluoroacetate or sulfonate. Cycloaddition of the allyl cation proceeds best with a cyclic diene, particularly for intermolecular reactions. Thus, cyclohexadiene and methylallyl cation gave the bicyclo[3.2.2]nonadiene **187** (3.125). Many intramolecular examples are known,[129] such as the formation of the

[125] F. Neumann, C. Lambert and P. v. R. Schleyer, *J. Am. Chem. Soc.*, **120** (1998), 3357.
[126] T. Kauffmann, *Angew. Chem. Int. Ed.*, **13** (1974), 627.
[127] W. H. Pearson and F. E. Lovering, *J. Am. Chem. Soc.*, **117** (1995), 12 336.
[128] H. M. R. Hoffmann, *Angew. Chem. Int. Ed. Engl.*, **23** (1984), 1; A. Hosomi and Y. Tominaga, in *Comprehensive Organic Synthesis*, ed. B. M. Trost and I. Fleming, vol. 5 (Oxford: Pergamon Press, 1991), p. 593; M. Harmata, *Tetrahedron*, **53** (1997), 6235; J. H. Rigby and F. C. Pigge, *Org. Reactions*, **51** (1997), 351.
[129] M. Harmata, *Acc. Chem. Res.*, **34** (2001), 595.

cycloadducts **189** and **190** from the allyl alcohol **188** and triflic anhydride, Tf_2O [$(CF_3SO_2)_2O$] (3.126). In this latter sequence the trimethylsilyl group serves both to stabilize the allyl cation and as a trigger for the subsequent formation of the exocyclic methylene group.

$$(3.125)$$

$$(3.126)$$

The 2-oxyallyl cation has found a number of applications in organic synthesis. These species can be produced from α,α'-dibromoketones, from α-halo-trialkylsilyl enol ethers or from allyl sulfones and a Lewis acid. For example, the 2-oxyallyl cation **192** can be prepared from the dibromide **191** and its cycloaddition with furan gave the adduct **193**, used in a synthesis of nonactic acid (3.127). These reactions may take a concerted or a stepwise course, depending on the nature of the diene and the allyl cation and the reaction conditions.

$$(3.127)$$

A convenient method for the formation of 2-oxyallyl cations has been developed from ketones and applied to both inter- and intramolecular cycloaddition reactions. The ketone is converted to the intermediate α-chloroketone, from which the 2-oxyallyl cation is formed with lithium perchlorate. For example, intramolecular cycloaddition of the 2-oxyallyl cation generated from the cyclic ketone **194**, gave the *exo* adduct **195** as the major stereoisomer (3.128).

$$\text{(3.128)}$$

58%

194   **195**

3.4 1,3-Dipolar cycloaddition reactions

The 1,3-dipolar cycloaddition reaction, like the Diels–Alder reaction, is a 6π electron pericyclic reaction, but it differs from the Diels–Alder reaction in that the 4π component, called the 1,3-dipole, is a three-atom unit containing at least one heteroatom and which is represented by a zwitterionic octet structure. The 2π component, here called the dipolarophile (rather than the dienophile), is a compound containing a double or triple bond. The product of the reaction is a five-membered heterocyclic compound.

A typical example is the well-known reaction between ozone (the 1,3-dipole) and an alkene to give an ozonide, formed by rearrangement of the initially formed cycloadduct (3.129). The ozonide can be broken down under reducing or oxidizing conditions to give two carbonyl compounds (see Section 5.4).

$$\text{(3.129)}$$

ozonide

A considerable number of 1,3-dipoles containing various combinations of carbons and heteroatoms is possible and many of these have been made and their reactions with dipolarophiles studied. All 1,3-dipoles contain 4π electrons in three parallel p-orbitals and some of the more commonly encountered classes are shown in Figure 3.130.[130]

[130] R. Huisgen, *J. Org. Chem.*, **41** (1976), 403; R. Huisgen, in *1,3-Dipolar Cycloaddition Chemistry*, ed. A. Padwa, vol. 1 (New York: Wiley, 1984), p. 1; A. Padwa, in *Comprehensive Organic Synthesis*, ed. B. M. Trost and I. Fleming, vol. 4 (Oxford: Pergamon Press, 1991), p. 1069; P. A. Wade, in *Comprehensive Organic Synthesis*, ed. B. M. Trost and I. Fleming, vol. 4 (Oxford: Pergamon Press, 1991), p. 1111; *Synthetic Applications of 1,3-Dipolar Cycloaddition Chemistry Toward Heterocycles and Natural Products*, ed. A. Padwa and W. H. Pearson (New Jersey: Wiley, 2003).

$$
\underset{\substack{\text{azides}}}{\overset{\underset{\displaystyle R}{\overset{+}{\text{N}}}}{\text{N}}=\text{N}=\overset{-}{\text{N}}}
\quad
\underset{\substack{\text{diazoalkanes}}}{\overset{\underset{\displaystyle R}{\overset{+}{}}}{}=\text{N}=\overset{-}{\text{N}}}
\quad
\underset{\substack{\text{nitrile oxides}}}{R\overset{+}{=\!\!\!\equiv\!\!\!=}\text{N}\!-\!\overset{-}{\text{O}}}
\quad
\underset{\substack{\text{nitrones}}}{}
\quad
\underset{\substack{\text{azomethine ylides}}}{}
\tag{3.130}
$$

Some 1,3-dipoles, such as azides and diazoalkanes, are relatively stable, isolable compounds; however, most are prepared *in situ* in the presence of the dipolarophile. Cycloaddition is thought to occur by a concerted process, because the stereochemistry (*E* or *Z*) of the alkene dipolarophile is maintained (*trans* or *cis*) in the cycloadduct (a stereospecific aspect). Unlike many other pericyclic reactions, the regio- and stereoselectivities of 1,3-dipolar cycloaddition reactions, although often very good, can vary considerably; both steric and electronic factors influence the selectivity and it is difficult to make predictions using frontier orbital theory.

Nitrile oxides are conveniently generated *in situ* by dehydration of primary nitro compounds (with phenylisocyanate or ethyl chloroformate or di-tert-butyl dicarbonate) or from α-chloro-oximes (by treatment with a base). The nitrile oxide reacts with an alkene to form an isoxazoline or with an alkyne to give a heteroaromatic isoxazole (3.131).[131] Nitrile oxides are prone to undergo dimerization, although this can be minimized by maintaining a low concentration of the dipole in the presence of the dipolarophile.

$$
\tag{3.131}
$$

The product isoxazolines formed in these reactions are useful synthetic intermediates, for on reductive cleavage (with, for example, H_2, Pd/C or $LiAlH_4$) they give 1,3-amino-alcohols and on hydrolytic reduction (with, for example, Raney nickel in aqueous acid) they give β-hydroxy ketones (which can be dehydrated to α,β-unsaturated ketones) (3.132). For example, in a synthesis of the amino sugar D-lividosamine, the major isoxazoline cycloadduct **196** (formed with the expected 5-substituted regiochemistry and with high stereoselectivity) was reduced with

[131] P. Caramella and P. Grünanger, in *1,3-Dipolar Cycloaddition Chemistry*, ed. A. Padwa, vol. 1 (New York: Wiley, 1984), p. 291; C. J. Easton, C. M. M. Hughes, G. P. Savage and G. W. Simpson, *Adv. Het. Chem.*, **60** (1994), 261.

LiAlH$_4$ to give a mixture ($\sim$4:1) in favour of the *syn* amino alcohol **197** (3.133).[132]
Hydrolysis of the acetal protecting groups led to the desired glycoside.

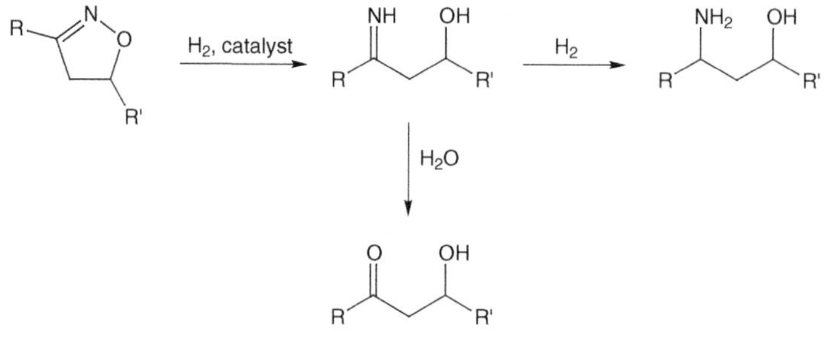

(3.132)

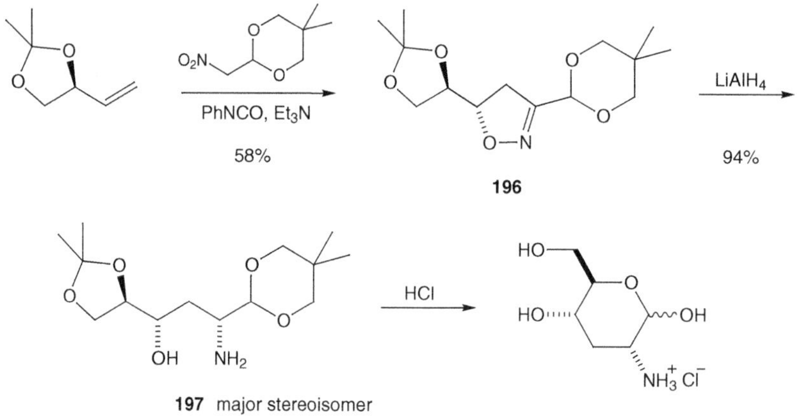

(3.133)

197 major stereoisomer

   Intramolecular 1,3-dipolar cycloaddition reactions take place readily and some
useful applications of this chemistry in synthesis have been reported. This is illustrated in the synthesis of the antitumor agents sarcomycin **201**, R=H (3.134) and
calicheamicin (3.135). In the former synthesis, dehydration of the nitro-alkene
198 gave the isoxazoline **199** *via* the intermediate nitrile oxide. Hydrogenolysis
with Raney nickel in aqueous acetic acid then led to the β-hydroxy ketone **200**,
which was dehydrated to the ethyl ester of sarcomycin **201**, R=Et.[133] In a synthesis of the much more complex calicheamicin, intramolecular cycloaddition of the
nitrile oxide, generated from the oxime **202** by *in situ* chlorination with sodium
hypochlorite and elimination, gave the two isoxazolines **203** and **204**. The isoxazoline **203** was later oxidized to the isoxazole, which was ring-opened (using aqueous
Mo(CO)$_6$) to give an amino aldehyde, used in a second ring-forming step to prepare
the required enediyne aglycon of calicheamicin.[134]

[132] V. Jäger and R. Schohe, *Tetrahedron*, **40** (1984), 2199.
[133] A. P. Kozikowski and P. D. Stein, *J. Am. Chem. Soc.*, **104** (1982), 4023.
[134] A. L. Smith, E. N. Pitsinos, C.-K. Hwang, Y. Mizuno, H. Saimoto, G. R. Scarlato, T. Suzuki and K. C. Nicolaou,
J. Am. Chem. Soc., **115** (1993), 7612.

$$(3.134)$$

198   199   200   201

$$(3.135)$$

202   203   3.6 : 1   204

1,3-Dipolar cycloaddition reactions occur readily even with 'non-activated' dipolarophiles, such as isolated alkenes. This contrasts with the Diels–Alder reaction, particularly for intermolecular reactions, in which an 'activated' alkene as the dienophile is required. Like the Diels–Alder reaction, [3+2] cycloaddition reactions of 1,3-dipoles are reversible, although in most cases it is the kinetic product that is isolated. For the intermolecular cycloaddition of nitrile oxides or nitrones, two of the most frequently used 1,3-dipoles, to monosubstituted or 1,1-disubstituted alkenes (except highly electron-deficient alkenes), the oxygen atom of the 1,3-dipole becomes attached to the more highly substituted carbon atom of the alkene double bond. Hence the 5-substituted isoxazolidine **206** is generated from the cycloaddition of the cyclic nitrone **205** with propene (3.136).[135] Reductive cleavage of the cycloadduct then gave the alkaloid sedridine. In this cycloaddition reaction the '*exo*' product is favoured.

$$(3.136)$$

205   206   sedridine

A related example, using the monosubstituted cyclic nitrone **207**, gave the cycloadduct **208** (3.137).[136] This cycloaddition reaction is highly stereoselective for the *trans*-2,6-disubstituted piperidine. Reduction of the N–O bond and dehydroxylation then led to the ant-venom constituent solenopsin. Cycloaddition of cyclic or acyclic nitrones with alkenes or alkynes leads to isoxazolidines or isoxazolines

[135] J. J. Tufariello and S. A. Ali, *Tetrahedron Lett.* (1978), 4647.
[136] D. A. Adams, W. Carruthers, P. J. Crowley and M. J. Williams, *J. Chem. Soc., Perkin Trans. 1* (1989), 1507; S. Chackalamannil and Y. Wang, *Tetrahedron*, **53** (1997), 11 203.

respectively,[137] both useful ring systems for further transformations. In particular, many alkaloid ring systems can be accessed efficiently by using a 1,3-dipolar cycloaddition reaction as the key step.[138]

(3.137)

207   **208**   solenopsin

Recent advances have been made in the enantioselective cycloaddition of nitrones and alkenes.[139] By using a chiral auxiliary attached to the nitrone or the alkene, moderate to good levels of asymmetric induction have been reported. A number of metal complexes with chiral ligands catalyse the cycloaddition reaction of nitrones, particularly for dipolarophiles containing two carbonyl groups for bidentate co-ordination to the metal.[140] An alternative approach, using α,β-unsaturated aldehydes and chiral secondary amines has been successful (3.138).[141] The *endo* product is the major stereoisomer in these cycloaddition reactions and the catalysis is thought to proceed *via* the reactive intermediate iminium ion **210**, with addition of the nitrone to the face of the alkene opposite the benzyl substituent.

(3.138)

98%   94% ee   94 : 6

209   **210**

[137] P. N. Confalone and E. M. Huie, *Org. Reactions*, **36** (1988), 1.
[138] G. Broggini and G. Zecchi, *Synthesis* (1999), 905.
[139] K. V. Gothelf and K. A. Jørgensen, *Chem. Commun.* (2000), 1449; M. Frederickson, *Tetrahedron*, **53** (1997), 403. For a review on asymmetric 1,3-dipolar cycloaddition reactions, see K. V. Gothelf and K. A. Jørgensen, *Chem. Rev.*, **98** (1998), 863.
[140] S. Kanemasa, Y. Oderaotoshi, J. Tanaka and E. Wada, *J. Am. Chem. Soc.*, **120** (1998), 12 355; S. Kanemasa, *Synlett* (2002), 1371.
[141] W. S. Jen, J. J. M. Wiener and D. W. C. MacMillan, *J. Am. Chem. Soc.*, **122** (2000), 9874.

Intramolecular cycloaddition reactions of nitrones have been used widely in synthesis. The required unsaturated nitrones can be obtained by oxidation of *N*-alkenyl-hydroxylamines or by condensation of an aldehyde with an *N*-substituted hydroxylamine. Thus the *cis* bicyclic isoxazolidine **212** was obtained by reaction of 5-heptenal with *N*-methylhydroxylamine, by way of the intermediate nitrone **211** (3.139).

$$(3.139)$$

211   **212**

Tethering the alkene to the carbon atom of the nitrone allows the preparation of *cis*-1,2-disubstituted cycloalkanes such as **212**. Examples in which the alkene is tethered to the nitrogen atom of the nitrone are also common. Thus, addition of formaldehyde to the hydroxylamine **213** promoted formation of the intermediate nitrone and hence the cycloadduct **214** (3.140).[142] Subsequent transformations led to the alkaloid luciduline. This synthesis illustrates a useful feature of the 1,3-dipolar cycloaddition reaction of nitrones, in that it provides an alternative to the Mannich reaction as a route to β-amino-ketones, via reductive cleavage of the N–O bond in the isoxazolidine and oxidation of the 1,3-amino-alcohol product. In another example of such an intramolecular cycloaddition reaction, the bridged bicyclic product **217**, used in a synthesis of indolizidine 209B, was formed by addition of an aldehyde to the hydroxylamine **215**, followed by heating the intermediate nitrone **216** (3.141).[143]

$$(3.140)$$

213   **214**   luciduline

[142] W. Oppolzer and M. Petrzilka, *Helv. Chim. Acta*, **61** (1978), 2755; E. G. Baggiolini, H. L. Lee, G. Pizzolato and M. R. Uskovic, *J. Am. Chem. Soc.*, **104** (1982), 6460.

[143] A. B. Holmes, A. L. Smith, S. F. Williams, L. R. Hughes, Z. Lidert and C. Swithenbank, *J. Org. Chem.*, **56** (1991), 1393. See also, C.-H. Tan and A. B. Holmes, *Chem. Eur. J.*, **7** (2001), 1845; E. C. Davison, M. E. Fox, A. B. Holmes, S. D. Roughley, C. J. Smith, G. M. Williams, J. E. Davies, P. R. Raithby, J. P. Adams, I. T. Forbes, N. J. Press and M. J. Thompson, *J. Chem. Soc., Perkin Trans. 1* (2002), 1494.

215   R = (CH₂)₃OAc   216   217

Another type of dipolar cycloaddition reaction occurs between an azomethine ylide 1,3-dipole and an alkene or alkyne dipolarophile. Cycloaddition leads to a pyrrolidine or dihydropyrrole ring respectively and the latter can be converted easily to a pyrrole ring. The reaction is valuable for organic synthesis as these ring systems are common in natural products and biologically active compounds. The azomethine ylide is generated *in situ* and not isolated. Most common are reactions of stabilized azomethine ylides, in which an electron-withdrawing group is present on one of the carbon atoms of the 1,3-dipole. These dipoles can be prepared by a number of methods. For example, thermal ring-opening of an aziridine (by a conrotatory pathway) generates an azomethine ylide. In the presence of a dipolarophile, cycloaddition of the azomethine ylide occurs stereospecifically.[144] Good levels of regioselectivity are also obtained. Thus, cycloaddition of the azomethine ylide **219**, generated from the aziridine **218**, gave predominantly the pyrrolidine **220** (3.142).[145]

218   219   220

Intramolecular 1,3-dipolar cycloaddition reactions normally proceed efficiently to give bicyclic products and these reactions do not require the presence of an electron-withdrawing group on the dipolarophile. Thus, in an approach to the alkaloid sarain A, the aziridine **221** was heated to generate the intermediate azomethine ylide **222**, and hence, after intramolecular cycloaddition, the pyrrolidine **223** (3.143).[146] An alternative method for the formation of the required azomethine ylide, and which avoids the need for the prior synthesis of an aziridine ring, uses the simple condensation of an aldehyde and a primary or secondary amine. Thus, in another approach to sarain A, addition of formaldehyde to the amine **224** resulted in the formation of the cycloadduct **226** (3.144).[147] Notice that in both cases the

[144] R. Huisgen, W. Scheer and H. Huber, *J. Am. Chem. Soc.*, **89** (1967), 1753.
[145] P. DeShong, D. A. Kell and D. R. Sidler, *J. Org. Chem.*, **50** (1985), 2309; P. DeShong and D. A. Kell, *Tetrahedron Lett.*, **27** (1986), 3979.
[146] O. Irie, K. Samizu, J. R. Henry and S. M. Weinreb, *J. Org. Chem.*, **64** (1999), 587.
[147] D. J. Denhart, D. A. Griffith and C. H. Heathcock, *J. Org. Chem.*, **63** (1998), 9616.

stereochemistry of the alkene (Z) determines the relative stereochemistry of the product.

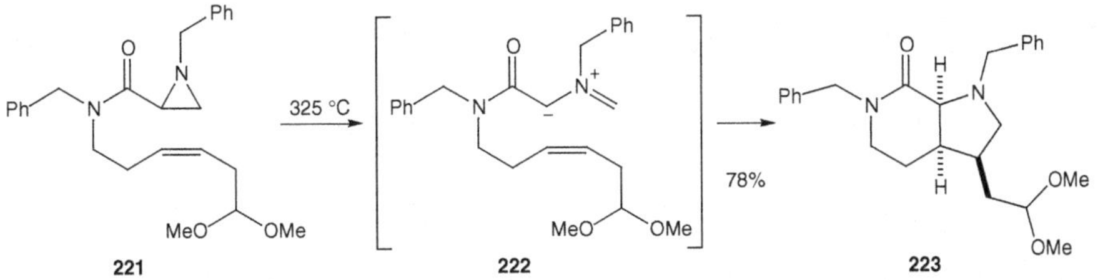

(3.143)

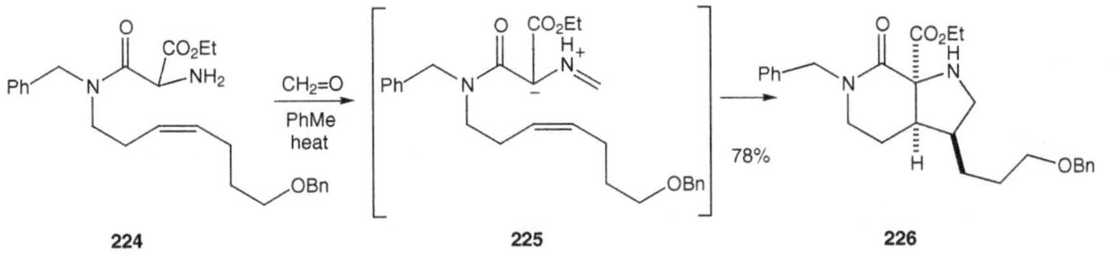

(3.144)

Addition of an aldehyde to an alkenyl-amine can provide the required alkenyl-azomethine ylide (such as **225**) for intramolecular cycloaddition. An alternative is the addition of an amine to an alkenyl-aldehyde. Thus, addition of *N*-methyl glycine ethyl ester to the aldehyde **227** gave the intermediate azomethine ylide **228** and hence the cycloadduct **229** (3.145).[148]

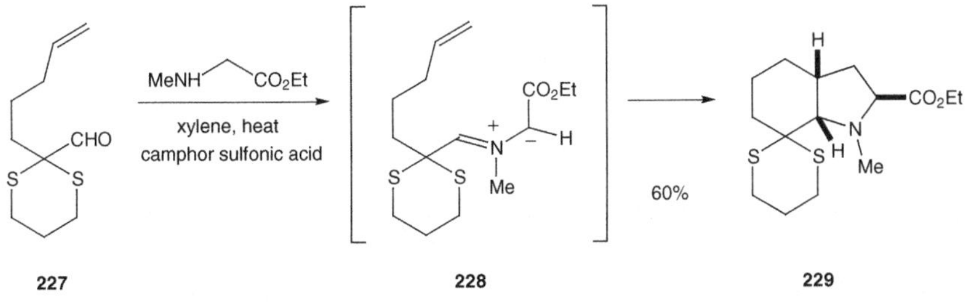

The condensation of a secondary amine and an aldehyde leads to an iminium ion, which can undergo deprotonation to give the required azomethine ylide. If a primary amine is used then an imine rather than an iminium ion is generated. Imines can be converted to azomethine ylides by heating or by addition of a metal salt, such as silver acetate or lithium bromide, in the presence of a base, such as

[148] P. N. Confalone and R. A. Earl, *Tetrahedron Lett.*, **27** (1986), 2695; I. Coldham, K. M. Crapnell, J. D. Moseley and R. Rabot, *J. Chem. Soc., Perkin Trans. 1* (2001), 1758.

triethylamine. In such cases the pyrrolidine product, obtained on cycloaddition with an alkene dipolarophile, is unsubstituted on the nitrogen atom and this is convenient for further functionalization as desired. Thus, formation of the azomethine ylide **231** from the imine **230** in the presence of methyl acrylate resulted in the formation of the pyrrolidine **232** (3.146).[149]

(3.146)

The azomethine ylides described above all bear an ester or other electron-withdrawing group in order to promote ylide formation. Non-stabilized azomethine ylides are less common, but can be produced, for example by decarboxylation or desilylation. The parent azomethine ylide **233** can be conveniently prepared by desilylation of an α-amino-silane, in which a leaving group (alkoxide, cyanide, benzotriazole, etc.) is present at an α′ position (3.147).[150] Formation of the non-stabilized ylide **233** in the presence of an alkene or alkyne dipolarophile leads to the 2,5-unsubstituted pyrrolidine or dihydropyrrole product.

(3.147)

A type of 1,3-dipole that has received considerable recent interest is the carbonyl ylide. One method for its formation makes use of carbenoid chemistry (see Section 4.2). Cyclization of an electrophilic rhodium carbenoid onto a nearby carbonyl group provides access to the carbonyl ylide.[151] Cycloaddition with an alkyne or alkene dipolarophile then gives the dihydro- or tetrahydrofuran product. For example, the carbonyl ylide **235**, formed from the diazo compound **234** and rhodium(II) acetate, reacts with dimethyl acetylenedicarboxylate to give the bridged dihydrofuran **236** (3.148).

[149] D. A. Barr, R. Grigg, H. Q. N. Gunaratne, J. Kemp, P. McMeekin and V. Sridharan, *Tetrahedron*, **44** (1988), 557; O. Tsuge, S. Kanemasa and M. Yoshioka, *J. Org. Chem.*, **53** (1988), 1384.
[150] E. Vedejs and F. G. West, *Chem. Rev.*, **86** (1986), 941.
[151] A. Padwa and M. D. Weingarten, *Chem. Rev.*, **96** (1996), 223; G. Mehta and S. Muthusamy, *Tetrahedron*, **58** (2002), 9477.

$$\text{(3.148)}$$

234   235   236

1,3-Dipolar cycloaddition reactions are not restricted to the use of alkene or alkyne dipolarophiles. Many hetero-dipolarophiles, particularly aldehydes and imines, undergo successful 1,3-dipolar cycloaddition with a range of 1,3-dipoles. The chemistry therefore provides access to a variety of five-membered heterocyclic compounds and compounds derived therefrom. Recent developments have focused on asymmetric dipolar cycloaddition reactions in the presence of a chiral catalyst, or the application of the chemistry to the preparation of biologically active compounds.

3.5 The ene reaction

The ene reaction involves the thermal reaction of an alkene bearing an allylic hydrogen atom (the ene component) with a compound containing an activated multiple bond (the enophile). A new σ-bond is formed between the unsaturated centres with migration of the allylic hydrogen atom to the other terminus of the enophile multiple bond (3.149).[152] The ene reaction, although not strictly a cycloaddition reaction, resembles the Diels–Alder reaction in having a cyclic six-electron transition state, but with two electrons of the allylic C—H σ-bond in place of two π-electrons of the diene. As expected, therefore, the activation energy is greater than the Diels–Alder reaction and high temperatures are generally required to effect the ene reaction. However, many ene reactions are catalysed by Lewis acids and proceed under relatively mild conditions, often with improved stereoselectivity.[2,153] It is possible to invoke frontier orbital theory to explain these reactions, with the Lewis acid catalyst exerting its effect by lowering the energy of the LUMO of the enophile.

$$\text{(3.149)}$$

The best type of ene components are 1,1-disubstituted alkenes, although other substitution patterns are possible, particularly with reactive enophiles. A typical

[152] H. M. R. Hoffmann, *Angew. Chem. Int. Ed. Engl.*, **8** (1969), 556; B. B. Snider, in *Comprehensive Organic Synthesis*, ed. B. M. Trost and I. Fleming, vol. 5 (Oxford: Pergamon Press, 1991), p. 1.
[153] B. B. Snider, *Acc. Chem. Res.*, **13** (1980), 426.

example is the ene reaction of maleic anhydride with 2-methylpropene, which occurs at a temperature of 170 °C (3.150). The reaction of methyl acrylate with 2-methylpropene requires a temperature of approximately 230 °C although, in the presence of a Lewis acid such as ethylaluminium dichloride, adducts from reactive alkenes can be obtained at 25 °C (3.151). With unsymmetrical enophiles such as methyl acrylate, the Lewis acid-catalysed ene reaction is often completely regioselective, with the new σ bond between the unsaturated centres forming at the more electron-deficient β-carbon atom of the enophile and at the more electron-rich alkene carbon atom of the ene component.

(3.150)

(3.151)

In most cases, the thermal ene reaction is thought to proceed by a concerted mechanism with a cyclic transition state, although this may be unsymmetrical, with one of the new σ bonds more highly developed than the other. The ene reaction often leads to predominantly the *endo* adduct (in which there is greater overlap of the two components). Thus, different major diastereomers are formed in the reaction of maleic anhydride with *E*-2-butene and *Z*-2-butene, both of which arise from a preference for the *endo* transition state. Further evidence for the concerted nature of the reaction comes from the observation that the new C—C and C—H bonds are formed *cis* to each other. The *cis* addition is exemplified in the reaction of 1-heptene with dimethyl acetylenedicarboxylate to give the adduct **237** (3.152). In this reaction the adduct is formed with the two ester groups on the same side of the alkene such that the hydrogen atom and the allyl residue add to the same side of the triple bond of the enophile.

(3.152)

237

A concerted mechanism is not universal, however, and it appears that some catalysed reactions, particularly those in which a carbonyl compound acts as the enophile, proceed by a stepwise mechanism involving a zwitterionic intermediate. Hetero-enophiles are common substrates in the ene reaction. For example, the reaction of an aldehyde and an alkene in the presence of a Lewis acid provides a convenient route to some homoallylic alcohols.[154] Thus, using dimethylaluminium chloride as the Lewis acid, limonene reacts selectively with acetaldehyde to give the alcohol **238** (3.153). This process is thought to occur by a stepwise mechanism, with co-ordination of the Lewis acid to the carbonyl oxygen atom and initial C—C bond formation to give an intermediate tertiary carbocation which undergoes proton transfer. Likewise, other enophiles such as diethyl azodicarboxylate ($EtO_2CN=NCO_2Et$), nitroso compounds,[155] singlet oxygen[156] and selenium dioxide (see Section 6.1) react readily with a variety of alkenes. The thermal or Lewis acid-catalysed imino-ene reaction gives rise to homoallylic amines.[157] For example, the ene reaction of *N*-sulfonyl imines, such as **239**, with alkenes gives adducts that are readily converted into γ,δ-unsaturated α-amino-acids (3.154).

$$(3.153)$$

65%   **238**

$$(3.154)$$

90%

239 $Ts = p\text{-}MeC_6H_4SO_2$

Various asymmetric ene reactions have been reported and particular success has been achieved with the carbonyl ene reaction of glyoxylate esters and chiral Lewis acid catalysts.[158] For example, 2,2'-binaphthol (BINOL) complexes of titanium(IV) salts[159] or bisoxazoline complexes of copper(II) salts (such as the complex **134**)[160] display excellent enantioselection in this ene reaction. Thus, the homoallylic alcohol (*R*)-**240** was formed from methylglyoxylate with very high optical purity

[154] B. B. Snider, in *Comprehensive Organic Synthesis*, ed. B. M. Trost and I. Fleming, vol. 2 (Oxford: Pergamon Press, 1991), p. 527.
[155] W. Adam and O. Krebs, *Chem. Rev.*, **103** (2003), 4131.
[156] M. Prein and W. Adam, *Angew. Chem. Int. Ed.*, **35** (1996), 477.
[157] R. M. Borzilleri and S. M. Weinreb, *Synthesis* (1995), 347.
[158] L. C. Dias, *Curr. Org. Chem.*, **4** (2000), 305; K. Mikami and M. Shimizu, *Chem. Rev.*, **92** (1992), 1021.
[159] K. Mikami, M. Terada and T. Nakai, *J. Am. Chem. Soc.*, **112** (1990), 3949; Y. Yuan, X. Zhang and K. Ding, *Angew. Chem. Int. Ed.*, **42** (2003), 5478.
[160] D. A. Evans, S. W. Tregay, C. S. Burgey, N. A. Paras and T. Vojkovsky, *J. Am. Chem. Soc.*, **122** (2000), 7936.

using the catalyst formed from (R)-2,2′-binaphthol and $TiCl_2(Oi-Pr)_2$ (3.155). The (S)-enantiomer of binaphthol can be used to give the (S)-enantiomer of the homoallylic alcohol. The metal–chiral ligand complex is thought to co-ordinate to the aldehyde carbonyl oxygen atom (and possibly also to the ester group) of the glyoxylate ester and the resulting activated enophile is then set up for the pericyclic reaction to occur preferentially from the less-hindered face of the carbonyl group. The chemistry has been applied to the synthesis of the terpene ipsdienol (3.156).

Interestingly, the asymmetric ene reaction with glyoxylate esters displays a strong positive non-linear effect,[161] such that 2,2′-binaphthol catalyst of only about 50% ee gives rise to product homoallylic alcohols with >90% ee! The non-linear relationship between the optical purity of the catalyst and that of the product can be ascribed to aggregation of the catalyst. Thus, if in solution the complex BINOL-TiX$_2$ is in equilibrium with dimeric species containing $(R)\cdot(R)$-, $(R)\cdot(S)$- and $(S)\cdot(S)$-titanium complexes, and if the $(R)\cdot(S)$-form is more stable and less reactive, this leaves the enantiopure complex as the predominant catalytic species.

---

[161] For a review on non-linear effects in asymmetric synthesis, see C. Girard and H. B. Kagan, *Angew. Chem. Int. Ed.*, **37** (1998), 2922.

Intramolecular ene reactions take place readily, even with compounds containing a normally unreactive enophile.[162] The cyclization is particularly effective for the formation of five-membered rings, although six-membered rings can also be formed from appropriate unsaturated compounds. The geometrical constraints imposed on the transition state often result in highly selective reactions. For example, the homoallylic alcohol **242** with an axial alcohol substituent was the exclusive product from the cyclization of the unsaturated aldehyde **241** (3.157).[163] In a synthesis of the neuroexcitatory amino acid $(-)$-α-kainic acid **245**, the diene **243**, itself prepared from (S)-glutamic acid, was cyclized to the pyrrolidine derivative **244** with almost complete stereoselectivity (3.158).[164]

$$(3.157)$$

$$(3.158)$$

The scope of the ene reaction has been extended by the discovery that allylic metal reagents (e.g. metals Mg, Zn, Li, Ni, Pd, Pt) take part readily by migration of the metal atom (instead of a hydrogen atom) and formation of a new carbon–metal bond.[165] For example, addition of crotyl magnesium chloride to trimethylsilylethene gave, after protonation of the intermediate Grignard species **246**, the alkene **247** (3.159).[166] Similarly, addition of allyl zinc bromide to 1-trimethylsilyl-1-octyne gave, after addition of iodine, the diene **248** (3.160).[167] The addition of allyl zinc bromide to vinyllithium or vinylmagnesium species proceeds readily to give *gem*-dimetallic species.[168] Although these particular reactions display excellent regio- and stereoselectivity, the intermolecular metallo-ene reaction is often unselective

[162] W. Oppolzer and V. Snieckus, *Angew. Chem. Int. Ed. Engl.*, **17** (1978), 476.
[163] N. H. Andersen, S. W. Hadley, J. D. Kelly and E. R. Bacon, *J. Org. Chem.*, **50** (1985), 4144.
[164] W. Oppolzer and K. Thirring, *J. Am. Chem. Soc.*, **104** (1982), 4978.
[165] W. Oppolzer, in *Comprehensive Organic Synthesis*, ed. B. M. Trost and I. Fleming, vol. 5 (Oxford: Pergamon Press, 1991), p. 29.
[166] H. Lehmkuhl, K. Hauschild and M. Bellenbaum, *Chem. Ber.*, **117** (1984), 383.
[167] E. Negishi and J. A. Miller, *J. Am. Chem. Soc.*, **105** (1983), 6761.
[168] I. Marek and J.-F. Normant, *Chem. Rev.*, **96** (1996), 3241.

and product yields are commonly poor, except with certain (e.g. metal-substituted) alkene or alkyne enophiles.

$$(3.159)$$

246   77%   **247**

$$(3.160)$$

83%

248   Z : E 85 : 15

Intramolecular metallo-ene reactions have received more interest for the synthesis of natural products.[169] For example, in a synthesis of the sesquiterpene $\Delta^{9(12)}$-capnellene, the allylic Grignard reagent **249** was warmed to promote cyclization and the new cyclopentylmethyl Grignard reagent was quenched with acrolein to give the cyclopentane **250** (3.161). High stereochemical control is characteristic of such reactions with the formation of the *cis*-1,2-disubstituted cyclopentane. Quaternary carbon centres can be formed without difficulty (as in **250**). The 'magnesio-ene' reaction was repeated with the allylic Grignard reagent derived from **250** to construct the second five-membered ring of the natural product.

$$(3.161)$$

249   57%   **250** $\Delta^{9(12)}$-capnellene

In the above example, the enophile is tethered to the terminal carbon atom of the ene component and this leads to a 1,2-disubstituted cyclic product. If the enophile is tethered to the central carbon atom of the allyl metal, then the ene reaction gives an *exo*-methylene-substituted cyclic product. A key step in a synthesis of khusimone involved the metallo-ene reaction of the allyl Grignard **251**, followed by trapping with carbon dioxide and hydrolysis to give the carboxylic acid **252** (3.162).

[169] W. Oppolzer, *Angew. Chem. Int. Ed. Engl.*, **28** (1989), 38.

$$(3.162)$$

251   85%   **252**

The lithium ene cyclization[170] occurs at a lower temperature than the magnesium analogue,[171] although it has seen little use because of problems with the high reactivity and basicity of the resulting organolithium species. Ene reactions with unsaturated zinc species as ene components alleviate these problems and have found application for the formation of carbocyclic and heterocyclic compounds. Thus, transmetallation with zinc bromide of the organolithium species derived from deprotonation of the alkyne **253**, resulted in cyclization via a chair-shaped conformation **254** (3.163).[172] The resulting organozinc species **255** can be trapped with a proton or iodine, or transmetallated to copper or palladium and coupled with unsaturated halides.

$$(3.163)$$

253   **254**   **255**

The intramolecular ene reaction with an enol as the ene component is known as the Conia reaction. Very high temperatures are normally required and more efficient is the use of a cobalt catalyst[173] or the use of a zinc enolate.[174] Thus, formation of the zinc enolate **257** from the ester **256** promoted cyclization at room temperature to give, after quenching the intermediate organozinc species with a proton, the *cis* pyrrolidine **258** (3.164).

$$(3.164)$$

256   **257**   60%   **258**

[170] D. Cheng, K. R. Knox and T. Cohen, *J. Am. Chem. Soc.*, **122** (2000), 412.
[171] D. Cheng, S. Zhu, Z. Yu and T. Cohen, *J. Am. Chem. Soc.*, **123** (2001), 30.
[172] C. Meyer, I. Marek, G. Courtemanche and J.-F. Normant, *J. Org. Chem.*, **60** (1995), 863.
[173] P. Cruciani, R. Stammler, C. Aubert and M. Malacria, *J. Org. Chem.*, **61** (1996), 2699.
[174] E. Lorthiois, I. Marek and J. F. Normant, *J. Org. Chem.*, **63** (1998), 2442.

The metallo-ene reactions described so far require a stoichiometric amount of the metal salt. Catalytic ene reactions with sub-stoichiometric amounts of nickel, palladium or platinum have been reported and are of value in terms of reducing the amount of metal required and in terms of reaction simplicity and ease. The use of a sub-stoichiometric amount of a metal is possible if the metal is regenerated during the reaction. This is feasible if the product from the ene reaction undergoes β-elimination or reductive elimination. Thus, the metallo-ene reaction of the allyl palladium species **259** generates a new palladium species **260**, which undergoes β-elimination to release the palladium(0) catalyst to continue the cycle (3.165).[165]

$$(3.165)$$

3.6 [3,3]-Sigmatropic rearrangements

Of all the sigmatropic rearrangements, the [3,3]-sigmatropic rearrangement has been used most in organic synthesis. The reaction has found particular use for the stereocontrolled preparation of carbon–carbon bonds. In the course of the reaction both a new carbon–carbon single bond and a new carbon–carbon double bond are formed. The high levels of stereoselectivity arise as a result of a highly ordered six-membered-ring transition state, which (unless constrained conformationally) prefers a chair shape. The reaction involves the interconversion of 1,5-dienes and the all-carbon system is known as the Cope rearrangement, whereas with an allyl vinyl ether the reaction is termed a Claisen rearrangement (3.166).[175]

$$(3.166)$$

[175] S. J. Rhoads and N. R. Raulins, *Org. Reactions*, **22** (1975), 1.

3.6.1 The Cope rearrangement

The Cope rearrangement[176] has found many synthetic applications, particularly when modified as the anionic oxy-Cope variant. The parent Cope rearrangement can be promoted by heating the 1,5-diene neat or in a high-boiling solvent such as xylene or decalin. The [3,3]-sigmatropic rearrangement of 1,5-hexadienes is a reversible process and the position of equilibrium depends on the substitution pattern and on the relative strain of the two 1,5-dienes. The thermodynamic stability of an alkene increases with increasing substitution or with increasing conjugation and the equilibrium therefore normally lies in favour of the more-substituted, more-conjugated product. Thus, heating the 1,5-diene **261** gave the new 1,5-diene **262**, in which two carbonyl groups come into conjugation with one of the new alkene π-bonds on rearrangement (3.167).

(3.167)

A disadvantage of the Cope rearrangement is the high temperature that is often required. Catalysis of the Cope rearrangement has been possible in some cases, particularly with metal salts such as palladium(II) chloride.[177] For example, treatment of the 1,5-diene **263** with [PdCl$_2$(PhCN)$_2$] as a catalyst at room temperature for 24 h gave the 1,5-dienes **264** and **265** (3.168). The reaction is even faster in benzene as the solvent. In contrast, the thermal Cope rearrangement of the 1,5-diene **261** requires temperatures in the region of 177 °C and is less stereoselective. Unfortunately, this type of reaction is restricted to 1,5-dienes that are substituted at either C-2 or C-5.

(3.168)

A good substrate for the Cope rearrangement is a *cis*-1,2-divinylcyclopropane, since ring-opening of the strained three-membered ring occurs on rearrangement.

[176] R. K. Hill, in *Comprehensive Organic Synthesis*, ed. B. M. Trost and I. Fleming, vol. 5 (Oxford: Pergamon Press, 1991), p. 785.

[177] L. E. Overman, *Angew. Chem. Int. Ed. Engl.*, **23** (1984), 579; R. P. Lutz, *Chem. Rev.*, **84** (1984), 205.

The *cis*-divinylcyclopropane unit rearranges readily by a concerted pericyclic process and gives rise to a cycloheptadiene ring.[178] The divinylcyclopropane **266**, for
example, rearranges at only 15 °C (3.169). Unlike conventional [3,3]-sigmatropic
rearrangements, this reaction proceeds by a boat-shaped six-membered-ring transition state **267** (the chair-shaped transition state would lead to the highly strained
E,E-cycloheptadiene and would be much higher in energy). The reaction provides
a useful entry to seven-membered rings and has been applied to a number of syntheses of natural products. For example, in a synthesis of the diterpene scopadulcic
acid A, stereoselective enolization of the ketone **269** to give the Z-silyl enol ether
270 is followed by rearrangement (which occurs prior to warming to room temperature) and hydrolysis (HCl—H$_2$O) to give the cyclohexenone **271** (3.170).[179]
The rearrangement is stereospecific and requires the Z-silyl enol ether to give the
desired diastereomer of the cyclohexenone. The related *E*-silyl enol ether can be
formed preferentially using Et$_3$N and Me$_3$SiOTf.

(3.169)

266   **267**   **268**

269   **270**   74%   **271**

R =

(3.170)

When the 1,5-diene is substituted by a hydroxy or alkoxy group at C-3 and/or
C-4, the [3,3]-sigmatropic shift is known as the oxy-Cope rearrangement. The
product enol or enol ether converts readily to the corresponding aldehyde or ketone.
High temperatures are normally required, but the variant has found use for the
synthesis of δ,ε-unsaturated carbonyl compounds, 1,6-dicarbonyl compounds and
for substrates in which the equilibrium would otherwise lie on the side of the
starting 1,5-diene. For example, the 1,5-diene **272** rearranges on heating to give the

[178] T. Hudlicky, R. Fan, J. W. Reed and K. G. Gadamasetti, *Org. Reactions*, **41** (1992), 1; E. Piers, in *Comprehensive Organic Synthesis*, ed. B. M. Trost and I. Fleming, vol. 5 (Oxford: Pergamon Press, 1991), p. 971.
[179] M. E. Fox, C. Li, J. P. Marino and L. E. Overman, *J. Am. Chem. Soc.*, **121** (1999), 5467.

ten-membered ring **273**, whereas the equilibrium for the Cope rearrangement lies in favour of the six-membered ring (3.171).

(3.171)

272   **273**

In addition to the high temperatures required, another problem with the oxy-Cope rearrangement of a 3-hydroxy-1,5-diene is fragmentation by a competing retro-ene process. An alternative and now common method for promoting the oxy-Cope rearrangement is the addition of a base to deprotonate the alcohol.[180] Very large increases in the rate of these rearrangements are obtained by using the potassium alkoxides, rather than the hydroxy compounds themselves. The variant is often referred to as the anionic oxy-Cope rearrangement.[181] Deprotonation of the alcohol with, for example, potassium hydride gives the potassium alkoxide, which has a higher-energy ground state than the alcohol and undergoes rearrangement directly at room temperature or on mild heating.

The anionic oxy-Cope rearrangement, like the parent Cope rearrangement, prefers the chair-shaped transition state. The ratio of stereoisomers of the product depends on the orientation of the substituents in the transition state. A substituent at C-3 (or C-4) of the 1,5-diene generally prefers the less-hindered equatorial position and this leads to the *E* alkene isomer of the product. The degree of stereocontrol across the new carbon–carbon single bond can also be very high and the preferred diastereomer can be related to the alkene geometry of the starting 1,5-diene. Thus the chair-shaped transition state results in a preference for different diastereomers of the product from the *E,E*- and *E,Z*-isomers of the starting 1,5-diene (3.172).[182] In addition, as a consequence of the ordered transition state, chirality transfer across the allylic system is possible.

(3.172)

[180] D. A. Evans and A. M. Golob, *J. Am. Chem. Soc.*, **97** (1975), 4765.
[181] S. R. Wilson, *Org. Reactions*, **43** (1993), 93; L. A. Paquette, *Tetrahedron*, **53** (1997), 13971.
[182] S.-Y. Wei, K. Tomooka and T. Nakai, *Tetrahedron*, **49** (1993), 1025.

The ease of the anionic oxy-Cope rearrangement and its high level of stereo-control make this reaction a popular and valuable synthetic method. For example, a key step of a synthesis of the sesquiterpene juvabione, made use of the stereocontrolled rearrangement of the potassium salt of the 3-hydroxy-1,5-diene **274** to give the cyclohexanone **275** (3.173).[183] The diastereoselectivity across the new carbon–carbon single bond reflects the preference for a chair-shaped transition state with the methoxy group in the pseudoequatorial position. In another example, the germacrane sesquiterpenes can be accessed readily using an anionic oxy-Cope rearrangement (3.174).[184]

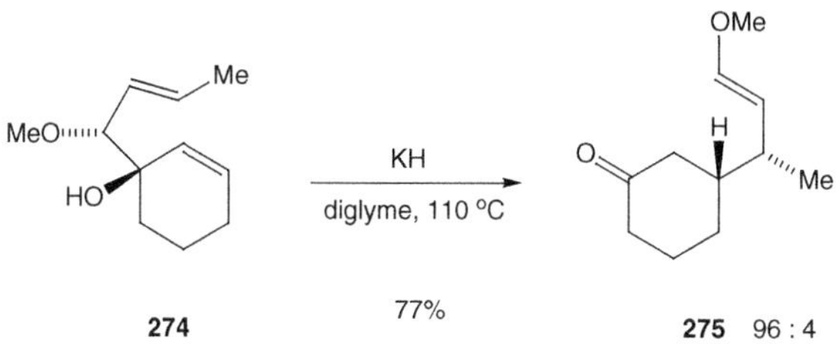

(3.173)

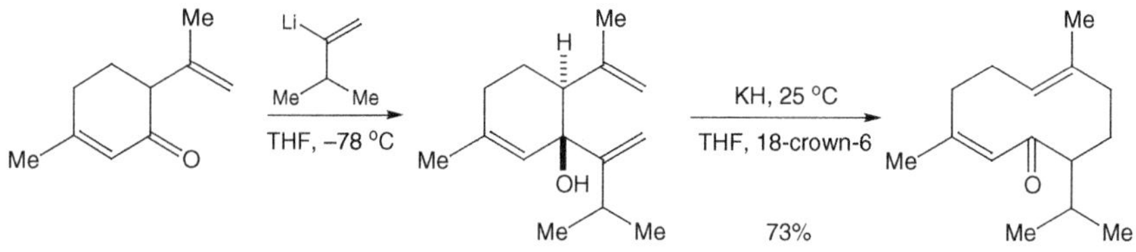

(3.174)

In addition to the oxy-Cope and anionic oxy-Cope rearrangements, an important variant is the aza-Cope rearrangement of *N*-butenyl-iminium ions (3.175). This rearrangement occurs under mild conditions, but suffers as a synthetic method because of its reversibility. However, with a hydroxy group attached to the butenyl chain (R=OH), the reaction is driven in the forward direction by capture of the rearranged iminium ion in an intramolecular Mannich reaction, to provide an excellent synthesis of substituted pyrrolidines.[185]

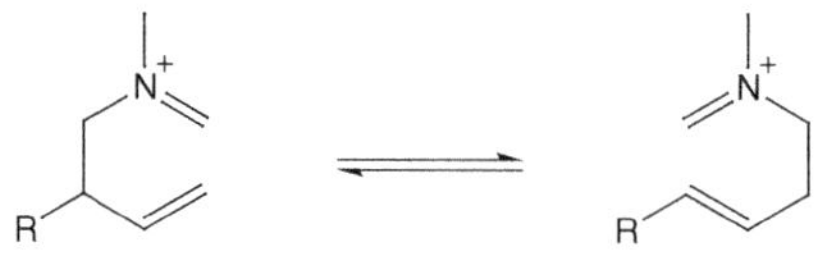

(3.175)

[183] D. A. Evans and J. V. Nelson, *J. Am. Chem. Soc.*, **102** (1980), 774.
[184] W. C. Still, *J. Am. Chem. Soc.*, **99** (1977), 4186; W. C. Still, *J. Am. Chem. Soc.*, **101** (1979), 2493.
[185] L. E. Overman, M. Kakimoto, M. E. Okazaki and G. P. Meier, *J. Am. Chem. Soc.*, **105** (1983), 6622; E. J. Jacobsen, J. Levin and L. E. Overman, *J. Am. Chem. Soc.*, **110** (1988), 4329.

The required iminium ion can be obtained readily by the condensation of an aldehyde with a butenylamine. For example, heating the butenylamine **276** with pyridine-3-carboxaldehyde and an acid catalyst (camphorsulfonic acid, CSA), gave the acetyl nicotine derivative **277** (3.176). The initial iminium ion **278** rearranges to the new iminium ion **279**, which is irreversibly trapped in an intramolecular Mannich reaction to give the pyrrolidine **277**.

$$(3.176)$$

When the hydroxy and amino groups are neighbouring substituents on a ring, an interesting conversion takes place to give a bicyclic pyrrolidine derivative, in which the original ring is expanded by one carbon atom. Thus, a key step in a synthesis of the *Amaryllidaceae* alkaloid pancracine involved the aza-Cope–Mannich reaction sequence (3.177).[186] In this case, condensation of the amino alcohol **280** with aqueous formaldehyde did not lead directly to rearrangement but to the oxazolidine **281**, which was treated with the Lewis acid BF$_3$·OEt$_2$ to promote the iminium ion formation, rearrangement and Mannich reaction. A key step in a synthesis of the alkaloid strychnine has also made use of this chemistry, by addition of paraformaldehyde to the unsaturated amine **282** (3.178).[187]

[186] L. E. Overman and J. Shim, *J. Org. Chem.*, **58** (1993), 4662.
[187] S. G. Knight, L. E. Overman and G. Pairaudeau, *J. Am. Chem. Soc.*, **117** (1995), 5776.

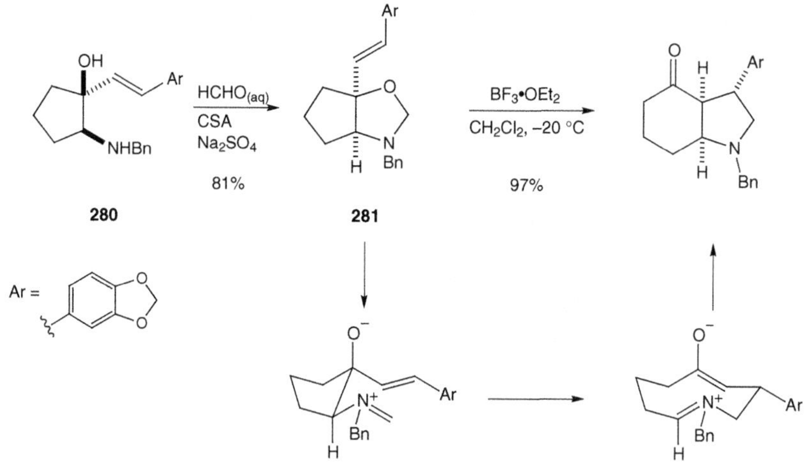

(3.177)

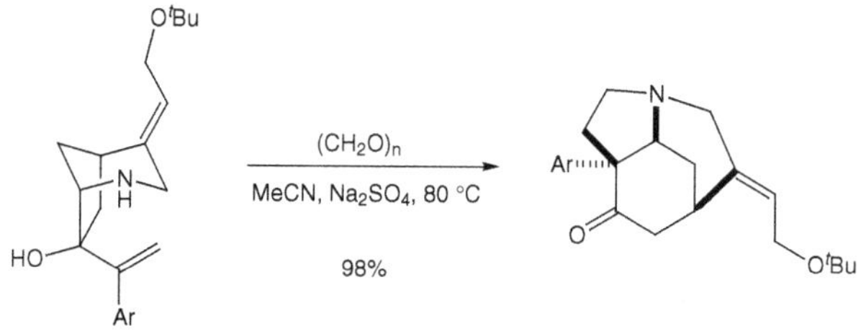

(3.178)

3.6.2 The Claisen rearrangement

The Claisen rearrangement of allyl vinyl ethers provides an excellent stereoselective route to γ,δ-unsaturated carbonyl compounds from allylic alcohols.[175,188] Like the Cope rearrangement (Section 3.6.1), the reaction involves a [3,3]-sigmatropic rearrangement, and takes place by a concerted mechanism through a cyclic six-membered transition state (3.179). Its value in synthesis stems from the ability to form a carbon–carbon bond at the expense of a carbon–oxygen bond, and from the fact that it is highly stereoselective, leading predominantly to the *E*-configuration of the new double bond and to the controlled stereochemical disposition of substituents on the single bond. A chair conformation is preferred for the cyclic transition state with the substituent R^1 (3.179) in the less-hindered pseudoequatorial position.

---

[188] F. E. Ziegler, *Chem. Rev.*, **88** (1988), 1423; P. Wipf, in *Comprehensive Organic Synthesis*, ed. B. M. Trost and I. Fleming, vol. 5 (Oxford: Pergamon Press, 1991), p. 827.

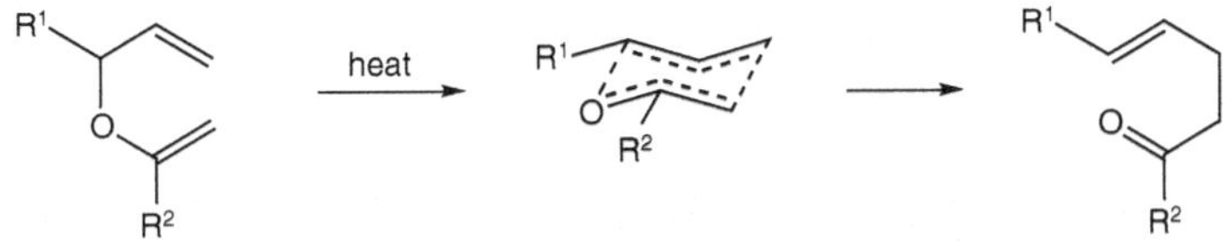

$$R^2 = H, \text{ alkyl, OR, OSiR}_3, \text{ NR}_2$$

(3.179)

The allyl vinyl ethers used in the reaction are prepared directly from allylic alcohols by acid-catalysed ether exchange. For example, reaction of the allylic alcohol **283** with ethyl vinyl ether and Hg(OAc)$_2$, followed by heating, gave the γ,δ-unsaturated aldehyde **284** (3.180). An alternative procedure, sometimes referred to as the Johnson–Claisen rearrangement, involves heating the allylic alcohol with an orthoester in the presence of a weak acid (propionic acid is often used).[188] A synthesis of the insect pheromone **286** from the allylic alcohol **285** and trimethyl orthoacetate illustrates this procedure and the high selectivity in favour of the *E*-alkene product (3.181). In the Johnson–Claisen rearrangement, a mixed orthoester is formed first and loses methanol to form a ketene acetal which rearranges to the γ,δ-unsaturated ester. In a similar way, unsaturated carboxylic amides can be obtained by heating allylic alcohols with the dialkyl acetal of an *N*, *N*-dialkylamide.

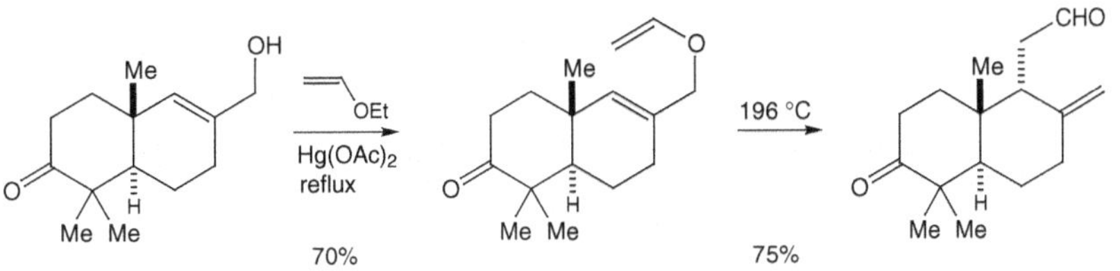

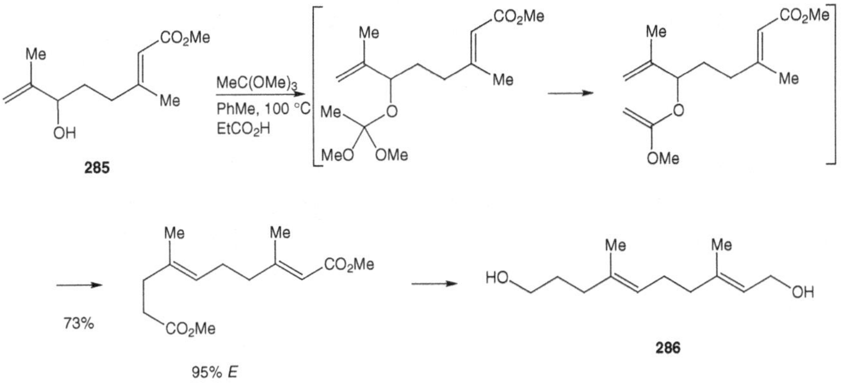

(3.180)

(3.181)

The relative stereochemistry across the new carbon–carbon single bond is established as a result of the chair-like transition state and depends on the geometry of the double bonds in the starting material. Thus, rearrangement of the *E,E*- (or *Z,Z*-) diene gave (≥95%) the diastereomer **287** (the *syn* diastereomer in the extended conformation), whereas the *E,Z*-dienes gave (≥95%) the diastereomer **288** (the *anti* diastereomer) (3.182).

(3.182)

In contrast, a boat-shaped transition state may be favoured if the allyl unit is constrained conformationally, for example as part of a ring system. The rearrangement of the vinyl ether of the cyclic allylic alcohol **289** occurs to give the unsaturated lactone **290** with complete stereoselectivity (3.183). In this reaction, the configuration at the new chiral centre of the cyclohexene ring is controlled directly by the configuration of the original allylic hydroxy group, and the configuration at the new centre of the lactone ring is controlled by the (boat) conformation of the transition state. The stereochemistry of **290** indicates that the reaction must have proceeded entirely through a boat-like transition state.

(3.183)

The defined transition state for the [3,3]-sigmatropic rearrangement allows 1,3-transfer of chirality, which proceeds suprafacially across the allyl unit. Therefore, the configuration at the new allylic carbon atom is related directly to that of the starting alcohol. In acyclic substrates, either configuration of the new chiral centre may be obtained by changing the configuration of the allylic double bond in the starting material. Hence, different stereoisomers can be obtained starting from the *E*- or *Z*-allylic alcohols. A key step in a synthesis of (+)-15-(*S*)-prostaglandin A$_2$ used the Claisen rearrangement from the chiral allylic alcohol **291** to give the ester **292** (3.184).[189] In this example the chirality at the alcohol centre in **291** is transferred across the allyl unit.

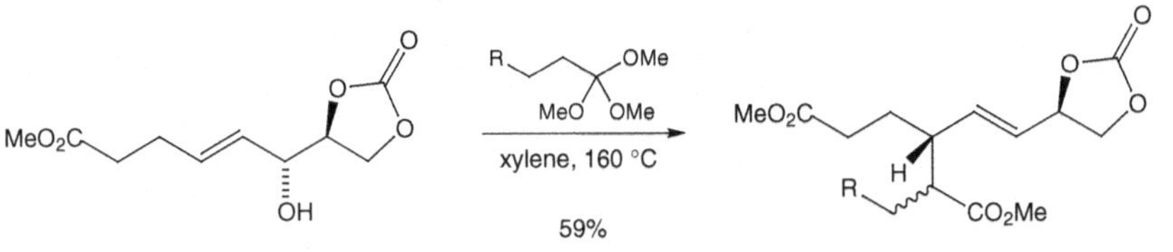

(3.184)

291   **292**

A disadvantage of the above procedures is that relatively high temperatures are required and that acid-sensitive substrates are not tolerated. A reduction in the temperature required to effect the Claisen rearrangement may be possible under aqueous conditions[190] or in the presence of a catalyst.[177] Various metal salts have been investigated, with successful Claisen rearrangements in the presence of $[PdCl_2(PhCN)_2]$,[191] aluminium(III) or other metal catalysts.[192] Some examples are illustrated in Schemes 3.185–3.187. If the aluminium catalyst, $(ArO)_3Al$, consists of an axially chiral binaphthol ligand, then asymmetric Claisen rearrangement is possible.[193]

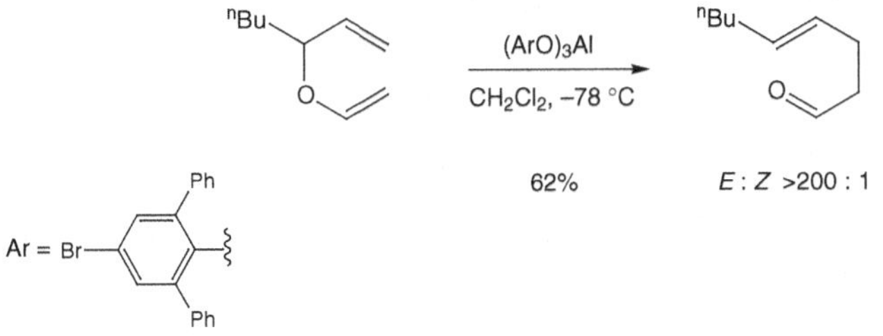

(3.185)

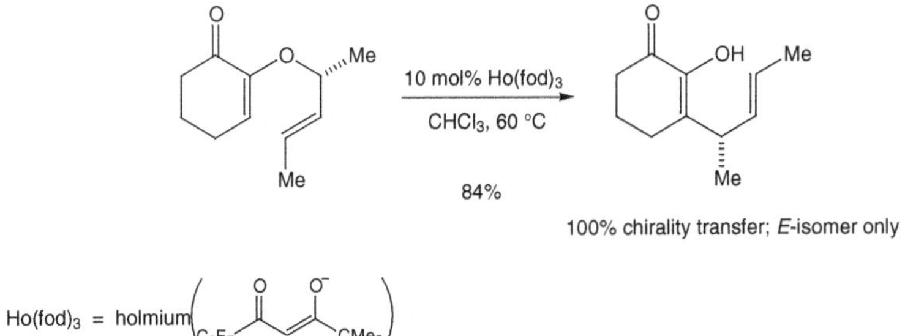

62%   E : Z >200 : 1

(3.186)

84%

100% chirality transfer; *E*-isomer only

[189] G. Stork and S. Raucher, *J. Am. Chem. Soc.*, **98** (1976), 1583.
[190] B. Ganem, *Angew. Chem. Int. Ed.*, **35** (1996), 937; J. J. Gajewski, *Acc. Chem. Res.*, **30** (1997), 219.
[191] M. Sugiura and T. Nakai, *Chem. Lett.* (1995), 697.
[192] S. Saito and K. Shimada and H. Yamamoto, *Synlett* (1996), 720; B. M. Trost and G. M. Schroeder, *J. Am. Chem. Soc.*, **122** (2000), 3785; M. Hiersemann and L. Abraham, *Eur. J. Org. Chem.* (2002), 1461.
[193] K. Maruoka, S. Saito and H. Yamamoto, *J. Am. Chem. Soc.*, **117** (1995), 1165.

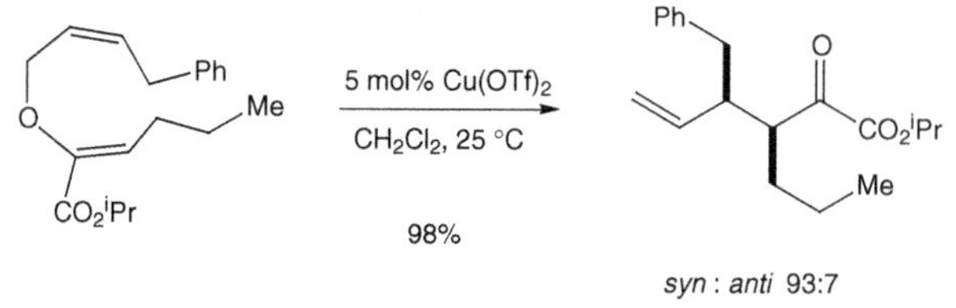

$$(3.187)$$

98%

syn : anti 93:7

An alternative and popular method for effecting the Claisen rearrangement has been developed by Ireland.[194] The allylic alcohol is first acylated to give a carboxylic ester, which is deprotonated to give the corresponding enolate. The problem of aldol side-products is avoided by formation of the silyl ketene acetal from the enolate prior to rearrangement. The Ireland–Claisen rearrangement of the silyl ketene acetal proceeds at low temperatures to give, after hydrolysis of the silyl ester, a γ,δ-unsaturated carboxylic acid.

The acyclic Ireland–Claisen rearrangement proceeds through a chair-shaped transition state and rearrangement of the *E*-enolate can lead to a different diastereomer from rearrangement of the *Z*-enolate. Careful choice of enolization conditions to favour one enolate geometry is therefore important for diastereocontrol of the rearrangement. Enolization at low temperature in THF followed by trapping with the silyl chloride favours the *E*-silyl ketene acetal, whereas in the presence of the co-solvent HMPA or DMPU the *Z*-silyl ketene acetal is favoured (3.188). The diastereomer formed also depends, of course, on the alkene geometry of the starting allylic alcohol, and opposite product diastereomers are formed using the same enolization conditions but changing from the *E*- to the *Z*-allylic alcohol.

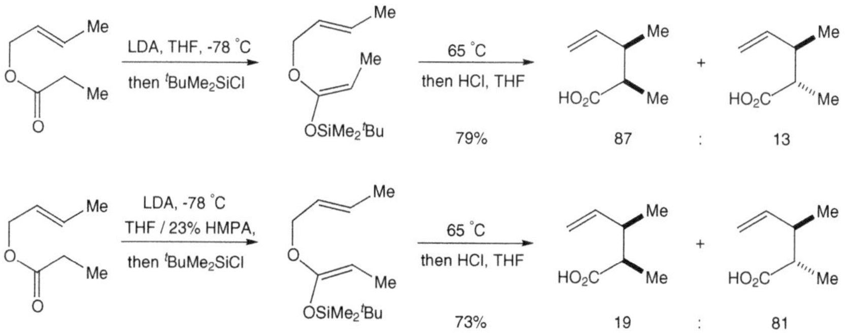

$$(3.188)$$

The selectivity in the enolization can be explained by the extent of co-ordination of the two different solvent systems. In the less-co-ordinating THF alone, association of the lithium cation with the ester carbonyl group is important and the R group prefers the less-hindered pseudoequatorial position distant from the LDA, thereby favouring **293** and leading to the *E*-enolate (note that LDA is drawn here

---

[194] R. E. Ireland and R. H. Mueller, *J. Am. Chem. Soc.*, **94** (1972), 5897; R. E. Ireland, R. H. Mueller and A. K. Willard, *J. Am. Chem. Soc.*, **100** (1976), 2868; S. Pereira and M. Srebnik, *Aldrichimica Acta*, **26** (1993), 17; Y. Chai, S. Hong, H. A. Lindsay, C. McFarland and M. C. McIntosh, *Tetrahedron*, **58** (2002), 2905.

as a monomer but is dimeric in THF) (3.189). In the presence of HMPA (a strongly lithium-co-ordinating molecule), the association of the lithium cation is less important and the allyloxy group (OR') becomes more sterically demanding, thereby favouring **294** (or a more open-chain transition structure) leading to the Z-enolate.

| *E*-enolate | **293** | **294** | *Z*-enolate |

(3.189)

With a cyclic substrate, for example in which three or more atoms of the allyl vinyl ether are constrained in a ring, then the boat-shaped transition state may be favoured. Formation of the silyl ketene acetal from the lactone **295** and rearrangement on warming gave the carboxylic acid **296** (3.190).[195] The reaction occurs via a boat-shaped transition state and was used in a synthesis of the sesquiterpene widdrol.

(3.190)

295   **296**

An interesting variant of the ester enolate Claisen rearrangement uses α-amino esters of allylic alcohols to give allyl glycine derivatives. In these examples, it is not necessary to prepare the silyl ketene acetal and rearrangement of the metal (normally zinc) enolate takes place on warming.[196] For example, deprotonation of the substrate **297** and warming to room temperature promoted rearrangement to the amino-acid derivative **299** (3.191). The rearrangement occurs with >98% selectivity to give the 2*R* epimer and the *E* alkene geometry via a chair-like transition state **298** with a chelated enolate.

(3.191)

297   **298**   **299**

[195] S. Danishefsky, R. L. Funk and J. F. Kerwin, *J. Am. Chem. Soc.*, **102** (1980), 6889.
[196] U. Kazmaier, *Liebigs Ann./Recl.* (1997), 285.

Of increasing interest and importance for stereoselective synthesis is the development of asymmetric processes using achiral substrates and a chiral catalyst. Such asymmetric induction in some pericyclic processes, such as the Diels–Alder reaction, has been studied extensively; however, the asymmetric Claisen rearrangement has received much less attention.[197] Very high selectivities in the Claisen rearrangement of α-amino esters in the presence of quinine or quinidine have been reported.[198] Another solution involves chiral boron enolate chemistry (3.192).[199] With the base Et$_3$N in toluene, the *E*-boron enolate is favoured and asymmetric induction using the chiral boron reagent **300** occurs to give predominantly one *anti* diastereomer. However, with Hünig's base (i-Pr$_2$NEt) in CH$_2$Cl$_2$, the *Z*-boron enolate is favoured and a very high diastereo- and enantioselectivity for one of the *syn* diastereomers is obtained.

(3.192)

The chiral boron reagent **300** has been used successfully for the aromatic Claisen rearrangement.[200] The aromatic Claisen rearrangement involves the [3,3]-sigmatropic rearrangement of allyl aryl ethers with migration of the allyl group (with allylic transposition) to the *ortho* position of the aromatic ring.[175,201] The

[197] U. Nubbemeyer, *Synthesis* (2003), 961; H. Ito and T. Taguchi, *Chem. Soc. Rev.*, **28** (1999), 43; D. Enders, M. Knopp and R. Schiffers, *Tetrahedron: Asymmetry*, **7** (1996), 1847.
[198] U. Kazmaier and A. Krebs, *Angew. Chem. Int. Ed. Engl.*, **34** (1995), 2012.
[199] E. J. Corey and D.-H. Lee, *J. Am. Chem. Soc.*, **113** (1991), 4026.
[200] H. Ito, A. Sato and T. Taguchi, *Tetrahedron Lett.*, **38** (1997), 4815.
[201] C. J. Moody, *Adv. Heterocycl. Chem.*, **42** (1987), 203.

rearrangement is normally promoted by heating the allyl aryl ether to 150–200 °C, although a lower temperature may be employed when the reaction is carried out in the presence of a protic or Lewis acid catalyst.[202] For example, heating the ether **301** gave the alcohol **302**, used in a synthesis of the aromatic unit of the antitumor agent calicheamicin γ_1^I (3.193).[203]

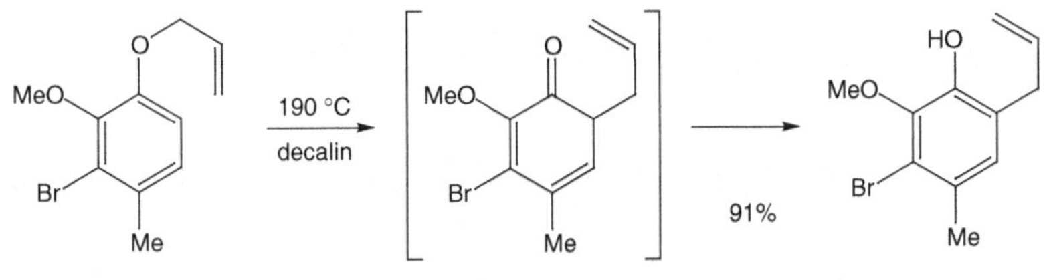

(3.193)

An important type of aromatic Claisen rearrangement occurs in the Fischer indole synthesis.[204] The Fischer indole synthesis involves the condensation of an arylhydrazine with an aldehyde or ketone to give an arylhydrazone, which, in the presence of a catalyst undergoes rearrangement and elimination of ammonia to give the indole ring. One of many different protic or Lewis acid catalysts can be used. For example, Woodward's synthesis of strychnine commenced with the condensation of the ketone **303** and phenylhydrazine in the presence of polyphosphoric acid to give the indole **304** (3.194).

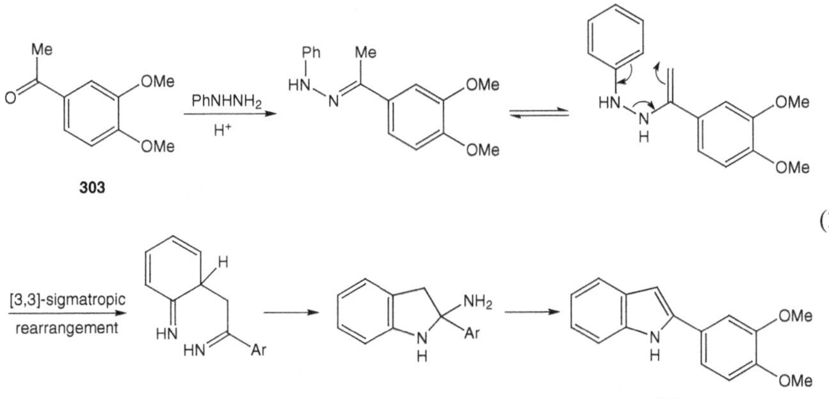

(3.194)

Examples of the aza-Claisen rearrangement of allyl vinyl amines and the thia-Claisen rearrangement of allyl vinyl sulfides have been reported. An efficient Lewis acid-catalysed Claisen rearrangement of zwitterionic *N*-acyl allylic amines

[202] B. M. Trost and F. D. Toste, *J. Am. Chem. Soc.*, **120** (1998), 815; P. Wipf and S. Rodríguez, *Adv. Synth. Catal.*, **344** (2002), 434.

[203] Y.-Z. Hu and D. L. J. Clive, *J. Chem. Soc., Perkin Trans. 1* (1997), 1421.

[204] B. Robinson, *Chem. Rev.*, **69** (1969), 227.

provides access to γ,δ-unsaturated carboxylic amides (3.195).[205] The rearrangement can be promoted by addition of an acid chloride to a tertiary allylic amine. The presence of a Lewis acid, such as TiCl$_4$ increases the yield and diastereoselectivity.

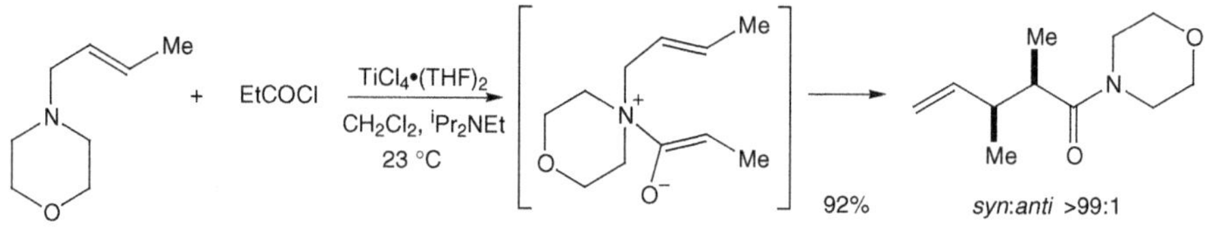

(3.195)

A popular [3,3]-sigmatropic rearrangement that provides a convenient method for the preparation of allylic amines makes use of the rearrangement of allylic alcohols via their imidates (sometimes called the Overman reaction).[206] The trichloroacetimidate has found most use and can be prepared from the allylic alkoxide and trichloroacetonitrile. Subsequent thermal rearrangement gives the allylic trichloroacetamide (3.196). The rearrangement is subject to the usual high stereoselectivity for the *E*-alkene product and occurs with chirality transfer as observed in other [3,3]-sigmatropic processes. Hydrolysis of the product amide (e.g. using aqueous NaOH) provides the allylic amine. Catalysis of the rearrangement, particularly by palladium(II) complexes is possible.[207] An example of the use of this rearrangement in a synthesis of the alkaloid pancratistatin is illustrated in Scheme 3.197.[208]

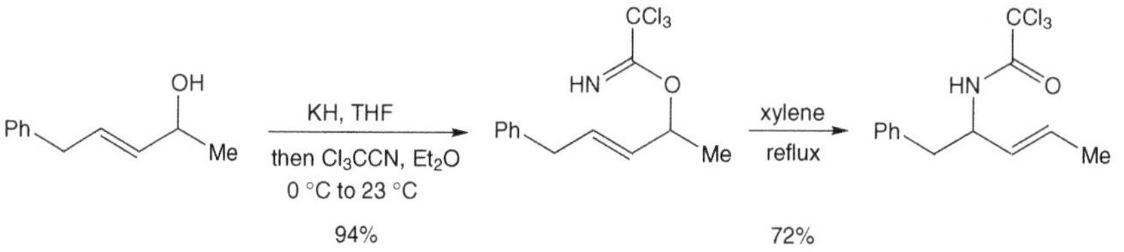

(3.196)

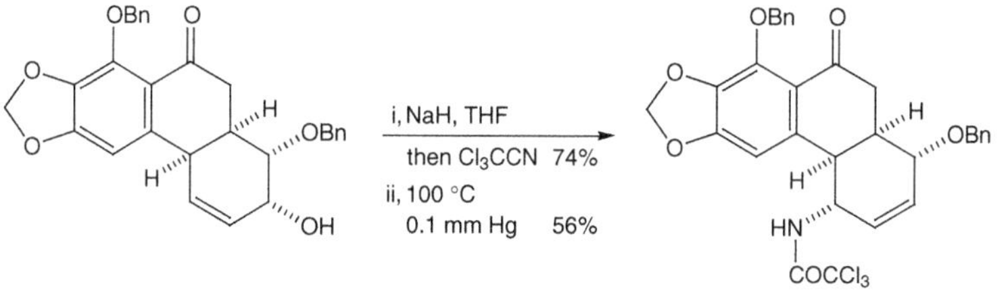

(3.197)

[205] T. P. Yoon, V. M. Dong and D. W. C. MacMillan, *J. Am. Chem. Soc.*, **121** (1999), 9726.
[206] L. E. Overman, *Acc. Chem. Res.*, **13** (1980), 218.
[207] T. K. Hollis and L. E. Overman, *J. Organomet. Chem.*, **576** (1999), 290; C. E. Anderson and L. E. Overman, *J. Am. Chem. Soc.*, **125** (2003), 12 412.
[208] S. Danishefsky and J. Y. Lee, *J. Am. Chem. Soc.*, **111** (1989), 4829.

3.7 [2,3]-Sigmatropic rearrangements

The [2,3]-sigmatropic rearrangement is the allylic variant of the [1,2]-sigmatropic rearrangement of sulfonium or ammonium ylides or α-metalated ethers. The new σ-bond forms at the end of the allylic system by a concerted process, with simultaneous cleavage of the allylic–heteroatom bond (3.198).[209] The heteroatom X is commonly a sulfur or nitrogen atom (as part of an ylide), in which case the reaction is termed a [2,3]-Stevens rearrangement, or an oxygen atom (as part of an ether), in which case the reaction is termed a [2,3]-Wittig rearrangement. The reaction occurs suprafacially across the allyl unit through a five-membered ring envelope-shaped transition state. Examples in which the allyl unit is replaced by a benzyl, propargyl or allenyl unit are known.

$$(3.198)$$

As a synthetic method, the most important examples involve the formation of a new carbon–carbon bond (in which Y is a carbon atom) at the expense of a carbon–heteroatom (C−X) bond. When X is an oxygen atom, the [2,3]-Wittig rearrangement proceeds to give a homoallylic alcohol product.[210] The carbanion is normally generated by direct deprotonation α- to the oxygen atom using a base such as *n*-BuLi or LDA. In such cases, the deprotonation must be regioselective and can be directed by an anion-stabilizing group such as an alkynyl, aryl or acyl group. If an alkene is used such that the substrate is a diallyl ether, then the base removes the most acidic proton on the less substituted allyl group to give the more stable carbanion. In the absence of an anion-stabilizing substituent, regioselective lithiation can be carried out by tin–lithium exchange or reductive lithiation of an *O,S*-acetal.

In a formal synthesis of brefeldin A, treatment of the allyl propargyl ether **305** with the base *n*-BuLi promoted deprotonation α- to the alkyne, followed by rearrangement to give predominantly the homoallylic alcohol **306** (3.199).[211] Likewise, deprotonation and rearrangement of the macrocyclic substrate **307** promoted rearrangement to the homoallylic alcohol **308**, used in a synthesis of the diterpene kallolide B (3.200).[212] Regioselective deprotonation of the diallyl ether **309** and

[209] R. Brückner, in *Comprehensive Organic Synthesis*, ed. B. M. Trost and I. Fleming, vol. 6 (Oxford: Pergamon Press, 1991), p. 873; R. W. Hoffmann, *Angew. Chem. Int. Ed. Engl.*, **18** (1979), 563.
[210] T. Nakai and K. Mikami, *Org. Reactions*, **46** (1994), 105; J. A. Marshall, in *Comprehensive Organic Synthesis*, ed. B. M. Trost and I. Fleming, vol. 3 (Oxford: Pergamon Press, 1991), p. 975.
[211] K. Tomooka, K. Ishikawa and T. Nakai, *Synlett* (1995), 901.
[212] J. A. Marshall, G. S. Bartley and E. M. Wallace, *J. Org. Chem.*, **61** (1996), 5729.

rearrangement led to the homoallylic alcohol **310** as the major product (3.201).[213]
The corresponding rearrangement of the *E*-alkene isomer of **309** was less diastereoselective.

(3.199)

305   **306**

(3.200)

307   **308**

(3.201)

309   **310**   97 : 3   **311**

The [2,3]-sigmatropic rearrangement is often highly stereoselective, with a
marked preference for the formation of the *E*-alkene product. The diastereoselectivity across the new carbon–carbon bond is more difficult to predict. With an
alkynyl, alkenyl or aryl anion-stabilizing group, the diastereoselectivity can be high,
particularly from the *Z*-alkene substrate, which favours the *syn* product. Use of the
E-alkene substrate often favours the *anti* product, although in these cases the degree
of stereoselection is normally lower. The diastereoselectivity reflects the preference
for the hydrocarbon group (G) to adopt the *exo* orientation in the envelope-like transition state (3.202). However, when the anion-stabilizing group is an acyl group
(e.g. carbonyl or oxazoline), such that an enolate intermediate is formed, then the
endo transition state is often favoured. Thus, rearrangement of the *E*-allylic ether
312 gave predominantly the *syn* diastereomer **313** (and the corresponding *Z*-isomer
gave predominantly the *anti* diastereomer **314**) (3.203).

[213] D. J.-S. Tsai and M. M. Midland, *J. Am. Chem. Soc.*, **107** (1985), 3915; D. J.-S. Tsai and M. M. Midland,
J. Org. Chem., **49** (1984), 1842.

Chirality transfer across the allyl unit is common. The products **310** and **311** are formed with no loss of optical purity on rearrangement of the enantiomerically enriched substrate **309** (i.e. substrate **309**, 91% ee, gave alcohols **310** and **311**, both 91% ee) (3.201).

G = hydrocarbon
in *exo* orientation

(3.202)

G = acyl
in *endo* orientation

$$\text{(3.203)}$$

312   LDA, THF, −78 °C   80%   **313**   84 : 16   **314**

In a few cases, a preference for the Z-alkene product is observed. For example, in a synthesis of the California red scale pheromone **317**, R=COCH$_3$, formation of the organolithium species **316** was accomplished by tin–lithium exchange from the stannane **315** (3.204).[214] Subsequent [2,3]-sigmatropic rearrangement at low temperature gave the homoallylic alcohol **317**, consisting predominantly (96%) of the Z-isomer. The use of tin–lithium or sulfur–lithium exchange to give a chiral organolithium species has allowed the determination that the [2,3]-sigmatropic rearrangement proceeds with complete inversion of configuration at the carbanion centre.

315   BuLi, THF, −78 °C   **316**   83%   **317**, R=H   (3.204)

Some examples of the related thia-Wittig or aza-Wittig rearrangement, in which a sulfur or nitrogen atom is located in place of the ether oxygen atom have been

[214] W. C. Still and A. Mitra, *J. Am. Chem. Soc.*, **100** (1978), 1927.

reported.[215] Much more common is the rearrangement of sulfonium or ammonium ylides. The [2,3]-sigmatropic rearrangement of such ylides is referred to as the [2,3]-Stevens rearrangement.[216] The ylides are normally prepared by deprotonation or desilylation of the sulfonium or ammonium salts or by reaction of the sulfide or amine with a carbene. Like the [2,3]-Wittig rearrangement, the reaction involves a five-membered envelope-shaped transition state. The reaction is normally stereoselective in favour of the *E*-alkene product, although mixtures of diastereomers across the new carbon–carbon single bond are typical, especially from acyclic ylides.

Sulfonium and ammonium salts are prepared readily by alkylation of sulfides or amines. For example, allylation of the amine **318** gave the intermediate ammonium salt **319** (3.205).[217] Deprotonation to the ylide with sodium hydride was followed by [2,3]-sigmatropic rearrangement at room temperature to give the penicillin derivative **320**. In a synthesis of γ-cyclocitral, the sulfonium salt **321** was prepared by alkylation of 1,3-dithiane and was converted to the ylide **322** and hence the rearranged product **323** (3.206).

$$(3.205)$$

$$(3.206)$$

[215] K. Brickmann and R. Brückner, *Chem. Ber.*, **126** (1993), 1227; C. Vogel, *Synthesis* (1997), 497; J. C. Anderson, S. C. Smith and M. E. Swarbrick, *J. Chem. Soc., Perkin Trans. 1* (1997), 1517.

[216] *Nitrogen, Oxygen and Sulfur Ylide Chemistry, A Practical Approach in Chemistry*, ed. J. S. Clark (Oxford: Oxford University Press, 2002); I. E. Markó, in *Comprehensive Organic Synthesis*, ed. B. M. Trost and I. Fleming, vol. 3 (Oxford: Pergamon Press, 1991), p. 913.

[217] G. V. Kaiser, C. W. Ashbrook and J. E. Baldwin, *J. Am. Chem. Soc.*, **93** (1971), 2342.

Addition of a carbene to an allylic sulfide or amine provides a method for direct ylide formation. Intermolecular reactions are typically performed with CH_2I_2 and Et_2Zn or with an α-diazocarbonyl compound and $[Rh_2(OAc)_4]$ or a copper salt as a catalyst.[218] For example, the allylic sulfide **324** was converted to the homoallylic sulfide **325** using this chemistry (3.207). Note that the rearrangement occurs suprafacially across the allyl unit. Intramolecular trapping of the carbene with the heteroatom, followed by *in situ* [2,3]-sigmatropic rearrangement provides a method to access heterocyclic compounds. Cyclic ethers and amines have been prepared in this way, such as the tetrahydropyran **326**, used in a synthesis of decarestrictine L (3.208).[219]

(3.207)

(3.208)

326 *trans:cis* 77:23

Ring expansion or ring contraction in the [2,3]-Stevens rearrangement of cyclic ylides provides an alternative method for the formation of heterocyclic products.[220]

If one of the substituents of the ammonium ylide is a benzyl group, then rearrangement to the *ortho* position of the aromatic ring is possible and this process is termed a Sommelet–Hauser rearrangement.[221] The ylide can be formed by proton abstracton with a base such as $NaNH_2$ in liquid ammonia, or by desilylation with CsF. For example, the product **328** is formed on rearrangement of the ylide generated from the ammonium salt **327** (3.209).[222] Depending on the substitution pattern and the conditions of the reaction, the Sommelet–Hauser rearrangement competes with the [1,2]-Stevens rearrangement, in which the ylide fragments to benzyl and

[218] Z. Kosarych and T. Cohen, *Tetrahedron Lett.*, **23** (1982), 3019; M. P. Doyle, W. H. Tamblyn and V. Bagheri, *J. Org. Chem.*, **46** (1981), 5094.

[219] J. S. Clark and G. A. Whitlock, *Tetrahedron Lett.*, **35** (1994), 6381; M. C. Pirrung and J. A. Werner, *J. Am. Chem. Soc.*, **108** (1986), 6060; E. J. Roskamp and C. R. Johnson, *J. Am. Chem. Soc.*, **108** (1986), 6062; J. S. Clark, P. B. Hodgson, M. D. Goldsmith and L. J. Street, *J. Chem. Soc., Perkin Trans. 1* (2001), 3312.

[220] E. Vedejs, *Acc. Chem. Res.*, **17** (1984), 358.

[221] S. H. Pine, *Org. Reactions*, **18** (1970), 403.

[222] T. Tanaka, N. Shirai, J. Sugimori and Y. Sato, *J. Org. Chem.*, **57** (1992), 5034.

α-amino radicals which then recombine, in this case to give the product **329**.

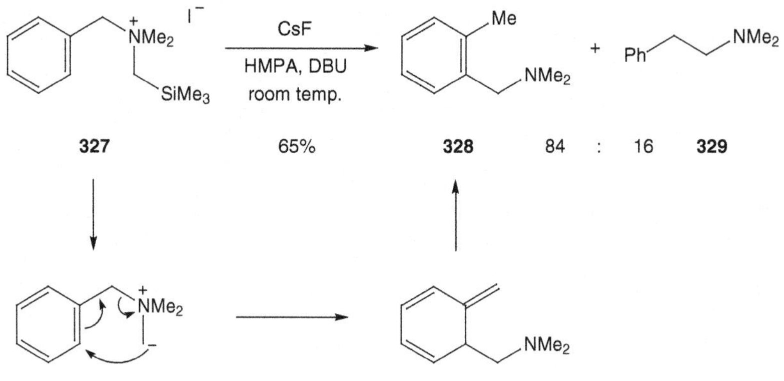

(3.209)

The above [2,3]-sigmatropic rearrangements generate a new carbon–carbon bond by making use of a carbanion α- to the heteroatom. A useful [2,3]-sigmatropic rearrangement in which a new carbon–heteroatom bond is formed is the sulfoxide–sulfenate rearrangement.[223] On warming, allyl sulfoxides (normally prepared by oxidation of allylic sulfides) are partly converted in a reversible reaction into rearranged allyl sulfenates. The equilibrium is usually much in favour of the sulfoxide, but if the mixture is treated with a thiophile (such as trimethyl phosphite) then the oxygen–sulfur bond of the sulfenate is cleaved to give an allylic alcohol. Even if the sulfenate is present in low equilibrium concentration, its removal by reaction with the thiophile results in conversion of the sulfoxide to the rearranged allylic alcohol in high yield. The rearrangement step occurs through a five-membered cyclic transition state and is stereoselective, leading, in the acyclic series, to predominantly the *E*-allylic alcohol (3.210).

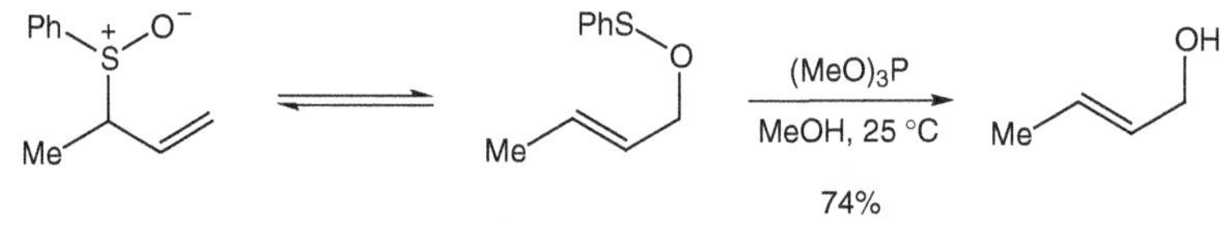

(3.210)

One property of the sulfoxide group is its ability to stabilize an adjacent carbanion. The combination of alkylation of the sulfoxide and its subsequent rearrangement leads to the synthesis of substituted allylic alcohols. For example, formation of the sulfenate **330** promotes rearrangement to the allyl sulfoxide **331** (3.211). Alkylation of the sulfoxide **331** gave the new sulfoxide **332** and rearrangement

[223] D. A. Evans and G. C. Andrews, *Acc. Chem. Res.*, **7** (1974), 147.

gave the alkylated allylic alcohol **333**.

$$(3.211)$$

Warming the corresponding allylic amine *N*-oxide (prepared from the tertiary amine and an oxidizing agent such as H_2O_2 or mCPBA) promotes the [2,3]-sigmatropic rearrangement to a hydroxylamine product in what is known as the [2,3]-Meisenheimer rearrangement (3.212). The *N–O* bond in the hydroxylamine can be cleaved, for example with zinc in acetic acid and ultrasound, to give the allylic alcohol product.

$$(3.212)$$

3.8 Electrocyclic reactions

Electrocyclic reactions are a class of pericyclic reactions in which a conjugated polyene interconverts with an unsaturated cyclic compound containing one less carbon–carbon double bond than the polyene.[224] The reactions can be promoted thermally or photochemically and take place with a very high degree of stereoselectivity.

A common type of electrocyclic reaction is the ring-opening of a cyclobutene to a butadiene.[225] The stereochemistry of the new alkene(s) in the diene can be interpreted on the basis of the Woodward–Hoffmann rules. For a four electron component, thermal ring-opening occurs by a conrotatory process (both terminal *p*-orbitals moving clockwise or anticlockwise), whereas the photochemical reaction

[224] E. N. Marvell, *Thermal Electrocyclic Reactions* (New York: Academic Press, 1980).
[225] T. Durst and L. Breau, in *Comprehensive Organic Synthesis*, ed. B. M. Trost and I. Fleming, vol. 5 (Oxford: Pergamon Press, 1991), p. 675.

occurs by a disrotatory process (3.213). This stereospecificity can be explained by the necessity to overlap orbitals of like sign in the highest occupied molecular orbital (HOMO) of the polyene.

(3.213)

In the thermal ring-opening of cyclobutenes, substituents tend to prefer an 'outward' motion to give the *E*-alkene, although a π-acceptor such as a carbonyl group can undergo preferential 'inward' motion. Hence ring-opening of the aldehyde **335**, formed by oxidation of the alcohol **334**, occurs to give the diene **336** with >97% isomeric purity, in which the aldehyde rather than the alkyl group has rotated 'inward' (3.214).[226]

An important electrocyclic reaction is the ring-opening of benzocyclobutenes to give *o*-quinodimethanes. The resulting diene is an excellent substrate for reaction with a dienophile in a Diels–Alder reaction (see Section 3.1.2). For example, in a synthesis of the steroid estrone, the benzocyclobutane **337**, prepared by a cobalt-mediated cyclotrimerization, was converted on heating to the *o*-quinodimethane

[226] F. Binns, R. Hayes, K. J. Hodgetts, S. T. Saengchantara, T. W. Wallace and C. J. Wallis, *Tetrahedron*, **52** (1996), 3631.

338, which undergoes cycloaddition to the tetracycle **339** (3.215).[227]

$$(3.215)$$

The electrocyclic reaction of a 1,3,5-hexatriene to give a cyclohexadiene provides an entry to unsaturated six-membered rings.[228] The central alkene double bond of the triene must possess the *Z* geometry for successful electrocyclization. A consideration of orbital symmetry (using the HOMO of the triene) allows the prediction that the thermal six-electron process occurs by a disrotatory pathway, whereas the photochemical reaction occurs by a conrotatory pathway. Indeed, thermal and photochemical induced cyclizations, for example in early work on the rearrangement of precalciferol (previtamin D), give complementary stereochemical results. Further confirmation of the thermal disrotatory electrocyclic reaction has been gained by heating the trienes **340** and **342** (3.216). The *E,Z,E*-triene **340** is converted to the *cis* product **341** with >99.5% diastereomeric purity, indicating a completely stereospecific disrotatory electrocyclization. The isomeric *E,Z,Z*-triene **342** is converted to the *trans* product **343**. In this latter case the product **344** (derived from the diene **343** by a 1,5-hydrogen shift) is also produced and the triene **342** has been found to interconvert readily with the *Z,Z,Z*-isomer by consecutive 1,7-hydrogen shifts.

$$(3.216)$$

Problems of competing hydrogen shifts and the difficulty of preparing the required triene as a single geometrical isomer, particularly in acyclic substrates, has limited the use of this reaction in synthesis. When the central double bond is

[227] R. L. Funk and K. P. C. Vollhardt, *J. Am. Chem. Soc.*, **102** (1980), 5253.
[228] W. H. Okamura and A. R. De Lera, in *Comprehensive Organic Synthesis*, ed. B. M. Trost and I. Fleming, vol. 5 (Oxford: Pergamon Press, 1991), p. 699.

part of a ring, then it is locked in the required Z geometry. The rearrangement of 1,2-divinyl-aromatic or heteroaromatic compounds has provided a useful entry to polycyclic aromatic compounds. For example, in a synthesis of the carbazole hyellazole **346**, the divinyl-indole **345** was heated to promote electrocyclization, followed by dehydrogenation with palladium on charcoal to give hyellazole (3.217).[229]

(3.217)

A common type of six electron electrocyclic reaction occurs in the photochemical reaction of 1,2-diaryl alkenes.[230] The parent substrate, stilbene can be converted to phenanthrene, a process that involves conrotatory electrocyclization under photochemical conditions and subsequent oxidation of the product to the polycyclic aromatic structure (3.218).

(3.218)

There are many examples of this type of reaction with both aromatic and heteroaromatic substrates. For successful electrocyclization, the central alkene must have Z geometry, however, as the action of light on stilbenes promotes E–Z isomerization, it is possible to start with either geometrical isomer of the substrate, or indeed a mixture of isomers. In a synthesis of cervinomycin A, photochemical electrocyclization of the mixture of E- and Z-diaryl alkenes **347** gave the polycyclic aromatic compound **348** after oxidation with iodine (3.219).[231]

(3.219)

[229] S. Kano, E. Sugino, S. Shibuya and S. Hibino, *J. Org. Chem.*, **46** (1981), 3856.
[230] F. B. Mallory and C. W. Mallory, *Org. Reactions*, **30** (1984), 1; W. H. Laarhoven, *Org. Photochem.*, **10** (1989), 163.
[231] G. Mehta, S. R. Shah and Y. Venkateswarhu, *Tetrahedron*, **50** (1994), 11 729.

Electrocyclic reactions are not limited to neutral polyenes. The cyclization of a pentadienyl cation to a cyclopentenyl cation offers a useful entry to five-membered carbocyclic compounds. One such reaction is the Nazarov cyclization of divinyl ketones.[232] Protonation or Lewis acid complexation of the oxygen atom of the carbonyl group of a divinyl ketone generates a pentadienyl cation. This cation undergoes electrocyclization to give an allyl cation within a cyclopentane ring. The allyl cation can lose a proton or be trapped, for example by a nucleophile. Proton loss occurs to give the thermodynamically more stable alkene and subsequent keto–enol tautomerism leads to the typical Nazarov product, a cyclopentenone (3.220).

$$(3.220)$$

In cases that provide a mixture of alkene regioisomers or in which the less-substituted alkene is desired, control of the position of the new alkene is possible using a trialkylsilyl group to direct its introduction. Desilylation is generally preferred over deprotonation, and the known β-cation stabilizing effect of a silyl group helps to reduce side reactions resulting from the intermediate allyl cation. The silicon-directed Nazarov cyclization has been made use of twice in a synthesis of the sesquiterpene $\Delta^{9(12)}$-capnellene **349** (3.221).[233]

$$(3.221)$$

349

The Nazarov cyclization is a four-electron cyclization and occurs thermally by a conrotatory process. The stereochemical outcome across the new carbon–carbon bond is often obscured by the loss of a proton at one of these centres during the cyclopentenone formation. If, however, the proton loss occurs *exo* to the five-membered ring or if the allyl cation is quenched by a nucleophile, then the stereochemistry can be observed. For example, trapping the allyl cation by reduction with

[232] K. L. Habermas, S. E. Denmark and T. K. Jones, *Org. Reactions*, **45** (1994), 1; S. E. Denmark, in *Comprehensive Organic Synthesis*, ed. B. M. Trost and I. Fleming, vol. 5 (Oxford: Pergamon Press, 1991), p. 751.
[233] G. T. Crisp, W. J. Scott and J. K. Stille, *J. Am. Chem. Soc.*, **106** (1984), 7500.

triethylsilane reveals the *trans* arrangement (arising from a conrotatory cyclization) of the two phenyl groups in the product **350** (3.222).[234]

(3.222)

350

Problems (answers can be found on page 472)

1. Explain why intermolecular Diels–Alder cycloaddition reactions usually fail with unactivated dienophiles such as ethene.

2. Diels–Alder reactions with nitroethene offer a method to carry out the equivalent of cycloaddition with ethene, such as in a synthesis of frondosin B, below. Draw the structure of the Diels–Alder adduct **1**.

R = Me

R = H, frondosin B

3. Explain the formation of the cycloadduct **2**, used in a synthesis of hybocarpone.

2

4. Draw the structures of the intermediates and hence explain the formation of the diazaindoline **3**. Draw the structure of the product **4**, used in a synthesis of a selection of *amaryllidaceae* alkaloids.

[234] S. Giese and F. G. West, *Tetrahedron*, **56** (2000), 10 221.

3

5. Draw the structure of the cycloadduct **5** and explain why the preparation of this compound is best carried out as a one-pot procedure, rather than by isolation of the diene and separate heating with the dienophile in toluene.

6. Draw the structure of the major stereoisomer of the cycloadduct **6**. (Hint: use the Felkin–Anh model to explain the stereochemistry.)

7. Draw the structure of the intermediate **7** and explain its formation.

8. Explain the formation of the isoxazoline **8**, formed in the following cycloaddition reaction.

9. Explain the formation of the pyrrolidine **10**, prepared by heating a mixture of the aldehyde **9** and *N*-methyl-glycine.

10. Draw a mechanism to explain the transformation given below.

11. Suggest reagents for the conversion of the alcohol **11** to the ester **12**.

12. Explain the formation of the aldehyde **14** on treatment of the ether **13** with potassium hydride. (Hint: two consecutive sigmatropic rearrangements are involved.)

13   **14**

13. Explain the formation of the enone **16**, from the triene **15**.

15   **16**

4

Radical and carbene chemistry

The previous chapters have concentrated on ionic or pericyclic reactions that give rise to new carbon–carbon single or double bonds. Reactive carbon- or heteroatom-centred radicals and carbenes allow alternative strategies for organic synthesis that have been used extensively. Radicals and carbenes are neutral, electron-deficient species that are not commonly isolable (although a few stable examples exist). Carbon-centred radicals are trivalent with a single non-bonding electron, whereas carbenes are divalent with two non-bonding electrons (4.1). Their ease of formation combined with their high reactivity yet tolerance to many functional groups, and their contrasting behaviour with many ionic species has promoted much use of these intermediates in synthesis. This chapter deals with salient aspects of their chemistry.

$$R_3C^- \qquad R_3C^{\bullet} \qquad R_3C^+ \qquad R_2C\colon \qquad (4.1)$$

carbanion   carbon radical   carbocation   carbene
                                   (or carbenium ion)

4.1 Radicals

Radicals can be generated by homolysis of weak σ-bonds. Homolysis is effected by photochemical, thermal or redox (electron transfer) methods. A common method to initiate a radical reaction is to warm a peroxide such as benzoyl peroxide or azobisisobutyronitrile (AIBN) **1** (4.2). The radical $\cdot C(CN)Me_2$ generated from AIBN is rather unreactive, but is capable of abstracting a hydrogen atom from weakly bonded molecules such as tributyltin hydride (4.3). The resulting tributyltin radical reacts readily with alkyl halides, selenides and other substrates to form a carbon-centred radical.

$$(4.2)$$

$$\text{Bu}_3\text{SnH} \quad + \quad \overset{\bullet}{\underset{}{\diagup}}\text{CN} \quad \longrightarrow \quad \text{Bu}_3\text{Sn}^\bullet \quad + \quad \overset{H}{\underset{}{\diagup}}\text{CN} \quad (4.3)$$

Most carbon-centred radicals are reactive and combine readily with a neutral species by abstraction or addition, or undergo elimination to generate a new radical species. A radical chain reaction is therefore set up and can lead to useful function-alized products.[1] Alternatively, radical–radical combination is possible, leading to a neutral product and this can form the termination step or indeed the key bond-forming step, as in the pinacol reaction of ketyl radicals (see Section 2.9).

Most of the useful radical reactions in synthetic chemistry involve a chain mechanism, in which radical species are continually regenerated and trapped. Such propagation steps are illustrated for reduction of a substrate RX (4.4). The feasibility of this sequence depends on the relative reaction rates which themselves are determined by the structures of the radicals (including that used to initiate the reaction). In reactions such as this, the trialkyltin radical is sometimes referred to as the chain carrier as it is continuously regenerated to propagate the cycle.

$$\text{R}{-}\text{X} \qquad {}^\bullet\text{SnBu}_3 \qquad \longrightarrow \qquad \text{R}^\bullet \quad + \quad \text{XSnBu}_3$$

$$(4.4)$$

$$\text{R}^\bullet \qquad \text{H}{-}\text{SnBu}_3 \qquad \longrightarrow \qquad \text{R}{-}\text{H} \quad + \quad {}^\bullet\text{SnBu}_3$$

4.1.1 Radical abstraction reactions

Scheme 4.4 can be regarded as a radical abstraction reaction, as the intermediate carbon-centred radical abstracts a hydrogen atom from the trialkyltin hydride (aided by the relatively weak H—Sn bond). Many examples of this process for dehalogenation of alkyl iodides or bromides in particular have been reported. For

[1] *Radicals in Organic Synthesis*, ed. P. Renaud and M. P. Sibi (New York: Wiley, 2001); J. Fossey, D. Lefort and J. Sorba, *Free Radicals in Organic Chemistry* (New York: Wiley, 1995); W. B. Motherwell and D. Crich, *Free Radical Chain Reactions in Organic Synthesis* (London: Academic Press, 1992); D. P. Curran, in *Comprehensive Organic Synthesis*, ed. B. M. Trost and I. Fleming, vol. 4 (Oxford: Pergamon Press, 1991), pp. 715, 779; B. Giese, *Radicals in Organic Synthesis: Formation of Carbon–Carbon Bonds* (Oxford: Pergamon Press, 1986).

example, a synthesis of the alkaloid epibatidine **3** made use of the radical debromination of the bromide **2** (4.5).[2] Other substrates such as tertiary or activated secondary nitro compounds can be reduced. Alkyl selenides are excellent substrates
for preparing carbon-centred radicals. Thus, in a synthesis of tylonolide hemiacetal,
de-hydroxylation of the alcohol **4** was accomplished via its selenide **5** (4.6).[3] Note
that the primary alcohol is converted more easily to the selenide than the secondary
alcohol, thereby allowing selective removal of one hydroxy group.

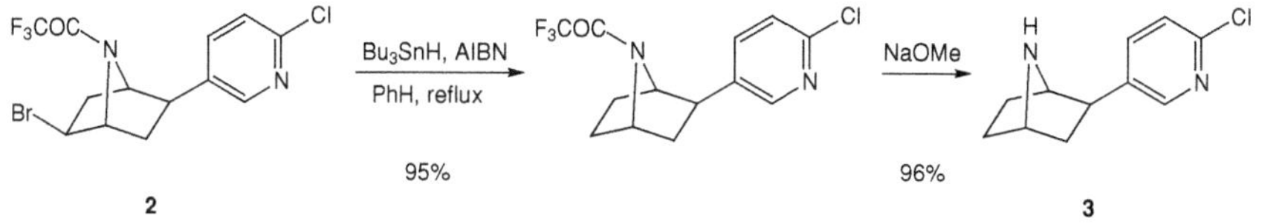

(4.5)

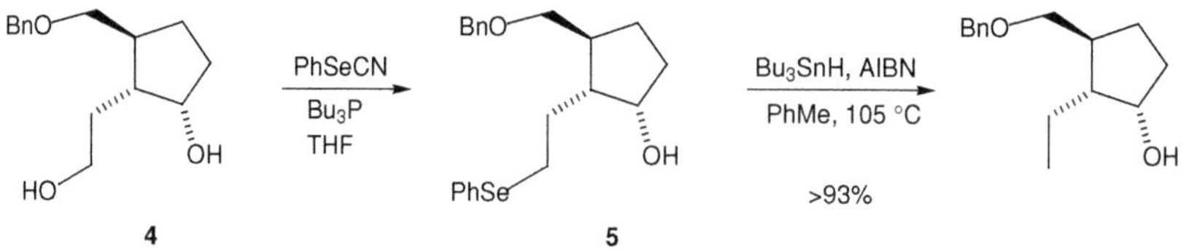

(4.6)

An alternative method for dehydroxylation via thiocarbonyl derivatives is popular. Thioacylation of the alcohol gives a thioester **6** (R' = SMe, OPh, imidazolyl,
etc.), which can be reduced under radical conditions (4.7).[4] The tributyltin radical
attacks the sulfur atom of the thiocarbonyl group to give a new radical **7**, which
fragments to give the desired carbon-centred radical R• and a carbonyl compound.
The radical R• then abstracts a hydrogen atom from tributyltin hydride, releasing
further tributyltin radical and giving the reduced product R−H.

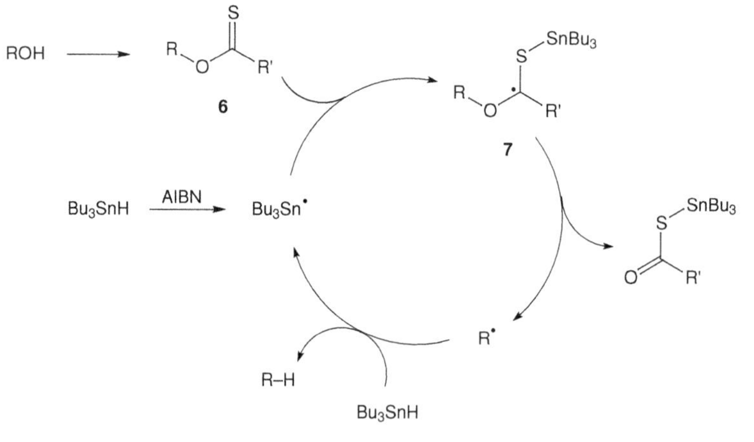

(4.7)

[2] E. J. Corey, T.-P. Loh, S. AchyuthaRao, D. C. Daley and S. Sarshar, *J. Org. Chem.*, **58** (1993), 5600.
[3] P. A. Grieco, J. Inanaga, N.-H. Lin and T. Yanami, *J. Am. Chem. Soc.*, **104** (1982), 5781.
[4] D. H. R. Barton and S. W. McCombie, *J. Chem. Soc., Perkin Trans. 1* (1975), 1574; W. Hartwig, *Tetrahedron*, **39** (1983), 2609; D. Crich and L. Quintero, *Chem. Rev.*, **89** (1989), 1413.

Radical dehydroxylation is most effective for secondary alcohols, including those derived from carbohydrates, in which traditional methods such as tosylation (or mesylation) and LiAlH₄ reduction often fail. The reaction tolerates many different functional groups, as illustrated in the reduction of the thiocarbonyl compound **8** (4.8).[5]

In addition to dehydroxylation, a useful protocol for decarboxylation has been developed.[6] The procedure was introduced by Barton, using thiohydroxamic esters **9**, prepared from activated carboxylic acids (RCOX) and the sodium salt of *N*-hydroxypyridine-2-thione. Simple thermolysis or photolysis of the esters (homolysis of the N—O bond) results in the production of alkyl radicals R•, which can attack the sulfur atom of the thiocarbonyl group to propagate the fragmentation (4.9).

In the presence of a hydrogen-atom source, such as tributyltin hydride or a thiol (R′SH), the alkyl radical is reduced and the reaction is propagated by the chain carrier (Bu₃Sn• or R′S•). Thus, in a synthesis of a segment of the immunosuppressant FK-506, decarboxylation was effected by heating the thiohydroxamic ester **10** with tert-butyl thiol (4.10).[7]

[5] L. A. Paquette and J. A. Oplinger, *J. Org. Chem.*, **53** (1988), 2953.
[6] D. H. R. Barton, D. Crich and W. B. Motherwell, *Tetrahedron*, **41** (1985), 3901.
[7] P. Kocienski, M. Stocks, D. Donald and M. Perry, *Synlett* (1990), 38.

The use of the thiohydroxamic ester has the advantage that the intermediate alkyl radical can be generated in the absence of tributyltin hydride (or other hydrogen-atom source). Therefore, in the presence of a suitable radical trap, the alkyl radical can be functionalized rather than simply reduced. Thus, in the presence of CCl_4, $BrCCl_3$ or CHI_3, the carboxylic acid RCO_2H can be decarboxylated and halogenated to give the alkyl halide RCl, RBr or RI. In the presence of oxygen gas, a hydroperoxide ROOH or alcohol product ROH can be formed.

An alternative method for forming a carbon-centred radical that can be trapped with a variety of neutral molecules such as a hydrogen-atom source (to give an alkane) or molecular oxygen (to give an alcohol) derives from organomercury compounds. The organomercury compounds can be prepared from Grignard reagents or by addition of mercury salts to alkenes. The resulting alkyl mercury halide or acetate can be reduced with $NaBH_4$ to give an alkyl mercury hydride, which fragments to give the radical species (4.11).[8] The radical abstracts a hydrogen atom from the alkyl mercury hydride to continue the cycle. In the presence of oxygen, the radical is trapped to give a new carbon–oxygen bond.[9] Thus, in an approach to the allosamidin disaccharides, the alkyl mercury acetate **12**, formed from the alkene **11** by amino-mercuration, was converted to the alcohol **13** (4.12).[10] The intermediate alkyl radical is not configurationally stable (it has considerable sp^2 character) but reacts with oxygen on the less hindered (convex) face of the molecule.

$$R-HgX \xrightarrow{NaBH_4} [R-HgH] \longrightarrow R^\bullet \longrightarrow R-H \qquad (4.11)$$

(4.12)

11   **12**   **13**

R = SiMe₂ᵗBu

Radicals are reactive species that readily abstract a hydrogen atom from metal hydrides. In some cases, in particular with substrates that meet certain structural and geometrical requirements, intramolecular $C-H$ abstraction can take place. In this way, a new radical can be generated at an unactivated position, thereby allowing the introduction of functional groups at this position. The geometrical requirements dictate that the most frequently observed intramolecular hydrogen transfers are 1,5-shifts, corresponding to specific attack on a hydrogen atom attached to a

[8] J. Barluenga and M. Yus, *Chem. Rev.*, **88** (1988), 487.
[9] C. L. Hill and G. M. Whitesides, *J. Am. Chem. Soc.*, **96** (1974), 870.
[10] W. D. Shrader and B. Imperiali, *Tetrahedron Lett.*, **37** (1996), 599.

carbon atom 5 atoms from the initial radical (4.13).[11] Homolytic cleavage of the
Y—X bond gives a radical Y• (normally nitrogen- or oxygen-centred), which is followed by hydrogen atom transfer. The resulting (more stable) carbon radical reacts
with a neutral molecule or with a radical X'•, which may or may not be identical
with X•.

(4.13)

One such example is the Hofmann–Löffler–Freytag reaction, which provides a
method for the synthesis of pyrrolidines from *N*-halogenated amines. The reaction is
effected by warming a solution of the halogenated amine in strong acid (e.g. H_2SO_4
or CF_3CO_2H), or by irradiation of the acid solution with ultra-violet light. The
initial product of the reaction is the δ-halogenated amine, but this is not generally
isolated, and by basification of the reaction mixture it is converted directly to
the pyrrolidine (4.14). Both *N*-bromo- and *N*-chloro-amines have been used as
substrates, although the *N*-chloro-amines usually give slightly better yields. The *N*-
chloro-amines can be obtained from the amines by the action of sodium hypochlorite
or *N*-chlorosuccinimide.

(4.14)

Thermal or photochemical dissociation of the *N*-chloro-ammonium salt, formed
by protonation of the *N*-chloro-amine, is thought to give the reactive ammonium
radical species (4.15). This abstracts a suitably situated hydrogen atom to give
the corresponding carbon radical. This in turn abstracts a chlorine atom from
another molecule of the *N*-chloro-ammonium salt, thus propagating the chain and
at the same time forming the δ-chloro amine, from which the cyclic amine is
obtained.

(4.15)

The first example of this type of reaction was reported by Hofmann in 1883. In
the course of a study of the reactions of *N*-bromo-amides and *N*-bromo-amines, he

---

[11] G. Majetich and K. Wheless, *Tetrahedron*, **51** (1995), 7095; H. Togo and M. Katohgi, *Synlett* (2001), 565.

treated *N*-bromo-coniine **14** with hot sulfuric acid and obtained, after basification, a tertiary base that was later identified as δ-coneceine (4.16). Further examples of the reaction were reported later by Löffler, including a synthesis of the alkaloid nicotine (4.17). Many other cyclizations leading to simple pyrrolidines and to more complex polycyclic structures have since been reported.

(4.16)

14   δ-coneceine

(4.17)

nicotine

The radical nature of the reaction is supported by a number of factors, including the fact that the reaction does not proceed in the dark at room temperature and that it is initiated by heat, light or iron(II) salts and inhibited by oxygen. The hydrogen abstraction step must be intramolecular in order to explain the specificity of reaction at the δ-carbon atom. Strong evidence for an intermediate carbon-centred radical that is trigonal is provided by the observation that the optically active *N*-chloro-amine **15**, on decomposition in acid on warming, gave the pyrrolidines **16** and **17** which were optically inactive (4.18). The intermediacy of δ-chloro-amines has been confirmed by their isolation in a few cases.

(4.18)

43%

15   **16** 22 : 78 **17**

As with other radical reactions, secondary hydrogen atoms react more readily than primary as the resulting secondary radical is more stable. Thus, in the reaction of *N*-chloro-amine **18**, attack by the nitrogen-centred radical on the δ-methyl group would lead to *N*-pentylpyrrolidine, whereas attack on the δ'-methylene would result in the formation of *N*-butyl-2-methylpyrrolidine (4.19). Only the latter compound was formed. Tertiary hydrogen atoms react very readily, but the resulting tertiary halides do not normally proceed to give cyclic amine products.

$$
\text{18} \xrightarrow[\text{heat}]{\text{H}_2\text{SO}_4} \qquad\qquad (4.19)
$$

18

An application of the Hofmann–Löffler–Freytag reaction is found in the synthesis of the steroidal alkaloid derivative dihydro-conessine **19** (4.20).[12] In this synthesis, the pyrrolidine ring is constructed by attack on the unactivated C-18 angular methyl group of the precursor by a suitably placed nitrogen radical. The ease of this reaction is a result of the fact that in the rigid steroid framework, the C-18 angular methyl group and C-20 side chain carrying the nitrogen radical are suitably disposed in space to allow easy formation of the six-membered transition state necessary for 1,5-hydrogen atom transfer.

$$
\xrightarrow[\text{ii, NaOH}]{\text{i, H}_2\text{SO}_4,\ h\nu} \qquad 79\% \qquad\qquad (4.20)
$$

19

A modification of the Hofmann–Löffler–Freytag reaction that avoids the harshly acidic conditions described above has been developed. The *N*-iodo compound is generated by reaction with iodine and iodobenzene diacetate.[11] Warming or irradiating the reaction mixture promotes the formation of the nitrogen-centred radical and hence subsequent remote hydrogen atom abstraction. The reaction is particularly effective with carboxylic amides, sulfonamides or phosphoramidates, as illustrated in the transannular cyclization to give the indolizidine **20** and in the formation of the bicyclic product **21** (4.21, 4.22).[13]

[12] E. J. Corey and W. R. Hertler, *J. Am. Chem. Soc.*, **81** (1959), 5209.
[13] R. L. Dorta, C. G. Francisco and E. Suárez, *J. Chem. Soc., Chem. Commun.* (1989), 1168; C. G. Francisco, A. J. Herrera and E. Suárez, *Tetrahedron: Asymmetry*, **11** (2000), 3879.

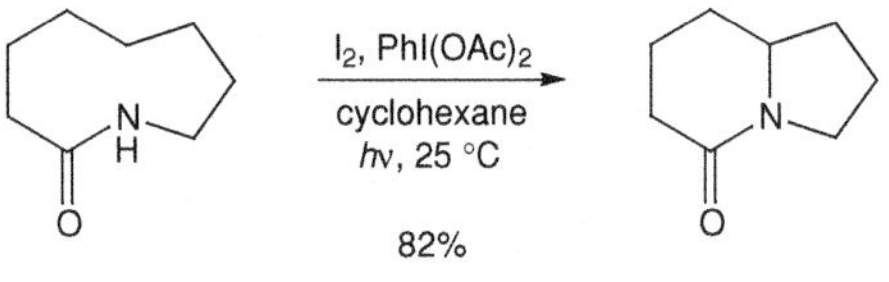

$$(4.21)$$

20

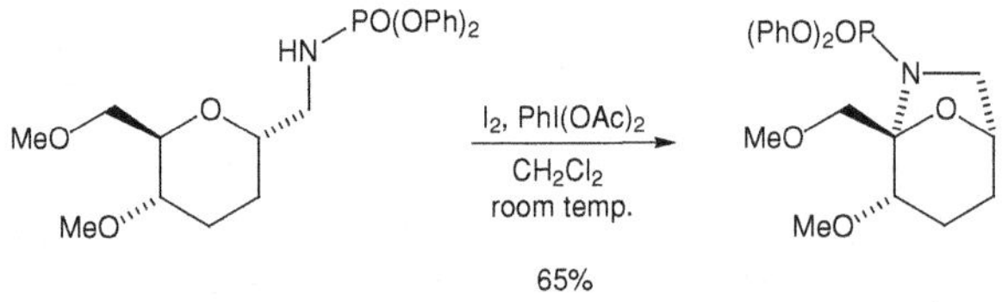

$$(4.22)$$

21

Oxygen-centred radicals can also be used for remote functionalization.[14] Heating or photolysis of an organic nitrite (RO—N=O) gives an alkoxy radical and nitrogen monoxide. Subsequent intramolecular hydrogen atom abstraction is followed by capture of nitrogen monoxide by the carbon radical and formation of a nitroso-alcohol, which may be isolated as the dimer or rearrange, where possible, to an oxime (4.23). The nitroso or oxime products may be further transformed into other functional groups such as carbonyl compounds, amines or nitrile derivatives. The photolytic conversion of organic nitrites into nitroso compounds has become known as the Barton reaction and the sequence has found most use in the synthesis of steroid derivatives.

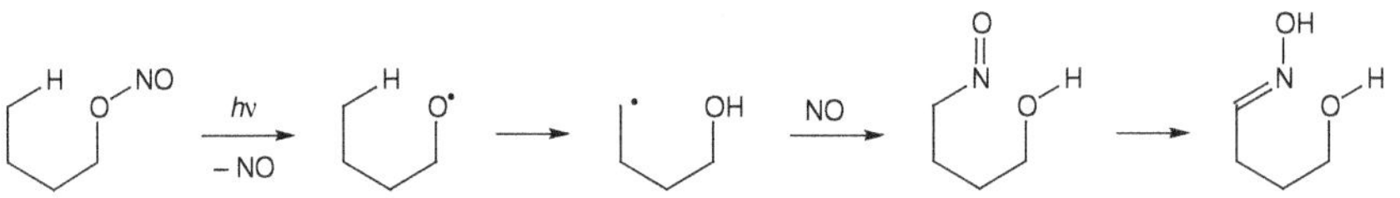

$$(4.23)$$

The nitrite can be prepared from the alcohol and has weak absorption bands in the region 320–380 nm. Irradiation using a Pyrex filter to limit the wavelengths to greater than 300 nm, thus avoiding side-reactions induced by more-energetic lower-wavelength radiation, brings about the dissociation of the nitrite.

A classic example of the Barton reaction is the key step in a synthesis of the acetate **24** of aldosterone, a hormone of the adrenal cortex (4.24).[15] Photolysis of the nitrite **22** provided the oxime **23**, which on hydrolysis with nitrous acid gave aldosterone-21-acetate directly. In this case the yield is limited in part by competing

---

[14] Z. Cekovic, *Tetrahedron*, **59** (2003), 8073.
[15] D. H. R. Barton and J. M. Beaton, *J. Am. Chem. Soc.*, **83** (1961), 4083.

attack of the alkoxy radical at the C-19 methyl group instead of C-18, which led to the side-product **25** (4.25).

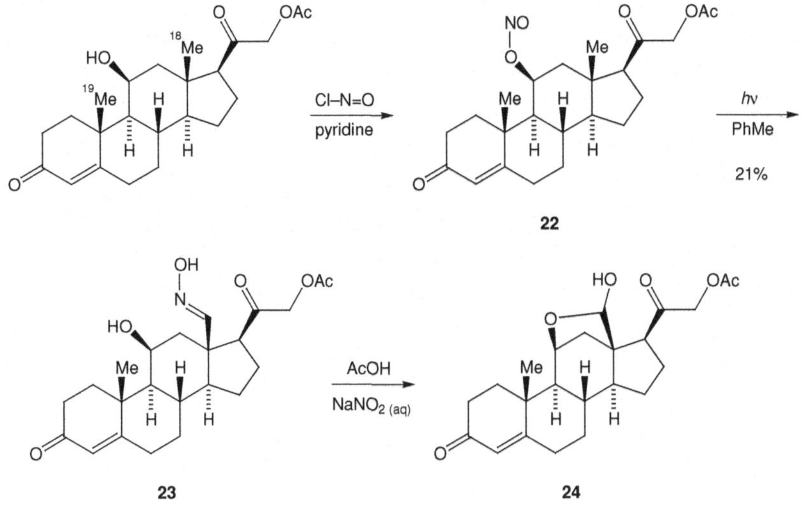

22

(4.24)

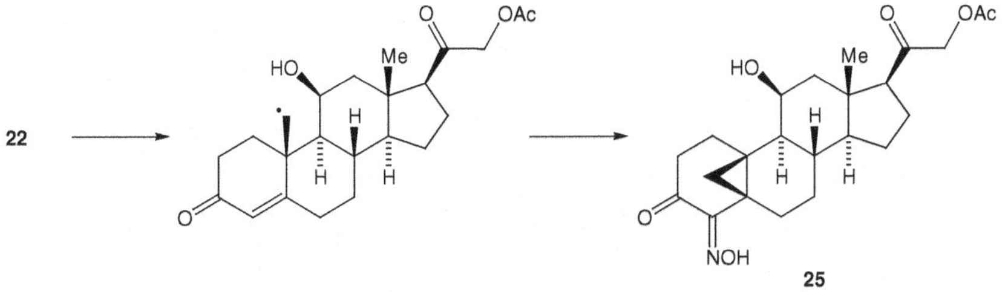

23   **24**

22 ⟶

(4.25)

25

In another application of this chemistry, photolysis of the nitrite **26** was the key step in Corey's synthesis of perhydrohistrionicotoxin (4.26).[16] The oxime **27** could be converted to the spirocyclic lactam **28** on Beckmann rearrangement.

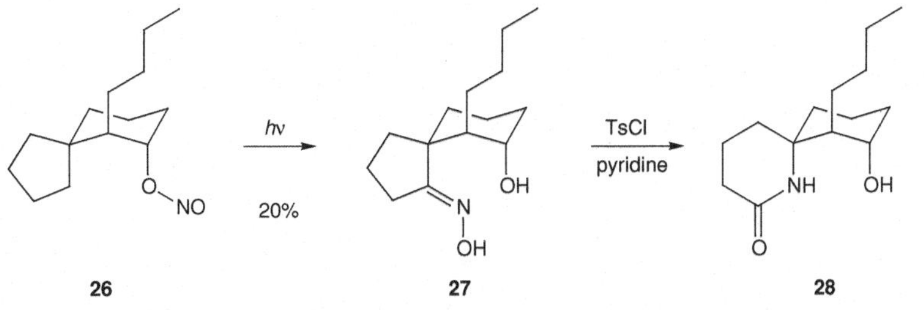

(4.26)

26   **27**   **28**

A modified procedure leads to the introduction of a hydroxyl group at the site of an unactivated C–H bond, a reaction that is common in nature but that is not easily

[16] E. J. Corey, J. F. Arnett and G. N. Widiger, *J. Am. Chem. Soc.*, **97** (1975), 430.

effected in the laboratory. If photolysis of the nitrite is carried out in the presence of oxygen, the product of rearrangement is a nitrate (rather than a nitroso compound or an oxime). The nitrate can be converted into the corresponding alcohol by mild reduction. The reaction is believed to take a pathway in which the initial carbon radical is captured by oxygen instead of nitric oxide.

The generation of alkoxy radicals that can undergo intramolecular hydrogen abstraction can also be achieved by photolysis of hypohalites. Photolysis of a hypochlorite (RO—Cl) gives a 1,4-chloro-alcohol, formed as expected by abstraction of a hydrogen atom attached to the δ-carbon atom. The 1,4-chloro-alcohol can be converted readily to a tetrahydrofuran product (4.27). The hydrogen abstraction reaction proceeds through a six-membered cyclic transition state as in the photolysis of nitrites.

A competing reaction is β-cleavage of the alkoxy radical to form a carbonyl compound and a carbon radical (4.28). The extent of this reaction varies with the structure of the substrate and will predominate if 1,5-hydrogen atom abstraction is unfavourable.

$$(4.27)$$

$$(4.28)$$

A convenient method for generating the alkoxy radical is by fragmentation of hypoiodites prepared *in situ* from the corresponding alcohol. This can be accomplished by treatment of the alcohol with iodine and lead tetraacetate or mercury(II) oxide, or with iodine with iodobenzene diacetate. For example, irradiation of the alcohol **29** under these latter conditions gave a high yield of the tetrahydrofuran **30** (4.29).[17] In another application of this chemistry, for the specific deprotection of benzyl ethers, irradiation of the alcohol **31** and *N*-iodosuccinimide (NIS) gave the cyclic acetal **32** (4.30).[18]

[17] P. de Armas, J. I. Concepción, C. G. Francisco, R. Hernández, J. A. Salazar and E. Suárez and *J. Chem. Soc., Perkin Trans. 1* (1989), 405.
[18] J. Madsen, C. Viuf and M. Bols, *Chem. Eur. J.*, **6** (2000), 1140.

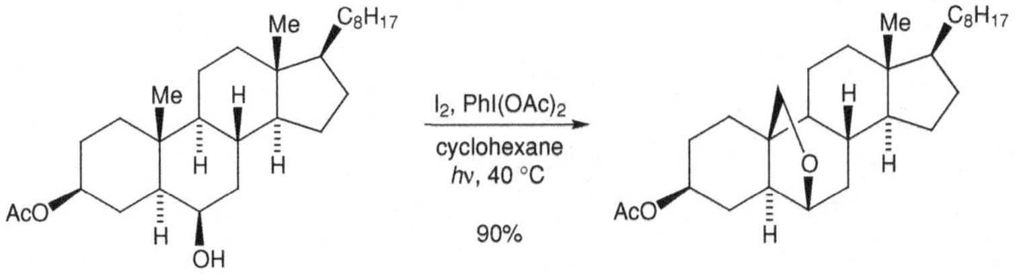

(4.29)

29   **30**

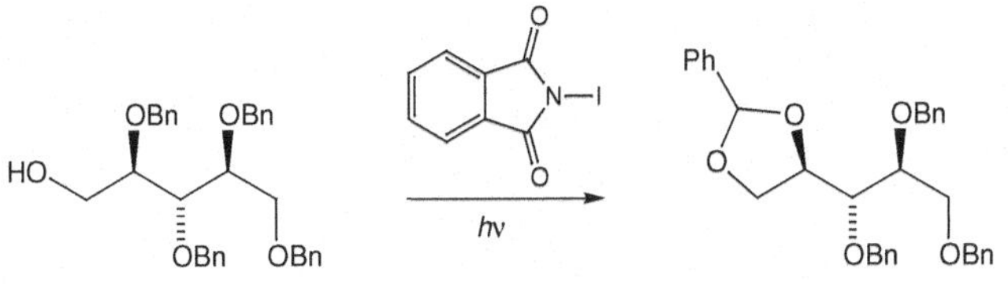

(4.30)

31   **32**

There are an increasing number of examples of the use of alkoxy radicals in β-cleavage processes (4.28). Thus, in a synthesis of 8-deoxyvernolepin, treatment of the hemiacetal **33** with iodine and iodobenzene diacetate and irradiation gave the lactone **34** (4.31).[19] The alkoxy radical generated from **33** is not set up for 1,5-hydrogen abstraction and undergoes β-cleavage to the lactone. The cleavage reaction is regioselective and might have been expected to take place to give the more-stable secondary carbon radical; however, the ease of elimination of the tributyltin radical promotes fragmentation on the side of the primary carbon atom.

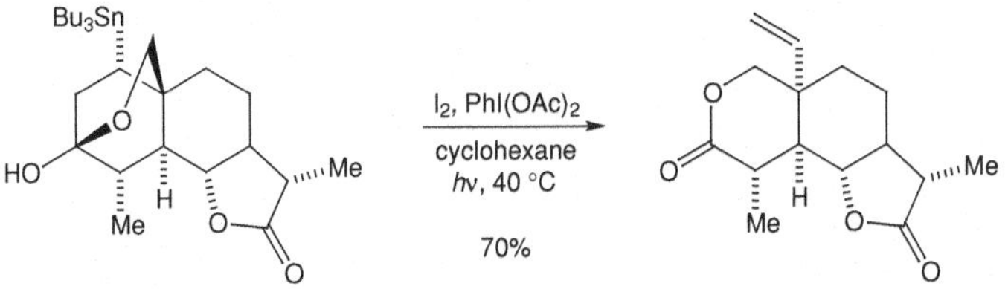

(4.31)

33   **34**

If the β-cleavage reaction gives rise to a carbon radical located α- to an oxygen or nitrogen atom, then the resulting oxonium or iminium ion can be trapped with a nucleophile. An example of this process with intramolecular trapping to give an azasugar ring system is depicted in Scheme 4.32.[20]

[19] R. Hernández, S. M. Velázquez, E. Suárez and M. S. Rodríguez, *J. Org. Chem.*, **59** (1994), 6395.
[20] C. G. Francisco, R. Freire, C. C. González, E. I. León, C. Riesco-Fagundo and E. Suárez, *J. Org. Chem.*, **66** (2001), 1861.

(4.32)

4.1.2 Radical addition reactions

In the presence of a double or triple bond, a radical species can undergo an addition reaction. It has been known for many years that alkyl radicals add to the double bond of alkenes with the formation of a new carbon–carbon bond. The reaction, of course, forms the basis of important industrial processes for making polymers. In organic synthesis, the use of radical chemistry has some advantages over ionic reactions. For example, radicals rarely undergo molecular rearrangements or elimination reactions, sometimes encountered in ionic reactions involving carbocations or carbanions. Further, the rate of reaction of alkyl radicals with many functional groups (hydroxy, ester, halogen) is slow compared with their rates of addition to carbon–carbon double bonds, so that it is often possible to bring about the addition without the need for protection and deprotection of functional groups.

Radical addition reactions involve a number of propagation steps and the relative rates of the reactions of the intermediate radicals and the concentration of any neutral species that permits abstraction, such as tributyltin hydride, becomes critical to the outcome. For successful reaction, the initial alkyl radical must undergo addition faster than it abstracts a hydrogen atom from a hydrogen donor, but the adduct radical must react faster with the hydrogen donor than with the double bond, which would lead to polymerization. In addition, the propagation steps must compete favourably with chain-terminating radical combination. The rate of addition of radicals to alkenes depends on the nature of the substituent groups on the double bond and for alkyl radicals is greater in the presence of electron-withdrawing substituents. The orientation of addition is influenced by steric and electronic factors, and addition of alkyl radicals to monosubstituted alkenes takes place predominantly at the unsubstituted carbon atom of the double bond (4.33).

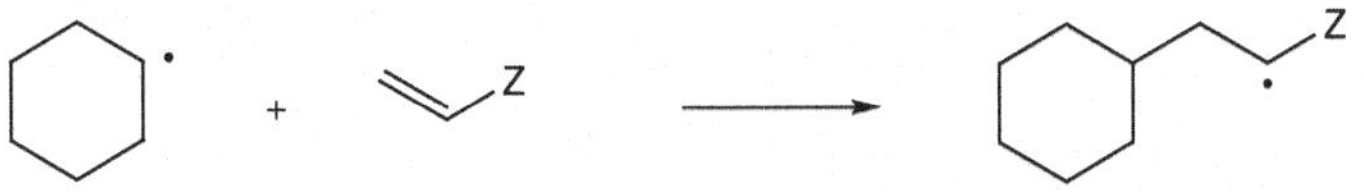

(4.33)

Z	k (relative)
H	1
Ph	65
CO_2Me	450
CHO	2300

The alkyl radicals used in these reactions may be generated in a number of ways, as described previously in this chapter, by using, for example, light, a peroxide or trialkyltin hydride and azobisisobutyronitrile (AIBN). Thus, in a synthesis of the pheromone *exo*-brevicomin **37**, the radical addition product **36** was obtained from the iodide **35** and methyl vinyl ketone using the tributyltin hydride method (4.34).[21]

Intermolecular addition of an alkyl radical to an electron-deficient alkene bearing a chiral auxiliary or even in the presence of an external chiral ligand is possible.[22] High selectivities have been achieved in the presence of Lewis acid catalysts.[23] The Lewis acid promotes chelation, for example of two carbonyl groups, thereby reducing the conformational freedom and favouring addition to one face of the alkene (4.35).

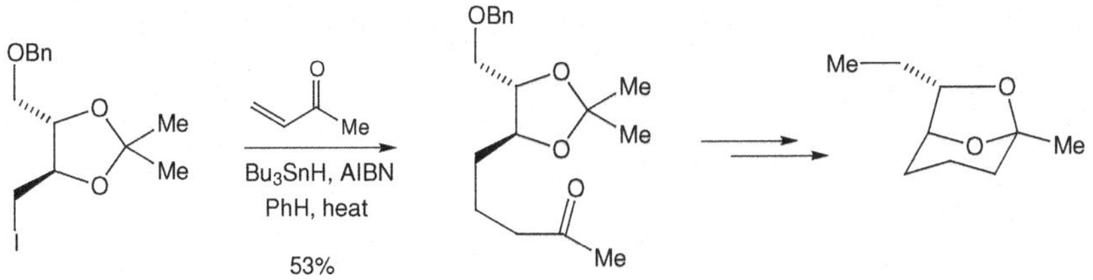

(4.34)

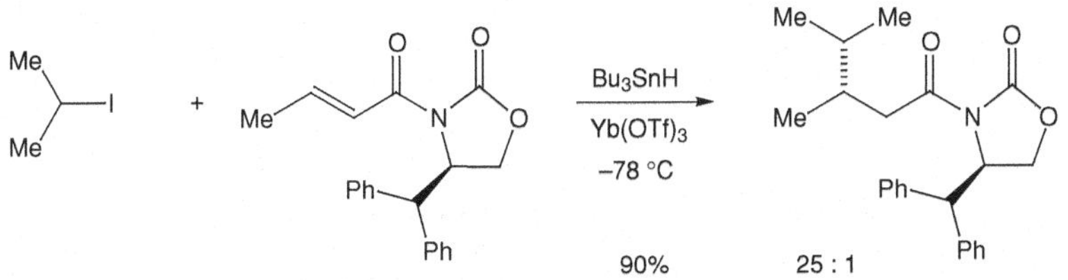

(4.35)

Successful intermolecular radical addition reactions depend on a number of factors. The tributyltin radical must allow formation of the carbon-centred radical

[21] B. Giese and R. Rupaner, *Synthesis* (1988), 219.
[22] M. P. Sibi and N. A. Porter, *Acc. Chem. Res.*, **32** (1999), 163; M. P. Sibi, S. Manyem and J. Zimmerman, *Chem. Rev.*, **103** (2003), 3263.
[23] P. Renaud and M. Gerster, *Angew. Chem. Int. Ed.*, **37** (1998), 2562.

faster than hydrostannylation of the alkene. Therefore reactive substrates such as alkyl iodides are suitable precursors. The first-formed carbon radical must add to the alkene faster than hydrogen atom abstraction from tributyltin hydride and this is favoured as described above by using a nucleophilic alkyl radical and an unhindered electron-deficient alkene. The new carbon radical generated after the first addition reaction, being electrophilic, reacts only slowly with further electron-deficient alkene. Therefore this radical is reduced to give the product and further tributyltin radical to continue the cycle (4.36). The whole process is therefore controlled by the electronic nature of the carbon radicals which dictates their rate of reaction with the alkene and can allow reasonable yields of intermolecular addition products in suitable cases.

$$R\text{--I} \quad + \quad Bu_3Sn^{\bullet} \quad \longrightarrow \quad R^{\bullet} \quad + \quad Bu_3SnI$$

$$R^{\bullet} \quad + \quad \overset{Z}{\diagdown} \quad \longrightarrow \quad R\diagup\diagdown\diagup^{Z}_{\bullet} \tag{4.36}$$

$$R\diagup\diagdown\diagup^{Z}_{\bullet} \quad + \quad Bu_3SnH \quad \longrightarrow \quad R\diagup\diagdown\diagup^{Z} \quad + \quad Bu_3Sn^{\bullet}$$

The problem of competing hydrogen atom abstraction and the difficulty in removing trialkyltin halide residues, which are toxic, has led to the development of modified methods for radical chemistry.[24] A procedure using tributyltin hydride as a catalyst, with regeneration of the tin hydride by reduction of tributyltin chloride, has shown some success. Thus, using only 20 mol% (0.2 molar equivalents) of tributyltin chloride in the presence of the reducing agent sodium borohydride has allowed good yields of coupled products such as **38** to be obtained (4.37).[25]

$$\underset{}{\bigcirc}\text{--I} \quad + \quad \overset{}{\diagup}\diagdown^{CN} \quad \xrightarrow[\substack{1.3\ NaBH_4 \\ EtOH,\ 25\ °C}]{0.2\ Bu_3SnCl} \quad \bigcirc\diagup\diagdown\diagup^{CN} \tag{4.37}$$

$$\text{95\%} \qquad\qquad\qquad \textbf{38}$$

A useful intermolecular radical reaction that avoids tributyltin hydride and excess alkene makes use of the ready β-elimination of tin or sulfur radicals. Addition of the carbon-centred radical to the γ-position of an unhindered allyl stannane gives an intermediate radical that eliminates a tin radical (4.38). The product is therefore the result of overall allyl addition and the released tin radical reacts

[24] P. A. Baguley and J. C. Walton, *Angew. Chem. Int. Ed.*, **37** (1998), 3072; A. Studer and S. Amrein, *Synthesis* (2002), 835. For organoboranes as a source of radicals, see C. Ollivier and P. Renaud, *Chem. Rev.*, **101** (2001), 3415.
[25] B. Giese, J. A. González-Gómez and T. Witzel, *Angew. Chem. Int. Ed. Engl.*, **23** (1984), 69.

with the starting material R–X to generate further carbon radical. By this method cyclohexyl bromide was converted into allylcyclohexane in 88% yield and the bromide **39** gave the allyl derivative **40** (4.39).[26] The reaction is effective even with alkyl radicals, which are 'nucleophilic', although additions to more electrophilic allyl stannanes or sulfides bearing an electron-withdrawing group such as CO_2Et in the β-position are also known.[27]

$$\text{(4.38)}$$

$$\text{(4.39)}$$

39   76%   **40**

Stereoselective allylation of secondary radicals is possible when a suitable steric bias is present.[28] For example, the thiocarbonyl compound **41** reacts to give exclusively the *exo* allylated product **42**, in which allyl tributylstannane approaches from the less-hindered convex face of the cyclic radical (4.40).[26] In acyclic substrates high stereoselectivity can be achieved by chelation with a Lewis acid.[23] For example, allylation of the selenide **43** is much more stereoselective in the presence of trimethylaluminium, in which the aluminium alkoxide chelates to the carbonyl group to give the species **44**, such that the approach of the allyl stannane is directed to the less hindered face (4.41).

$$\text{(4.40)}$$

41   **42**

$$\text{(4.41)}$$

43							
	no additive	98%		63	:	37	
	Me₃Al	97%		95	:	5	

44

Addition–elimination reactions are not restricted to allylic stannanes or sulfides. The vinyl stannane **45** acts as a suitable radical acceptor, leading to α,β-unsaturated

[26] G. E. Keck and J. B. Yates, *J. Am. Chem. Soc.*, **104** (1982), 5829; G. E. Keck, E. J. Enholm, J. B. Yates and M. R. Wiley, *Tetrahedron*, **41** (1985), 4079.
[27] D. H. R. Barton and D. Crich, *J. Chem. Soc., Perkin Trans. 1* (1986), 1613; J. E. Baldwin, R. M. Adlington, D. J. Birch, J. A. Crawford and J. B. Sweeney, *J. Chem. Soc., Chem. Commun.* (1986), 1339.
[28] M. P. Sibi and T. R. Rheault, *J. Am. Chem. Soc.*, **122** (2000), 8873.

carboxylic esters after elimination of the tributyltin radical (4.42).[29] Mixtures of *E* and *Z* geometrical isomers of the products are often formed. A more recent example, using radical addition α- to a sulfone then β-elimination (to give $EtSO_2^{\bullet}$ and hence SO_2 and $Et^{\bullet}$ to continue the chain process) is illustrated in Scheme 4.43.[30]

(4.42)

45

(4.43)

64%   85   :   15

lauroyl peroxide =
$(C_{11}H_{23}COO)_2$

Formation of a radical adjacent to a three-membered ring, such as a cyclopropane or epoxide, promotes rapid fragmentation of the strained ring.[31] The rate of this process is particularly fast ($k = 1.3 \times 10^8$ s^{-1} at 25 °C) and the preparation of a substrate containing a cyclopropane ring has been used frequently as a test for a radical intermediate at an adjacent reacting carbon centre (although in fact organometallic species can also cause rapid ring-opening of adjacent cyclopropanes). Ring-opening of the cyclopropane occurs to give the more-stable radical intermediate. For example, treatment of the iodide **46** with tributyltin hydride gave only the product **47** (with no loss of enantiopurity), resulting from selective ring-opening to the more stable benzylic radical intermediate (4.44).[32]

(4.44)

46   94%   47

Alternative radical chain addition reactions that avoid tributyltin hydride include the use of tris(trimethylsilyl)silane[33] $[(Me_3Si)_3SiH]$ or the use of thiohydroxamic

[29] J. E. Baldwin and D. R. Kelly, *J. Chem. Soc., Chem. Commun.* (1985), 682.
[30] F. Bertrand, B. Quiclet-Sire and S. Z. Zard, *Angew. Chem. Int. Ed.*, **38** (1999), 1943.
[31] P. Dowd and W. Zhang, *Chem. Rev.*, **93** (1993), 2091; A. Gansäuer, T. Lauterbach and S. Narayan, *Angew. Chem. Int. Ed.*, **42** (2003), 5556.
[32] Y. Takekawa and K. Shishido, *J. Org. Chem.*, **66** (2001), 8490.
[33] C. Chatgilialoglu, *Acc. Chem. Res.*, **25** (1992), 188.

esters (see Scheme 4.9). The alkyl radical, generated by N–O cleavage and decarboxylation, adds to the electron-deficient alkene to give a new radical, which then reacts with the thiohydroxamic ester to give the coupled product and hence propagate the cycle. Thus, the radical generated from the thiohydroxamic ester **48** adds to methyl vinyl ketone to give the product **49** (4.45).[34] Oxidation of such products with *meta*-chloroperoxybenzoic acid gives the sulfoxide and heating promotes elimination to give the α,β-unsaturated ketone.

$$(4.45)$$

68%

48

49

Reduction of an alkylmercury halide or acetate with a borohydride provides an alternative method for accessing a carbon radical species (see Scheme 4.11). The alkylmercury compound can be prepared by one of a number of methods, such as from the corresponding Grignard reagent or from addition of $Hg(OAc)_2$ to an alkene. For example, intramolecular amido-mercuration of the alkene **50**, followed by formation of the alkyl radical and addition to methyl acrylate provides a route to the alkaloid δ-coniceine (4.46).[35]

$$(4.46)$$

50

64%

δ-coniceine

A different method to access carbon radicals makes use of the one-electron reducing agent samarium diiodide, SmI_2. From a primary iodide, RI, the intermediate carbon radical $R^\bullet$ is converted to the organometallic species $RSmI_2$ by addition of a second electron from the SmI_2. Addition of the organosamarium species to ketones (activated by the samarium Lewis acid) gives tertiary alcohols.[36] From an aldehyde or ketone, addition of SmI_2 gives an intermediate ketyl radical that can be reduced to give an alcohol with further SmI_2 in the presence of a proton source or couple with itself in a pinacol reaction to give a diol or can add to an alkene. Thus, mixing a ketone and ethyl acrylate with SmI_2 gives the intermolecular radical addition product which cyclizes to give the lactone **51** (4.47).[37] For example, octan-2-one

[34] D. H. R. Barton and J. C. Sarma, *Tetrahedron Lett.*, **31** (1990), 1965.
[35] S. Danishefsky, E. Taniyama and R. R. Webb, *Tetrahedron Lett.*, **24** (1983), 11.
[36] A. Krief and A.-M. Laval, *Chem. Rev.*, **99** (1999), 745; P. Girard, J. L. Namy and H. B. Kagan, *J. Am. Chem. Soc.*, **102** (1980), 2693.
[37] S. Fukuzawa, A. Nakanishi, T. Fujinami and S. Sakai, *J. Chem. Soc., Perkin Trans. 1* (1988), 1669; for reductions with samarium diiodide, see G. A. Molander, *Org. Reactions*, **46** (1994), 211.

gave the lactone **51**, R $= n\text{-}C_6H_{13}$, R$' =$ Me (71%). Activated (electron deficient) alkenes are most suitable and the reaction can often be accelerated in the presence of the additive hexamethylphosphoramide (HMPA).

(4.47)

51

Addition reactions to alkenes are by far the most common type of radical carbon–carbon bond forming reaction. However, there are also examples of the addition of radicals to carbon monoxide or isonitriles.[38] Under appropriate conditions, normally high pressure, an alkyl radical will add to carbon monoxide faster than to an alkene (or hydrogen atom abstraction). The resulting acyl radical can then be reduced or add to an alkene. For example, the radical generated from 1-iodo-octane adds to carbon monoxide followed by the allyl stannane **52** in a three-component coupling process (4.48).

(4.48)

Alkyl radical additions to heteroatoms are less common, but there are important transformations in which the heteroatom is a halogen, sulfur, oxygen or even nitrogen atom.[1] Addition of an alkyl radical to an azide provides a method for formation of a carbon–nitrogen bond and the use of sulfonyl azides for this purpose has been developed recently.[39] The alkyl radical is best generated by using a peroxide initiator (such as lauroyl peroxide) or from tributyltin radicals generated initially from hexabutylditin and di-tert-butylhyponitrite. On heating di-tert-butylhyponitrite the tert-butoxy radical is released, which reacts with hexabutylditin (4.49). The resulting tributyltin radical then reacts with the alkyl halide or thiocarbonyl compound to propagate the radical reaction. These methods avoid metal hydrides, which would reduce the carbon radical by competing with the slower addition to the azide. For example, conversion of the iodide **53** under radical conditions to the corresponding

[38] I. Ryu, N. Sonoda and D. P. Curran, *Chem. Rev.*, **96** (1996), 177.
[39] C. Ollivier and P. Renaud, *J. Am. Chem. Soc.*, **123** (2001), 4717.

azide can be accomplished effectively with benzenesulfonyl azide (as expected, addition to the azide from the less-hindered face is preferred) (4.50).

$$^{t}BuO-N=N-O^{t}Bu \xrightarrow{80\ ^{\circ}C} 2\ ^{t}BuO^{\bullet} + N_2 \tag{4.49}$$

$$^{t}BuO^{\bullet} + Bu_3SnSnBu_3 \longrightarrow {}^{t}BuOSnBu_3 + Bu_3Sn^{\bullet}$$

$$\begin{array}{ccc} & \xrightarrow[\substack{Bu_3SnSnBu_3 \\ {}^{t}BuON=NO^{t}Bu \\ PhH,\ 80\ ^{\circ}C}]{3\ equiv.\ PhSO_2N_3} & + & \tag{4.50} \end{array}$$

53 55% 84 : 16

Trapping the carbon-centred radical intramolecularly, particularly with an alkene, has been used extensively in organic synthesis.[1,40] This valuable strategy has found many important applications for the preparation of carbocyclic and heterocyclic natural products. Intramolecular reactions are inherently more favourable than the corresponding intermolecular reactions due to entropic factors and are typically fast processes that provide a σ-bond at the expense of a π-bond.

Most favourable are reactions that form a five-membered ring by an *exo* (rather than an *endo*) mode of addition.[41] Hence, the kinetically controlled cyclization of a 5-hexenyl radical leads mainly to the five- rather than the six-membered ring (4.51). This can be explained by invoking an early transition state, in which stereoelectronic factors are most crucial. For geometric reasons, the interconnecting chain is not long enough to favour approach of the radical to the terminal carbon atom of the double bond along the ideal trajectory. Formation of the five-membered ring, during which suitable orbital overlap can be achieved, is therefore favoured, even though this requires the generation of the less stable primary radical.

17% 81% 2%

[40] B. Giese, B. Kopping, T. Göbel, J. Dickhaut, G. Thoma, K. J. Kulicke and F. Trach, *Org. Reactions*, **48** (1996), 301.

[41] For guidelines for ring formation see J. E. Baldwin, *J. Chem. Soc., Chem. Commun.* (1976), 734.

For successful ring formation, the rate of cyclization must be faster than that of hydrogen atom abstraction. As cyclization is a unimolecular process, whereas intermolecular abstraction is bimolecular, conducting the reaction at low concentration can be beneficial. The rate of cyclization of the 5-hexenyl radical has been measured to be $k = 2.3 \times 10^5\,\mathrm{s}^{-1}$ at $25\,^{\circ}\mathrm{C}$, corresponding to a half-life $t_{1/2} = (\ln 2)/k = 3 \times 10^{-6}\,\mathrm{s}$, indicating a very rapid cyclization. Substituents within the chain tend to enhance the rate of cyclization (Thorpe–Ingold effect), unless located at C-5, in which there is steric hindrance to cyclization. A chair-like transition state has been proposed and this model can be used to account for the observed stereoselectivity with substituted hexenyl radicals. For example, the 2- and 3-substituted hexenyl radicals **54** and **55** cyclize to give different major stereoisomers of 1,3-dimethylcyclopentane, indicating a preference for the substituent to adopt the less hindered pseudoequatorial position in the transition state (4.52).

(4.52)

Various methods for the generation of carbon-centred radicals have been described earlier in this chapter and these are generally applicable for subsequent intramolecular reaction. Commonly, tributyltin hydride is used to initiate carbon radical formation. For example, treatment of the bromide **56** with tributyltin hydride and AIBN gave the cyclopentanes **57** and **58** (4.53). In such cyclizations, the 4-substituent plays a dominant role in the stereoselection (to favour the *trans* arrangement between the substituent at this position and the new chiral centre), although the cyclization of the isomeric *E*-alkene occurs with no diastereoselectivity.[42]

---

[42] For a review on intramolecular radical conjugate addition reactions, see W. Zhang, *Tetrahedron*, **57** (2001), 7237.

(4.53)

Cyclization using an alkenyl radical is efficient and leads to a product containing an alkene in a defined position, which may be used in further chemical transformations. Thus, the bromide **59** was cyclized to the indane derivative **60** without the need to protect the hydroxyl or cyano groups (4.54). In this example cyclization is high yielding even though it leads to the formation of a quaternary carbon centre. Substrates in which the alkenyl bromide can exist as *E*- or *Z*-geometrical isomers converge on the same product alkene stereochemistry due to the rapid isomerization of alkenyl radicals.

(4.54)

Another route to alkenyl radicals is by addition of radicals to alkynes. An application of this procedure, which serves as a model for the synthesis of the CD ring system of cardiac aglycones was reported by Stork and co-workers (4.55).[43] The initial alkyl radical, formed selectively from the bromide **61** attacks the alkyne regioselectively to give an intermediate alkenyl radical, which reacts further with the alkene of the cyclohexene to give the product **62**. A mixture of alkene stereoisomers is produced owing to the ease of *E*–*Z* alkenyl radical isomerization.

Two (or more) consecutive reactions are often termed a tandem or cascade process and allow the rapid formation of complex polycyclic structures.[44] In the case illustrated in Scheme 4.55, two carbon–carbon bonds, two rings and a quaternary carbon centre are generated in a single step.

[43] G. Stork and R. Mook, *J. Am. Chem. Soc.*, **105** (1983), 3720.
[44] A. J. McCarroll and J. C. Walton, *Angew. Chem. Int. Ed.*, **40** (2001), 2225.

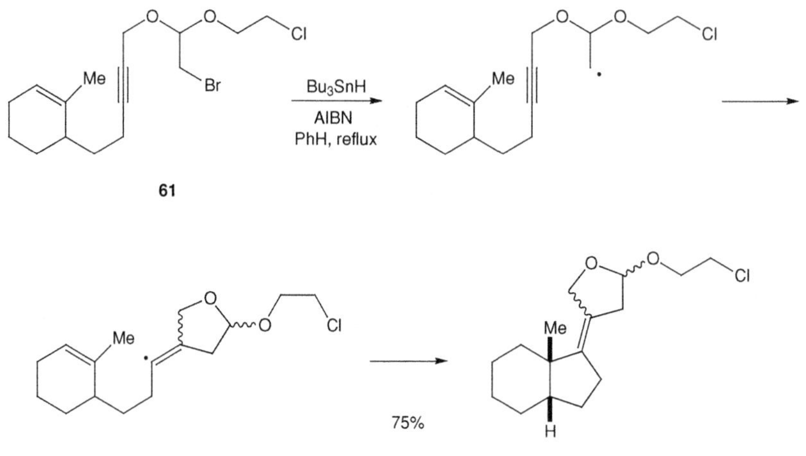

(4.55)

Many natural products contain bicyclic or polycyclic ring systems, and tandem radical cyclization reactions provide an efficient approach for their synthesis. This strategy is illustrated by a synthesis of the triquinane hirsutene, outlined in Scheme 4.56.[45] Treatment of the iodide **63** with tributyltin hydride results in the direct formation of the natural product via two consecutive radical cyclization reactions. The initial alkyl radical undergoes a 5-*exo*-*trig* cyclization to give a *cis*-fused bicyclic radical, which undergoes a 5-*exo*-*dig* cyclization to give a further *cis*-fused ring system. The relative stereochemistry is dictated by the stereochemistry of the initial substrate **63**. The vinyl radical abstracts a hydrogen atom from the tributyltin hydride to release the product and further tributyltin radical to propagate the cycle.

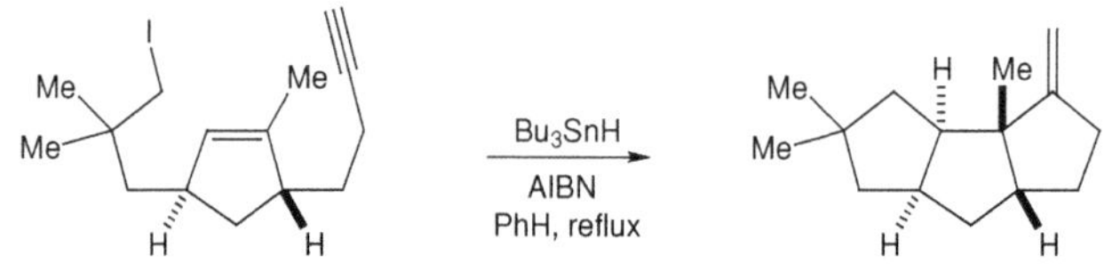

(4.56)

In some cases it is possible to intercept the final carbon-centred radical with an external radical trap. Such tandem processes involving cyclization followed by intermolecular trapping provide a rapid entry to highly functionalized cyclic compounds. Attempts to use tributyltin hydride in such processes is often thwarted by competing hydrogen atom abstraction from the metal hydride. An early and excellent example of the power of this strategy is provided by a synthesis of (+)-prostaglandin $F_{2\alpha}$ (PGF$_{2\alpha}$). Treatment of the iodide **64** with tributyltin chloride and sodium cyanoborohydride (in order to minimize the amount of tin hydride

[45] D. P. Curran and D. M. Rakiewicz, *Tetrahedron,* **41** (1985), 3943.

present) in the presence of excess 2-(trimethylsilyl)-oct-1-en-3-one, resulted in the formation of the product **65** (4.57).[46] In this reaction, the first-formed radical reacts faster with the intramolecular alkene than with the enone; however, the cyclic radical is trapped intermolecularly by the enone to give a stabilized α-keto-α-silyl radical, which abstracts a hydrogen atom from tributyltin hydride. A modification involves intermolecular trapping with 1-tributylstannyl-oct-1-en-3-one.[47]

$$(4.57)$$

The order of the reactions can be reversed in appropriate substrates, such that an intermolecular radical reaction occurs first, to give a new radical that is set up for an intramolecular reaction.[48] An efficient synthesis of the antitumor agent camptothecin made use of an initial intermolecular radical addition to phenyl isocyanide, followed by radical cyclization onto the pendant alkyne and a second cyclization onto the phenyl group (4.58).[49]

$$(4.58)$$

[46] G. Stork, P. M. Sher and H.-L. Chen, *J. Am. Chem. Soc.*, **108** (1986), 6384.
[47] G. E. Keck and D. A. Burnett, *J. Org. Chem.*, **52** (1987), 2958.
[48] T. R. Rheault and M. P. Sibi, *Synthesis* (2003), 803.
[49] H. Josien, S.-B. Ko, D. Bom and D. P. Curran, *Chem. Eur. J.*, **4** (1998), 67.

Some examples of intermolecular addition of carbon-centred radicals, followed by β-elimination of tin or sulfur radicals were provided in Schemes 4.38–4.43 and this strategy is effective in intramolecular processes. Thus, in a synthesis of the antitumor agent CC-1065, the aryl radical generated from the bromide **66** underwent cyclization and subsequent β-elimination to give the indoline **67** (4.59).[50] An advantage of this type of elimination procedure is that it provides a new alkene in a defined position that is suitable for further elaboration. The β-elimination of a sulfur radical has found other applications, such as in syntheses of the alkaloid morphine and the neuroexcitatory amino-acid kainic acid.[51]

$$(4.59)$$

Cyclization onto an enone or enoate provides an alternative method to access a cyclic product with suitable functionality for further elaboration. Thus, a key step in a synthesis of the alkaloid gelsemine made use of the radical cyclization after homolytic C–S bond cleavage of the sulfide **68** (4.60).[52] A second radical cyclization was subsequently used to set up the indolinone ring of the natural product.

$$(4.60)$$

[50] D. L. Boger and R. S. Coleman, *J. Am. Chem. Soc.*, **110** (1988), 4796; D. L. Boger, R. J. Wysocki and T. Ishizaki, *J. Am. Chem. Soc.*, **112** (1990), 5230.

[51] K. A. Parker and D. Fokas, *J. Am. Chem. Soc.*, **114** (1992), 9688; M. D. Bachi and A. Melman, *J. Org. Chem.*, **62** (1997), 1897.

[52] S. Atarashi, J.-K. Choi, D.-C. Ha, D. J. Hart, D. Kuzmich, C.-S. Lee, S. Ramesh and S. C. Wu, *J. Am. Chem. Soc.*, **119** (1997), 6226.

Alkyl, alkenyl, aryl and acyl radicals can all be used in cyclization reactions. Acyl radicals can be generated by addition of alkyl radicals to carbon monoxide, or more conveniently from acyl selenides, and undergo a variety of radical reactions.[53] A synthesis of the sesquiterpene (−)-kamausallene made use of the radical cyclization from the acyl selenide **69** (4.61).[54] Tris(trimethylsilyl)silane and triethylborane in air were used to promote the reaction, which is highly selective (32:1) in favour of the *cis* stereoisomer **70**, as expected from a chair-like transition state. Best yields in the cyclization reactions of acyl radicals are found with electron-deficient alkenes, indicating the nucleophilic character of acyl radicals.

$$(4.61)$$

The carbon-centred radical species required for these cyclizations can be generated in one of many different ways. Although the most common procedure uses organic halides and treatment with tributyltin hydride and AIBN, there are cases in which it is advantageous to use alternative methods. Successful cyclizations can sometimes be achieved by treating the halide with a cobalt(I) complex[55] or with samarium diiodide.[56] In these cases, cyclization is followed by reincorporation of the metal and this can be very useful as it allows further functionalization. An example is the cyclization of ketyl radicals, generated by one-electron reduction of aldehydes or ketones with samarium diiodide.[56] Treating the ketone **71** with samarium diiodide, followed by addition of an electrophile provides the substituted cyclopentanols **72** (4.62). Very high stereoselectivities can be achieved in these cyclization reactions and the chemistry allows the formation of a variety of substituted products. A synthesis of the diterpenoid grayanotoxin III made use of such a cyclization reaction (4.63).[57] The synthesis also involved a second cyclization with samarium diiodide in a pinacol reaction.

[53] C. Chatgilialoglu, D. Crich, M. Komatsu and I. Ryu, *Chem. Rev.*, **99** (1999), 1991.

[54] P. A. Evans, V. S. Murthy, J. D. Roseman and A. L. Rheingold, *Angew. Chem. Int. Ed.*, **38** (1999), 3175.

[55] G. Pattenden, *Chem. Soc. Rev.*, **17** (1988), 361.

[56] G. A. Molander and C. R. Harris, *Chem. Rev.*, **96** (1996), 307; G. A. Molander and C. R. Harris, *Tetrahedron*, **54** (1998), 3321.

[57] T. Kan, S. Hosokawa, S. Nara, M. Oikawa, S. Ito, F. Matsuda and H. Shirahama, *J. Org. Chem.*, **59** (1994), 5532.

$$(4.62)$$

E$^+$ = PhSSPh, E = SPh 77%
E$^+$ = Ac$_2$O, E = COMe 74%

$$(4.63)$$

One-electron oxidation of carbonyl compounds provides another entry to radical species, suitable for carbon–carbon bond formation.[58] Best results have been obtained using β-dicarbonyl compounds with oxidation by manganese triacetate. The resulting carbon radical, generated at the α-position, is electrophilic and reacts best with electron-rich π-systems. Cyclization can take place by an *exo* or an *endo* mode of addition, the regioselectivity being influenced predominantly by the stability of the resulting radical (more-substituted radicals are more stable). Hence, in a synthesis of *O*-methylpodocarpic acid, the β-keto-ester **73** was treated with Mn(OAc)$_3$ to give the product **74** (4.64).[59] The intermediate α-keto radical undergoes 6-*endo-trig* rather than 5-*exo-trig* cyclization, as this leads to the more-stable tertiary radical. Subsequent cyclization onto the aromatic ring, followed by oxidation of the resulting radical species, leads to the tricyclic product **74**.

$$(4.64)$$

[58] B. B. Snider, *Chem. Rev.*, **96** (1996), 339.
[59] Q. Zhang, R. M. Mohan, L. Cook, S. Kazamis, D. Peisach, B. M. Foxman and B. B. Snider, *J. Org. Chem.*, **58** (1993), 7640.

The alkyl mercury method is convenient for the formation of radical species from precursor alkenes. The example in Scheme 4.65 illustrates the use of this procedure and highlights one of the advantages of radical reactions in synthesis. The new carbon–carbon bond generated in the formation of the bicyclic ketone **76** would be difficult to prepare by other methods. For example, the corresponding carbanion, rather than undergo an intramolecular Michael reaction, would simply undergo β-elimination of the acetoxy anion (to generate the alkene **75**). The β-elimination of acetoxy radicals does not take place. Radicals containing alkoxy or amino groups on the β-carbon atom can also be used in this way (although tin and sulfur radicals do β-eliminate).

(4.65)

75   **76**

Radical reactions are not restricted to cyclizations onto alkenes or alkynes. Increasingly popular is the use of an imine or imine derivative, such as an oxime or hydrazone.[60] Most examples involve 5- or 6-*exo-trig* cyclization to give cyclopentane or cyclohexane ring systems. Thus, treatment of the bromide **77** with tributyltin hydride gave the cyclopentane **78** (4.66). The stereoselectivity of the cyclization is in line with that expected on the basis of a chair-like transition state (compare with Scheme 4.53).

(4.66)

77   **78**

A useful variant of this chemistry involves the radical cyclization onto an *N*-aziridinyl hydrazone. Fragmentation of the intermediate nitrogen-centred radical to release nitrogen gas and an alkene (typically styrene or stilbene) results in the formation of a new carbon radical at the original hydrazone carbon atom. Thus, in a synthesis of the sesquiterpene α-cedrene, the radical species **80**, formed from the thiocarbonyl compound **79**, cyclizes onto the hydrazone to give the nitrogen-centred

[60] G. K. Friestad, *Tetrahedron*, **57** (2001), 5461.

radical **81** (4.67).[61] Fragmentation provides the new radical **82** which is set up for a second radical cyclization to give, after hydrogen atom abstraction from tributyltin hydride and hydrolysis, the product **83**. The *N*-aziridinyl hydrazone acceptor therefore provides a method to form a radical at the same carbon atom as that attacked in the first cyclization reaction. This complements the more conventional radical cyclization onto an alkene, in which the new radical is generated one carbon atom from where the first carbon–carbon bond was formed.

(4.67)

The majority of reported intramolecular radical reactions generate a five-membered ring product by a 5-*exo-trig* cyclization. Some examples of the disfavoured 5-*endo-trig* process are known, particularly with substrates that result after cyclization in a stabilized radical.[62] For example, the α-acyl radical formed from the chloride **84** cyclizes onto the alkene to give the tertiary radical **85** (stabilized by the α-amido group) which abstracts a hydrogen atom from Bu₃SnH to give the product **86** (4.68). In contrast, treatment of the chloride **87** with tributyltin hydride gave only the β-lactam **89** (4-*exo-trig* cyclization) via the benzyl radical intermediate **88** (4.69).

[61] H.-Y. Lee, S. Lee, D. Kim, B. K. Kim, J. S. Bahn and S. Kim, *Tetrahedron Lett.*, **39** (1998), 7713.
[62] H. Ishibashi, T. Sato and M. Ikeda, *Synthesis* (2002), 695.

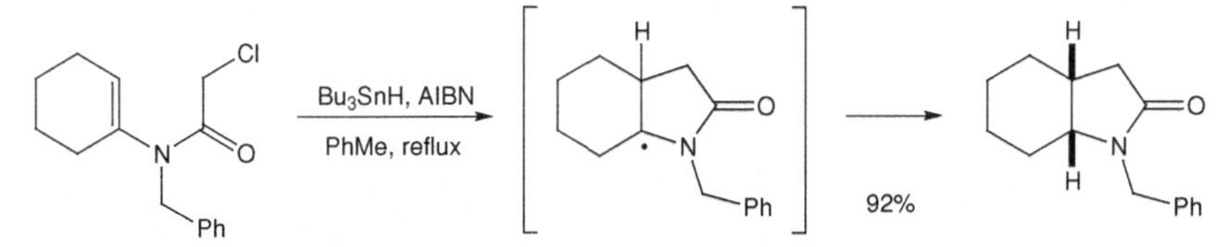

$$(4.68)$$

84   85   86

92%

$$(4.69)$$

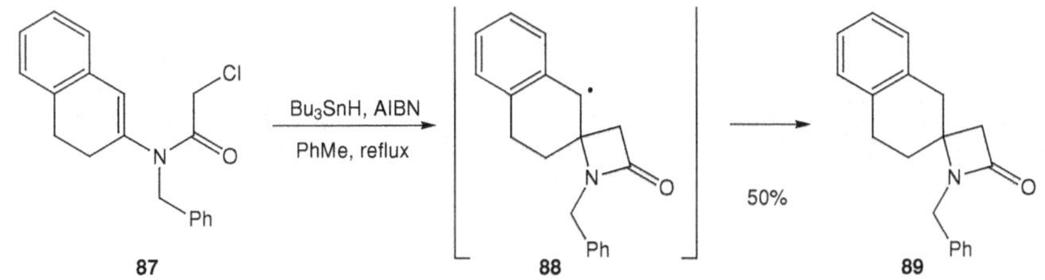

87   88   89

50%

Radical reactions that result in medium or large rings are less common than those that give five-membered rings. The rate of radical cyclization to give a six-membered or larger ring is considerably slower than that to a five-membered ring, and reactions such as hydrogen atom abstraction can compete with the desired ring formation. Such competing reactions can be offset by careful choice of substrate, in which conformational constraints and/or activation of the double bond enhance the rate of cyclization.

Cyclization reactions to prepare medium-sized (seven- to nine-membered) rings are often low yielding, although a significant and increasing number of successful radical reactions have been reported.[63] Thus, the use of a reactive aryl radical and the formation of a stabilized benzylic α-amido radical promotes the ready formation of the alkaloid lennoxamine by a 7-*endo-trig* reaction (4.70). The ketyl radical generated by addition of tributyltin hydride or samarium diiodide to the aldehyde **90** undergoes 7-*exo-trig* cyclization onto the oxime as part of a synthesis of balanol (4.71).

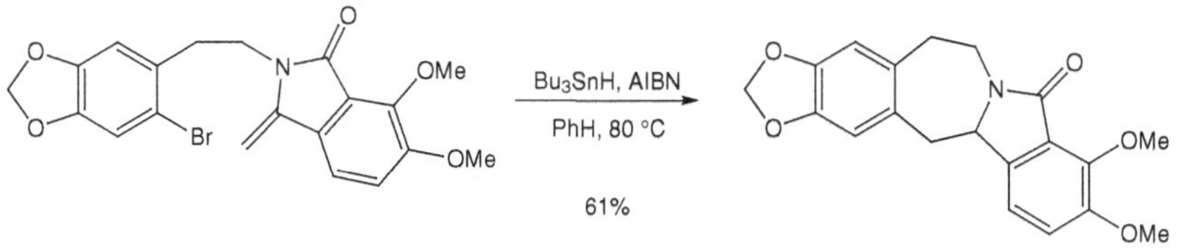

$$(4.70)$$

61%

lennoxamine

[63] L. Yet, *Tetrahedron*, **55** (1999), 9349.

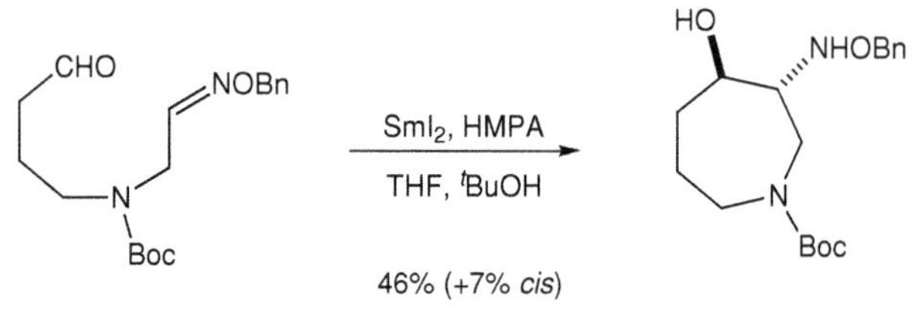

$$(4.71)$$

46% (+7% cis)

90

The formation of eight-membered rings by radical cyclization has been found to occur using α-acyl radical species. The reaction occurs at a rate that is faster remarkably even than 5-*exo-trig* cyclization. Hence, the eight-membered lactone **92** was formed, rather than a five-, six- or seven-membered lactone, on treatment of the bromide **91** with tributyltin hydride (4.72).

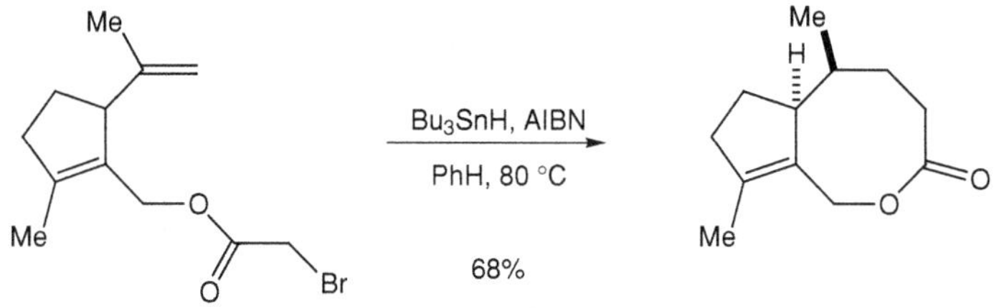

$$(4.72)$$

91   **92**

With appropriate choice of substrate and conditions it is possible to prepare large-sized (10–20 membered) rings using radical cyclization.[64] Such macrocyclizations are normally best accomplished by an *endo* cyclization of a carbon-centred radical onto a terminal, electron-deficient alkene under high dilution conditions. Thus, treatment of the iodide **93** with tributyltin hydride promoted *endo* cyclization to give the 14-membered ring ketone **94** (4.73). Significant amounts of uncyclized (hydrogen atom abstraction) products such as **95** are formed in such reactions, although these may be suppressed by using photolytic or other radical-generating conditions.

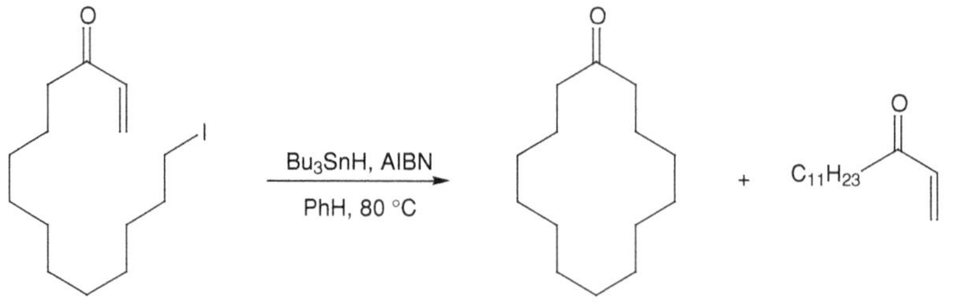

$$(4.73)$$

93   **94** 63%   **95** 22%

[64] S. Handa and G. Pattenden, *Contemp. Org. Synth.*, **4** (1997), 196.

4.2 Carbenes

A carbene is a neutral intermediate containing divalent carbon, in which the carbon atom is covalently bonded to two other groups and has two valency electrons distributed between two non-bonding orbitals. If the two electrons are spin-paired the carbene is a singlet; if the spins of the electrons are parallel it is a triplet.

A singlet carbene is believed to have a bent sp^2 hybrid structure, in which the paired electrons occupy the vacant sp^2-orbital. A triplet carbene may be either a bent sp^2 hybrid with an electron in each unoccupied orbital, or a linear sp hybrid with one electron in each of the unoccupied p-orbitals (4.74). Structures in between the last two are also possible.

| lowest singlet | triplet | triplet | (4.74) |

The results of experimental observations and molecular orbital calculations indicate that many carbenes have a nonlinear triplet ground state. Exceptions are the dihalocarbenes and carbenes with oxygen, nitrogen or sulfur atoms attached to the bivalent carbon, all of which are singlets. The singlet and triplet states of a carbene do not necessarily show the same chemical behaviour. For example, addition of singlet carbenes to olefinic double bonds to form cyclopropane derivatives is more stereoselective than addition of triplet carbenes.

A variety of methods is available for the generation of carbenes, but for synthetic purposes they are usually obtained by thermal, photolytic or transition metal catalysed decomposition of diazoalkanes, or by α-elimination of HX from a haloform CHX_3 or of halogen from a *gem*-dihalide by action of an organolithium or a metal (4.75). In many of these reactions it is doubtful whether a 'free' carbene is actually formed. It is more likely that the carbene is complexed with a metal or held in a solvent cage with a salt, or that the reactive intermediate is, in fact, an organometallic compound and not a carbene. Such organometallic or complexed intermediates which, while not 'free' carbenes, give rise to products expected of carbenes are usually called metallocarbenes or carbenoids.

$$N_2CH-CO_2Et \xrightarrow{\text{heat}} {:}CH-CO_2Et \quad + \quad N_2$$

$$CHCl_3 \xrightarrow{\text{base}} {:}CCl_3^- \longrightarrow {:}CCl_2 \quad + \quad Cl^- \qquad (4.75)$$

$$R_2CBr_2 \xrightarrow{\text{BuLi}} R_2CBrLi \longrightarrow R_2C{:} \quad + \quad LiBr$$

Carbenes produced by photolysis of diazoalkanes are highly energetic species and often react indiscriminately. Thermal decomposition of diazoalkanes can be catalysed by certain transition metal (particularly copper or rhodium) salts, to produce less energetic and more selective carbenes. The active species in such a reaction is thought to be the metallocarbene rather than the 'free' carbene. Another convenient and widely used route to alkylcarbenes is the thermal or photolytic decomposition of the lithium or sodium salts of toluene-*p*-sulfonylhydrazones. The diazoalkane is first formed (by elimination of the toluenesulfinate anion) and decomposes under the reaction conditions to give the carbene (4.76). This process is often referred to as the Bamford–Stevens reaction.

$$(4.76)$$

Carbenes, in general, are very reactive electrophilic species. Their activity depends to some extent on the method and conditions of preparation, on the nature of the substituent groups and also on the presence or absence of metals or metallic salts. Carbenes undergo a variety of reactions, including insertion into C—H bonds, addition to multiple bonds and skeletal rearrangements.[65] The simplest carbene, methylene itself, attacks primary, secondary and tertiary C—H bonds indiscriminately. However, alkyl carbenes can undergo selective intramolecular C—H, N—H or O—H insertion to provide useful synthetic transformations.[66] In general, no intermolecular reactions are observed when intramolecular insertion is possible.

Intramolecular insertion reactions can allow transformations that would otherwise be difficult to achieve. Geometrically rigid structures favour insertions, but this is not a necessity. For example, diazocamphor was converted into cyclocamphanone **96** in high yield, yet the open-chain diazo compound **97** also readily reacted to give the cyclopentanone derivative **98** (4.77).[67]

[65] S. D. Burke and P. A. Grieco, *Org. Reactions*, **26** (1979), 361; A. Padwa and K. E. Krumpe, *Tetrahedron*, **48** (1992), 5385; T. Ye and M. A. McKervey, *Chem. Rev.*, **94** (1994), 1091.

[66] D. J. Miller and C. J. Moody, *Tetrahedron* (1995), 10 811.

[67] D. F. Taber, *J. Am. Chem. Soc.*, **108** (1986), 7686; D. F. Taber and S. C. Malcolm, *J. Org. Chem.*, **66** (2001), 944.

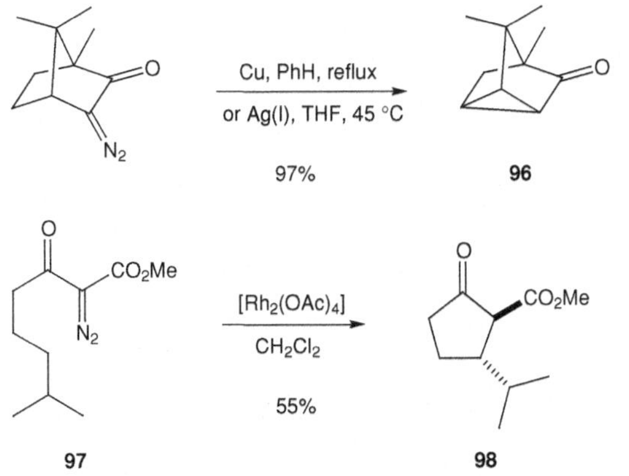

(4.77)

97% 96

97 98

[Rh₂(OAc)₄] structure: 97 → 98 (55%)

Cyclopentanones and other five-membered ring-containing compounds can be prepared readily by intramolecular C—H insertion of carbenes derived from diazo-carbonyl compounds and rhodium(II) acetate. Thus, treatment of the diazoketone **99** with rhodium(II) acetate gave the C—H insertion product **100**, used in a synthesis of the toxin muscarine (4.78).

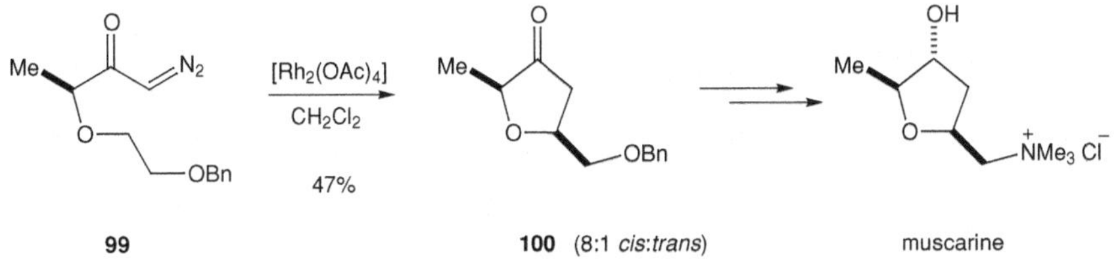

(4.78)

99   100 (8:1 *cis:trans*)   muscarine

The ability of rhodium or copper complexes to promote carbene formation allows the study of asymmetric reactions with chiral ligands attached to the metal centre.[68] Some highly enantioselective transformations are possible in certain cases. For example, the lactone **101** was formed with high optical purity using the complex [Rh₂(5S-MEPY)₄] (4.79). In the absence of competing intramolecular reactions, intermolecular C—H insertion is possible and such reactions are also amenable to asymmetric induction. Thus, high enantioselectivity in the insertion into a C—H bond of cyclohexane has been reported (4.80).

[68] H. M. L. Davies and R. E. J. Beckwith, *Chem. Rev.*, **103** (2003), 2861; C. A. Merlic and A. L. Zechman, *Synthesis* (2003), 1137; D. C. Forbes and M. C. McMills, *Curr. Org. Chem.*, **5** (2001), 1091; H. M. L. Davies and E. G. Antoulinakis, *J. Organomet. Chem.*, **617** (2001), 47; M. P. Doyle and D. C. Forbes, *Chem. Rev.*, **98** (1998), 911; G. A. Sulikowski, K. L. Cha and M. M. Sulikowski, *Tetrahedron: Asymmetry*, **9** (1998), 3145.

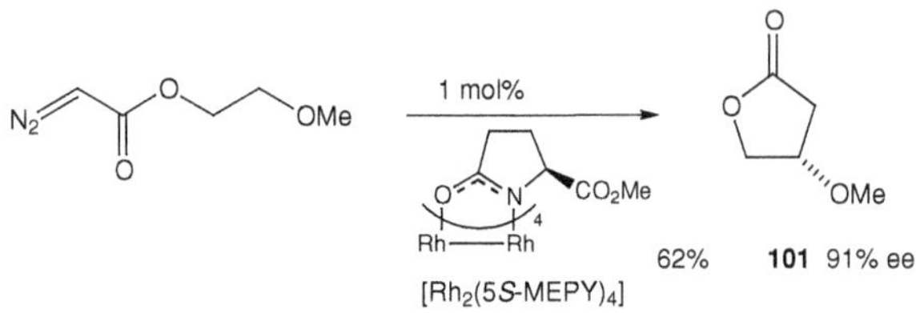

(4.79)

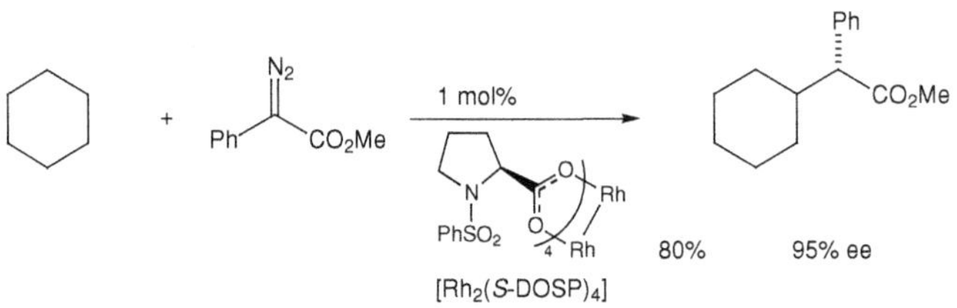

(4.80)

Insertion reactions of alkylidene carbenes offer a useful entry to cyclopentene ring systems (4.81).[69] Insertion is most effective with dialkyl-substituted alkylidene carbenes (R = alkyl), since rearrangement of the alkylidene carbene to the alkyne occurs readily when R = H or aryl. A number of methods have been used to access alkylidene carbenes. One of the most convenient uses a ketone and the anion of trimethylsilyl diazomethane. Addition of the anion to the ketone and elimination gives an intermediate diazoalkene, which loses nitrogen to give the alkylidene carbene. For example, a synthesis of the antibiotic (−)-malyngolide started from the ketone **102** (4.82). The insertion reaction takes place with retention of configuration at the C−H bond.

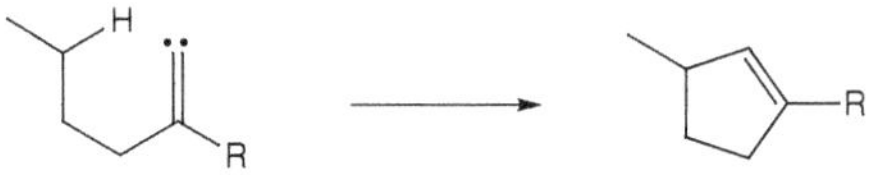

(4.81)

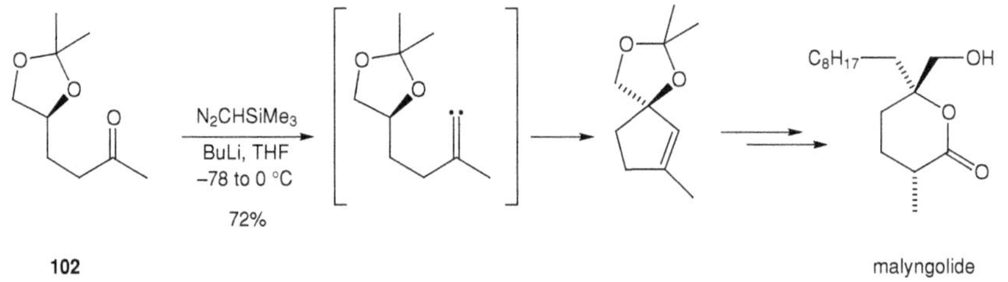

(4.82)

102   malyngolide

An alternative approach to alkylidene carbenes uses the deprotonation or halogen–lithium exchange of vinyl halides.[70] Hence, treatment of the vinyl chloride **103** with potassium hexamethyldisilazide (KHMDS) resulted in the formation of the cyclopentene **104** via the intermediate alkylidene carbene (4.83).[71] The carbene

[69] W. Kirmse, *Angew. Chem. Int. Ed. Engl.*, **36** (1997), 1164.
[70] M. Braun, *Angew. Chem. Int. Ed.*, **37** (1998), 430.
[71] D. F. Taber and T. D. Neubert, *J. Org. Chem.*, **66** (2001), 143.

undergoes selective C—H insertion with retention of configuration. The cyclopentene **104** was used in a synthesis of the alkaloid (−)-mesembrine.

$$(4.83)$$

Insertion of carbenes into other bonds, particularly O—H and N—H bonds, has found worthwhile application in organic synthesis.[66] Both inter- and intramolecular reactions are possible to give ethers, amines or amides. Intramolecular O—H insertion of the carbene, generated from the diazoketone **105**, gave rise to the seven-membered cyclic ether **106** (4.84). In a synthesis of the β-lactam antibiotic thienamycin, intramolecular N—H insertion of the carbene, formed from the diazoketone **107**, was a key step to give the β-lactam **108** (4.85).

$$(4.84)$$

$$(4.85)$$

The most common application of carbenes in synthesis is in the formation of three-membered rings by addition to multiple bonds. This is a typical reaction of all carbenes that do not undergo intramolecular insertion. Generation of the carbene in the presence of an alkene gives a cyclopropane product.[65,72] Addition of halocarbenes to alkenes is a stereospecific *cis* reaction, but this is not necessarily the case with all carbenes. Hence Z-2-butene **109** gives the cyclopropane **110**, in which the two methyl groups remain *cis* to one another (4.86). The stereospecificity

[72] For reviews on the synthesis of cyclopropanes, see H. Lebel, J.-F. Marcoux, C. Molinaro and A. B. Charette, *Chem. Rev.*, **103** (2003), 977; W. A. Donaldson, *Tetrahedron*, **57** (2001), 8589.

of the reaction can be attributed to the fact that these carbenes are singlets. A singlet carbene allows concerted addition to the alkene since the two new σ-bonds of the cyclopropane can be formed without changing the spin of any of the electrons involved. Addition of a triplet carbene, on the other hand, would be stepwise, proceeding through a triplet diradical intermediate, with the possibility of rotation about one of the bonds before spin inversion and closure to the cyclopropane could take place. Carbenes are normally generated as the singlet even if they have a triplet ground state. Therefore, depending on the conditions, cyclopropanation can occur stereospecifically if the rate of reaction with the alkene is faster than the conversion to the triplet ground state.

$$(4.86)$$

109   **110**

The concerted cyclopropanation reaction of singlet carbenes can be classified as a symmetry allowed pericyclic process. It is sometimes referred to as a cheletropic reaction. The carbene is thought to adopt a sideways approach in order to develop a bonding interaction between the HOMO of the alkene and the LUMO of the carbene.

Addition of carbenes to aromatic systems leads to ring-expanded products. Methylene itself, formed by photolysis of diazomethane, adds to benzene to form cycloheptatriene in 32% yield; a small amount of toluene is also formed by an insertion reaction. The cycloheptatriene is formed by a Cope rearrangement of the intermediate cyclopropane (a norcaradiene). More satisfactory is the reaction of benzene with diazomethane in the presence of copper salts, such as copper(I) chloride, which gives cycloheptatriene in 85% yield (4.87). The reaction is general for aromatic systems, substituted benzenes giving mixtures of the corresponding substituted cycloheptatrienes.

$$(4.87)$$

A valuable cyclopropanation reaction is the Simmons–Smith reaction of alkenes with diiodomethane and zinc–copper couple or diethyl zinc. This is a versatile reaction and has been applied with success to a wide variety of alkenes.[73] Many

---

[73] H. E. Simmons, T. L. Cairns, S. A. Vladuchick and C. M. Hoiness, *Org. Reactions*, **20** (1973), 1; P. Helquist, in *Comprehensive Organic Synthesis*, ed. B. M. Trost and I. Fleming, vol. 4 (Oxford: Pergamon Press, 1991), p. 951; W. B. Motherwell and C. J. Nutley, *Contemp. Org. Synth.*, **1** (1994), 219; A. B. Charette and A. Beauchemin, *Org. Reactions*, **58** (2001), 1.

functional groups are unaffected, making possible the formation of a variety of cyclopropane derivatives. For example, dihydrosterculic acid was obtained from methyl oleate **111** (4.88). The reaction is stereospecific and takes place by *cis* addition to the alkene. The reactive intermediate is thought to be an iodomethylenezinc iodide [ICH$_2$ZnI] or [Zn(CH$_2$I)$_2$] complex, which reacts with the alkene in a bimolecular process to give the cyclopropane and zinc iodide.

$$(4.88)$$

111

The Simmons–Smith reaction is influenced by a suitably situated hydroxy group in the alkene substrate. With allylic and homoallylic alcohols or ethers, the rate of the reaction is greatly increased and, in five- and six-membered cyclic allylic alcohols, the product in which the cyclopropane ring is *cis* to the hydroxy group is formed stereoselectively (4.89). These effects are ascribed to co-ordination of the oxygen atom to the zinc, followed by transfer of methylene to the same face of the adjacent double bond.

$$(4.89)$$

Asymmetric cyclopropanation reactions have been developed by using diiodomethane and diethyl zinc in the presence of a chiral Lewis acid. A particularly effective chiral Lewis acid, introduced by Charette, is the dioxaborolane **112**, which induces high levels of optical purity in the resultant cyclopropanes derived from allylic alcohols (4.90).[74] This methodology has been used in natural product synthesis, such as in the preparation of the antifungal agent FR-900848 (4.91).[75]

$$(4.90)$$

112   98%   93% ee

[74] A. B. Charette and J.-F. Marcoux, *Synlett* (1995), 1197; A. B. Charette, H. Juteau, H. Lebel and C. Molinaro, *J. Am. Chem. Soc.*, **120** (1998), 11 943; for use of the TADDOL chiral ligand, see A. B. Charette, C. Molinaro and C. Brochu, *J. Am. Chem. Soc.*, **123** (2001), 12168.

[75] A. G. M. Barrett and K. Kasdorf, *J. Am. Chem. Soc.*, **118** (1996), 11 030.

$$ (4.91) $$

FR-900848

Although the Simmons–Smith reaction has found considerable use in organic synthesis, it is not readily applicable to the formation of highly substituted cyclopropanes, since 1,1-diiodoalkanes (other than diiodomethane) are not readily available. Substituted zinc carbenoids can be prepared from aryl or α,β-unsaturated aldehydes (or ketones) with zinc metal, and these species can be trapped with an alkene to give substituted cyclopropanes.[76] The addition of chromium carbenes (see Section 1.2.2) to alkenes can be used to effect cyclopropanation to give substituted cyclopropanes.[77] Thus, addition of excess 1-hexene to the chromium carbene **113** gave the cyclopropane **114** as a mixture of diastereomers, with the isomer **114** predominating (4.92).[78]

$$ (4.92) $$

113   **114** 72% de

A popular method for the formation of substituted cyclopropanes is the condensation of a diazocarbonyl compound and an alkene in the presence of a metal catalyst.[79] Most common is the use of rhodium acetate, although copper and other metal salts are effective. The reaction is normally stereospecific with respect to the alkene geometry, however the stereoselectivity (*trans:cis* ratio) is rarely high. Typical is the reaction of ethyl diazoacetate with styrene, which gives the cyclopropanes

---

[76] W. B. Motherwell, *J. Organomet. Chem.*, **624** (2001), 41.

[77] M. P. Doyle, in *Comprehensive Organometallic Chemistry II*, ed. E. W. Abel, F. G. A. Stone and G. Wilkinson, vol. 12 (Oxford: Elsevier, 1995), p. 387; see also Chapter 1, Reference 136.

[78] J. Barluenga, S. López, A. A. Trabanco, A. Fernández-Acebes and J. Flóres, *J. Am. Chem. Soc.*, **122** (2000), 8145.

[79] H. M. L. Davies and E. G. Antoulinakis, *Org. Reactions*, **57** (2001), 1; H. M. L. Davies, in *Comprehensive Organic Synthesis*, ed. B. M. Trost and I. Fleming, vol. 4 (Oxford: Pergamon Press, 1991), p. 1031.

115 and **116** in high overall yield (4.93).

$$Ph\diagup\hspace{-0.3em}\diagdown \quad + \quad N_2\diagup\hspace{-0.3em}\diagdown CO_2Et \xrightarrow[93\%]{[Rh_2(OAc)_4]} \underset{115}{Ph^{\text{''''}}\triangle CO_2Et} \quad + \quad \underset{116}{Ph\triangle CO_2Et} \tag{4.93}$$

$$1.4 \quad : \quad 1$$

The cyclopropanation reaction with metal salts is readily amenable to asymmetric induction in the presence of a chiral ligand, and some excellent enantioselectivities have been achieved.[68,80] Particularly effective are the bisoxazoline ligands, developed by Pfaltz, Masamune and Evans. Both enantiomers of the chiral ligand are available and the reaction is amenable to a wide variety of different alkenes. For example, very high selectivity in favour of the cyclopropane **118** was achieved using only small amounts of the bisoxazoline ligand **117** and copper triflate (4.94).

$$\diagup\hspace{-0.3em}\diagdown\hspace{-0.3em}\diagup \quad + \quad N_2\diagup\hspace{-0.3em}\diagdown CO_2Et \xrightarrow[0.1\ mol\%]{0.1\ mol\%\ CuOTf} \triangle\text{''''}CO_2Et \tag{4.94}$$

$$91\% \ \textbf{118} \ 99\% \ ee$$

117

Intramolecular cyclopropanation reactions give access to bicyclic compounds, with, most commonly, the three-membered ring fused to a five- or six-membered ring. Metal-catalysed decomposition of the diazocarbonyl compound **119** gives the *exo*-bicyclo[3.1.0]hexanone **120** (4.95). The reaction is stereospecific, and the Z-isomer of **119** gives the *endo* diastereomer of the product.

$$\xrightarrow[58\%]{CuSO_4} \tag{4.95}$$

119   **120**

Asymmetric intramolecular cyclopropanation reactions have been reported with a variety of metal–ligand complexes,[80] the best choice of complex depending on the structure of the substrate. The bicyclic lactone **121** was formed with very high enantioselectivity using the chiral rhodium complex [Rh$_2$(5S-MEPY)$_4$] (4.96) (compare

---

[80] M. P. Doyle and M. N. Protopopova, *Tetrahedron*, **54** (1998), 7919; V. K. Singh, A. DattaGupta and G. Sekar, *Synthesis* (1997), 137.

with Scheme 4.79). This reaction was used in a recent synthesis of the chiral cyclopropane unit in the antifungal antibiotic ambruticin S.[81] However, copper(I) complexes with a bisoxazoline ligand, such as **117**, although less effective for the formation of **121** (20% ee), are more suitable for other substrates.

$$(4.96)$$

Intramolecular reactions of carbenes with alkenes have been exploited in synthesis. The sesquiterpene cycloeudesmol was prepared using, as a key step, the intramolecular cyclopropanation of the diazoketone **122** (4.97). The cyclopropanation reaction occurs stereoselectively to give the tricyclic product **123**, which was subsequently converted into the natural product. A synthesis of sesquicarene was achieved using the copper(I)-catalysed decomposition of the diazo compound **125**, itself prepared by oxidation of the hydrazone **124** (4.98).

$$(4.97)$$

$$(4.98)$$

The cyclopropanation reaction is not restricted to the formation of cyclopropane-containing products. Many different ring-opening reactions of cyclopropanes are known and the methodology therefore provides a useful carbon–carbon bond-forming reaction that has potential for a variety of targets. For example, vinyl cyclopropanes undergo thermal rearrangement to give cyclopentenes.[82] Thus, in a synthesis of isocomenic acid, the tricyclic compound **126** was prepared by

[81] T. A. Kirkland, J. Colucci, L. S. Geraci, M. A. Marx, M. Schneider, D. E. Kaelin and S. F. Martin, *J. Am. Chem. Soc.*, **123** (2001), 12432.

[82] For a recent asymmetric example, see H. M. L. Davies, B. Xiang, N. Kong and D. G. Stafford, *J. Am. Chem. Soc.*, **123** (2001), 7461.

intramolecular cyclopropanation followed by thermal rearrangement (4.99). The formation of divinyl cyclopropanes allows subsequent Cope rearrangement (see Section 3.6.1) to give seven-membered rings.[83]

It should be noted that if cyclopropanation is unfavourable, then other reactions such as insertion reactions or Wolff rearrangement (see below) may take place. A useful reaction of diazocarbonyl compounds (not involving carbenes) is an acid-catalysed cyclization by electrophilic attack on suitably situated carbon–carbon double bonds or aromatic rings. Thus, the unsaturated diazoketone **127** readily gave the fused cyclopentenone **128** and the bridged tricyclic compound **130** was obtained from the diazoketone **129**, with the generation of a quaternary carbon centre (4.100). The reaction is believed to proceed by initial protonation of the diazoketone (or complexation if a Lewis acid is used), followed by nucleophilic displacement of nitrogen from the resultant diazonium ion by the double bond or aromatic ring.

(4.100)

An important reaction of diazoketones is the Wolff rearrangement.[84] The reaction is the key step in the well-known Arndt–Eistert method for converting a carboxylic

[83] H. M. L. Davies, *Curr. Org. Chem.*, **2** (1998), 463; H. M. L. Davies, *Tetrahedron*, **49** (1993), 5203.
[84] W. Kirmse, *Eur. J. Org. Chem.* (2002), 2193; G. B. Gill, in *Comprehensive Organic Synthesis*, ed. B. M. Trost and I. Fleming, vol. 3 (Oxford: Pergamon Press, 1991), p. 887.

acid into its next higher homologue. Typically, an acid chloride or mixed anhydride is formed from the carboxylic acid and is treated with diazomethane to give the diazoketone. The Wolff rearrangement is normally effected by heat, photolysis or with a metal salt, often a silver(I) salt (4.101 and 4.102).

$$RCOCl \xrightarrow{CH_2N_2} \cdots \longrightarrow [\cdots] \longrightarrow RCH=C=O \xrightarrow{R'OH} RCH_2CO_2R' \quad (4.101)$$

$$\xrightarrow[\text{ii, CH}_2N_2]{\text{i, EtOCOCl, Et}_3N} \quad \xrightarrow[\text{Et}_3\text{N, MeOH}]{PhCO_2Ag} \quad (4.102)$$

76%   89%

The R group migrates with retention of configuration and the resulting ketene can be trapped by water, an alcohol, a thiol or an amine to give the product carboxylic acid, ester, thioester or amide. With cyclic diazoketones, the rearrangement leads to ring contraction. Such reactions have been used to prepare derivatives of strained small-ring compounds such as bicyclo[2.1.1]hexanes (4.103).

$$\xrightarrow[h\nu]{MeOH} \quad (4.103)$$

The addition of a diazocarbonyl compound to an alkene with metal catalysis is an effective method for the formation of cyclopropanes, as discussed above. However, direct addition to aldehydes, ketones or imines is normally poor.[85] Epoxide or aziridine formation can be promoted by trapping the carbene with a sulfide to give an intermediate sulfur ylide, which then adds to the aldehyde or imine.[86] For example, addition of tetrahydrothiophene to the rhodium carbenoid generated from phenyldiazomethane gave the ylide **131**, which adds to benzaldehyde to give the *trans* epoxide **132** in high yield (4.104). On formation of the epoxide, the sulfide is released and hence the sulfide (and the rhodium complex) can be used in substoichiometric amounts.

[85] See, however, M. P. Doyle, W. Hu and D. J. Timmons, *Org. Lett.*, **3** (2001), 933; H. M. L. Davies and J. DeMeese, *Tetrahedron Lett.*, **42** (2001), 6803.

[86] V. K. Aggarwal, H. Abdel-Rahman, L. Fan, R. V. H. Jones and M. C. H. Standen, *Chem. Eur. J.*, **2** (1996), 1024; V. K. Aggarwal, M. Ferrara, C. J. O'Brien, A. Thompson, R. V. H. Jones and R. Fieldhouse, *J. Chem. Soc., Perkin Trans. 1* (2001), 1635; V. K. Aggarwal and J. Richardson, *Chem. Commun.* (2003), 2644.

(4.104)

131   **132**

Preparing the diazoalkane *in situ* by warming the sodium salt of the corresponding toluene-*p*-sulfonylhydrazone (Bamford–Stevens reaction, see Scheme 4.76) avoids isolation of the potentially explosive diazo compound. In the presence of a chiral sulfide the methodology has been applied to the asymmetric synthesis of epoxides (4.105).[87]

(4.105)

PhCHO   82%   94% ee

Ylide formation can be used to good effect for a variety of transformations.[88] With an allylic sulfide, the resulting sulfonium ylide undergoes [2,3]-sigmatropic rearrangement (4.106). This type of rearrangement (see Section 3.7) can be effected with allylic amines, ethers and even allylic halides.

In the presence of a carbonyl group, a carbene can form a carbonyl ylide (oxonium ylide) and this species undergoes 1,3-dipolar cycloaddition reactions (see Section 3.4). Intramolecular capture of a metallocarbene, generated from a diazocarbonyl compound is a useful method for the formation of a cyclic carbonyl ylide. Cycloaddition is effective with alkene, carbonyl or other dipolarophiles and leads to bridged bicyclic compounds. An example is the formation of the carbonyl ylide **133** using rhodium(II) and its intramolecular cycloaddition to give the bridged compound **134** (4.107).[89] Improved conditions were found using [Rh$_2$(OCOCF$_3$)$_4$] (87% yield of **134**), that avoid competing Wolff rearrangement or dipolar cycloaddition of the diazo group across the alkenyl π-bond.

---

[87] V. K. Aggarwal, E. Alonso, I. Bae, G. Hynd, K. M. Lydon, M. J. Palmer, M. Patel, M. Porcelloni, J. Richardson, R. A. Stenson, J. R. Studley, J.-L. Vasse and C. L. Winn, *J. Am. Chem. Soc.*, **125** (2003), 10 926.

[88] A. Padwa and S. F. Hornbuckle, *Chem. Rev.*, **91** (1991), 263; M. P. Doyle, in *Comprehensive Organometallic Chemistry II*, ed. E. W. Abel, F. G. A. Stone and G. Wilkinson, vol. 12 (Oxford: Elsevier, 1995), p. 421; A.-H. Li, L.-X. Dai and V. K. Aggarwal, *Chem. Rev.*, **97** (1997), 2341; D. M. Hodgson, F. Y. T. M. Pierard and P. A. Supple, *Chem. Soc. Rev.*, **30** (2001), 50.

[89] A. Padwa, L. Precedo and M. A. Semones, *J. Org. Chem.*, **64** (1999), 4079.

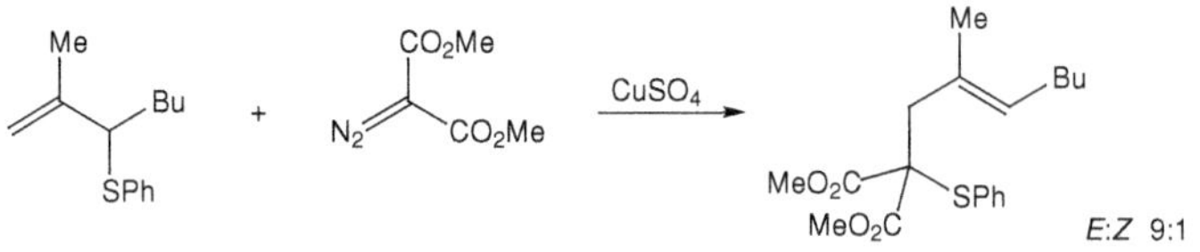

(4.106)

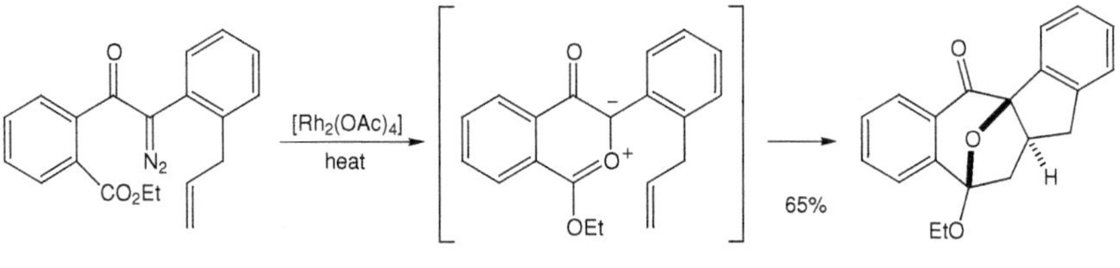

(4.107)

133   134

Problems (answers can be found on page 476)

1. Suggest reagents and conditions for the decarboxylation of the carboxylic acid **1** to give **2**.

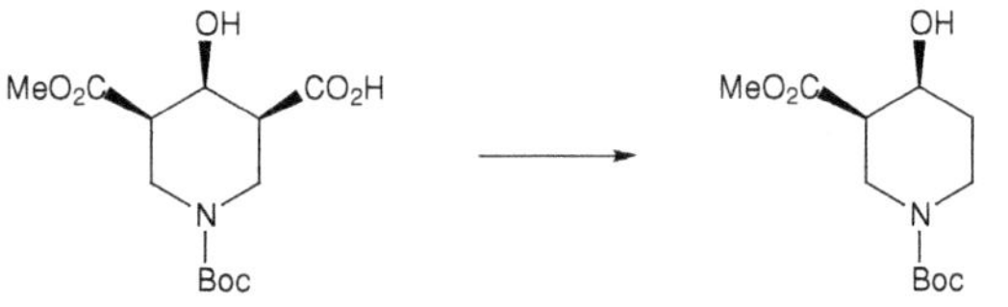

1   2

2. Explain the formation of the alcohol **4** from the epoxide **3**.

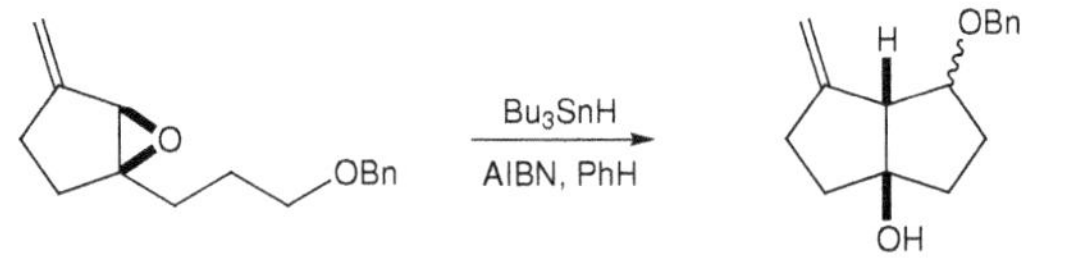

3   4

3. Explain (using the benzoyl radical and *N*-hydroxy-phthalimide) the formation of the ketone **5**.

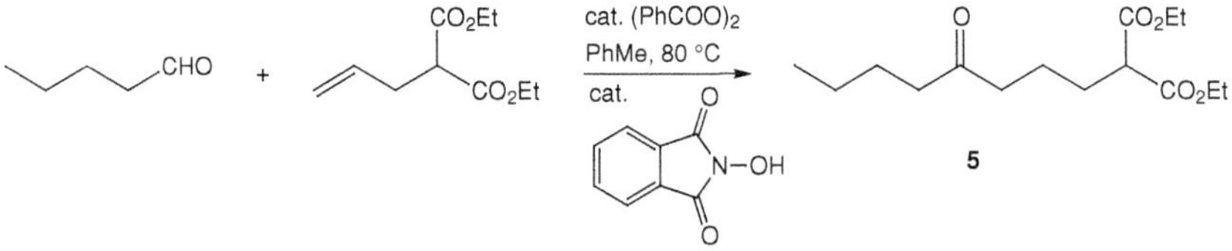

4. Draw the structures of the intermediate radicals to explain the formation of the three ketones **7–9** on treatment of the iodide **6** with a trialkyltin hydride.

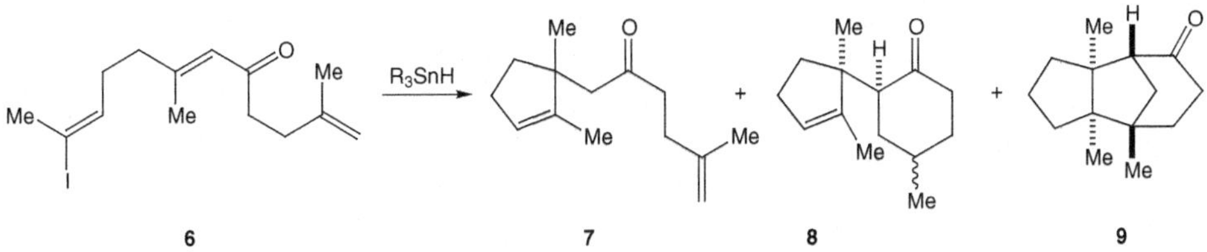

6   7   8   9

5. Suggest a mechanism for the formation of the cyclohexanones **10** that involves a phenylthio radical catalyst.

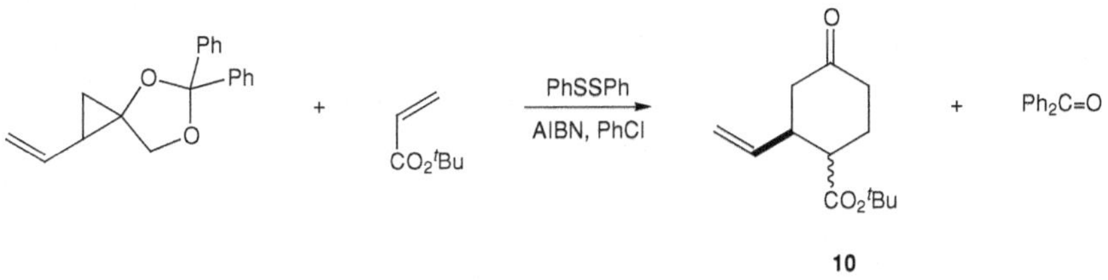

10

6. Draw the structure of the products resulting from Wittig alkenylation of the ketone **11**, and subsequent treatment with sodium hexamethyldisilazide (NaHMDS). Note that the base NaHMDS promotes the formation of the alkylidene carbene.

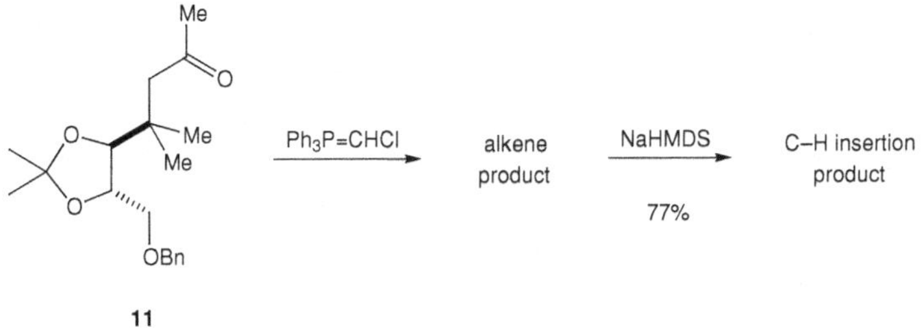

11

7. Draw the structure of the cyclopropane intermediate and explain the formation of the cycloheptadiene in the following reaction. Explain the stereochemistry of the product by drawing an appropriate three dimensional conformation of the transition state.

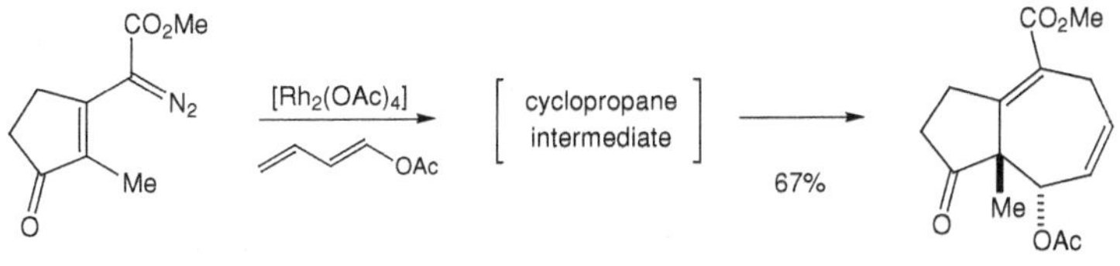

8. Explain the formation of the lactam product **13** on treatment of the diazoketone **12** with silver(I) benzoate.

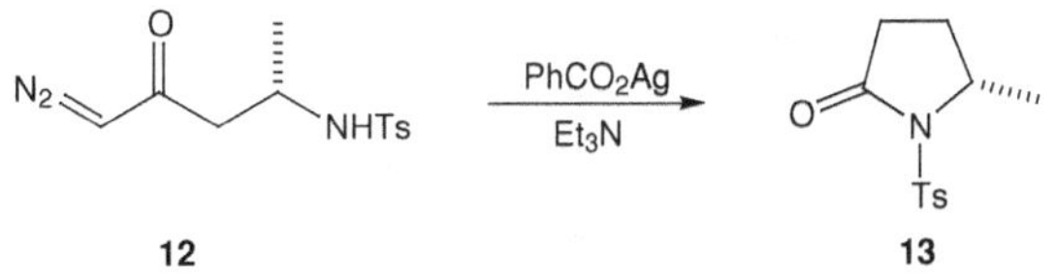

12   13

9. A synthesis of the pine beetle pheromones *exo*- and *endo*-brevicomin makes use of the following reaction. Draw the structure of the intermediate ylide and explain the regiochemistry of the cycloaddition reaction (don't worry about the stereochemistry – a mixture of the *exo* and *endo* isomers is formed).

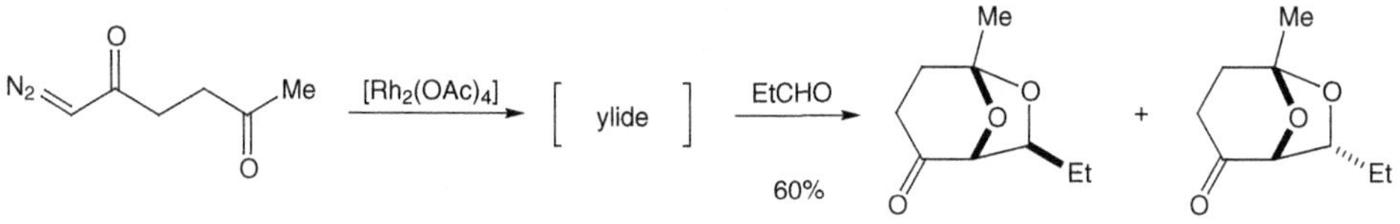

5

Functionalization of alkenes

Alkenes are very useful in synthesis, owing to their ready conversion to many different functional groups. As a result of their importance, Chapter 2 was devoted to methods for their preparation. You should have noticed that we have already encountered many reactions of alkenes. For example, nucleophilic additions (such as conjugate addition) and palladium-catalysed reactions are covered in Chapter 1, pericyclic reactions (such as the Diels–Alder reaction) are described in Chapter 3, and addition reactions with radicals and carbenes are given in Chapter 4. This chapter covers other functionalization reactions, including hydroboration, epoxidation, aziridination, dihydroxylation, oxidative cleavage and palladium-catalysed oxidation. Commonly, the carbon–carbon double bond is converted to a single bond with the incorporation of one or two heteroatoms at the original alkene carbon atom(s). Examples of the oxidation of alkenes are included here, whereas the oxidation of other functional groups are provided in Chapter 6. Reduction of alkenes are described in Chapter 7.

5.1 Hydroboration

Organoboranes are obtained by addition of borane or alkyl boranes to alkenes (or alkynes). Borane itself can be prepared by reaction of boron trifluoride etherate with sodium borohydride. Borane exists as a dimer, but solutions containing an electron donor, such as an ether, amine or sulfide, allow adduct formation. The complexes $BH_3 \cdot THF$ and the borane–dimethyl sulfide complex $BH_3 \cdot SMe_2$ are commercially available and provide a convenient source of borane. The dimethyl sulfide complex is more stable than $BH_3 \cdot THF$ and has the additional advantage that it is soluble in a variety of organic solvents, such as diethyl ether and hexane.

The most important synthetic application of borane is for the preparation of alkyl boranes by addition to alkenes, a process known as hydroboration (5.1).[1] Borane and its derivatives can also be used for reduction (see Section 7.3). The hydroboration reaction has been applied to a large number of alkenes of widely differing structures. In nearly all cases the addition proceeds rapidly at room temperature, and only the most hindered alkenes do not react.

$$
\begin{array}{ccc}
\diagdown\diagup C{=}C\diagup\diagdown & + & H{-}B\diagup\diagdown \longrightarrow \overset{H}{\diagdown}C{-}C\overset{B{-}}{\diagup} & (5.1)
\end{array}
$$

Hydroboration occurs by a concerted process and takes place through a four-membered cyclic transition state, formed by addition of a polarized B–H bond (boron is the more positive) to the alkene double bond (5.2). This is supported by the fact that the reaction is stereospecific, with *syn* addition of the boron and hydrogen atoms. The reaction can also be stereoselective, with hydroboration taking place preferentially on the less hindered side of the double bond. Stereospecific addition of borane to a 1-alkylcycloalkene such as 1-methylcyclohexene, gives, after oxidation of the organoborane product (see Scheme 5.21), almost exclusively the *trans* alcohol product (5.3).

$$
\begin{array}{c}
\overset{\delta-}{H}{-}{-}{-}\overset{\delta+}{B}\diagup \\
\diagdown C{=}C\diagup\diagdown
\end{array}
\qquad (5.2)
$$

$$(5.3)$$

85%

Hydroboration of mono- and disubstituted alkenes with borane gives rise typically to a trialkylborane product. However, trisubstituted alkenes normally give a dialkylborane and tetrasubstituted alkenes form only the monoalkylboranes (5.4). The extent of hydroboration may also be controlled by the stoichiometry of alkene and borane. This has been exploited in the preparation of a number of mono- and dialkylboranes that are less reactive and more selective than borane itself. Important in this respect are the so-called disiamylborane **1** (name derived from

---

[1] K. Smith and A. Pelter, in *Comprehensive Organic Synthesis*, ed. B. M. Trost and I. Fleming, vol. 8 (Oxford: Pergamon Press, 1991), p. 703; for a review of organoboron chemistry, see M. Vaultier and B. Carboni, in *Comprehensive Organometallic Chemistry II*, ed. E. W. Abel, F. G. A. Stone and G. Wilkinson, vol. 11 (Oxford: Elsevier, 1995), p. 191.

the di-*s*-isoamyl group), thexylborane **2** (derived from *t*-hexyl) and 9-BBN **3** (9-borabicyclo[3.3.1]nonane) (5.5), formed by addition of borane to 2-methyl-2-butene, 2,3-dimethyl-2-butene and 1,5-cyclo-octadiene respectively. These partially alkylated boranes may themselves be used to hydroborate less-hindered alkenes.

$$[(CH_3)_2CHCH(CH_3)]_2BH$$

1

(5.4)

$$[(CH_3)_2CHC(CH_3)_2]BH_2$$

2

$$\equiv \quad HB$$

(5.5)

3

Hydroboration is readily effected with alkenes containing many types of functional group. Where the other functional group is not reduced by borane, hydroboration generally proceeds without difficulty and even some functional groups that react only slowly with borane may be tolerated, for example carboxylic esters. Easily reduced carbonyl groups of aldehydes and ketones, however, must be protected as their acetals, and carboxylic acids as esters.

Addition of borane to an unsymmetrical alkene could, of course, give rise to two different products by addition of boron at either end of the double bond. It is found in practice, however, that in the absence of strongly polar neighbouring substituents, the reactions are highly selective and give predominantly the isomer in which boron is attached to the less highly substituted carbon atom (5.6). Typically, the less-hindered end of the alkene double bond is also the more electron rich and therefore interacts better with the electron-deficient boron atom. However, the selectivity is diminished with increasing electronegativity of the substituent.

EtO   Cl $CH_3CH{=}CHC(CH_3)_3$ (5.6)

6% 94%   2% 98%   19% 81%   40% 60%   58% 42%

With 1,2-disubstituted alkenes, however, there is little discrimination in reactions with borane itself. In addition, the regioselectivity in the hydroboration of terminal alkenes, although high, is not complete. Further, there is little difference in the rate of reaction of borane with differently substituted double bonds, so that it is rarely possible to achieve selective hydroboration of one double bond in the presence of others. These difficulties can be overcome by hydroboration with substituted boranes such as disiamylborane **1**, thexylborane **2** or 9-BBN **3**. Such reagents are less reactive and more selective than borane. For example, 1-hexene is hydroborated with 94% addition of the boron atom to C–1 using borane, 99% using disiamylborane **1** and 99.9% using 9-BBN **3**. Unsymmetrical 1,2-disubstituted alkenes, in which borane shows almost no regioselectivity, may be hydroborated with high selectivity using a bulky borane reagent, as illustrated with 4-methyl-2-pentene **4** (5.7). Even greater discrimination is shown by 9-BBN (0.2:99.8 ratio of organoborane products), whereas borane gives a 43:57 mixture.

$$(5.7)$$

Disiamylborane **1** and 9-BBN **3** are more sensitive to the structure of the alkene than is borane itself. Terminal alkenes react more rapidly than internal alkenes and Z-alkenes more rapidly than *E*-alkenes. This sometimes allows the selective hydroboration of one double bond in a diene or triene. Thus, the non-conjugated diene **5** is readily hydroborated to give the organoborane **6**, with no hydroboration of the more-hindered alkene (5.8).

$$(5.8)$$

Hydroboration of dienes with borane itself usually leads to the formation of polymers. The monoalkylborane thexylborane **2** can promote the hydroboration of dienes to give cyclic or bicyclic organoboranes. 1,5-Hexadiene, for example, is converted mainly into the boracycloheptane **7** (5.9). Thexylborane has been used to make trialkylboranes containing three different alkyl groups by stepwise addition to two different alkenes (5.10). Ideally the first alkene is relatively unreactive (otherwise two groups would add to the borane at this stage) and so the procedure is less well suited to making trialkylboranes containing two different primary alkyl groups.

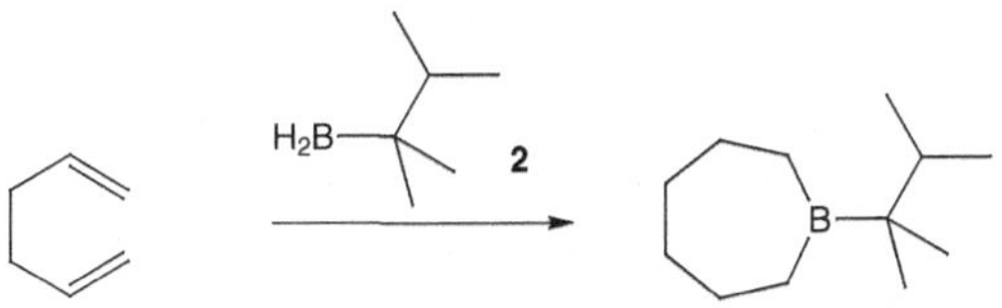

(5.9)

7

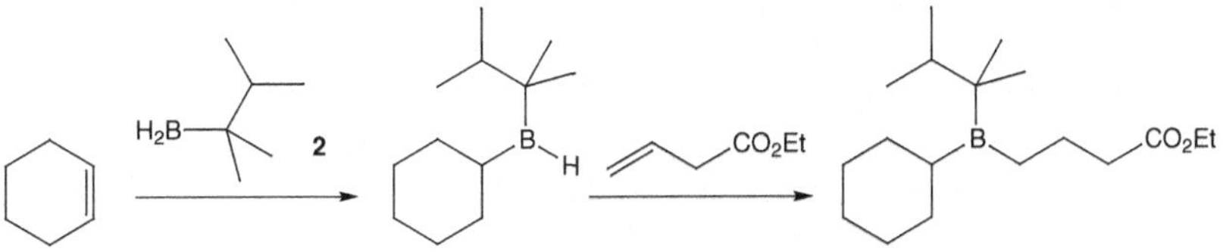

(5.10)

Halogenated boranes react with the usual regioselectivity, with addition of the boron atom to the less-substituted end of the carbon–carbon double bond, although they do not always show the same reactivity as other borane derivatives. For example, the far higher reactivity of dibromoborane towards more-substituted alkenes, in contrast to disiamyborane, makes possible the selective hydroboration of the double bonds in 2-methyl-1,5-hexadiene (5.11).

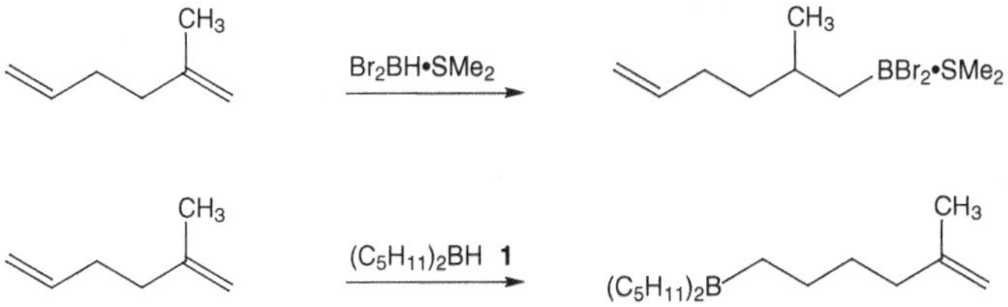

(5.11)

Catecholborane **8**, which has two boron–oxygen bonds, is much less reactive as a hydroborating agent. Typically, high temperatures are required for hydroboration with catecholborane. However, in the presence of a catalyst, most commonly a rhodium complex, high reactivity and selectivity can be achieved at room temperature or below.[2] Wilkinson's catalyst, $Rh(PPh_3)_3Cl$, is the most frequently used. For example, hydroboration of 1-hexene with catecholborane requires heating to 90 °C, whereas in the presence of $Rh(PPh_3)_3Cl$ the reaction occurs at room temperature in just 5 min.

The regioselectivity in the metal-catalysed hydroboration of alkyl-substituted alkenes is the same as that with the non-catalysed system, in which the boron atom becomes attached to the less-hindered carbon atom of the alkene. However,

---

[2] I. Beletskaya and A. Pelter, *Tetrahedron*, **53** (1997), 4957; K. Burgess and M. J. Ohlmeyer, *Chem. Rev.*, **91** (1991), 1179.

the mechanism for the hydroboration is different, occurring via a H—Rh(III)–B(catechol) species, formed by oxidative addition of the catalyst into the hydrogen–boron bond. Monosubstituted alkenes react fastest, with tri- and tetrasubstituted alkenes unreactive towards catecholborane and Wilkinson's catalyst. This allows the selective hydroboration of less substituted alkenes, as illustrated during a synthesis of calyculin A (5.12).[3]

$$\text{(5.12)}$$

The different mechanisms for the catalysed and non-catalysed reactions can be beneficial for controlling the stereochemical outcome of the hydroboration reaction. Hence, hydroboration of the allylic alcohol **9** occurs with opposite stereoselectivity under the two different reaction conditions (5.13).[4]

$$\text{(5.13)}$$

9	**8**, Rh(PPh₃)₃Cl	79%	81	:	19
	9-BBN	91%	17	:	83

Remarkably, catalytic hydroboration of arylethenes (styrenes) using a cationic rhodium complex occurs with unconventional regioselectivity. Essentially complete selectivity for the secondary alcohol (formed after oxidation of the organoborane) is obtained by using the complex [Rh(COD)₂]BF₄ in the presence of a phosphine ligand (5.14). This contrasts with that obtained in the absence of a catalyst, or in the presence of neutral transition-metal catalysts.

$$\text{(5.14)}$$

BH₃•THF	82	:	18
8, Rh(PPh₃)₃Cl	90	:	10
8, [Rh(COD)₂]BF₄, PPh₃	<1	:	>99

[3] D. A. Evans and J. R. Gage, *J. Org. Chem.*, **57** (1992), 1958.
[4] D. A. Evans, G. C. Fu and A. H. Hoveyda, *J. Am. Chem. Soc.*, **114** (1992), 6671.

Hydroboration of alkenes with an optically active alkylborane, or with a chiral ligand for the metal-catalysed hydroboration, followed by oxidation has been used in the asymmetric synthesis of optically active secondary alcohols. Good results have been obtained with di-isopinocampheylborane, $(Ipc)_2BH$ **10**, which is prepared readily in either enantiomeric form by reaction of borane with (+)- or (−)-α-pinene under appropriate conditions.[5] The dialkylborane **10** is most effective for asymmetric hydroboration of *cis*-disubstituted alkenes. For example, reaction of 2,3-dihydrofuran with (−)-$(Ipc)_2BH$ occurs with perfect enantioselectivity to give, after conversion to the diethyl boronate and oxidation, the chiral alcohol **12** (5.15).[6]

The dialkylborane **10** is normally, however, poorly selective with *trans-* or trisubstituted alkenes. In such cases, the mono-isopinocampheylborane, $(Ipc)BH_2$ **11**, may be effective. For example, the cyclohexanol **13** was obtained as an 85:15 ratio of enantiomers using the hydroborating agent **11**, and was hydrolysed to give cryptone (5.16).[7]

The reversal in the regioselectivity of the hydroboration of styrenes using a cationic rhodium complex (see Scheme 5.14), to provide the secondary (rather than primary) organoborane, allows the study of the asymmetric hydroboration of mono-(aryl) substituted alkenes (styrenes). A variety of chiral phosphine ligands can be used to good effect, a popular choice being (*R*)-2,2′-bis(diphenylphosphino)-1, 1′-binaphthyl (BINAP). Just 0.02 molar equivalents of the rhodium catalyst and

---

[5] H. C. Brown and P. V. Ramachandran, *J. Organomet. Chem.*, **500** (1995), 1.
[6] H. C. Brown and J. V. N. Vara Prasad, *J. Am. Chem. Soc.*, **108** (1986), 2049.
[7] R. C. Hawley and S. L. Schreiber, *Synth. Commun.*, **20** (1990), 1159.

this ligand promote the highly enantioselective hydroboration of styrene (5.17).[8] Asymmetric hydroboration of β-substituted styrenes (ArCH=CHR) with the chiral ligand BINAP is often poorly enantioselective. In such cases, the ligand **14** can provide a solution (5.18).[9]

$$
\text{Ph} \diagup \!\!\!\!\! = \quad \xrightarrow[\substack{\text{2 mol\% [Rh(COD)}_2\text{]BF}_4 \\ \text{2 mol\% (}R\text{)-BINAP} \\ \text{DME, }-78\ ^\circ\text{C} \\ \text{then H}_2\text{O}_2\text{, NaOH}}]{\textbf{8}} \quad \underset{\text{91\%}\quad\text{96\% ee}}{\text{Ph}\diagup\!\!\!\text{CH}_3}
$$

(*R*)-BINAP $\qquad$ (5.17)

$$
\xrightarrow[\substack{\text{1 mol\% [Rh(COD)}_2\text{]BF}_4 \\ \text{1 mol\% } \textbf{14} \\ \text{PhMe, room temp.} \\ \text{then H}_2\text{O}_2\text{, NaOH}}]{\textbf{8}} \quad \underset{\text{78\%}\quad\text{96\% ee}}{}
$$

14 $\qquad$ (5.18)

5.1.1 Reactions of organoboranes

The usefulness of the hydroboration reaction in synthesis arises from the fact that the alkylboranes formed can be converted by further reaction into a variety of other products.[10] On hydrolysis (protonolysis), for example, the boron atom is replaced by a hydrogen atom. Particularly important is the oxidation to alcohols. Some of the transformations are described below.

Protonolysis is best effected with an organic carboxylic acid and provides a convenient method for the reduction of a carbon–carbon multiple bond. Heating the alkylborane with propionic acid is used most commonly, although alkenylboranes are more reactive and can undergo rapid protonolysis with acetic acid at room temperature. The reaction takes place through a cyclic transition state with retention of configuration at the carbon atom concerned (5.19). Therefore hydroboration of an alkyne (*syn* addition) followed by protonolysis leads to the Z-alkene product (see Section 2.6). An advantage of the hydroboration–protonolysis procedure is that it can sometimes be used for the reduction of double or triple bonds in compounds that contain other easily reducible groups. Allyl methyl sulfide, for example, is converted into methyl propyl sulfide (5.20).

[8] T. Hayashi, Y. Matsumoto and Y. Ito, *Tetrahedron: Asymmetry*, **2** (1991), 601.
[9] H. Doucet, E. Fernandez, T. P. Layzell and J. M. Brown, *Chem. Eur. J.*, **5** (1999), 1320.
[10] E. Negishi and M. J. Idacavage, *Org. Reactions*, **33** (1985), 1.

$$R_2B\!\!-\!\!OCOEt \quad + \quad RH \qquad (5.19)$$

$$\qquad (5.20)$$

78%

Oxidation of organoboranes to alcohols is usually effected with alkaline hydrogen peroxide.[11] The reaction is of wide applicability and many functional groups are unaffected by the reaction conditions, so that a variety of substituted alkenes can be converted into alcohols by this procedure. Several examples have been given above. A valuable feature of the reaction is that it results in the overall addition of water to the double (or triple) bond, with a regioselectivity opposite to that from acid-catalysed hydration. This follows from the fact that, in the hydroboration step, the boron atom adds to the less-substituted carbon atom of the multiple bond. Terminal alkynes, for example, give aldehydes in contrast to the methyl ketones obtained by mercury-assisted hydration.

The oxidation reaction involves migration of an alkyl group from boron to oxygen, in an intermediate borate species. All three alkyl groups on the boron atom can undergo this reaction. Hydrolysis of the resulting $B(OR)_3$ derivative releases the desired alcohol product (5.21).

$$\qquad (5.21)$$

The alkyl group migrates with retention of stereochemistry at the migrating carbon centre. Since the hydroboration reaction occurs by a *syn* addition pathway, subsequent oxidation results in *syn* addition of the elements of water across the

[11] G. W. Kabalka and H. C. Hedgecock, *J. Org. Chem.*, **40** (1975), 1776; G. Zweifel and H. C. Brown, *Org. Reactions*, **13** (1963), 1.

double bond. Thus, hydroboration–oxidation of 1-methylcyclohexene gives *trans*-2-methylcyclohexanol (see Scheme 5.3). Such stereospecific aspects can occur in conjunction with the stereoselective addition of borane to the less-hindered face of the alkene. The alkene β-pinene undergoes stereoselective hydroboration to give, after oxidation, *cis*-myrtanol (5.22). In a synthesis of a fragment of rifamycin, the allylic alcohol **15** was converted selectively into the *meso*-triol **16**, in which one centre controls the formation of the new chiral centres (by way of diastereomeric transition states in which allylic 1,3-strain is avoided) (5.23).[12] In each case, *syn* addition of B–H to the less hindered face is followed by oxidation of the carbon–boron bond with retention of configuration. Asymmetric hydroboration–oxidation gives rise to optically active alcohol products, as described in Schemes 5.15–5.18.

$$(5.22)$$

cis-myrtanol

$$(5.23)$$

15   75%   **16**

Hydroboration of **15** is thought to occur *via* the conformation with the allylic hydrogen atom eclipsing the alkene (compare with reactions of allylic derivatives in Section 1.1.8)

Tr = CPh₃

Oxidation of the organoborane produced by hydroboration can be accomplished by using oxidants other than hydrogen peroxide, such as amine oxides, sodium perborate, sodium percarbonate, Oxone® (2KHSO₅·KHSO₄·K₂SO₄) or even molecular oxygen. For example, on a large scale, or when oxidation with hydrogen peroxide is slow or gives side products, then Oxone® has been found to

¹² W. C. Still and J. C. Barrish, *J. Am. Chem. Soc.*, **105** (1983), 2487.

be a valuable alternative.[13] Transformation of the alkene **17** to the alcohol **18** is best accomplished under these conditions (5.24).

$$(5.24)$$

17   **18**

Direct oxidation of primary trialkylboranes to aldehydes and of secondary trialkylboranes to ketones, without isolation of the alcohol, is possible with pyridinium chlorochromate (PCC) or aqueous chromic acid (formed from $Na_2Cr_2O_7$ and H_2SO_4).[14] For example, 1-octene was converted directly to octanal using disiamylborane, followed by oxidation with PCC (5.25).

$$(5.25)$$

71%

Replacement of the boron atom for a nitrogen atom is possible with a suitable aminating agent. Treatment of a trialkylborane with a chloramine, prepared *in situ* by oxidation of ammonia or an amine with sodium hypochlorite, provides a method to form a carbon–nitrogen bond (5.26).[15] The transformation of an alkene to an amine by overall addition of ammonia is much less straightforward than hydration and this methodology, although used less frequently than oxidation with peroxide, provides a solution to this problem.

$$(5.26)$$

91%

If catecholborane **8** is used for hydroboration, then conversion of the boronic ester product to a trialkylborane with a Grignard reagent or a dialkylzinc species is needed prior to amination. Thus, asymmetric hydroboration of styrene, followed by addition of methyl magnesium chloride then hydroxylamine-*O*-sulfonic acid gave

[13] K. S. Webb and D. Levy, *Tetrahedron Lett.*, **36** (1995), 5117; D. H. Brown Ripin, W. Cai and S. J. Brenek, *Tetrahedron Lett.*, **41** (2000), 5817.
[14] H. C. Brown and C. P. Garg, *Tetrahedron*, **42** (1986), 5511; H. C. Brown, S. U. Kulkarni, C. G. Rao and V. D. Patil, *Tetrahedron*, **42** (1986), 5515.
[15] G. W. Kabalka, K. A. R. Sastry, G. W. McCollum and H. Yoshioka, *J. Org. Chem.*, **46** (1981), 4296; G. W. Kabalka, G. W. McCollum and S. A. Kunda, *J. Org. Chem.*, **49** (1984), 1656.

α-methylbenzylamine with high optical purity (5.27).[16] Secondary amines can be formed by using a primary amine and sodium hypochlorite, or by reaction of the organoborane with an alkyl or aryl azide.[17]

$$Ph \diagup\diagdown \xrightarrow[\substack{\text{1 mol\% [Rh(COD)(acac)]OTf} \\ \text{1 mol\% } \textit{ent.} \text{ 14, THF} \\ \text{then MeMgCl, THF}}]{\substack{\text{8}}} Ph \overset{BMe_2}{\underset{CH_3}{|}} \xrightarrow[54\%]{NH_2OSO_3H} Ph \overset{NH_2}{\underset{CH_3}{|}} \quad (5.27)$$

87% ee

Intramolecular reaction of the organoborane with an azide provides a cyclic amine product.[18] For example, in a synthesis of the cyclic hexapeptide echinocandin D, stereoselective hydroboration of the alkene **19** was followed by cyclization to give the substituted pyrrolidine **20** (5.28).[19]

$$(5.28)$$

72%

19   **20**

A useful transformation in synthesis is the reaction of an organoborane to form a carbon–carbon bond. One method to achieve this is by addition of carbon monoxide, followed by oxidation, which can be directed to give primary, secondary or tertiary alcohols, aldehydes or ketones under appropriate conditions.[10,20] On heating, many organoboranes absorb one molecule of carbon monoxide at atmospheric pressure to form intermediates that are oxidized to tertiary alcohols by alkaline hydrogen peroxide. The reaction is of wide applicability, even for bulky alkyl groups. For example, tricyclohexylcarbinol R_3C-OH, $R=c-C_6H_{11}$, is obtained from cyclohexene in 85% yield.

The reaction involves migration of the alkyl groups from the boron to the carbon atom of carbon monoxide. This in an intramolecular process, as shown by the fact that carbonylation of an equimolar mixture of triethylborane and tributylborane gave, after oxidation, only triethylcarbinol and tributylcarbinol; no 'mixed'

[16] E. Fernandez, K. Maeda, M. W. Hooper and J. M. Brown, *Chem. Eur. J.*, **6** (2000), 1840.
[17] H. C. Brown, A. M. Salunkhe and B. Singaram, *J. Org. Chem.*, **56** (1991), 1170.
[18] B. Carboni and M. Vaultier, *Bull. Soc. Chim. Fr.*, **132** (1995), 1003.
[19] D. A. Evans and A. E. Weber, *J. Am. Chem. Soc.*, **109** (1987), 7151.
[20] H. C. Brown, *Acc. Chem. Res.*, **2** (1969), 65.

alcohols were formed. A stepwise pathway, involving three successive intramolecular transfers, has been proposed (5.29).

$$R_3B \rightleftharpoons R_3B-C\equiv O \longrightarrow R_2B-\underset{\underset{O}{\|}}{C}-R \longrightarrow RB-CR_2 \longrightarrow OB-CR_3 \xrightarrow[HO^-]{H_2O_2} HO-CR_3$$

$$(5.29)$$

If the carbonylation is conducted in the presence of a small amount of water, migration of the third alkyl group is inhibited. Oxidation of the hydrate produced then gives the ketone (5.30). Yields obtained are generally high, and the sequence provides a convenient synthetic route to ketones. For example, cyclopentene was converted into dicyclopentyl ketone in 90% yield.

$$R_3B \xrightarrow[H_2O]{CO} RB-CR_2 \xrightarrow[HO^-]{H_2O_2} \underset{O}{R}\overset{R}{\underset{\|}{\diagup}}R \qquad (5.30)$$

The method can be extended to the synthesis of unsymmetrical ketones by using 'mixed' organoboranes prepared from thexylborane **2** or thexylchloroborane. The thexyl (tert-hexyl) group shows an exceptionally low aptitude for migration. Carbonylation of trialkylboranes thexylB(R)R′ in the presence of water, followed by oxidation, leads to high yields of the ketone RCOR′. Because of the bulky nature of the thexyl group, carbonylation of these compounds requires more vigorous conditions than usual and generally has to be carried out under pressure. Functional groups in the alkene rarely interfere with the reaction, and the procedure can be used to synthesize a ketone from almost any two (unhindered) alkenes (5.31).[21]

$$(5.31)$$

---

[21] H. C. Brown and E. Negishi, *J. Am. Chem. Soc.*, **89** (1967), 5285.

Dienes similarly yield cyclic ketones and bicyclic ketones can be prepared by this method. For example, the thermodynamically disfavoured *trans*-perhydroindanone **21** is formed on stereospecific hydroboration of 1-vinylcyclohexene, followed by carbonylation (5.32). Likewise, *trans*-1-decalone is obtained from 1-allylcyclohexene. The stereoselectivity of the reactions, leading exclusively to the *trans* fused compounds, is a consequence of the mechanism of the hydroboration, which requires *syn* addition of the B—H group to the double bond of the alkene.

$$\text{(5.32)}$$

The carbonylation reaction can be adapted to the preparation of aldehydes and primary alcohols. In the presence of certain hydride reducing agents, such as $LiAlH(OMe)_3$, the rate of reaction of carbon monoxide with organoboranes is greatly increased. The intermediate acylborane (R_2BCOR) is reduced and the product, on oxidation with buffered hydrogen peroxide gives an aldehyde. Alkaline hydrolysis gives the corresponding primary alcohol (containing one more carbon atom than the original alkene and therefore differing from the product obtained on direct oxidation of the original organoborane). Hydroboration of an alkene with 9-borabicyclo[3.3.1]nonane **3**, followed by reaction of the alkyl–B(9-BBN) derivative with carbon monoxide and $LiAlH(OMe)_3$ takes place with preferential migration of the alkyl group. High yields of aldehydes containing a variety of functional groups have been obtained from substituted alkenes by this method.

A useful carbon–carbon bond-forming process from organoboranes uses the anion of dichloromethane (or a derivative, such as dichloromethyl methyl ether).[22] Addition of butyllithium to dichloromethane generates dichloromethyllithium, which adds to the organoborane. Migration of an alkyl group is accompanied by loss of chloride ion. The reaction is effective with boronic esters and the product α-chloroboronic esters can be oxidized or reacted with a nucleophile to displace the chloride (followed, if desired, by iterative addition of $LiCHCl_2$). For example, asymmetric hydroboration then addition of $LiCHCl_2$ and oxidation gave optically active 2-phenylpropanoic acid (5.33).[23] In a similar way, organoboranes react with the anion of α-halo-ketones and α-halo-esters, to give the corresponding α-alkyl

[22] D. S. Matteson and D. Majumdar, *Organomet.*, **2** (1983), 1529.
[23] A. Chen, L. Ren and C. M. Crudden, *J. Org. Chem.*, **64** (1999), 9704.

carbonyl compounds in which the halogen atom has been replaced by an alkyl group from the boron atom (5.34).[24]

$$(5.33)$$

$$(5.34)$$

Examples described so far make use of nucleophilic addition to the organoborane, in which the nucleophile bears a leaving group. An alternative strategy involves transmetallation of the organoborane to a new organometallic species capable of reaction with an electrophile. Conversion of the organoborane to an organozinc species is possible using a dialkylzinc reagent. Subsequent addition of copper cyanide gives an organocopper–zinc species that can be used in a wide variety of reactions (see Section 1.1.7). For example, hydroboration of 1-phenylcyclohexene, transmetallation and reaction with allyl bromide gave *trans*-1-allyl-2-phenylcyclohexane (5.35).[25]

$$(5.35)$$

In the presence of palladium(0), organoboranes undergo coupling reactions with unsaturated halides (or triflates). This type of *B*-alkyl Suzuki reaction is a useful method for the synthesis of substituted alkenes (see Section 1.2.4, Scheme 1.205).[26] For example, hydroboration of the alkene **22** and coupling with the alkenyl iodide **23** was used in a synthesis of the alkaloid halichlorine (5.36).

[24] H. C. Brown and M. M. Rogic, *J. Am. Chem. Soc.*, **91** (1969), 2146.
[25] A. Bourdier, C. Darcel, F. Flachsmann, L. Micouin, M. Oestreich and P. Knochel, *Chem. Eur. J.*, **6** (2000), 2748.
[26] S. R. Chemler, D. Trauner and S. J. Danishefsky, *Angew. Chem. Int. Ed.*, **40** (2001), 4544.

$$(5.36)$$

Homolytic cleavage of the carbon–boron bond of a trialkylborane can be pro-moted by oxygen.[27] The so-formed alkyl radical can be used in synthesis (for radical reactions, see Section 4.1). Indeed, triethylborane in air can be used to gen-erate radicals from precursors such as alkyl iodides or selenides. Hydroboration followed by addition of an α,β-unsaturated aldehyde or ketone leads to trans-fer of an alkyl group from the boron atom via an alkyl radical intermediate. The reaction takes place by addition of the alkyl radical to the conjugated system to form an enol borinate, hydrolysis of which gives the aldehyde or ketone product (5.37).

$$(5.37)$$

A problem with this methodology is that only one of the three alkyl groups is transferred to the unsaturated carbonyl compound. A solution to this uses the radical generated from the boronic ester, itself derived from hydroboration with catecholb-orane **8**. Treatment of the boronic ester with oxygen and 1,3-dimethyl-hexahydro-2-pyrimidinone (DMPU) in the presence of the α,β-unsaturated aldehyde or ketone gives the desired radical addition product, with transfer of the *B*-alkyl group.[28] Thus, cyclohexene was converted to 1-cyclohexyl-3-pentanone **24** using this chemistry (5.38).

[27] H. C. Brown and M. M. Midland, *Angew. Chem. Int. Ed. Engl.*, **11** (1972), 692; K. Nozaki, K. Oshima and K. Utimoto, *J. Am. Chem. Soc.*, **109** (1987), 2547; C. Ollivier and P. Renaud, *Chem. Rev.*, **101** (2001), 3415.
[28] C. Ollivier and P. Renaud, *Chem. Eur. J.*, **5** (1999), 1468.

$$(5.38)$$

94%

5.2 Epoxidation and aziridination

This section, and those following in this chapter, can be classified as presenting oxidation reactions of alkenes. The formation of a three-membered ring, particularly epoxidation, is an extremely valuable transformation in organic synthesis. Epoxides and even aziridines are present in a number of natural products and biologically active compounds. Of crucial importance, however, is their use as building blocks in organic synthesis; their ring-opening reactions allow the formation of a wide variety of substituted alcohol- and amine-containing compounds.

5.2.1 Epoxidation

Reaction of an alkene with an oxidising agent, such as a peroxy-acid, leads to the formation of an epoxide ring.[29] A number of peroxy-acids can be used, although the most common is *meta*-chloroperoxybenzoic acid (mCPBA). This is a fairly stable, white solid, that is commercially available. The reaction is believed to take place by electrophilic attack of the peroxy-acid on the double bond (5.39).

$$(5.39)$$

[29] A. S. Rao, in *Comprehensive Organic Synthesis*, ed. B. M. Trost and I. Fleming, vol. 7 (Oxford: Pergamon Press, 1991), p. 357.

In accordance with this mechanism, the rate of epoxidation is increased by electron-withdrawing groups in the peroxy-acid or electron-donating groups on the double bond. Terminal mono-alkenes react slowly with most peroxy-acids and the rate of reaction increases with the degree of alkyl substitution. 1,2-Dimethyl-1,4-cyclohexadiene for example, reacts preferentially at the more electron-rich tetra-substituted double bond and the diene **25** reacts selectively at the disubstituted double bond (5.40). On the other hand, conjugation of the alkene double bond with other unsaturated groups reduces the rate of epoxidation because of delocalization of the π-electrons. α,β-Unsaturated acids and esters require the strong reagent trifluoroperoxyacetic acid, or mCPBA at an elevated temperature, for successful oxidation. With α,β-unsaturated ketones, reaction is complicated by competing Baeyer–Villiger oxidation of the carbonyl group (see Section 6.3). Epoxides of α,β-unsaturated carbonyl compounds are best made by the action of nucleophilic reagents such as hydrogen peroxide or tert-butyl hydroperoxide in alkaline solution.

$$(5.40)$$

Epoxidations with peroxy-acids are highly stereoselective and take place by *cis* addition to the double bond of the alkene. For example, oleic acid **26** gave *cis*-9,10-epoxystearic acid **27**, whereas elaidic acid **28** gave the isomeric *trans*-epoxide **29** (5.41).

$$(5.41)$$

Epoxidation of alkenes normally occurs with approach of the peroxy-acid from the less-hindered side of the double bond. However, where there is a polar substituent, particularly in the allylic position, this may influence the direction of attack by the peroxy-acid. Thus, whereas 2-cyclohexenyl acetate gives a mixture consisting predominantly of the *trans*-epoxide (as expected with attack from the less-hindered side of the double bond), the free alcohol gives almost exclusively the *cis*-epoxide under the same conditions (5.42). The stereoselectivity and the faster rate of reaction with the hydroxy compound result from hydrogen bonding of the reactants.

$$(5.42)$$

Highly selective epoxidation of acyclic allylic alcohols with mCPBA has been observed in certain cases. For example, the allylic alcohol **30** gave the epoxide **31** almost exclusively, through attack on the preferred conformation **32** (5.43). Epoxidation is directed by co-ordination of the reagent with the ethereal oxygen atom as well as the allylic hydroxy group. The conformation **32** has the allylic hydrogen atom in the plane of the alkene, in order to avoid steric interactions in the *cis*-substituted alkene. Related *cis* alkenes have been reported to epoxidize stereoselectively, although epoxidation of the isomeric *trans* alkenes is normally poorly selective.

$$(5.43)$$

The use of peroxy-acids such as mCPBA is not always ideal. Acid-sensitive groups, including sometimes the desired epoxides (especially when aryl-substituted) are not well tolerated and the peroxy-acid reagent may not be sufficiently reactive (especially for electron-deficient alkenes). In addition, mCPBA is explosive in pure form. Alternative reagents such as oxygen, hydrogen peroxide or bleach are attractive as they are cheap and give rise to inert by-products, although the rate of epoxidation is slow in the absence of activation. This has been solved by using the oxidizing agent and a catalyst, such as a metal salt or complex, for alkene epoxidation.

Reaction of alkenes with tert-butyl hydroperoxide (t-BuOOH) in the presence of a transition metal catalyst, for example, a vanadium(V), molybdenum(VI) or titanium(IV) complex, provides an excellent method for the preparation of epoxides.[30] The molybdenum catalysts are most effective for the epoxidation of isolated double bonds, and the vanadium or titanium catalysts are most effective for allylic alcohols. Even terminal alkenes can be epoxidized readily. For example, 1-decene was converted into its epoxide with t-BuOOH and $Mo(CO)_6$ on heating in 1,2-dichloroethane.

The oxidation of allylic alcohols has been studied thoroughly using a variety of catalysts.[31] The reactivity of the vanadium-tert-butyl hydroperoxide reagents towards the double bond of allylic alcohols makes possible selective epoxidation. Thus, reaction of geraniol with t-BuOOH and vanadium acetylacetonate [VO(acac)$_2$] gave the 2,3-epoxide **33** (5.44). With peroxy-acids, reaction takes place preferentially at the other double bond.

$$\text{(5.44)}$$

Vanadium catalysts have found particular advantage for stereoselective epoxidations. Thus, the acyclic allylic alcohol **34** is oxidized with high selectivity using t-BuOOH and vanadium acetylacetonate, whereas with mCPBA a nearly equal mixture of the diastereomeric epoxides was produced (5.45).

$$\text{(5.45)}$$

90 : 10

but mCPBA   52 : 48

[30] K. A. Jørgensen, *Chem. Rev.*, **89** (1989), 431.
[31] W. Adam and T. Wirth, *Acc. Chem. Res.*, **32** (1999), 703.

Homoallylic alcohols, in which the hydroxy group is further removed from the double bond also show the effect. For example, the homoallylic alcohol **35** gave mainly the epoxide **36** and with the cyclic alcohol **37**, the *syn* directing effect of the hydroxy group was even more marked than in the reaction with peroxy-acids (5.46). The precise course of these reactions is not entirely clear, but the rate accelerations and high stereoselectivity suggest the formation of an intermediate in which the hydroxy group and the peroxide are co-ordinated to the metal.

(5.46)

More environmentally friendly and cheap is the use of hydrogen peroxide as the oxidizing agent. This can be used in the presence of various metal catalysts.[32] A powerful catalyst is methyltrioxorhenium ($MeReO_3$), which can be used in amounts less than 1 mol% in the presence of hydrogen peroxide. The reaction works best with some added pyridine, or even better 3-cyanopyridine.[33] For example, *E*-3-decene gave a very high yield of the *trans*-epoxide **38** using the oxidant aqueous hydrogen peroxide and methyltrioxorhenium as the catalyst (5.47).

(5.47)

Epoxidation in the absence of a metal catalyst is possible.[34] Hydrogen peroxide, in the presence of a nitrile, aldehyde or ketone, or a relatively acidic alcohol (e.g. phenol), can effect epoxidation of an alkene. Peroxy-imidic acids RC(=NH)OOH, formed *in situ* by reaction of nitriles (RC≡N) with hydrogen peroxide, react under mildly alkaline or neutral conditions. For example, 2-allyl-cyclohexanone was readily converted into the corresponding epoxide with the alkaline reagent, whereas with peroxy-acetic acid Baeyer–Villiger ring-expansion intervenes. The perhydrate **39**,

[32] B. S. Lane and K. Burgess, *Chem. Rev.*, **103** (2003), 2457.
[33] H. Adolfsson, C. Copéret, J. P. Chiang and A. K. Yudin, *J. Org. Chem.*, **65** (2000), 8651.
[34] W. Adam, C. R. Saha-Möller and P. A. Ganeshpure, *Chem. Rev.*, **101** (2001), 3499.

formed by addition of hydrogen peroxide to hexafluoroacetone or its hydrate, is suitable for epoxidation of unactivated alkenes (5.48). For example, the acid-sensitive epoxide **40** is formed using the perhydrate **39** in the presence of the buffer disodium hydrogen phosphate (5.49).

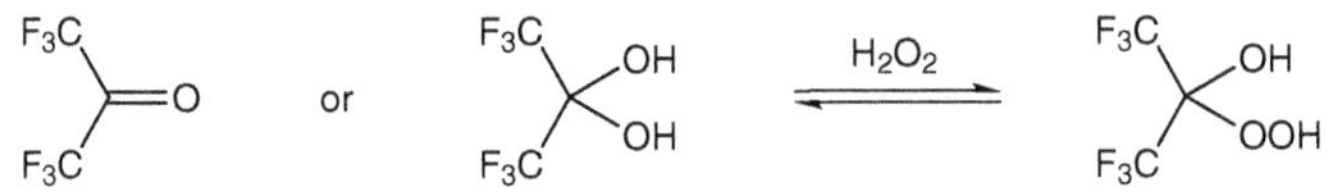

$$\text{(5.48)}$$

39

$$\text{(5.49)}$$

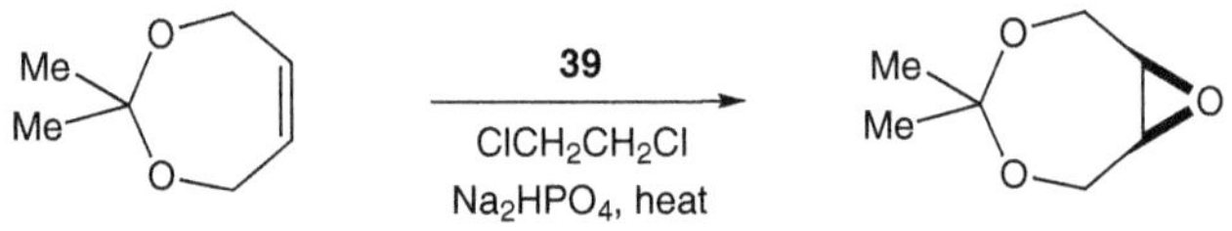

83%   **40**

An epoxidizing agent that has found widespread use is dimethyl dioxirane (DMDO). The reagent is generated from acetone and Oxone®, a source of potassium peroxomonosulfate ($KHSO_5$) (5.50). Epoxidation with DMDO occurs under mild, neutral conditions, without any nucleophilic component, which is ideal for preparing sensitive epoxides. For example, the enol ether **41** was epoxidized selectively using DMDO (5.51).[35]

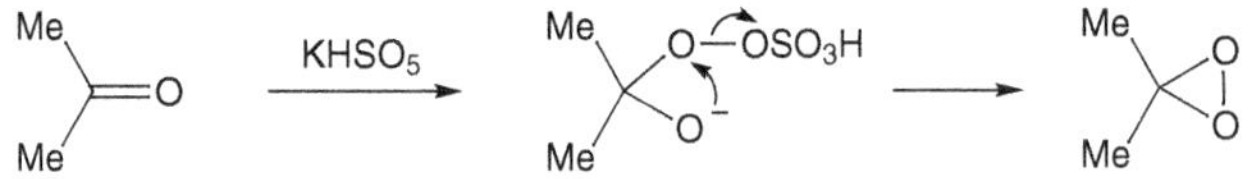

$$\text{(5.50)}$$

DMDO

$$\text{(5.51)}$$

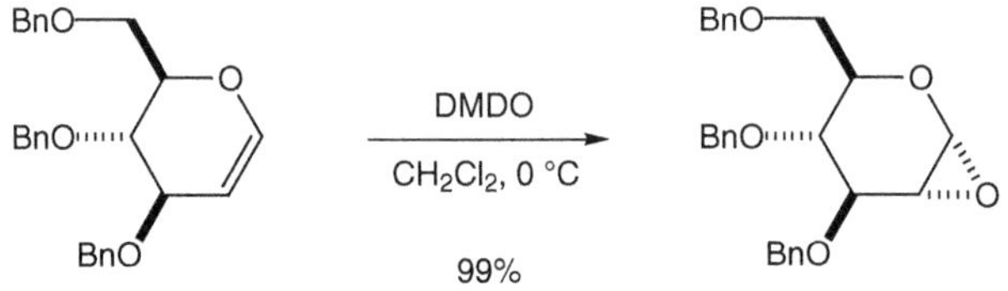

41   99%   20:1 α:β

Ketones other than acetone can be used for the formation of dioxiranes. Methyl(trifluoromethyl)dioxirane, formed from $KHSO_5$ and the more electrophilic ketone trifluoroacetone, is a reactive dioxirane. This reagent can be used for the epoxidation of electron-poor α,β-unsaturated carbonyl compounds. Trifluoroacetone can be used as a catalyst in combination with, for example, hydrogen peroxide

[35] R. L. Halcomb and S. J. Danishefsky, *J. Am. Chem. Soc.*, **111** (1989), 6661.

and acetonitrile at pH 11 to provide an efficient epoxidizing system (5.52).[36] Other fluorinated ketones or ketones bearing ammonium salts are also effective catalysts and can be used in sub-stoichiometric amounts.[34]

$$
\begin{array}{c}
\text{30 mol\% CF}_3\text{COMe} \\
\xrightarrow{\hspace{2cm}} \\
\text{H}_2\text{O}_2,\ \text{MeCN} \\
\text{K}_2\text{CO}_3,\ 0\ °\text{C}
\end{array}
\tag{5.52}
$$

89%

Two other classes of oxidizing agent are worthy of mention, namely oxaziridines and oxaziridinium salts. These are nitrogen analogues of dioxiranes, with a three-membered carbon–nitrogen–oxygen ring system. Oxidation of imines gives oxaziridines which are, in general, less reactive than dioxiranes. The most common type, *N*-sulfonyl-oxaziridines, react only slowly with alkenes to give epoxides.[37] *N*-Unsubstituted or *N*-acyl oxaziridines can in fact be used as aminating agents (although this chemistry is not yet well developed).[38]

Oxaziridinium salts **42** can be prepared by oxidation of iminium salts (5.53). They are more powerful oxidizing agents than oxaziridines and allow ready epoxidation of alkenes.[34] Typically, the reaction is performed all in one pot, with the iminium salt used as a catalyst. For example, epoxidation of *trans*-stilbene occurs with 0.1 equivalents of the iminium salt **43** and 2 equivalents of $KHSO_5$ at room temperature (5.54).[39]

$$
\begin{array}{c}
\xrightarrow{\begin{array}{c}\text{KHSO}_5\\ \text{MeCN, H}_2\text{O}\end{array}} \\
\textbf{42}
\end{array}
\qquad
\begin{array}{c}
\xrightarrow{\textbf{42}} \\
{}
\end{array}
\tag{5.53}
$$

43

10 mol% **43**, $KHSO_5$

MeCN, H_2O

89%

5.2.2 Asymmetric epoxidation

Asymmetric epoxidation ranks as one of the most reliable and selective methods for the formation of single enantiomer products. In particular, the asymmetric epoxidation of allylic alcohols with tert-butyl hydroperoxide (t-BuOOH),

[36] L. Shu and Y. Shi, *J. Org. Chem.*, **65** (2000), 8807.

[37] F. A. Davis and A. C. Sheppard, *Tetrahedron*, **45** (1989), 5703.

[38] J. Vidal, S. Damestoy, L. Guy, J.-C. Hannachi, A. Aubry and A. Collet, *Chem. Eur. J.*, **3** (1997), 1691; A. Armstrong, M. A. Atkin and S. Swallow, *Tetrahedron Lett.*, **41** (2000), 2247.

[39] A. Armstrong, G. Ahmed, I. Garnett, K. Goacolou and J. S. Wailes, *Tetrahedron*, **55** (1999), 2341.

a titanium(IV) metal catalyst and a tartrate ester ligand, called the *Sharpless asymmetric epoxidation* (or *Katsuki–Sharpless epoxidation*), is particularly valuable.[40] The reaction was developed in the early 1980s and was a major factor in the (shared) award of the Nobel Prize in chemistry to Professor Sharpless in 2001. The epoxidation system possesses two especially striking and useful features. It gives uniformly high asymmetric inductions throughout a range of substitution patterns in the allylic alcohol and, secondly, the absolute configuration of the epoxide produced can be predicted. Each enantiomer of the tartrate ligand delivers the epoxide oxygen atom to one face of the double bond, regardless of the substitution pattern. This is represented in Scheme 5.55, in which the allylic alcohol (which is required for co-ordination to the metal) is drawn with the hydroxymethyl group at the lower right. The oxygen atom is then delivered to the bottom face in the presence of L-(+)-diethyl tartrate (the natural isomer) and to the top face in the presence of D-(−)-diethyl tartrate. The enantiomeric (+)- and (−)-di-isopropyl tartrates are also common chiral ligands.

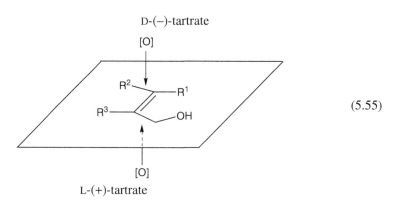

$$\text{(5.55)}$$

The epoxidation reaction is normally best carried out with only 5–10 mol% of the titanium catalyst in the presence of activated molecular sieves.[41] These conditions avoid the traditional use of stoichiometric catalyst and provide a mild and convenient method (although often at the expense of a slight reduction in enantioselectivity and rate of reaction). Numerous examples of highly enantioselective epoxidations of allylic alcohols by this procedure have been reported. For example, the allylic alcohol **44** was converted selectively into the epoxides **45** and **46** (5.56).

[40] T. Katsuki and V. S. Martin, *Org. Reactions*, **48** (1996), 1; R. A. Johnson and K. B. Sharpless, in *Catalytic Asymmetric Synthesis*, 2nd edition, ed. I. Ojima, chapter 6A (New York: Wiley–VCH, 2000), p. 231; or in *Comprehensive Organic Synthesis*, ed. B. M. Trost and I. Fleming, vol. 7 (Oxford: Pergamon Press, 1991), p. 389.

[41] Y. Gao, R. M. Hanson, J. M. Klunder, S. Y. Ko, H. Masamune and K. B. Sharpless, *J. Am. Chem. Soc.*, **109** (1987), 5765.

$$(5.56)$$

$$(+)\text{-DET} = (+)\text{-diethyl tartrate} =$$

A wide selection of substituted allylic alcohols are amenable to asymmetric epoxidation under these conditions. Allylic alcohols with E-geometry or unhindered Z-allylic alcohols are excellent substrates (5.57). However, branched Z-allylic alcohols, particularly those branched at C-4, exhibit decreased reactivity and selectivity.

$$(5.57)$$

The titanium-catalysed reaction is highly chemoselective for epoxidation of allylic alcohols. Thus, the dienol **47** gave only the epoxide **48** (5.58). The reaction is also tolerant of many different functional groups, including esters, enones, acetals, epoxides, etc.

$$C_5H_{11}\diagdown\!\!\!\diagup\!\!\!\diagdown\!\!\!\diagup\!\!\!\diagdown OH \xrightarrow[\substack{1.2\ \text{equiv. }(-)\text{-DIPT} \\ CH_2Cl_2,\ -25\ °C}]{\substack{{}^tBuOOH \\ 1\ \text{equiv. Ti(O}^iPr)_4}} C_5H_{11}\diagdown\!\!\!\diagup\!\!\!\diagdown\!\!\!\diagup\!\!\overset{\displaystyle}{\triangle}\!\!\diagdown OH \quad (5.58)$$

$$47 \qquad\qquad\qquad 90\% \qquad\qquad\qquad 48 \quad 86\%\ ee$$

DIPT = di-isopropyl tartrate

If the allylic alcohol contains an stereogenic centre, then two diastereomers can be formed. It is found that the Sharpless asymmetric epoxidation is a powerful reagent-controlled reaction that commonly over-rides any substrate control. For example, stereocontrolled access to polyols is possible by epoxidation of the allylic alcohol **49** (5.59).[42] In the absence of a chiral ligand, the chiral centre in the substrate **49** directs the oxidation to only a small degree (low diastereoselectivity in favour of epoxide **50**). This stereoselectivity is 'matched' with (−)-DIPT and 'mismatched' with (+)-DIPT, although both enantiomers of the chiral ligand far outweigh the influence of the substrate chirality, to provide either epoxide with high selectivity.

$$(5.59)$$

49	**50**	**51**
no ligand	2.3 : 1	
(−)-DIPT	90 : 1	
(+)-DIPT	1 : 22	

The rate of epoxidation of a chiral allylic alcohol will be different with the two enantiomers of the chiral ligand. Epoxidation of racemic secondary alcohols proceeds rapidly with only one of the enantiomers of the ligand, leaving the slower-reacting enantiomer of the secondary alcohol behind, produced, effectively, by a kinetic resolution. This slower-reacting enantiomer is the one in which the substituent R group hinders approach of the metal-bound tert-butyl hydroperoxide to its preferred face of the alkene (5.60). Hence, in the oxidation of the allylic alcohol **52** using the chiral ligand (+)-di-isopropyl tartrate, the (*S*)-enantiomer reacts about a hundred times faster than the (*R*)-enantiomer, leading to, predominantly, the epoxide **53** (5.61). If the reaction is run to only 55% completion (*e.g.* by limiting the amount of ${}^tBuOOH$), the allylic alcohol (*R*)-**52** is recovered with greater than 96% optical purity. In addition to being slower, the reaction of the (*R*)-alcohol with the (+)-tartrate is much less stereoselective (5.62).

[42] S. Y. Ko, A. W. M. Lee, S. Masamune, L. A. Reed, K. B. Sharpless and F. J. Walker, *Tetrahedron*, **46** (1990), 245.

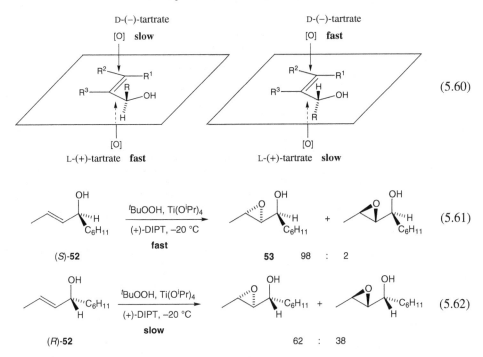

$$\text{(5.60)}$$

$$\text{(5.61)}$$

$$\text{(5.62)}$$

The kinetic resolution reaction can be used for the asymmetric synthesis of chiral secondary allylic alcohols or their corresponding epoxides. The yield of either is, of course, limited to 50%, starting from the racemic allylic alcohol, but the methodology has found widespread use in organic synthesis. For example, epoxidation of the racemic allylic alcohol **54** gave the epoxide **55**, used to prepare the anticoccidial antibiotic diolmycin A1 (5.63).[43]

$$\text{(5.63)}$$

If the epoxidation reaction is carried out with enantiomerically pure secondary allylic alcohol, then high yields of the desired epoxide can be obtained. Selective epoxidation of a single enantiomer of the bis-allylic alcohol **56** with (+)-DIPT gave the mono-epoxide anti-tumor agent laulimalide **57** (5.64).[44] No epoxidation of the mismatched allylic alcohol, nor of the other alkenes, was observed.

[43] T. Sunazuka, N. Tabata, T. Nagamitsu, H. Tomoda, S. Omura and A. B. Smith, *Tetrahedron Lett.*, **34** (1993), 6659.

[44] J. Mulzer and E. Öhler, *Angew. Chem. Int. Ed.*, **40** (2001), 3842; A. Ahmed, E. K. Hoegenauer, V. S. Enev, M. Hanbauer, H. Kaehlig, E. Öhler and J. Mulzer, *J. Org. Chem.*, **68** (2003), 3026; I. Paterson, C. De Savi and M. Tudge, *Org. Lett.*, **3** (2001), 3149.

$$(5.64)$$

The mechanism of epoxidation is believed to take place by co-ordination of the tartrate ligand to the metal catalyst (with displacement of two isopropoxide ligands). The equilibrium between $Ti(Oi-Pr)_4$ and $Ti(tartrate)(Oi-Pr)_2$ lies in favour of the tartrate complex because of the the higher binding constant of the chelating diol (bidentate ligand). Further displacement of the remaining two isopropoxide ligands by the allylic alcohol and the t-BuOOH then takes place. The dimeric species shown in Scheme 5.65 [for the (+)-tartrate ligand] has been proposed as the active catalyst. The alkene acts as a nucleophile to attack the equatorial oxygen atom of the peroxide from the alkene's lower face (as drawn). This is consistent with the model in Scheme 5.55.

For epoxidation of tertiary allylic alcohols, $Zr(Oi-Pr)_4$ has been found to be more efficient than the titanium catalysts.[45] This is probably because of the longer Zr–O bond, which facilitates complex formation between the metal and the tertiary allylic alcohol.

$$(5.65)$$

Allylic alcohols and their epoxides are valuable building blocks in organic synthesis, although the requirement for an allylic alcohol as the substrate in the Sharpless asymmetric epoxidation limits the type of alkene that can be used. If an enantiomerically pure epoxide of an unfunctionalized alkene is desired then other methodology is needed. A partial solution to this problem has been discovered by Jacobsen and Katsuki and co-workers using chiral manganese complexes.[46] Suitable substrates for highly enantioselective oxidation are conjugated *cis*-di- and

[45] A. C. Spivey, S. J. Woodhead, M. Weston and B. I. Andrews, *Angew. Chem. Int. Ed.*, **40** (2001), 769.

[46] E. N. Jacobsen, in *Comprehensive Organometallic Chemistry II*, ed. E. W. Abel, F. G. A. Stone and G. Wilkinson, vol. 12 (Oxford: Elsevier, 1995), p. 1097; T. Katsuki, in *Catalytic Asymmetric Synthesis*, 2nd edition, ed. I. Ojima, chapter 6B (New York: Wiley–VCH, 2000), p. 287; T. Katsuki, *Adv. Synth. Cat.*, **344** (2002), 131.

trisubstituted alkenes. For example, epoxidation of *cis*-β-methyl styrene gave the epoxide **59** with high optical purity using the manganese complex **58** and bleach as the oxidant (5.66). Unfortunately, epoxidation of the *trans* geometrical isomer takes place with low enantioselectivity. Epoxidation of styrene itself can be achieved using the catalyst **58** or related manganese–salen complexes, with mCPBA as the oxidant at low temperature (86% ee).

$$\text{Ph}\diagup\diagdown\text{Me} \xrightarrow[\text{4 mol\% }\textbf{58}]{\text{NaOCl, CH}_2\text{Cl}_2\text{, 4 °C}} \text{Ph}\diagup\!\!\triangledown\!\!\diagdown\text{Me}$$

81%

58

59 92% ee

(5.66)

The active oxidizing agent is a manganese–oxo species **60** (5.67). The alkene approaches the oxo–metal bond along the pathway that is least sterically congested. In the case of catalyst **58** (and hence **60**), this implies that the alkene approaches over the cyclohexane ring, to avoid the bulky tert-butyl groups. If the larger phenyl group is then placed on the side away from the axial hydrogen atom (the two nitrogen atoms are located in equatorial positions), epoxidation leads to the enantiomer **59** (5.67). The epoxidation reaction is not always stereospecific, such that *cis*-alkenes can produce some (or even predominantly) *trans*-epoxide products. This is thought to be a result of a radical intermediate **61** (stabilized by the conjugating substituent) that may rotate prior to formation of the second carbon–oxygen bond. Indeed, epoxidation of *cis*-β-methyl styrene (as shown in Schemes 5.66 and 5.67) gave the *cis* epoxide **59**, together with some *trans* epoxide (92:8 ratio).

60

61

59   (5.67)

The rate of Jacobsen–Katsuki epoxidation can be enhanced in the presence of additives such as pyridine *N*-oxide or related aromatic *N*-oxides. For example, in a synthesis of the potassium channel activator BRL-55834, only 0.1 mol% of the (*S*,*S*)-(salen)Mn(III)Cl catalyst **58** was required for efficient epoxidation of the chromene **62** in the presence of 0.1 mol% isoquinoline *N*-oxide (5.68). In the absence of the additive, at least 1 mol% of the catalyst **58** was required and the reaction was slower (12 h, 90% ee).

$$(5.68)$$

Finding increasing use in organic synthesis is the preparation of optically active epoxides using a hydrolytic kinetic resolution (HKR) protocol.[47] Formation of the racemic epoxide is followed by a kinetic resolution with the chiral salen-complexed cobalt catalyst **63**, to provide the recovered unreacted epoxide with very high enantiomeric excess (maximum 50% yield) and the diol. The reaction is especially good for terminal epoxides, compounds that are not commonly accessible by Sharpless or Jacobsen–Katsuki asymmetric epoxidation methods. The resolution is normally run with slightly more than 0.5 equivalents of the nucleophile water, in order to isolate the unreacted epoxide with very high optical purity. Substrates with a high selectivity factor ($k_{fast}/k_{slow} > 50$) can allow the formation of the product diol with reasonable yield and high optical purity (using 0.45 equivalents of water). Alternatively hydrolysis of the resolved epoxide gives the required diol. An important example of this process is the resolution of epichlorohydrin, which provides essentially enantiomerically pure epoxide using only 0.5 mol% of the catalyst (5.69). Many other terminal alkenes can be resolved using this procedure.

$$(5.69)$$

The methodology described above allows the asymmetric epoxidation of allylic alcohols or *cis*-substituted conjugated alkenes and the resolution of terminal epoxides. The asymmetric synthesis of *trans*-di- and trisubstituted epoxides can be achieved with the dioxirane formed from the fructose-derived ketone **64**, developed by Shi and co-workers.[48] The oxidizing agent potassium peroxomonosulfate

[47] S. E. Schaus, B. D. Brandes, J. F. Larrow, M. Tokunaga, K. B. Hansen, A. E. Gould, M. E. Furrow and E. N. Jacobsen, *J. Am. Chem. Soc.*, **124** (2002), 1307.

[48] M. Frohn and Y. Shi, *Synthesis* (2000), 1979.

(KHSO$_5$) is used most commonly, although hydrogen peroxide in acetonitrile is an alternative.[49] Formation of the dioxirane of the ketone (see Scheme 5.50) is followed by enantioselective epoxidation of the alkene. Two examples of the asymmetric epoxidation using the catalyst **64** are shown in Schemes 5.70 and 5.71.

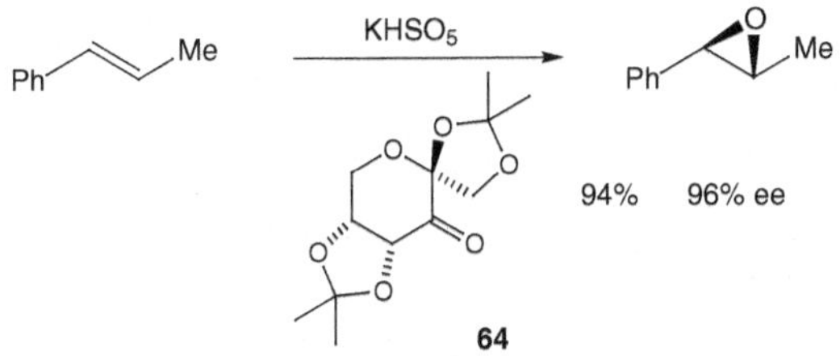

$$ (5.70) $$

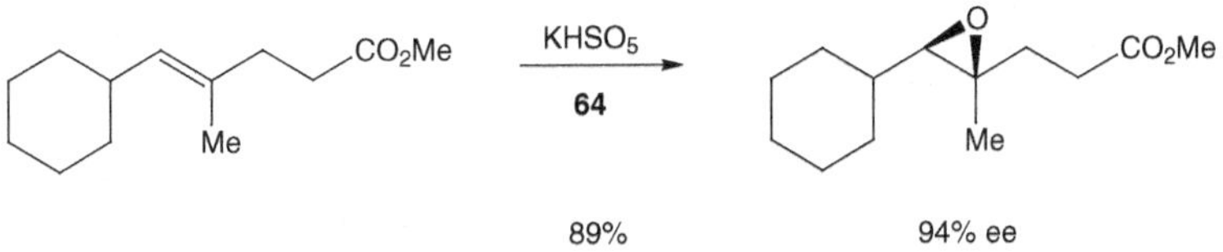

94%   96% ee

64

$$ (5.71) $$

89%   94% ee

The reactions described in this section so far have focused on the epoxidation of electron-rich alkenes. Epoxidation of electron-deficient alkenes is normally sluggish with the electrophilic oxidizing agents described above. However, nucleophilic oxidizing agents are well-suited for the epoxidation of α,β-unsaturated carbonyl compounds and other related electron-deficient alkenes. Good yields of the required epoxides are obtained using alkaline solutions of hydrogen peroxide or tert-butyl hydroperoxide. In these cases, conjugate addition of the peroxide onto the β-carbon atom of the α,β-unsaturated carbonyl compound is followed by cyclization to give the epoxide (5.72).

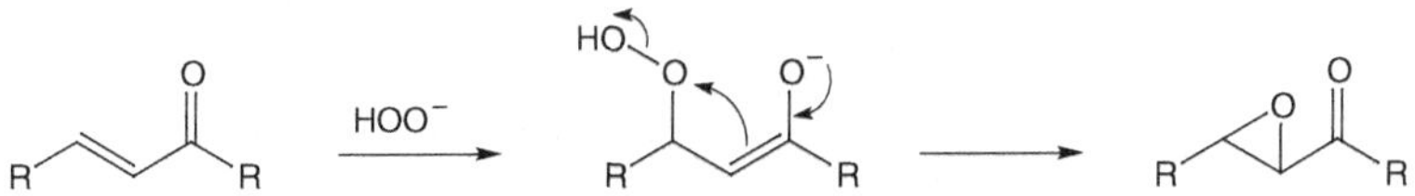

$$ (5.72) $$

Reaction with alkaline peroxide (or hypochlorite) and a chiral catalyst allows the asymmetric epoxidation of enones.[50] Excellent asymmetric induction has been achieved using metal–chiral ligand complexes, such as those derived from lanthanides and (*R*)- or (*S*)-BINOL. Alternatively, phase-transfer catalysis using ammonium salt derivatives of *Cinchona* alkaloids, or the use of polyamino acid

[49] L. Shu and Y. Shi, *Tetrahedron*, **57** (2001), 5213.
[50] M. J. Porter and J. Skidmore, *Chem. Commun.* (2000), 1215.

catalysts, such as poly-L-leucine, have proved effective. The epoxidations are typically most efficient with aromatic substrates such as chalcone **65** (5.73), although some examples of the epoxidation of alkyl-substituted enones or of α,β-unsaturated amides have been reported to be highly enantioselective.[51]

		(5.73)
*t*BuOOH, THF		
5 mol% La(O*i*Pr)₃, **66**		
4 Å mol. sieves	99%	96% ee
15 mol% Ph₃P=O		
KOCl, PhMe		
10 mol% **67**	96%	93% ee
−40 °C		
poly-L-leucine		
urea–H₂O₂	85%	>95% ee (opposite enantiomer)
DBU, THF		

(*R*)-BINOL **66**   **67**

5.2.3 Aziridination

Although less common than an epoxide in natural products and as an intermediate in organic synthesis, the aziridine ring is an important functional group. Aziridines can be prepared from epoxides by ring-opening with azide anion and cyclization with triphenylphosphine. This methodology provides a convenient, stereospecific way to access *N*-unsubstituted aziridines. For example, epoxidation of the alkene **68** occurred stereoselectively to give the *cis* epoxide **69**, which was converted to the aziridine **71** via azido-alcohol **70** with overall double inversion of stereochemistry (5.74).

[51] T. Nemoto, H. Kakei, V. Gnanadesikan, S. Tosaki, T. Ohshima and M. Shibasaki, *J. Am. Chem. Soc.*, **124** (2002), 14544.

Direct preparation of an aziridine from an alkene is possible by reaction of the alkene with a nitrene or metal nitrenoid species.[52] Nitrenes can be generated thermally or photochemically from azides, although their reaction with alkenes to give aziridines is often low yielding and is complicated by side reactions. Oxidation of *N*-amino-phthalimide or related hydrazine compounds (e.g. with Pb(OAc)$_4$ or by electrolysis) and reaction with an alkene has found some generality.[53] The metal-catalysed reaction of nitrenes with alkenes has received considerable study. A variety of metal catalysts can be used, with copper(II) salts being the most popular. For example, styrene was converted to its *N*-tosyl aziridine **72** by reaction with [*N*-(tosyl)imino]phenyliodinane (PhI=NTs) and copper(II) triflate (5.75).[54]

The reagents *N*-chloramine-T (TsNClNa) and *N*-bromamine-T (TsNBrNa) are convenient sources of nitrogen for aziridination. In the presence of copper(II)

[52] J. E. G. Kemp, in *Comprehensive Organic Synthesis*, ed. B. M. Trost and I. Fleming, vol. 7 (Oxford: Pergamon Press, 1991), p. 469.

[53] R. S. Atkinson, *Tetrahedron*, **55** (1999), 1519; G. Hilt, *Angew. Chem. Int. Ed.*, **41** (2002), 3586.

[54] D. A. Evans, M. M. Faul and M. T. Bilodeau, *J. Am. Chem. Soc.*, **116** (1994), 2742. For an account of the chemistry of iminoiodinanes, see P. Dauban and R. H. Dodd, *Synlett* (2003), 1571.

chloride and microwave irradiation, aziridination of styrene with TsNBrNa gave *N*-tosyl aziridine **72** in 70% yield.[55] An alternative to a metal salt is the use of a source of bromine as a catalyst. Reaction of alkenes with TsNClNa and phenyltrimethylammonium tribromide provides an efficient method to prepare aziridines.[56] With these reagents, styrene gave the *N*-tosyl aziridine **72** (68%), although a strength of the procedure is that it is amenable to a wide range of alkenes, as illustrated in Scheme 5.76. The reaction is thought to occur by attack of *N*-chloramine-T on the intermediate bromonium ion.

$$(5.76)$$

The asymmetric synthesis of aziridines can be achieved by a number of methods.[57] The best alkene substrates are typically α,β-unsaturated esters, styrenes or chromenes, with aziridination by PhI=NTs and a metal–chiral ligand complex.[58] For example, aziridination of tert-butyl cinnamate **73** occurs highly enantioselectively with copper(I) triflate and a bisoxazoline ligand (5.77).

$$(5.77)$$

[55] B. M. Chanda, R. Vyas and A. V. Bedekar, *J. Org. Chem.*, **66** (2001), 30; see also D. P. Albone, P. S. Aujla, P. C. Taylor, S. Challenger and A. M. Derrick, *J. Org. Chem.*, **63** (1998), 9569.

[56] J. U. Jeong, B. Tao, I. Sagasser, H. Henniges and K. B. Sharpless, *J. Am. Chem. Soc.*, **120** (1998), 6844.

[57] D. Tanner, *Angew. Chem. Int. Ed.*, **33** (1994), 599; H. M. I. Osborn and J. Sweeney, *Tetrahedron: Asymmetry*, **8** (1997), 1693; P. Müller and C. Fruit, *Chem. Rev.*, **103** (2003), 2905.

[58] D. A. Evans, M. M. Faul, M. T. Bilodeau, B. A. Anderson and D. M. Barnes, *J. Am. Chem. Soc.*, **115** (1993), 5328; Z. Li, K. R. Conser and E. N. Jacobsen, *J. Am. Chem. Soc.*, **115** (1993), 5326; H. Nishikori and T. Katsuki, *Tetrahedron Lett.*, **37** (1996), 9245; K. M. Gillespie, C. J. Sanders, P. O'Shaughnessy, I. Westmoreland, C. P. Thickitt and P. Scott, *J. Org. Chem.*, **67** (2002), 3450.

5.3 Dihydroxylation

The dihydroxylation of alkenes provides 1,2-diol products, present in a great many natural products and biologically active molecules. The transformation has, therefore, received considerable interest and methods are now well developed for catalytic, racemic and asymmetric dihydroxylation. The most common of these is *syn* (or *cis*) dihydroxylation, in which the two hydroxy groups are added to the same side of the double bond. The best methods involve reaction with osmium tetroxide; other reagents include potassium permanganate or iodine with silver carboxylates. To prepare the *anti* (or *trans*) diol product, ring-opening of epoxides is most convenient, although the Prévost reaction (iodine and silver acetate under anhydrous conditions) is also useful. The alkene may, of course, have either the *E*- or *Z*-configuration and *syn* dihydroxylation gives rise to isomeric diols. Thus, fumaric acid, on *syn* dihydroxylation gives racemic tartaric acid, whereas maleic acid gives *meso* tartaric acid (5.78).

(5.78)

5.3.1 Dihydroxylation with osmium tetroxide

The most popular method for dihydroxylation of alkenes uses osmium tetroxide.[59] This reagent can be used stoichiometrically, although its expense and toxicity have led to the development of catalytic variants. There has been considerable debate over the mechanism of the reaction, which has been postulated to proceed by a direct [3+2] cycloaddition, or *via* a [2+2] cycloaddition followed by a rearrangement, to give the intermediate osmate ester.[60] This osmium(VI) species can be oxidized or reduced and hydrolysed to release the diol product (5.79). The reaction is accelerated by tertiary amine and other bases, such as pyridine, which co-ordinate to the osmium metal.

---

[59] M. Schröder, *Chem. Rev.*, **80** (1980), 187; A. H. Haines, in *Comprehensive Organic Synthesis*, ed. B. M. Trost and I. Fleming, vol. 7 (Oxford: Pergamon Press, 1991), p. 437.

[60] For a recent discussion, see D. V. Deubel and G. Frenking, *Acc. Chem. Res.*, **36** (2003), 645.

$$R\diagup\hspace{-0.3em}\diagdown \xrightarrow{\text{OsO}_4} \underset{\text{osmate ester}}{R\diagup\hspace{-0.3em}\diagdown\text{O-Os}} \longrightarrow R\diagup\overset{\text{OH}}{\hspace{-0.3em}}\diagdown\text{OH} \qquad (5.79)$$

Many different co-oxidants can be used in conjunction with osmium tetroxide for the catalytic dihydroxylation reaction. The most popular is *N*-methylmorpholine *N*-oxide (NMO); the use of NMO with less than one equivalent of osmium tetroxide is often referred to as the Upjohn conditions.[61] Other oxidants, such as [K$_3$Fe(CN)$_6$], tert-butyl hydroperoxide, hydrogen peroxide or bleach are effective.[62] In these reactions, the intermediate osmate ester is oxidized to an osmium(VIII) species that is then hydrolysed with regeneration of osmium tetroxide to continue the cycle. For example, less than 1 mol% of osmium tetroxide is needed for the dihydroxylation of the alkene **74** (5.80).

$$\text{MeO}\diagup\hspace{-0.3em}\diagdown\hspace{-0.3em}\diagup\hspace{-0.3em}\diagdown\text{CO}_2\text{Et} \xrightarrow[\substack{\text{NMO, }^t\text{BuOH, H}_2\text{O}\\ \text{room temp.}\\ \text{then Na}_2\text{SO}_{3(aq)}}]{\text{0.2 mol\% OsO}_4} \text{MeO}\diagup\hspace{-0.3em}\diagdown\overset{\text{OH}}{\underset{\text{OH}}{\hspace{-0.3em}}}\text{CO}_2\text{Et} \qquad (5.80)$$

74   88%

NMO
$$\underset{\underset{\text{Me}}{\overset{+}{N}}\overset{O^-}{}}{\overset{O}{\diagup\hspace{-0.3em}\diagdown}}$$

The dihydroxylation reaction is very general, giving high yields of diol products from electron-rich or electron-poor alkenes. High levels of stereocontrol can often be obtained on dihydroxylation of alkenes bearing one or more chiral centre.[63] The large steric requirements of the reagent normally dictates that dihydroxylation with osmium tetroxide takes place predominantly from the less-hindered side of the double bond.

Dihydroxylation of allylic alcohols provides a route to 1,2,3-triols, a structural feature found in many natural products. The reaction with allylic alcohols or their

[61] V. Van Rheenen, R. C. Kelly and D. Y. Cha, *Tetrahedron Lett.* (1976), 1973.
[62] See, for example, S. Y. Jonsson, K. Färnegårdh and J.-E. Bäckvall, *J. Am. Chem. Soc.*, **123** (2001), 1365; G. M. Mehltretter, S. Bhor, M. Klawonn, C. Döbler, U. Sundermeier, M. Eckert, H.-C. Militzer and M. Beller, *Synthesis* (2003), 295.
[63] J. K. Cha and N.-S. Kim, *Chem. Rev.*, **95** (1995), 1761; J. K. Cha, W. J. Christ and Y. Kishi, *Tetrahedron*, **40** (1984), 2247.

corresponding ethers is highly stereoselective, giving preferentially the isomer in which the original hydroxy or alkoxy group and the adjacent newly introduced diol are *anti* to one another. Thus, 2-cyclohexen-1-ol gave almost exclusively the triol **75** and the acyclic allylic ether **77** gave the diol **78** in preference to its isomer **79** (5.81) and 5.82).[64] The dihydroxylation reaction takes place by preferential addition of osmium tetroxide to the face of the double bond opposite to the hydroxy or alkoxy group; with the substrate **77**, the reaction occurs in the least sterically encumbered conformation **80** (with an 'inside' allylic hydrogen atom).

$$(5.81)$$

$$(5.82)$$

In contrast, dihydroxylation of allylic alcohols using stoichiometric osmium tetroxide in the presence of tetramethylethylene diamine (TMEDA) as a ligand provides predominantly the *syn* product **76** (5.83).[65] The diamine TMEDA co-ordinates to the osmium atom of OsO$_4$, thereby increasing the electronegativity of the oxo ligands and favouring hydrogen bonding to the allylic hydroxy group. This hydrogen bonding directs the dihydroxylation and hence these conditions often provide complementary stereoselectivity in comparison with the conventional Upjohn conditions using catalytic osmium tetroxide and NMO.

$$(5.83)$$

[64] Note that the dihydroxylation reaction occurs by a *syn* addition of the two hydroxy groups to the same face of the alkene, such that the Z-alkene **77** gives the *anti* products **78**, **79**.
[65] T. J. Donohoe, *Synlett* (2002), 1223; T. J. Donohoe, K. Blades, P. R. Moore, M. J. Waring, J. J. G. Winter, M. Helliwell, N. J. Newcombe and G. Stemp, *J. Org. Chem.*, **67** (2002), 7946.

Asymmetric dihydroxylation has been developed into an extremely efficient and selective process for a wide variety of substituted alkenes.[66] Chiral amine ligands provide the required rate enhancement and asymmetric induction, by co-ordinating to the osmium atom. The most popular ligands are based on the naturally occurring cinchona alkaloids dihydroquinine (DHQ) and dihydroquinidine (DHQD). In particular, the ligands (DHQ)$_2$PHAL **81** and (DHQD)$_2$PHAL **82**, in which two of the alkaloids are connected to a phthalazine ring, have found widespread use. The dihydroxylation reaction can be carried out with osmium tetroxide as a catalyst (typically added in the lower oxidation state as the solid [K$_2$OsO$_4$·2H$_2$O]), and the favoured co-oxidant is [K$_3$Fe(CN)$_6$]. The additive methanesulfonamide MeSO$_2$NH$_2$ often enhances the rate of the reaction. Therefore, the overall reaction mixture normally contains a combination of these reagents, plus solvent, as illustrated in the asymmetric dihydroxylation of *E*-5-decene and ethyl *E*-oct-2-enoate (5.84).

$$(5.84)$$

(DHQ)$_2$PHAL **81**   (DHQD)$_2$PHAL **82**

The reagent combination with the ligand (DHQ)$_2$PHAL **81** is sometimes referred to as AD-mix α and with (DHQD)$_2$PHAL **82** as AD-mix β. The two ligands provide

[66] H. C. Kolb, M. S. Van Nieuwenhze and K. B. Sharpless, *Chem. Rev.*, **94** (1994), 2483; R. A. Johnson and K. B. Sharpless, in *Catalytic Asymmetric Synthesis*, 2nd edition, ed. I. Ojima, chapter 6D (New York: Wiley–VCH, 2000), p. 357; C. Bolm, J. P. Hildebrand and K. Muñiz, in *Catalytic Asymmetric Synthesis*, 2nd edition, ed. I. Ojima, chapter 6E (New York: Wiley–VCH, 2000), p. 399. For a comparison of asymmetric epoxidation, dihydroxylation, see C. Bonini and G. Righi, *Tetrahedron*, **58** (2002), 4981.

the opposite enantioselectivity and therefore either enantiomer of the diol product can be accessed readily. The ligands **81** and **82** are not enantiomeric, but do engender each mirror image environment around the metal, which co-ordinates to the quinuclidine nitrogen atom (the chiral centre adjacent to this nitrogen atom and that bearing the oxygen atom are enantiomeric in the two ligands). The alkene must approach in a preferred orientation and this has been dipicted by the mnemonic shown in Figure 5.85. With the largest alkene substituent in the lower left corner as drawn, the DHQ-based ligand system promotes dihydroxylation from the lower (α) face, whereas the top (β) face reacts with the DHQD-based ligand system.

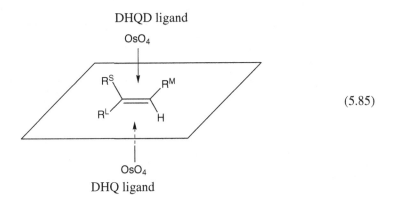

(5.85)

From this model it is clear that E-1,2-disubstituted alkenes (with the larger substituent in the lower left corner) are well suited to the dihydroxylation reaction. However, the selectivity with Z-alkenes, in which one of the substituents must occupy a more-hindered position, is often poor. Indeed, dihydroxylation of E-β-methylstyrene occurs with near perfect enantioselectivity (97% ee using the ligand **81**), whereas the isomeric Z-β-methylstyrene gives the diol product (opposite diastereomer) with a low enantiomer ratio (29% ee using the ligand **82**). The enantioselectivity in the dihydroxylation of Z-alkenes can be improved using other related ligands, in which the DHQ or DHQD is linked to a different aromatic unit, such as an N-carboxy-indoline. Improvements in the enantioselectivity of dihydroxylation of terminal alkenes have been obtained using an anthraquinone bridge [(DHQ)$_2$AQN or (DHQD)$_2$AQN].

The asymmetric dihydroxylation reaction is finding increasing use as a stereoselective method in organic synthesis. For example, in a synthesis of the bioactive acetogenin parviflorin, asymmetric dihydroxylation of the diene **83** gave, selectively, the triol **84** (5.86).[67] The dihydroxylation reaction was run to approximately two

---

[67] B. M. Trost, T. L. Calkins and C. G. Bochet, *Angew. Chem. Int. Ed.*, **36** (1997), 2632.

thirds conversion to avoid over-oxidation of the less-reactive terminal alkene.

$$\text{(5.86)}$$

83   71%   **84** 94% ee

Dihydroxylation of a chiral substrate can give rise to two diastereomeric diol products (see, for example, Scheme 5.82). A particular merit of the asymmetric dihydroxylation reaction is that reagent control often dominates over substrate control, such that either diol product can be obtained, depending on the choice of chiral ligand. For example, the intrinsic diastereofacial selectivity in the dihydroxylation of the alkene **85** was poor, but in favour of the diol **86** (5.87). The chiral ligand (DHQD)$_2$PHAL **82** reinforced this selectivity (double asymmetric induction, matched case), with both the ligand and the substrate favouring the diol **86**. In contrast, the ligand (DHQ)$_2$PHAL **81** promoted the preferential formation of the other diastereomer **87** (mismatched case). Therefore, despite the fact that the substrate favours reaction to give the diol **86**, reagent control dominates and the diol **87** can be prepared with high selectivity. Unfortunately, unlike the Sharpless asymmetric epoxidation reaction of racemic secondary allylic alcohols, the asymmetric dihydroxylation reaction does not promote efficient kinetic resolution.

$$\text{(5.87)}$$

85   **86**   **87**

Conditions	Yield	Ratio	
10 mol% OsO$_4$, 3 equiv. NMO, THF, H$_2$O	83%	76	: 24
8 mol% [K$_2$OsO$_4$·2H$_2$O], 3 equiv. [K$_3$Fe(CN)$_6$] (DHQD)$_2$PHAL, MeSO$_2$NH$_2$, ^{t}BuOH, H$_2$O	82%	98	: 2
8 mol% [K$_2$OsO$_4$·2H$_2$O], 3 equiv. [K$_3$Fe(CN)$_6$] (DHQ)$_2$PHAL, MeSO$_2$NH$_2$, ^{t}BuOH, H$_2$O	85%	5	: 95

In the same way as many other reactions, there are advantages in using a chiral ligand that is attached to a polymer, as the ligand can normally be recovered by simple filtration.[68] This makes for easier purification of the product and the possibility to recycle the catalyst. On the other hand, this approach typically suffers from the need to prepare the polymer-supported catalyst, the difficulty in its characterization and, crucially, often a reduction in the selectivity of the desired reaction.

---

[68] P. H. Toy and K. D. Janda, *Acc. Chem. Res.* (2000), 546; S. V. Ley, I. R. Baxendale, R. N. Bream, P. S. Jackson, A. G. Leach, D. A. Longbottom, M. Nesi, J. S. Scott, R. I. Storer and S. J. Taylor, *J. Chem. Soc., Perkin Trans. 1* (2000), 3815.

A number of different polymer-supported catalysts have been prepared for the asymmetric dihydroxylation of alkenes.[69] The enantioselectivities with these catalysts are now comparable with the conventional solution phase chemistry described above using chiral ligands such as (DHQD)$_2$PHAL. For example, dihydroxylation of the alkene *trans*-stilbene occurs with essentially complete enantioselectivity using the polymer-supported ligand **88** (5.88). This polyethyleneglycol (PEG)-linked ligand is soluble in organic solvents, a facet that often improves activity and selectivity in comparison with insoluble polymer-bound ligands, since the reaction is homogeneous. The ligand can be isolated by precipitation, which occurs on addition of a suitable solvent (in this case t-BuOMe), followed by filtration. The ligand retains its activity for subsequent dihydroxylation reactions.

$$\text{(5.88)}$$

91%   99% ee

88

5.3.2 Other methods of dihydroxylation

The formation of chiral 1,2-diols can be achieved by dihydroxylation of benzene derivatives by using appropriate enzymes or whole cell systems, in particular *Pseudomonas putida* (5.89).[70] This biotransformation reaction provides cyclohexadiene diols in essentially enantiopure form. Many of these diols are commercially available and can be converted to a variety of different enantiomerically enriched compounds. For example, protection of diol **90** as its acetonide and aziridination

[69] C. Bolm and A. Gerlach, *Eur. J. Org. Chem.* (1998), 21; P. Salvadori, D. Pini and A. Petri, *Synlett* (1999), 1181; B. M. Choudary, N. S. Chowdari, K. Jyothi and M. L. Kantam, *J. Am. Chem. Soc.*, **124** (2002), 5341.
[70] T. Hudlicky, D. Gonzalez and D. T. Gibson, *Aldrichimica Acta*, **32** (1999), 35.

(see Section 5.2.3) gave the aziridine **91**, which was converted to the alkaloid pan-cratistatin (5.90).[71]

$$(5.89)$$

$$(5.90)$$

Oxidation of alkenes with potassium permanganate provides an alternative method for dihydroxylation, avoiding the use of the toxic and expensive reagent osmium tetroxide.[72] The reaction needs careful control to avoid over-oxidation and best results are obtained in alkaline solution, using water or aqueous-soluble organic solvents (e.g. acetone, ethanol, t-BuOH). Use of acidic or neutral solutions gives α-hydroxy ketones or even cleavage products. The method is particularly suitable for dihydroxylation of unsaturated acids, which dissolve in the alkaline solution. Oxidation of other substrates, which are typically insoluble in the aqueous oxidizing medium, results in poor yields of the diol products. This can sometimes, however, be overcome using vigorous stirring with a phase-transfer catalyst, such as a quater-nary ammonium salt or a crown ether. For example, dihydroxylation of cyclooctene with aqueous alkaline permanganate in the presence of benzyltrimethylammonium chloride gave the *cis*-diol **92** in reasonable yield (5.91), whereas in the absence of the catalyst the yield was only 7%.

$$(5.91)$$

The reactions are believed to proceed through the formation of cyclic manganese esters and it is this which controls the *syn* (*cis*) addition of the two hydroxy groups. *Syn*-addition is shown by the conversion of maleic acid into *meso*-tartaric acid, and of fumaric acid into (±)-tartaric acid. Competition between ring-opening of the cyclic manganese ester by hydroxide ion (to give the diol) and further oxidation by permanganate accounts for the effect of pH on the distribution of products.

---

[71] T. Hudlicky, X. Tian, K. Königsberger, R. Maurya, J. Rouden and B. Fan, *J. Am. Chem. Soc.*, **118** (1996), 10752.

[72] A. J. Fatiadi, *Synthesis* (1987), 85.

Oxidation of alkenes with ruthenium tetroxide normally gives cleavage products (see Section 5.4). The reaction is believed to proceed through a diol intermediate, which is not normally isolated. However, use of a biphasic system slows the rate of cleavage of the diol and allows the conversion of a range of different alkenes to the diol product.[73] For example, dihydroxylation of cyclooctene to give the diol **92** (58%) was achieved using 7 mol% $RuCl_3 \cdot 3H_2O$ and sodium periodate ($NaIO_4$) in ethyl acetate–acetonitrile–water (3:3:1).

The formation of 1,2-diol products from alkenes can be achieved using Prévost's reagent – a solution of iodine in carbon tetrachloride together with an equivalent of silver(I) acetate or silver(I) benzoate.[74] Under anhydrous conditions, this oxidant yields directly the diacyl derivative of the *anti*-diol (Prévost conditions), while in the presence of water the monoester of the *syn*-diol is obtained (Woodward conditions). Thus, treatment of a *cis*-alkene with iodine and silver benzoate in boiling carbon tetrachloride under anhydrous conditions gives the *trans*-dibenzoate (5.92). With iodine and silver(I) acetate in moist acetic acid, however, the monoacetate of the *cis*-1,2-dihydroxy compound is formed.

(5.92)

The value of these reagents results from their specificity and the mildness of the reaction conditions. The reaction proceeds through the formation of an iodonium ion which, in the presence of carboxylate and silver ions, forms the resonance-stabilized cation **93** (5.93). Attack on the cation by the carboxylate anion in an S_N2 process gives the *trans*-diacyl compound. In the presence of water, however, a hydroxy acetal is formed; this breaks down to gives the *cis*-monoacylated diol. Note that with conformationally rigid molecules, or indeed with any alkene in which there is a preference for initial attack on one of the two faces of the double bond, the *cis*-diol obtained by the Woodward–Prévost method may not have the same configuration as that obtained with osmium tetroxide. Related procedures, that avoid the use of expensive silver salts, have been reported with, for example, iodine and thallium(I) acetate or bismuth(III) acetate.

[73] T. K. M. Shing, E. K. W. Tam, V. W.-F. Tai, I. H. F. Chung and Q. Jiang, *Chem. Eur. J.*, **2** (1996), 50.
[74] C. V. Wilson, *Org. Reactions*, **9** (1957), 332; C. Prévost, *C. R. Acad. Sci.*, **196** (1933), 1129; R. B. Woodward and F. B. Brutcher, *J. Am. Chem. Soc.*, **80** (1958), 209.

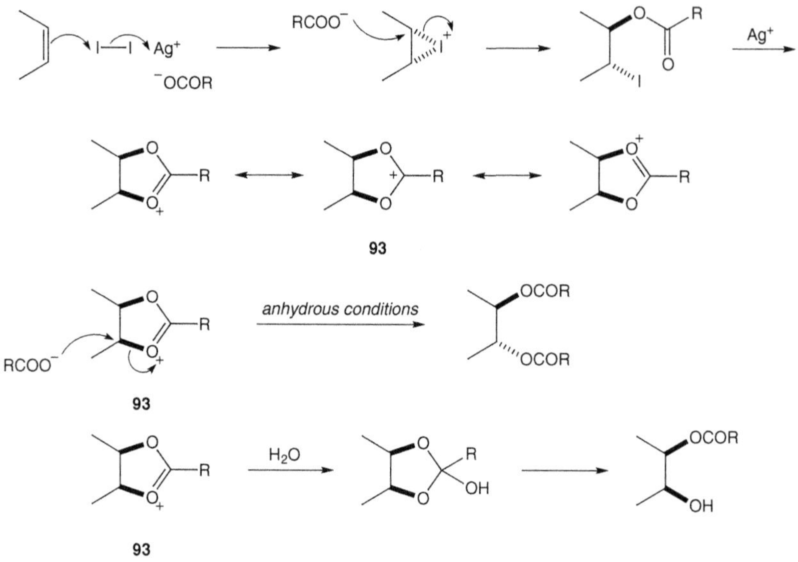

$$(5.93)$$

5.3.3 Amino-hydroxylation

Related to the dihydroxylation of alkenes with osmium tetroxide is the direct conversion of alkenes into 1,2-amino alcohols. Treatment of an alkene with osmium tetroxide in the presence of *N*-chloramine-T (TsNClNa) provides the 1,2-hydroxy toluene-*p*-sulfonamide (5.94).[75] The sulfonylimido osmium compound **94** is believed to be the active reagent, and is continuously regenerated during the reaction. The sulfonamide products can be converted to their free amines by cleavage of the tosyl group with sodium in liquid ammonia.

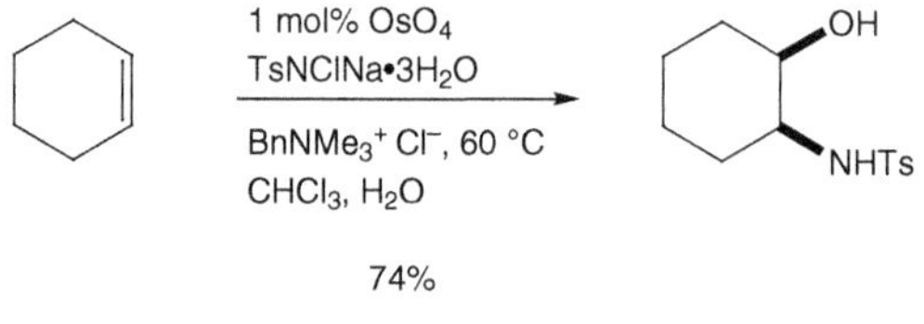

$$(5.94)$$

74%

Ts = $SO_2C_6H_4$-*p*-Me

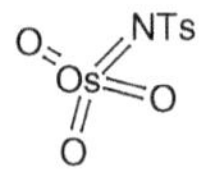

94

75 K. B. Sharpless, A. O. Chong and K. Oshima, *J. Org. Chem.*, **41** (1976), 177; E. Herranz and K. B. Sharpless, *Org. Synth.*, **61** (1980), 85.

The amino-hydroxylation reaction occurs stereoselectively with incorporation of the amino and hydroxy groups *syn* to one another. However, a serious problem is its poor regioselectivity with unsymmetrical alkenes. Despite this, there has been recent interest in the development of the asymmetric amino-hydroxylation reaction.[76] The chiral ligands (DHQ)$_2$PHAL **81** and (DHQD)$_2$PHAL **82** (see Scheme 5.84), effective for the asymmetric dihydroxylation reaction, are also suitable for the asymmetric amino-hydroxylation reaction. Various nitrogen-atom sources are possible, the best being the sodium salts of *N*-chloro-carbamates or *N*-bromo-carboxylic amides. For example, treatment of the styrene derivative **95** with *N*-chloro-benzylcarbamate sodium salt and a source of catalytic osmium tetroxide and the chiral ligand (DHQ)$_2$PHAL, gave a mixture of the two regioisomeric 1,2-hydroxy carbamates **96** and **97**, the former with high optical purity (5.95).[77] The structure of the ligand can have a dramatic effect on the regio- and enantioselectivity of the reaction; for example the ligand (DHQ)$_2$AQN (AQN = anthraquinone) results in predominantly the secondary, rather than the primary, alcohol product.

$$(5.95)$$

The asymmetric amino-hydroxylation reaction provides a very short synthesis of the side chain of the anticancer agent Taxol®. The substrate isopropyl cinnamate was converted to the chiral 1,2-hydroxy amide **98** as essentially a single enantiomer after recrystallization (5.96).[78] Hydrolysis then gave the required amine, as its hydrochloride salt.

$$(5.96)$$

[76] J. A. Bodkin and M. D. McLeod, *J. Chem. Soc., Perkin Trans. 1* (2002), 2733; D. Nilov and O. Reiser, *Adv. Synth. Catal.*, **344** (2002), 1169.

[77] K. L. Reddy and K. B. Sharpless, *J. Am. Chem. Soc.*, **120** (1998), 1207.

[78] M. Bruncko, G. Schlingloff and K. B. Sharpless, *Angew. Chem. Int. Ed. Engl.*, **36** (1997), 1483.

5.4 Oxidative cleavage

A useful reaction in organic synthesis is the oxidative cleavage of an alkene, in which the carbon–carbon single and double bonds are both broken, to give two carbonyl compounds.[79] The reaction may be accomplished by using ozone, and in this case is termed ozonolysis. Ozone is an electrophilic reagent that reacts with carbon–carbon double bonds by a 1,3-dipolar cycloaddition reaction (see Section 3.4) followed by a rearrangement reaction to form ozonides. The resulting ozonides can be cleaved oxidatively or reductively to carboxylic acids, ketones, aldehydes or alcohols; the nature of the products formed depends on the method used and on the structure of the alkene (5.97).

$$\text{(5.97)}$$

The reaction is normally carried out by passing a stream of oxygen containing 2%–10% of ozone into a solution or suspension of the compound in a suitable solvent, such as CH_2Cl_2 or methanol. Oxidation of the ozonide, without isolation, by hydrogen peroxide or other reagent, leads to carboxylic acids or ketones or both, depending on the degree of substitution of the alkene. Reductive decomposition of the crude ozonide leads to aldehydes and ketones (5.97), or to primary or secondary alcohols with the reducing agent sodium borohydride. Various methods of reduction have been used including catalytic hydrogenation and reduction with zinc and acids or with triethyl phosphite. In general, the yields of aldehydes with these reagents are not high. Reaction with dimethyl sulfide has been found to give excellent results and this appears to be the reagent of choice. Reduction takes place under neutral conditions and the reagent is highly selective; carbonyl groups or even alkynes, for example, elsewhere in the molecule are not affected (5.98). High yields of the aldehydes or ketones on addition of dimethyl sulfide to the ozonide hinges on the fact that peroxides are reduced rapidly and cleanly by sulfides.

$$\text{(5.98)}$$

91%

[79] D. G. Lee and T. Chen, in *Comprehensive Organic Synthesis*, ed. B. M. Trost and I. Fleming, vol. 7 (Oxford: Pergamon Press, 1991), p. 541.

The ozonolysis reaction has been the subject of considerable mechanistic study.[80] It is likely that in most cases the reaction proceeds by breakdown of the 1,3-dipolar cycloaddition product to a carbonyl oxide **99** and an aldehyde (or ketone) (5.99). The fate of the carbonyl oxide depends on the solvent and on its structure and the structure of the carbonyl compound. In an inert (non-participating) solvent, the carbonyl compound may react with the carbonyl oxide to form an ozonide **100**; otherwise the carbonyl oxide may dimerize to the peroxide **101** or give ill-defined polymers. In nucleophilic solvents such as methanol or acetic acid, hydroperoxides of the type **102** are formed.

Evidence for the intermediacy of a carbonyl oxide was found by Criegee in the ozonolysis of 2,3-dimethylbut-2-ene (5.100). In an inert solvent the cyclic peroxide and acetone were obtained. However, when formaldehyde was added to the reaction mixture, the known ozonide of 2-methylpropene was isolated. In the first case the intermediate carbonyl oxide dimerizes, but in the second it reacts preferentially with the reactive formaldehyde.

$$R_2C=CR_2 \xrightarrow{O_3} \cdots \longrightarrow \cdots \longrightarrow \cdots \tag{5.99}$$

99   **100**

101   **102**

$$ \tag{5.100}$$

Ozonolysis of cyclic alkenes gives dicarbonyl compounds. For example, treatment of cyclohexene in methanol with ozone, followed by addition of hydrogen peroxide gave adipic acid (5.101).[81] Particularly useful are variants of this process that lead to differentiated functional groups, thereby making subsequent selective reactions feasible. Ozonolysis at low temperature followed by addition of

[80] R. Criegee, *Angew. Chem. Int. Ed. Engl.*, **14** (1975), 745; R. L. Kuczkowski, *Chem. Soc. Rev.*, **21** (1992), 79; O. Horie and G. K. Moortgat, *Acc. Chem. Res.*, **31** (1998), 745.

[81] P. S. Bailey, *J. Org. Chem.*, **22** (1957), 1548.

p-toluenesulfonic acid gives an intermediate acetal–alkoxy-hydroperoxide, which on neutralization with $NaHCO_3$ and reduction with dimethylsulfide gives, in one pot, the acetal–aldehyde product (5.102).[82] Conversion to the acetal–ester or aldehyde–ester is also possible.

$$\text{(5.101)}$$

$$\begin{array}{c} O_3, MeOH \\ \xrightarrow{\hspace{2cm}} \\ \text{then } H_2O_2 \\ 85\% \end{array}$$

CO$_2$H
CO$_2$H

$$\text{(5.102)}$$

$$\begin{array}{c} O_3, -78\ ^\circ C \\ CH_2Cl_2, MeOH \\ \hline \text{then TsOH} \\ \text{room temp.} \end{array}$$

OMe
—OOH
—OMe
OMe

$$\begin{array}{c} NaHCO_3 \\ \hline \text{then Me}_2S \\ 93\% \end{array}$$

CHO
CH(OMe)$_2$

Ozonolysis of an enol ether provides a carboxylic ester, as one of the two carbonyl products. For example, the enol ether **103** (formed by Birch reduction – see Section 7.2) was converted to the ester–alcohol **104**, used in a synthesis of the Cecropia moth juvenile hormone (5.103).[83] This example illustrates the preferential oxidation of the more electron-rich alkene by the electrophilic ozone.

$$\text{(5.103)}$$

OMe
Me

$$\begin{array}{c} O_3, MeOH \\ \hline Me_2S, -78\ ^\circ C \\ \text{then NaBH}_4 \end{array}$$

CO$_2$Me
Me
OH

103   >52%   **104**

In the absence of a more electron-rich alkene within the same molecule, an α,β-unsaturated carbonyl compound will undergo ozonolysis. The reaction generally gives a product containing one less carbon atom than the starting material. Thus, the tricyclic α,β-unsaturated ketone **105** was converted to the keto–acid **106** with loss of a carbon atom (5.104).[84] It is thought that the ozonide intermediate fragments by release of electrons from the neighbouring hydroxy group (5.105). This is supported by related fragmentations that have been reported on ozonolysis of some allylic alcohols.

[82] S. L. Scheiber, R. E. Claus and J. Reagan, *Tetrahedron Lett.*, **23** (1982), 3867.
[83] E. J. Corey, J. A. Katzenellenbogen, N. W. Gilman, S. A. Roman and B. W. Erickson, *J. Am. Chem. Soc.*, **90** (1968), 5618.
[84] E. Wenkert, V. I. Stenberg and P. Beak, *J. Am. Chem. Soc.*, **83** (1961), 2320.

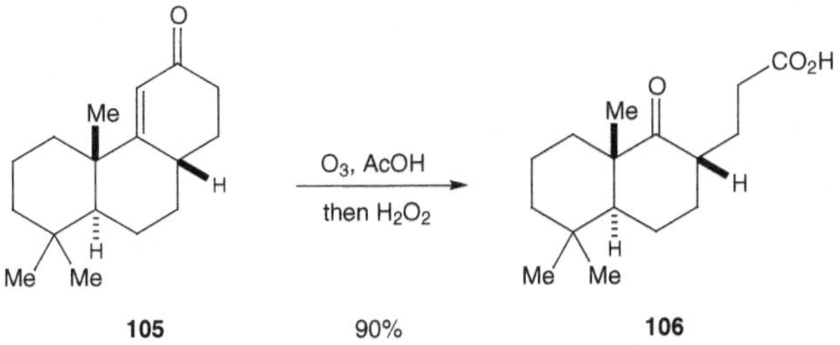

(5.104)

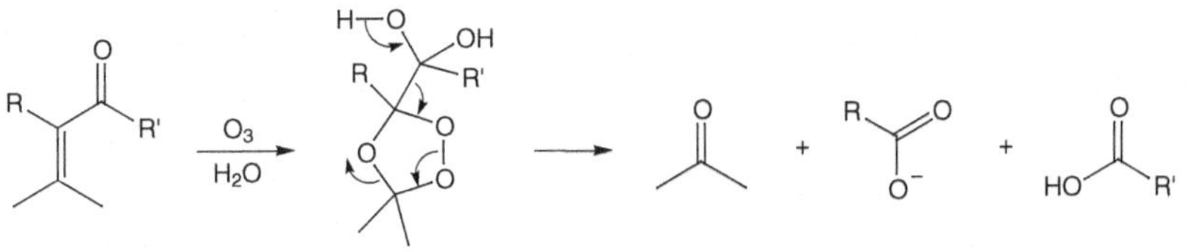

105 90% **106**

(5.105)

Oxidative cleavage of alkenes is not restricted to the use of ozone. In fact, some excellent alternative mild methods have been reported. A popular approach is the use of catalytic osmium tetroxide and sodium periodate ($NaIO_4$). As expected, the use of osmium tetroxide leads to the diol products, but in the presence of periodate, these are cleaved directly to give aldehydes and/or ketones (5.106).[85] Hence the method provides the same products as that from ozonolysis followed by reductive cleavage of the ozonide. The reaction is normally carried out in one step in aqueous dioxane, aqueous THF or other mixed aqueous solvent system, the water being required for hydrolysis of the intermediate osmate ester. Alternatively, dihydroxylation (OsO_4, NMO, t-BuOH, H_2O, THF) of the alkene followed directly by addition of $NaIO_4$ can be used. Cyclohexene, for example, was converted to adipaldehyde in 77% yield and the alkene **107** was converted to the aldehyde **108** (5.107). Another example, towards the sesquiterpene phytuberin, involved oxidative cleavage of the alkene **109** to give the aldehyde **110**, which was subsequently treated with base to promote intramolecular aldol condensation to give the cyclopentenone **111** (5.108).[86]

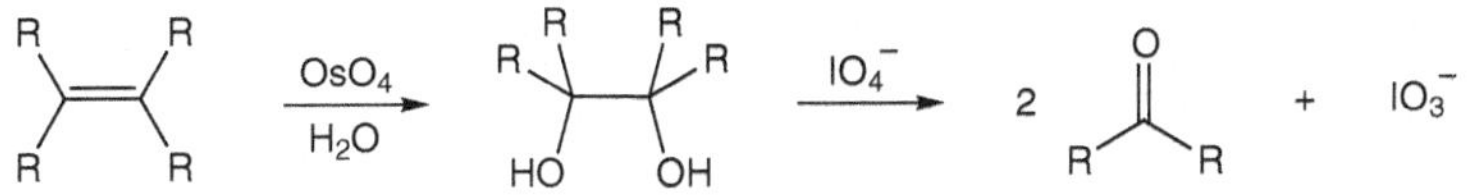

(5.106)

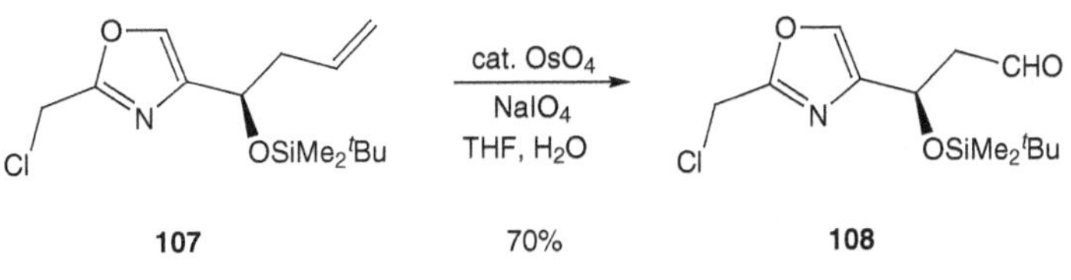

(5.107)

107 70% **108**

---

[85] R. Pappo, D. S. Allen, R. U. Lemieux and W. S. Johnson, *J. Org. Chem.*, **21** (1956), 478.
[86] F. Kido, H. Kitahara and A. Yoshikoshi, *J. Org. Chem.*, **51** (1986), 1478.

(5.108)

Osmium tetroxide can be used in substoichiometric amounts because the perio-date oxidizes the osmium back to the tetroxide. The periodate therefore plays two roles, one to cleave the diol and the other to re-oxidize the osmium(VI), although it does not itself react with alkenes or aldehydes. Reaction often occurs at the least-hindered alkene and therefore provides a selective method for the oxidative cleavage of, for example, a vinyl group in the presence of a di- or trisubstituted alkene. If potassium permanganate is used in place of osmium tetroxide, then any aldehyde products are usually oxidized to give the corresponding carboxylic acids.

Recently it has been reported that oxidative cleavage of alkenes with catalytic osmium tetroxide is possible in the absence of water using the co-oxidant Oxone® (KHSO$_5$).[87] In this case the diol is not formed and the intermediate osmate ester is oxidized by the Oxone® and fragments to regenerate osmium tetroxide and release the carbonyl products. For example, the alkene 1-nonene gave octanoic acid (90% yield) under these conditions.

Another excellent method for cleaving carbon–carbon double bonds is by action of ruthenium tetroxide, prepared using ruthenium trichloride (or ruthenium dioxide) as a catalyst in the presence of excess sodium periodate. Carboxylic acids are nor-mally produced from mono- or 1,2-disubstituted alkenes with this reagent, although it is possible to obtain the aldehyde products under neutral conditions or with NaHCO$_3$ and Oxone® as the co-oxidant.[88] Two examples of the use of ruthenium tetroxide for oxidative cleavage are shown in Schemes 5.109 and 5.110. Ruthenium tetroxide is a very powerful oxidizing agent and can be effective for the oxidation of alkenes that are resistant to other oxidants such as osmium tetroxide, potassium permanganate or ozone. It can also be used for the oxidative cleavage of aromatic rings.

(5.109)

[87] B. R. Travis, R. S. Narayan and B. Borhan, *J. Am. Chem. Soc.*, **124** (2002), 3824.
[88] P. H. J. Carlsen, T. Katsuki, V. S. Martin and K. B. Sharpless, *J. Org. Chem.*, **46** (1981), 3936; D. Yang and C. Zhang, *J. Org. Chem.*, **66** (2001), 4814.

$$\text{MeO}_2\text{C} \diagdown\diagdown\diagdown\diagdown\diagdown\diagdown\diagdown\diagup \quad \xrightarrow[\substack{\text{NaHCO}_3 \\ \text{MeCN, H}_2\text{O}}]{\text{cat. RuCl}_3, \text{Oxone}^{\circledR}} \quad \text{MeO}_2\text{C}\diagdown\diagdown\diagdown\diagdown\diagdown\diagdown\diagdown\diagdown\text{CHO}$$

94%

(5.110)

5.5 Palladium-catalysed oxidation of alkenes

The oxidation of ethene to ethanal by oxygen and a solution of a palladium(II) salt in aqueous hydrochloric acid is an important industrial process (the Wacker reaction). The palladium(II) is simultaneously reduced to the metal, but the reaction is made catalytic by addition of copper(II) chloride in the presence of air or oxygen, whereby the palladium is continuously re-oxidized to palladium(II) (5.111).

$$\text{H}_2\text{C}=\text{CH}_2 \;+\; \text{PdCl}_2 \;+\; \text{H}_2\text{O} \longrightarrow \text{CH}_3\text{CHO} \;+\; \text{Pd} \;+\; 2\,\text{HCl}$$

$$\text{Pd} \;+\; 2\,\text{CuCl}_2 \longrightarrow \text{PdCl}_2 \;+\; 2\,\text{CuCl}$$ (5.111)

$$2\,\text{CuCl} \;+\; 0.5\,\text{O}_2 \;+\; 2\,\text{HCl} \longrightarrow 2\,\text{CuCl}_2 \;+\; \text{H}_2\text{O}$$

The Wacker reaction has found most use for the oxidation of terminal alkenes to give methyl ketones.[89] It is believed to take place by an initial *trans* hydroxypalladation of the alkene to form an unstable complex that undergoes rapid β-elimination to the enol **112** (5.112). Hydropalladation then reductive elimination completes the overall process that involves transfer of hydride ion from one carbon to the other, via the palladium atom. The hydride migration is required to explain the observation that when the reaction is conducted in deuterium oxide, no deuterium is incorporated in the aldehyde produced.

$$\text{H}_2\text{O}\!: \diagdown\!\overset{\text{R}}{\diagup}\!\!\parallel\!\!\cdots\text{Pd}^{2+}\text{Cl}^-_2 \quad\longrightarrow\quad \underset{\text{Me}}{\overset{\text{O}}{\diagup}}\!\!\diagdown^{\text{R}} \;+\; \text{Pd} \;+\; 2\,\text{HCl}$$ (5.112)

$$\left[\text{HO}\diagdown\overset{\text{R}}{\underset{\text{Pd}^+\text{L}_n}{\diagdown}}\right] \longrightarrow \left[\text{HO}\diagdown\overset{\text{R}}{\diagup}\parallel\!\!\cdots\text{Pd}^+\text{L}_n\atop\text{H}\right] \longrightarrow \left[\text{HO}\diagdown\overset{\text{R}}{\underset{\text{H}}{\diagup}}\!\!\text{Pd}^+\text{L}_n\right]$$

112

Conversion of a terminal alkene to a methyl ketone is a useful transformation in organic synthesis. The reaction is typically carried out in aqueous DMF as solvent, using palladium(II) chloride as a catalyst (commonly 10 mol%) with copper(II)

---

[89] J. M. Takacs and X. Jiang, *Curr. Org. Chem.*, **7** (2003), 369; J. Tsuji, in *Comprehensive Organic Synthesis*, ed. B. M. Trost and I. Fleming, vol. 7 (Oxford: Pergamon Press, 1991), p. 449.

or copper(I) chloride and 1 atmosphere of oxygen. Copper(I) chloride is normally preferable as this avoids the formation of α-chlorinated ketones. Many different functional groups are tolerated and the reaction is selective for the oxidation of terminal alkenes in the presence of di- or trisubstituted alkenes. For example, only the terminal alkene is converted to a ketone on oxidation of the dienes **113** and **114** (5.113 and 5.114).

The Wacker reaction provides a method for the preparation of 1,4-dicarbonyl compounds, by formation of an enolate, allylation with an allyl halide, followed by palladium-catalysed oxidation of the terminal alkene. The product 1,4-dicarbonyl compounds can be treated with base to promote intramolecular aldol reaction (Robinson annulation – see Section 1.1.2) to give cyclopentenones. Thus, in a synthesis of pentalenene, Wacker oxidation of the 2-allyl ketone **115** gave the 1,4-diketone **116**, which was converted to the cyclopentenone **117** (5.115).[90]

[90] G. Mehta and K. S. Rao, *J. Am. Chem. Soc.*, **108** (1986), 8015.

Oxidation of 1,2-disubstituted alkenes occurs more slowly than that of terminal alkenes and a mixture of the two regioisomeric products is normally formed. With certain substrates, however, very high levels of regioselectivity have been obtained. For example, oxidation of the allylic ether **118** gave only the β-alkoxy ketone **119** (5.116).[91] The regioselectivity in oxidation reactions of unsymmetrical 1,2-disubstituted alkenes can be explained by electronic and neighbouring group effects, the latter involving co-ordination of a heteroatom or even an allylic hydrogen atom to the palladium atom in the intermediate.[92]

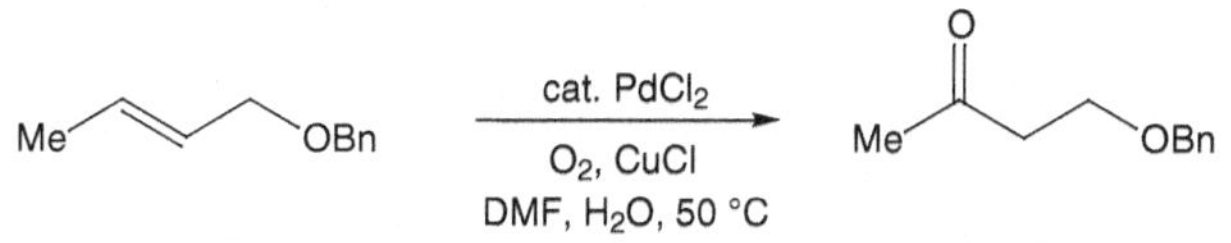

$$\text{(5.116)}$$

118   67%   **119**

Problems (answers can be found on page 479)

1. Explain the regio- and stereoselectivity in the formation of the organoborane **2** from the alkene **1**.

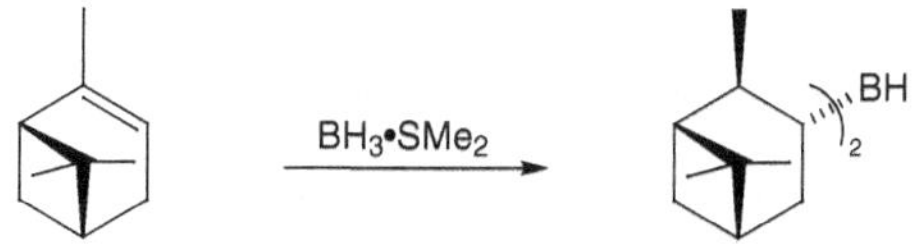

1   **2**

2. Explain the formation of the pyrrolidine **3**, R=H, used in a synthesis of the unnatural enantiomer of nicotine **3**, R=Me.

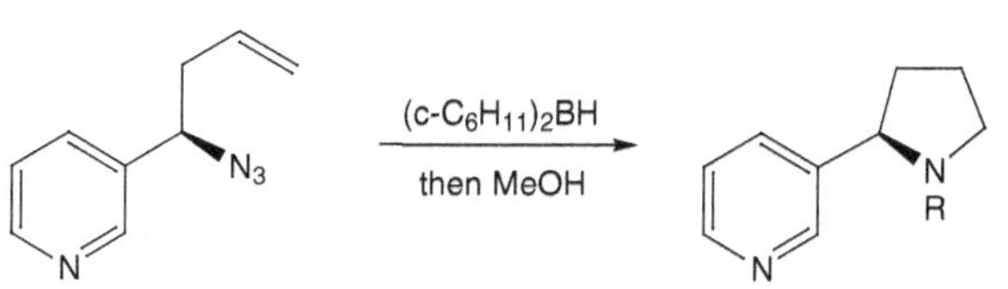

3, R=H

3. Draw the structures of the intermediates in the carbonylation of cyclohexene to give the aldehyde **4**.

---

[91] J. Tsuji, H. Nagashima and K. Hori, *Tetrahedron Lett.*, **23** (1982), 2679; see also S.-K. Kang, K.-Y. Jung, J.-U. Chung, E.-Y. Namkoong and T.-H. Kim, *J. Org. Chem.*, **60** (1995), 4678; H. Pellissier, P.-Y. Michellys and M. Santelli, *Tetrahedron*, **53** (1997), 10733.

[92] M. J. Gaunt, J. Yu and J. B. Spencer, *Chem. Commun.* (2001), 1844.

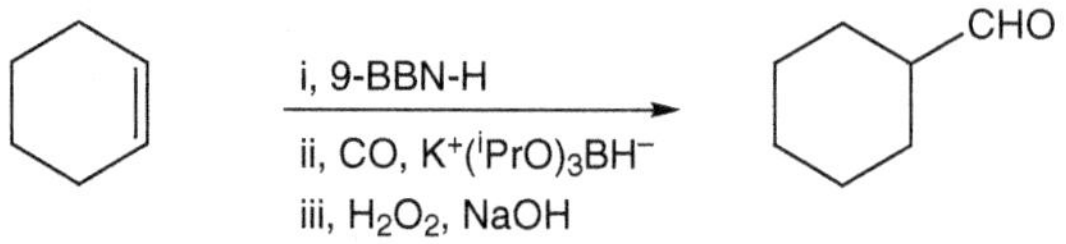

4

4. Suggest reagents for the asymmetric epoxidation of the allylic alcohol **5** to give the epoxide **6**.

5   6

5. Explain the selective formation of the epoxide **8** from 1,4-pentadien-3-ol **7**.

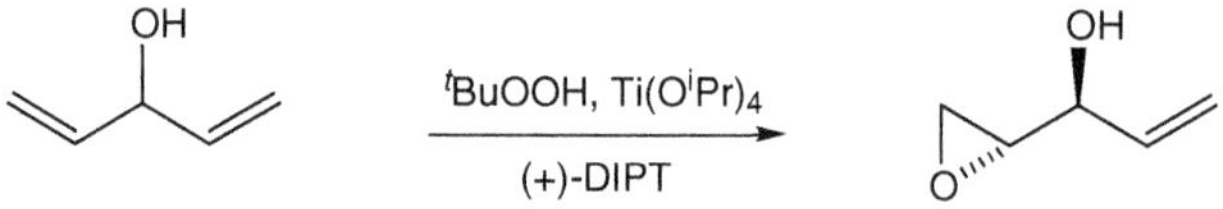

7   8

6. Suggest reagents and conditions for the asymmetric synthesis of the epoxide **9**, used in a synthesis of the HIV protease inhibitor indinavir.

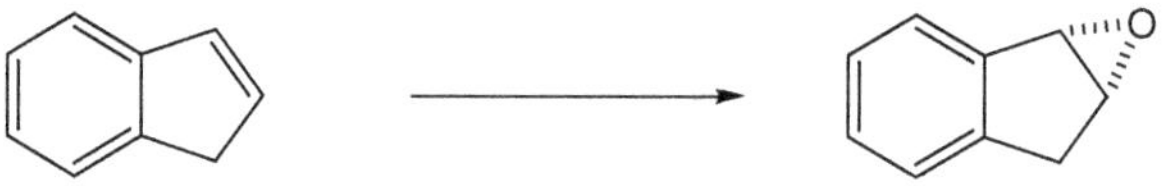

9

7. Suggest reagents and conditions for the synthesis of the two diastereomeric diols *syn* and *anti* **10**.

syn-**10**

anti-**10**

8. Suggest reagents and conditions for the formation of the *exo-cis*-diol and reagents and conditions for the formation of the *endo-cis*-diol product by oxidation of the alkene **11**.

11 **exo** OR **endo**

9. Suggest reagents, including an appropriate chiral ligand, and conditions for the formation of the chiral 1,2-amido-alcohol **12**, used in a synthesis of the antibiotic vancomycin.

12 87% ee

10. Unexpected products sometimes arise on ozonolysis of alkenes bearing allylic heteroatoms. Draw the structure of the ozonide from the reaction of the allylic alcohol **13** with ozone and suggest an explanation for the formation of the product **14**, used in a synthesis of grandisol.

13 **14**

11. Suggest reagents for the conversion of the alkene **15** to the ester **16**.

15 **16**

12. Draw the structures of the compounds **17** and **18**, and suggest reagents for the conversion of **17** to **18** and for **18** to the cyclopentenone **19**.

17 **18**

19

6

Oxidation

For practical purposes, most organic chemists mean by 'oxidation' either addition of oxygen to the substrate (such as epoxidation of an alkene), removal of hydrogen (such as the conversion of an alcohol to an aldehyde or ketone), or removal of one electron (such as the conversion of phenoxide anion to the phenoxy radical). Examples of oxidation reactions of alkenes have been described in Chapter 5, including epoxidation, aziridination, dihydroxylation and Wacker oxidation. This chapter therefore concentrates on oxidations of hydrocarbons, alcohols and ketones.

6.1 Oxidation of hydrocarbons

6.1.1 Alkanes

Under vigorous conditions strong oxidizing agents such as chromic acid and permanganate attack alkanes, but the reaction is of little synthetic use for usually mixtures of products are obtained in low yield. The reaction was traditionally used in the Kuhn–Roth estimation of the number of methyl groups in an unknown compound. This depends on the fact that a methyl group is rarely oxidized (the relative rates of oxidation of primary, secondary and tertiary C–H bonds are 1:110:7000) and hence the amount of ethanoic acid formed can be quantified.

The controlled oxidation of unactivated, saturated CH_3, CH_2 and CH groups is not uncommon in nature under the influence of oxidizing enzymes, but there are very few methods for effecting controlled reactions of this kind in the laboratory.

Oxidation of saturated hydrocarbons, the main feedstocks for the chemical industry, is extremely important. With simple substrates such as cyclohexane or adamantane, selective oxidation can be achieved, typically by using hydrogen peroxide or peroxycarboxylic acids in combination with strong acids or transition metal catalysts.[1] For example, oxidation of cyclohexane gave predominantly the ester **1** using an oxidant and trifluoroacetic acid (6.1). However, such forcing conditions are rarely used in the synthesis of more complex substrates, owing to the fact that mixtures of products are typically obtained.

$$
\underset{\substack{\text{30\% AcOOH, 1 mol\% RuCl}_3\,77\% \\ \text{or 30\% aq. H}_2\text{O}_2\quad 80\%}}{\xrightarrow{\text{CF}_3\text{CO}_2\text{H, CH}_2\text{Cl}_2}}
$$

OCOCF$_3$

1

(6.1)

adamantane

Some examples of selective oxidation of steroids are known, particularly using the Barton reaction (see Section 4.1) or using appropriate enzymes.[2] Biohydroxylation is less common with other substrates, but increasing levels of success in terms of yields and selectivities are being reported. For example, oxidation of *N*-benzyl-pyrrolidin-2-one using the biocatalyst *Sphingomonas* sp. HXN-200 in potassium phosphate buffer resulted in a highly regio- and enantioselective hydroxylation to give (*S*)-*N*-benzyl-4-hydroxy-pyrrolidin-2-one (together with some of the 3-hydroxy derivative) (6.2).[3]

$$
\xrightarrow[\quad 68\% \quad]{\textit{Sphingomonas} \text{ sp. HXN-200}}
$$

HO

>99% ee

(6.2)

6.1.2 Aromatic hydrocarbons

In the absence of activating hydroxy or amino substituents, benzene rings are attacked only slowly by oxidizing agents such as chromic acid or permanganate,

[1] A. E. Shilov and G. B. Shul'pin, *Chem. Rev.*, **97** (1997), 2879; D. H. R. Barton and D. Doller, *Acc. Chem. Res.*, **25** (1992), 504; P. Stavropoulos, R. Celenligil-Cetin and A. E. Tapper, *Acc. Chem. Res.*, **34** (2001), 745; N. C. Deno, E. J. Jedziniak, L. A. Messer, M. D. Meyer, S. G. Stroud and E. S. Tomezsko, *Tetrahedron*, **33** (1977), 2503; N. Komiya, S. Noji and S.-I. Murahashi, *Chem. Commun.* (2001), 65.
[2] H. L. Pellissier and M. Santelli, *Org. Prep. Proced. Int.*, **33** (2001), 1; S. M. Brown, in *Comprehensive Organic Synthesis*, ed. B. M. Trost and I. Fleming, vol. 7 (Oxford: Pergamon Press, 1991), p. 53.
[3] D. Chang, B. Witholt and Z. Li, *Org. Lett.*, **2** (2000), 3949.

but alkyl side chains are degraded with formation of benzoic acids (6.3). This is a useful method for the preparation of carboxylic acids. With side chains longer than methyl, initial attack takes place at the benzylic carbon atom, with the rate-determining step in these chromic acid oxidations being the cleavage of the benzylic C—H bond. This is suggested by the fact that tert-butylbenzene is very resistant to oxidation and ethylbenzene gives some acetophenone as well as benzoic acid.

$$\text{(6.3)}$$

The conversion of a methyl group attached to a benzene ring into an aldehyde can be achieved by oxidation with chromium trioxide in acetic anhydride in the presence of strong acid, or with a solution of chromyl chloride in carbon tetrachloride (the Étard reaction) (6.4). The success of the first reaction comes from the initial formation of the diacetate, which protects the aldehyde group against further oxidation. Cerium(IV) salts also readily oxidize aromatic methyl groups to aldehydes in acidic media. The aldehyde is not oxidized further and, in a polymethyl compound only one methyl group is oxidized under normal conditions. 1,3,5-Trimethylbenzene (mesitylene), for example, gave 3,5-dimethylbenzaldehyde quantitatively. The hypervalent iodine species *o*-iodoxy benzoic acid (IBX) is a useful oxidant and allows selective oxidation at the benzylic position on heating in DMSO.[4] Other reagents, such as tert-BuOOH and catalytic $RuCl_2(PPh_3)_3$ also oxidize alkyl benzenes.[5] For example, butylbenzene and *meta*-chloro-ethylbenzene can be oxidized readily to their corresponding aryl ketones (6.5).

$$\text{(6.4)}$$

[4] K. C. Nicolaou, P. S. Baran and Y.-L. Zhong, *J. Am. Chem. Soc.*, **123** (2001), 3183.
[5] S.-I. Murahashi, N. Komiya, Y. Oda, T. Kuwabara and T. Naota, *J. Org. Chem.*, **65** (2000), 9186.

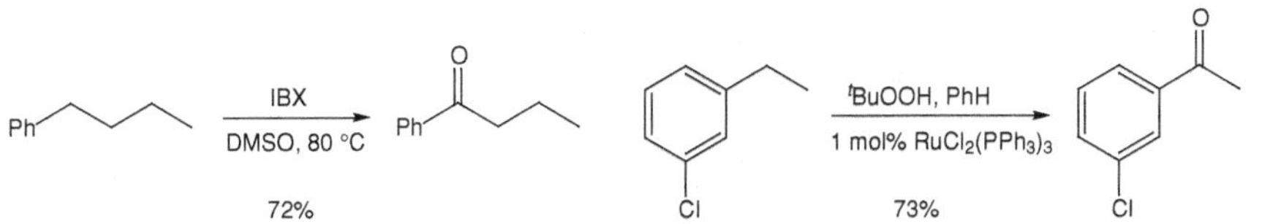

(6.5)

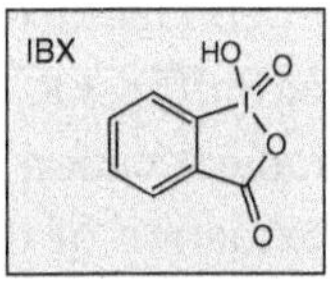

If a hydroxy or amino group is attached to the aromatic ring then this must first be protected, otherwise oxidation of the aromatic ring to a quinone may take place. This, of course, may be the desired transformation and quinones can be formed by using one of a variety of different oxidizing agents,[6] such as Fremy's salt [(KSO$_3$)$_2$NO],[7] CAN (ceric ammonium nitrate), DDQ (2,3-dichloro-5,6-dicyano-1,4-benzoquinone) or hypervalent iodine reagents.[8] The oxidation reaction can be achieved by using substituted phenols, anilines or derivatives to give *para-* or *ortho*-quinones, and can tolerate a wide variety of functional groups. Two examples, in the synthesis of juglone and avarone, are illustrated in Scheme 6.6. Common substrates include *para*-disubstituted dihydroxybenzenes or their ethers, although simple phenols can be employed. Thus, oxidation of the phenol **2** to the indolequinone **3** occurs even in the presence of the aldehyde, α,β-unsaturated ester and indole functionalities (6.7). The indolequinone **3** was converted to the anti-tumor agent EO9.

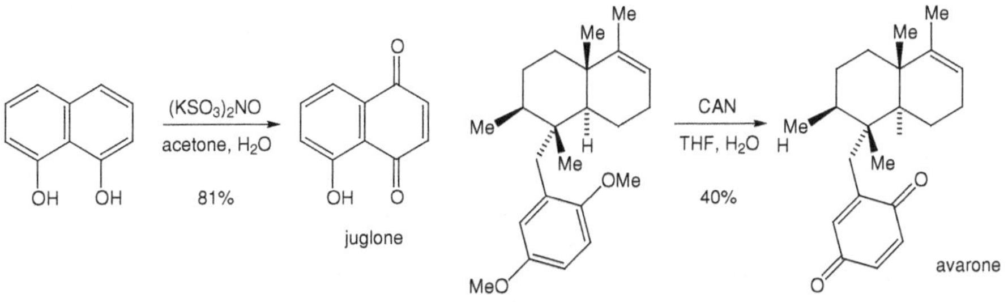

(6.6)

[6] S. Akai and Y. Kita, *Org. Prep. Proced. Int.*, **30** (1998), 605; P. J. Dudfield, in *Comprehensive Organic Synthesis*, ed. B. M. Trost and I. Fleming, vol. 7 (Oxford: Pergamon Press, 1991), p. 345.

[7] H. Zimmer, D. C. Lankin and S. W. Horgan, *Chem. Rev.*, **71** (1971), 229.

[8] R. M. Moriarty and O. Prakash, *Org. Reactions*, **57** (2001), 327.

$$(6.7)$$

2   **3**   **EO9**

Oxidation to quinones using hypervalent iodine reagents, in particular $PhI(OAc)_2$ or $PhI(OCOCF_3)_2$, has been finding increasing use in synthesis.[8] The intermediate radical cation can be trapped to give substituted dienone products. For example, oxidation of the dimethoxyaniline **4** in methanol gave the quinone monoacetal **5** and the phenol **6** gave the spirodienone **7** (6.8). Oxidation of the phenol **8** with polymer-supported $PhI(OAc)_2$ promoted cyclization of the nucleophilic arene onto the intermediate radical cation to give the spirodienone **9**, used in a synthesis of the alkaloid plicamine (6.9).[9]

$$(6.8)$$

4   **5**   **6**   **7**

$$(6.9)$$

8   **9**   plicamine

6.1.3 Alkenes

Oxidation of an alkene may take place at the double bond or at the adjacent allylic positions, and important synthetic reactions of each type are known. Reactions at the double bond, such as epoxidation and dihydroxylation, are described in Chapter 5. Allylic oxidation is of value in synthesis and provides a method to access allylic alcohols, α,β-unsaturated aldehydes or α,β-unsaturated ketones.[10] A common reagent for such transformations is selenium dioxide.[11] For example, with

[9] I. R. Baxendale, S. V. Ley, M. Nessi and C. Piutti, *Tetrahedron*, **58** (2002), 6285.
[10] P. C. B. Page and T. J. McCarthy, in *Comprehensive Organic Synthesis*, ed. B. M. Trost and I. Fleming, vol. 7 (Oxford: Pergamon Press, 1991), p. 83.
[11] N. Rabjohn, *Org. Reactions*, **24** (1976), 261.

selenium dioxide in water, 1-methylcyclohexene gave 2-methyl-2-cyclohexenone, although in ethanol a mixture of the α,β-unsaturated ketone and the alcohol 2-methyl-2-cyclohexenol was formed. The allylic alcohols can be oxidized easily to the α,β-unsaturated carbonyl compound if desired (see Section 6.2). For reaction in ethanol, the order of reactivity of allylic groups is $CH_2 > CH_3 > CH$, but this may not hold for reaction under other conditions or for all types of alkene. Generally, reactions are effected by using stoichiometric amounts of selenium dioxide, but very good yields of more easily purified products are often obtained with selenium dioxide as a catalyst in the presence of tert-BuOOH, which serves to re-oxidize the spent catalyst.

The oxidations are believed to involve an ene reaction (see Section 3.5) between the alkene and the hydrated form of the dioxide, followed by a [2,3]-sigmatropic rearrangement of the resulting allylseleninic acid and final hydrolysis of the Se(II) ester to the allylic alcohol (6.10). Further oxidation of the alcohol gives the α,β-unsaturated carbonyl compound.

(6.10)

A useful application of this reaction is in the oxidation of 1,1-dimethyl-alkenes to the corresponding *E*-allylic alcohols or aldehydes by selective attack on the *E*-methyl group. Thus, geranyl acetate gave a mixture of the *E,E*-alcohol **10** and the corresponding aldehyde (6.11). The aldehyde can become the major product using excess selenium dioxide. The high selectivity in these reactions is a consequence of the mechanism of the reaction. The initial ene reaction occurs by attack of the more electron-rich carbon atom of the alkene on the electrophilic selenium atom (with involvement of the least hindered allylic hydrogen atom) and is followed by the [2,3]-sigmatropic rearrangement, which establishes the *E*-stereochemistry by virtue of the preference for the substituent R group (6.10) to be in the pseudoequatorial position in the cyclic transition state.

(6.11)

Allylic oxidation can be carried out with reagents other than selenium dioxide, such as chromium trioxide–pyridine complex $CrO_3 \cdot py_2$ (Collins' reagent) or pyridinium chlorochromate $pyH^+CrO_3Cl^-$ (PCC). For example, oxidation to the lactone **11** was achieved on heating with PCC (6.12). Another useful allylic oxidation system uses a hydroperoxide or peroxycarboxylic ester in combination with a transition metal catalyst, such as a copper(I) salt, in what is referred to as the Kharasch–Sosnovsky reaction. For example, cyclohexene was oxidized to the allylic acetate **12** with copper(I) chloride and tert-BuOOH; asymmetric variants are possible using a chiral ligand for the metal (such as the bisoxazoline **13**) and a peroxycarboxylic ester as the oxidant (6.13).[12] The reaction is believed to proceed by hydrogen atom abstraction by a peroxy radical to give an allyl radical species that combines with the carboxylate via an allyl copper(III) carboxylate intermediate.

$$(6.12)$$

$$(6.13)$$

Allylic amination of alkenes can be carried out by selenium or sulfur reagents of the type Ts–N=S=N–Ts.[13] Reactions take place readily at room temperature and follow the sequence of ene reaction and [2,3]-sigmatropic rearrangement established for oxidations with selenium dioxide. The main problem with this chemistry is the difficulty in deprotecting the *N*-tosyl group from the product allylic amine. As a result, variations on these reagents have been reported, that allow easy subsequent deprotection. For example, in a synthesis of the alkaloid agelastatin A, treatment of the alkene **14** with the sulfur diimido reagent SES–N=S=N–SES (SES =

[12] M. B. Andrus and J. C. Lashley, *Tetrahedron*, **58** (2002), 845.
[13] K. B. Sharpless and T. Hori, *J. Org. Chem.*, **41** (1976), 176; M. Johannsen and K. A. Jørgensen, *Chem. Rev.*, **98** (1998), 1689.

β-trimethylsilylethylsulfonyl) gave the allylic amine **17** by way of the sulfonamide **16** (6.14).[14] The nitrogen–sulfur bond can be cleaved using trimethyl phosphite (MeO)$_3$P or sodium borohydride to give the product **17**. The SES group can be cleaved using a source of fluoride ion such as Bu$_4$NF or CsF in THF. Notice that the stereoselectivity of the allylic amination reaction arises from the first step (the ene reaction), which takes place on the less-hindered convex face of the substrate to give the intermediate **15**, that undergoes [2,3]-sigmatropic rearrangement to give the sulfonamide **16**, in which the new carbon–nitrogen bond is in the *exo* position.

$$(6.14)$$

SES = SO$_2$CH$_2$CH$_2$SiMe$_3$

An alternative to the direct electrophilic allylic amination of an alkene is the reaction of an allylic acetate or carbonate with a transition metal (typically a palladium or rhodium complex) to give a π-allyl metal species that reacts with a nitrogen nucleophile to give an allylic amine (see Section 1.2.4).

Another method for the conversion of an alkene into an allylic alcohol, but with a shift in the position of the double bond, proceeds from the corresponding β-hydroxyselenide. The β-hydroxyselenide can be obtained from the epoxide by reaction with phenylselenide anion or directly from the alkene by addition of phenylselenenic acid, phenylselenenyl chloride in aqueous MeCN, or by acid-catalysed reaction with *N*-phenylseleno-phthalimide.[15] The hydroxyselenide does not need to be isolated, but can be oxidized directly with tert-BuOOH to the unstable selenoxide, which spontaneously eliminates phenylselenenic acid to form the *E*-allylic alcohol. For example, 4-octene gave 5-octen-4-ol (6.15). Elimination takes place away from the hydroxy group to give the allylic alcohol; no more than traces

---

[14] D. Stien, G. T. Anderson, C. E. Chase, Y. Koh and S. M. Weinreb, *J. Am. Chem. Soc.*, **121** (1999), 9574.

[15] T. Hori and K. B. Sharpless, *J. Org. Chem.*, **43** (1978), 1689; A. Toshimitsu, T. Aoai, H. Owada, S. Uemura and M. Okano, *Tetrahedron*, **41** (1985), 5301; K. C. Nicolaou, D. A. Claremon, W. E. Barnette and S. P. Seitz, *J. Am. Chem. Soc.*, **101** (1979), 3704.

of the alternative keto products have been found in these reactions. With trisubstituted alkenes, addition of phenylselenenic acid is highly regioselective; citronellol methyl ether, for example, gave only allylic alcohol **18** by opening of the intermediate episelenonium ion at the more electropositive end (6.16).

$$(6.15)$$

88%

$$(6.16)$$

87%   **18**

6.2 Oxidation of alcohols

Many methods have been developed for the oxidation of primary and secondary alcohols. Oxidation of secondary alcohols normally gives rise to ketone products, whereas primary alcohols form aldehydes or carboxylic acids, depending on the reagent and conditions. Selective oxidation reactions have been developed that give these different types of products, even in the presence of other sensitive functionality. This section will describe, in turn, the different reagents used for the formation of aldehydes and ketones, before discussing the formation of carboxylic acids.

6.2.1 Chromium reagents

The most well-used of all the oxidizing agents are those based on chromium(VI),[16] typically prepared from chromium trioxide, CrO_3, or potassium dichromate, $K_2Cr_2O_7$. The oxidation of simple, particularly secondary alcohols can be accomplished using chromic acid, H_2CrO_4. The reaction is commonly effected with a solution of the alcohol in acetone and aqueous chromic acid in acetic or sulfuric acid (Jones' reagent). High yields of ketone products are usually obtained for substrates that are tolerant of strongly acidic, oxidizing conditions (6.17).

---

[16] F. A. Luzzio, *Org. Reactions*, **53** (1998), 1; S. V. Ley and A. Madin, in *Comprehensive Organic Synthesis*, ed. B. M. Trost and I. Fleming, vol. 7 (Oxford: Pergamon Press, 1991), p. 251.

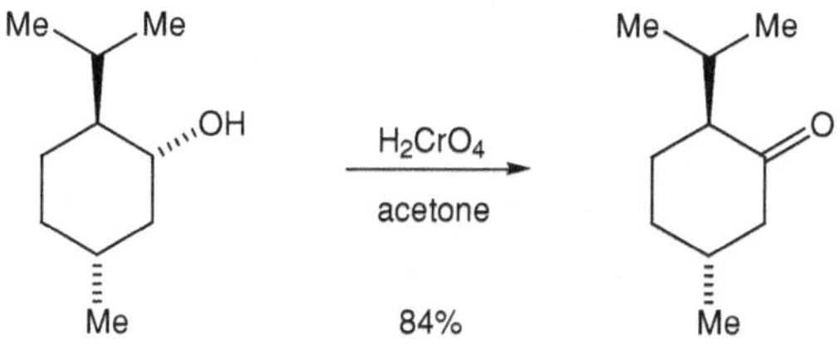

$$(6.17)$$

Oxidation of primary alcohols to aldehydes with acidic solutions of chromic acid is usually less satisfactory because the aldehyde is easily oxidized further to the carboxylic acid and, more importantly, because under the acidic conditions the aldehyde reacts with unchanged alcohol to form a hemiacetal which is oxidized rapidly to an ester.

In general, tertiary alcohols are unaffected by chromic acid, but tertiary 1,2-diols are cleaved readily to give ketones, provided they are capable of forming cyclic chromate esters. For oxidative cleavage of diols, see Section 5.4.

Oxidation of alcohols by chromic acid is believed to take place by initial formation of a chromate ester, followed by breakdown of the ester, as shown for isopropanol in Scheme 6.18. Proton abstraction with a base allows formation of the ketone product and generates a chromium(IV) species, which itself is thought to act as an oxidant to give more ketone product.[17]

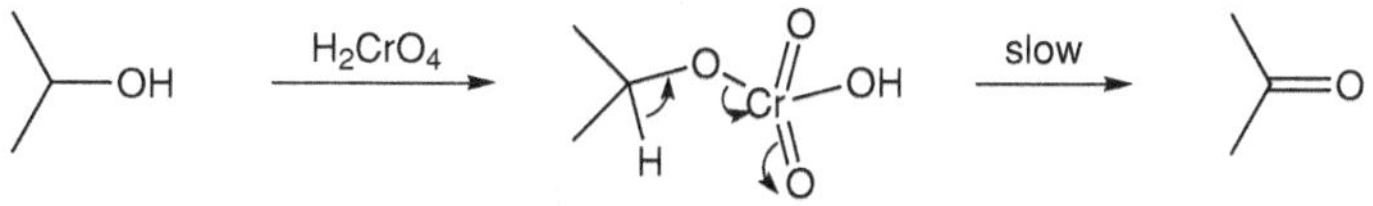

$$(6.18)$$

With unhindered alcohols, the initial reaction to form the chromate ester is fast, and the subsequent cleavage of the C—H bond is the rate-determining step. Where formation of the ester results in steric overcrowding, ester decomposition is accelerated because steric strain is relieved in going from reactant to product. In extreme cases, the initial esterification may become rate-determining. In the cyclohexane series, it is found that axial hydroxy groups are generally oxidized more rapidly than equatorial by a factor of about 3, presumably because of destabilizing 1,3-diaxial interactions in the axial chromate ester.

Oxidation with acid solutions of chromic acid is normally unsuitable for alcohols that contain acid-sensitive groups or other easily oxidizable groups, such as allylic or benzylic C—H bonds elsewhere in the molecule. However, the end-point of the oxidation can in many cases by observed by the persistence of the red colour of the chromic acid. Therefore the reaction can be monitored by dropwise addition of a solution of chromium trioxide in aqueous sulfuric acid to a cooled (0–20 °C) solution of the alcohol in acetone, thereby adding only the stoichiometric amount

[17] S. L. Scott, A. Bakac and J. H. Espenson, *J. Am. Chem. Soc.*, **114** (1992), 4205.

of the powerful oxidant. Over-oxidation is thus lessened or prevented, and selective oxidation of unsaturated secondary alcohols to unsaturated ketones without appreciable oxidation or rearrangement of double bonds can be achieved in good yield. In some cases it is possible to carry out oxidations using a chromium salt as a catalyst with a co-oxidant, such as tert-BuOOH.[18]

A useful, mild reagent for the oxidation of alcohols that contain acid-sensitive functional groups is the chromium trioxide–pyridine complex $CrO_3 \cdot py_2$, which is obtained readily by addition of chromium trioxide to pyridine. Reactions are best effected with a solution of the complex in CH_2Cl_2 – the so-called Collins' reagent – under anhydrous conditions.[16] Primary and secondary alcohols are converted into the carbonyl compounds in good yield, and acid-sensitive protecting groups are unaffected. For example, 1-heptanol gave heptanal in 80% yield and the alcohol **19** gave the aldehyde **20** (6.19). Oxidation of polyhydroxy compounds can sometimes be effected selectively at one position by protection of the other hydroxy groups, followed by subsequent deprotection. For example, protection of the 1,3-diol unit of the triol **21** and oxidation of the remaining alcohol gave the ketone **23**, which can be deprotected with dilute aqueous hydrochloric acid (6.20).

A disadvantage of Collins' original procedure is that a considerable excess of reagent is usually required to ensure rapid and complete oxidation of the alcohol, and a number of modifications have been introduced to overcome this. Excellent results have been obtained with pyridinium chlorochromate $pyH^+CrO_3Cl^-$ (PCC). When used in small excess in solution in CH_2Cl_2, it gives good yields of aldehydes and ketones from the corresponding alcohols.[16,19] However, the mildly acidic nature of PCC may preclude its use with acid-sensitive compounds. Another good reagent is pyridinium dichromate $(pyH^+)_2Cr_2O_7^{2-}$ (PDC), which in CH_2Cl_2 or DMF solution oxidizes alcohols to aldehydes or ketones in excellent yield and allylic and benzylic

[18] J. Muzart, *Chem. Rev.*, **92** (1992), 113.
[19] E. J. Corey and J. W. Suggs, *Tetrahedron Lett.* (1975), 2647; G. Piancatelli, A. Scettri and M. D'Auria, *Synthesis* (1982), 245.

alcohols to unsaturated carbonyl compounds. Some examples of the use of the reagents PCC and PDC are illustrated in Schemes 6.21 and 6.22.

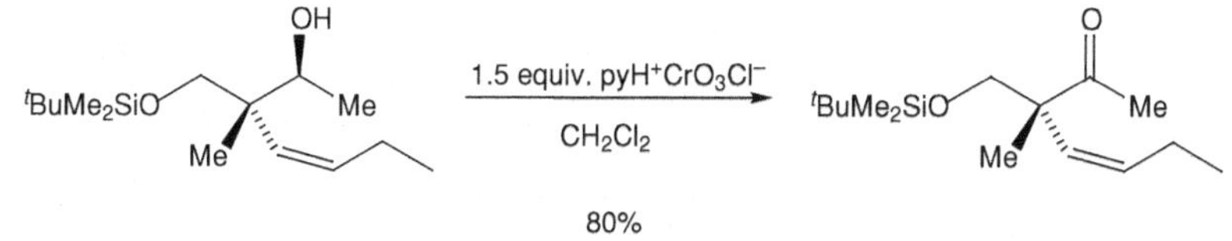

$$(6.21)$$

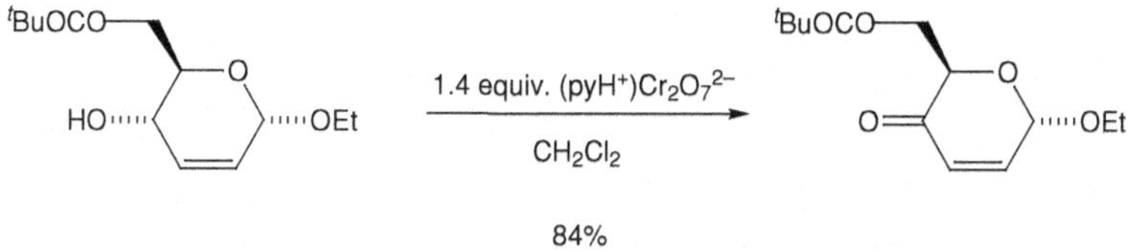

80%

$$(6.22)$$

84%

A useful application of chromium-based oxidants, especially pyridinium chlorochromate, is in the conversion of allylic tertiary alcohols to their transposed α,β-unsaturated ketones. For example, treatment of the allylic alcohol **24** with PCC gave the α,β-unsaturated ketone **25** (6.23). The reaction is thought to proceed by rearrangement of the chromate ester of the allylic alcohol to give a new allyl chromate ester that is oxidized to the ketone.

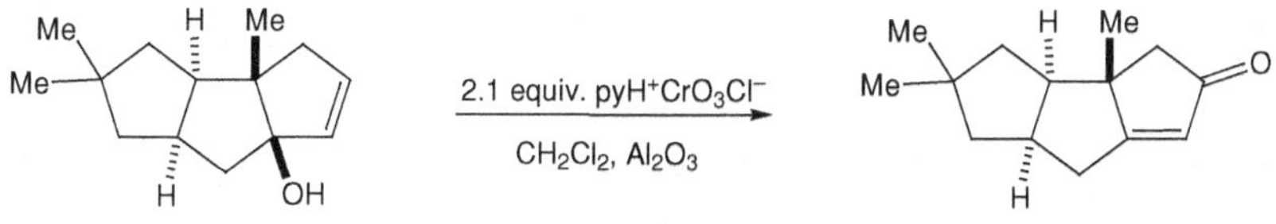

$$(6.23)$$

64%

24 **25**

6.2.2 *Oxidation* via *alkoxysulfonium salts*

One the most popular of all oxidations of alcohols to aldehydes or ketones involves the formation of intermediate alkoxysulfonium salts. A number of methods are available for the formation of the alkoxysulfonium salts, which are treated with a base to give the aldehyde or ketone.[20] The conditions of the reaction are mild, the reactions proceed rapidly and high yields of carbonyl compounds are generally obtained.

One of the earliest procedures involved reaction of the alcohol with dimethyl sulfoxide (DMSO) and dicyclohexylcarbodiimide (DCC) in the presence of a proton

[20] A. J. Mancuso and D. Swern, *Synthesis* (1981), 165; T. T. Tidwell, *Org. Reactions*, **39** (1990), 297; T. V. Lee, in *Comprehensive Organic Synthesis*, ed. B. M. Trost and I. Fleming, vol. 7 (Oxford: Pergamon Press, 1991), p. 291.

source. This method has been used to oxidize a number of sensitive compounds, such as 3′-*O*-acetylthymidine **26** (6.24).

$$(6.24)$$

26

A disadvantage of the carbodiimide route is that the product has to be separated from the dicyclohexylurea formed in the reaction. To overcome this, a number of other reagents have been used in conjunction with dimethyl sulfoxide, including acetic anhydride, trifluoroacetic anhydride, sulfur trioxide–pyridine complex, thionyl chloride and oxalyl chloride. Best results are normally obtained with oxalyl chloride in what is called the Swern oxidation.[21] By reaction with dimethyl sulfoxide and oxalyl chloride, followed by treatment of the resulting alkoxysulfonium salt with a base, usually triethylamine, many different alcohols have been converted into the corresponding carbonyl compounds in high yield under mild conditions. The Swern oxidation is one of the best methods for oxidizing alcohols; it is effective for almost all types of primary and secondary alcohol, including sensitive substrates such as allylic alcohols that give α,β-unsaturated aldehydes. In addition, no enolization typically takes place and therefore no loss of stereochemical integrity occurs in the formation of aldehydes that have an α-chiral centre. Some examples are given in Schemes 6.25–6.27.

$$(6.25)$$

$$(6.26)$$

[21] A. J. Mancuso, S.-L. Huang and D. Swern, *J. Org. Chem.*, **43** (1978), 2480.

$$(6.27)$$

The reaction is believed to proceed by way of the activated complex **28**, formed by spontaneous loss of carbon dioxide and carbon monoxide from the oxysulfonium salt **27** (6.28). Displacement of chloride by the alcohol gives the alkoxysulfonium salt **29**. This then undergoes proton abstraction by the base to form the ylide **30**, which fragments to the aldehyde or ketone by an intramolecular concerted process.

$$(6.28)$$

The reaction of dimethyl sulfoxide, oxalyl chloride and an alcohol is normally carried out at −78 or −60 °C, since the formation of the alkoxysulfonium salt **29** is rapid at this low temperature. After addition of the base triethylamine, the mixture may be warmed to −30 °C or higher to promote proton abstraction and fragmentation. The use of diisopropylethylamine instead of triethylamine as the base or addition of pH 7 phosphate buffer can, in the rare cases when it does occur, reduce the extent of enolization and therefore minimize any racemization or rearrangement of β,γ-double bonds.

An alternative method for the formation of aldehydes or ketones makes use of the complexes formed from a methyl sulfide with chlorine or N-chlorosuccinimide (NCS), in what is called the Corey–Kim oxidation.[22] With dimethyl sulfide the salt **28** is generated and reacts with the alcohol to give the alkoxysulfonium salts and hence, on treatment with a base, the carbonyl compound. This reaction has found particular application in the oxidation of 1,2-diols in which one alcohol is tertiary, to give α-hydroxy-aldehydes or ketones without rupture of the carbon–carbon bond. For example, the aldehyde **32** is formed in good yield from the diol **31** using dimethyl sulfide and NCS followed by addition of triethylamine (6.29). This transformation is thought to depend on the preferential five-membered transition

[22] E. J. Corey and C. U. Kim, *J. Am. Chem. Soc.*, **94** (1972), 7587.

state (**30**) leading to oxidation of the primary or secondary alcohol, rather than diol cleavage, which from the ylide requires a seven-membered transition state.

$$\text{(6.29)}$$

Related to these reactions is the oxidation of alkyl halides or tosylates to carbonyl compounds with dimethyl sulfoxide (or trimethylammonium N-oxide). The reaction is effected simply by warming the halide (normally the iodide) or sulfonate in DMSO (or Me_3NO), generally in the presence of a proton acceptor such as sodium hydrogen carbonate or a tertiary amine. Oxidation never proceeds beyond the carbonyl stage and other functional groups are unaffected. The reaction has been applied to benzyl halides, phenacyl halides, primary sulfonates and iodides and a limited number of secondary sulfonates. With substrates containing a secondary rather than primary halide or sulfonate elimination becomes an important side reaction and the oxidation is less useful with such compounds.

One drawback of these reactions is the formation of volatile, odorous dimethyl sulfide. Various solutions to this problem have been reported, such as the use of dodecyl methyl sulfoxide ($C_{12}H_{25}SOMe$) with oxalyl chloride followed by triethylamine.[23] The by-product dodecyl methyl sulfide is non-volatile and yields are only slightly reduced in comparison with the use of dimethyl sulfoxide.

6.2.3 Manganese reagents

A useful, mild reagent for the oxidation of primary and secondary alcohols to carbonyl compounds is manganese dioxide. This reagent has found most use as a highly specific oxidant for allylic and benzylic hydroxy groups, and reaction takes place under mild conditions (room temperature) in a neutral solvent (e.g. water, petroleum, acetone, DMF, CH_2Cl_2 or $CHCl_3$).[24] It avoids some of the problems of chromium reagents, which may promote epoxidation of the allylic alcohol or isomerization (Z to E geometry) of the double bond. For maximum activity

[23] K. Nishide, S. Ohsugi, M. Fudesaka, S. Kodama and M. Node, *Tetrahedron Lett.*, **43** (2002), 5177.
[24] A. J. Fatiadi, *Synthesis* (1976), 65.

it is best to prepare the manganese dioxide immediately prior to use. The best method appears to be by reaction of manganese(II) sulfate with potassium permanganate in alkaline solution; the hydrated manganese dioxide obtained is highly active.

For example, oxidation of the allylic alcohol **33** gave the α,β-unsaturated aldehyde **34**, used in a synthesis of the macrolactone bafilomycin A$_1$ (6.30). Chemoselective oxidation of allylic or benzylic alcohols can be achieved in the presence of aliphatic alcohols. Thus, in a synthesis of the alkaloid galanthamine, treatment of the diol **35** with manganese dioxide promoted selective oxidation of the benzylic alcohol to give the aldehyde **36** (6.31).

Hydroxy groups adjacent to triple bonds and cyclopropane rings are also easily oxidized, but under ordinary conditions saturated alcohols are not attacked (although they may be under more vigorous conditions). Manganese dioxide is used typically for the oxidation of allylic and benzylic alcohols, but it can be employed for the oxidation of unhindered primary aliphatic alcohols in refluxing toluene or CHCl$_3$ particularly with *in situ* removal of the aldehyde, for example with a Wittig reagent.[25] Thus, oxidation of the alcohol **37** with excess manganese dioxide in the presence of (carbomethoxymethylene)triphenylphosphorane resulted in the formation of the α,β-unsaturated ester **38** (6.32). The Wittig reagent traps the intermediate aldehyde to give the alkene (see Section 2.7). These *in situ* conditions are especially useful when the aldehyde is unstable or volatile. More hindered, including secondary alcohols can be oxidized and reacted *in situ* using the Swern oxidation (see section 6.2.2).

[25] L. Blackburn, X. Wei and R. J. K. Taylor, *Chem. Commun.* (1999), 1337.

$$(6.32)$$

Potassium permanganate adsorbed on a solid support is a useful alternative to manganese dioxide in the oxidation of allylic or benzylic alcohols.[26] Oxidation of allylic or benzylic primary alcohols with manganese dioxide in the presence of cyanide ions and an alcohol solvent promotes further oxidation to give the carboxylic ester.[27] The reaction proceeds through the aldehyde which reacts with cyanide to give the cyanohydrin. Oxidation of the cyanohydrin with manganese dioxide gives the acyl nitrile, which then reacts with the alcohol solvent to give the ester.

6.2.4 *Other metal-based oxidants*

A reagent for oxidizing primary and secondary alcohols to aldehydes and ketones under mild and essentially neutral conditions is silver carbonate precipitated on celite (Fetizon's reagent).[28] The reaction is carried out in refluxing solvent such as benzene and the product is recovered by simply filtering off the spent reagent and evaporating off the solvent. The reaction is highly chemoselective and other functional groups are normally unaffected. Under these conditions nerol, for example, is converted into neral in 95% yield and the allylic alcohol **39** gave the α,β-unsaturated ketone **40** (6.33). Highly hindered alcohols are not attacked, allowing selective oxidation in appropriate cases. Primary alcohols are oxidized more slowly than secondary, which are themselves much less reactive than benzylic or allylic alcohols; in acetone or methanol solution, selective oxidation of benzylic or allylic hydroxyl groups is easily effected.

$$(6.33)$$

[26] N. A. Noureldin and D. G. Lee, *Tetrahedron Lett.*, **22** (1981), 4889; M. Caldarelli, J. Habermann and S. V. Ley, *J. Chem. Soc., Perkin Trans. 1* (1999), 107.
[27] E. J. Corey, N. W. Gilman and B. E. Ganem, *J. Am. Chem. Soc.*, **90** (1968), 5616.
[28] A. McKillop and D. W. Young, *Synthesis* (1979), 401.

Treatment of a diol with silver carbonate normally promotes oxidation of only one of the hydroxyl groups. Butan-1,4-diols, pentan-1,5-diols and hexan-1,6-diols, with two primary hydroxyl groups, are converted into the corresponding lactones (6.34).[29] Initial oxidation to the aldehyde and cyclization gives an intermediate lactol, which is oxidized further to the lactone. When one of the hydroxyl groups is secondary then a mixture of the lactone and the hydroxy-ketone is often formed. However, good yields of the lactone from oxidation of the less-hindered primary alcohol can be obtained. Other diols give hydroxy-aldehydes or ketones depending on their structure. Thus, cyclohexan-1,2-diol gives 2-hydroxycyclohexanone and butan-1,3-diol forms 1-hydroxy-3-butanone, in line with the observation that secondary alcohols are generally more-readily oxidized than primary alcohols with this reagent (6.35). A number of other reagents have been reported to promote selective oxidation of secondary alcohols.[30]

$$\text{(6.34)}$$

65%

$$\text{(6.35)}$$

80%

A useful reagent for the oxidation of alcohols to aldehydes or ketones is tetrapropylammonium per-ruthenate (n-Pr$_4$N$^+$RuO$_4^-$, TPAP).[31] The reagent is soluble in a variety of organic solvents and is considerably milder than RuO$_4$ (see below), although still a powerful oxidant. It can be used as a catalyst with the co-oxidant N-methylmorpholine N-oxide (NMO). Oxidation with TPAP is often successful when other oxidants, such as DMSO/(COCl)$_2$ (the Swern oxidation, see above), fail or give low yields of the aldehyde or ketone. The reaction is best carried out in acetonitrile and can be accomplished at room temperature. For example, in a synthesis of the important anticancer agent Taxol®, oxidation of the alcohol **41** with TPAP as a catalyst gave the aldehyde **42** (6.36).

[29] M. Fetizon, M. Golfier and J.-M. Louis, *Tetrahedron*, **31** (1975), 171.
[30] J. B. Arterburn, *Tetrahedron*, **57** (2001), 9765.
[31] S. V. Ley, J. Norman, W. P. Griffith and S. P. Marsden, *Synthesis* (1994), 639.

(6.36)

41   **42**

Polymer-supported per-ruthenate reagents have some advantages over the solution-phase chemistry, not least in the ease of purification by simple filtration.[32] A sequence of polymer-supported reagents has been used in the synthesis of various natural products; for example, the alkaloid epibatidine was prepared using multi-step polymer-supported reagents, including the oxidation of 2-chloro-5-hydroxymethylpyridine with polymer-supported per-ruthenate (6.37).[33]

(6.37)

95%

Catalytic oxidation with a transition-metal catalyst and molecular oxygen (or H_2O_2) is another valuable method for oxidation of primary or secondary hydroxyl groups under mild and clean conditions.[34] The reaction is particularly effective for the oxidation of benzylic or allylic alcohols but, depending on the catalyst, may be successful for aliphatic alcohols. A variety of transition metals or their complexes can be used. For example, oxidation of primary aliphatic or allylic alcohols is possible using oxygen or even air (1 atmosphere pressure) as the oxidant with a palladium(II) catalyst supported on basic clay mineral (6.38).[35] No isomerization of the Z-alkene occurs and good yields of the aldehyde products are obtained. Oxidation at room temperature has been reported recently using Pd(OAc)$_2$ as a catalyst in the presence of Et$_3$N and 1 atmosphere of oxygen.[36] For example, under these conditions benzyl alcohol gave benzaldehyde in 84% yield.

[32] B. Hinzen, R. Lenz and S. V. Ley, *Synthesis* (1998), 977.
[33] J. Habermann, S. V. Ley and J. S. Scott, *J. Chem. Soc., Perkin Trans. 1* (1999), 1253.
[34] J. Muzart, *Tetrahedron*, **59** (2003), 5789.
[35] N. Kakiuchi, Y. Maeda, T. Nishimura and S. Uemura, *J. Org. Chem.*, **66** (2001), 6620.
[36] M. J. Schultz, C. C. Park and M. S. Sigman, *Chem. Commun.* (2002), 3034.

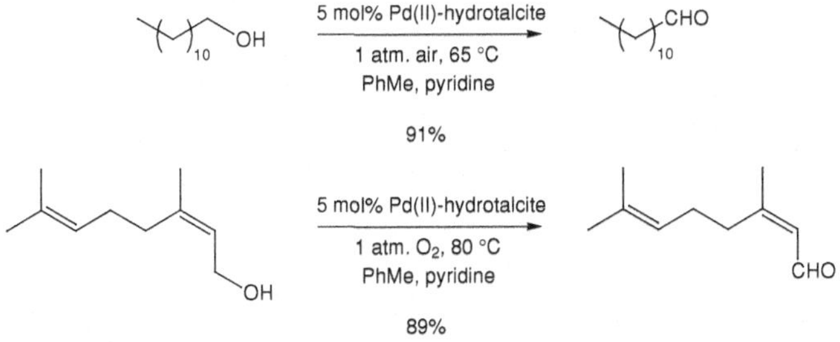

(6.38)

Kinetic resolution of some secondary allylic and benzylic alcohols has been shown to occur efficiently in the presence of the chiral ligand (−)-sparteine. For example, partial oxidation of the racemic alcohol **43** with a palladium(II) catalyst under an atmosphere of oxygen in the presence of (−)-sparteine occurs to give a mixture of the ketone **44** and recovered alcohol (*S*)-**43** (6.39).[37] Selective oxidation of the (*R*)-alcohol occurs with the chiral catalyst system.

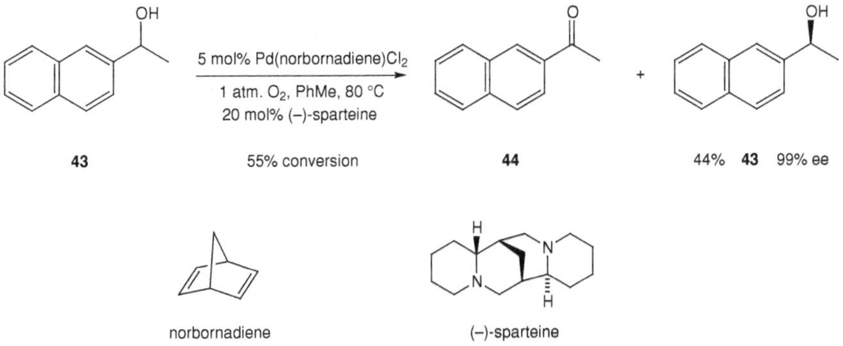

(6.39)

6.2.5 Other non-metal-based oxidants

A popular oxidizing agent that effects rapid oxidation of primary or secondary alcohols to aldehydes or ketones is the Dess–Martin reagent.[38] This is the hypervalent iodine(V) compound **46**, prepared from 2-iodoxybenzoic acid **45** (IBX) (6.40).[39] Both IBX and the Dess–Martin reagent are potentially explosive and should be handled with care.

[37] J. A. Mueller and M. S. Sigman, *J. Am. Chem. Soc.*, **125** (2003), 7005; D. R. Jensen, J. S. Pugsley and M. S. Sigman, *J. Am. Chem. Soc.*, **123** (2001), 7475; E. M. Ferreira and B. M. Stoltz, *J. Am. Chem. Soc.*, **123** (2001), 7725.

[38] D. B. Dess and J. C. Martin, *J. Org. Chem.*, **48** (1983), 4155, *J. Am. Chem. Soc.*, **113** (1991), 7277.

[39] R. E. Ireland and L. Liu, *J. Org. Chem.*, **58** (1993), 2899.

IBX **45**   **46**   (6.40)

The conditions for the oxidation of alcohols using the Dess–Martin reagent **46** are particularly mild and simple, typically using CH_2Cl_2 (or MeCN) as the solvent at room temperature. High yields of the carbonyl products are obtained, with no over-oxidation, and the neutral conditions make this method suitable for sensitive substrates. For example, of a selection of oxidizing agents, the Dess–Martin reagent was the only one found suitable for the oxidation of the alcohol **47** (6.41). The alcohol is thought to undergo ligand exchange with the periodinane **46** prior to the oxidation step. The reaction is effective for substrates in which the carbonyl products are sensitive to epimerization, elimination or rearrangement. In addition, amines and thioethers are tolerated. For example, oxidation of the alcohol **48** with the Dess–Martin reagent gave the ketone **49**, used in a synthesis of perhydrohistrionicotoxin (6.42).

47   70%   (6.41)

48   87%   **49**   (6.42)

The precursor, 2-iodoxybenzoic acid **45** (IBX), although less soluble in organic solvents than the Dess–Martin reagent, is more tolerant of moisture and promotes clean oxidation of alcohols in the solvent DMSO,[40] or on heating in other solvents

---

[40] M. Frigerio, M. Santagostino, S. Sputore and G. Palmisano, *J. Org. Chem.*, **60** (1995), 7272; T. Wirth, *Angew. Chem. Int. Ed.*, **40** (2001), 2812.

such as ethyl acetate.[41] For example, oxidation of 3-indolemethanol in DMSO at room temperature gave the aldehyde **50** in excellent yield (6.43). The conditions are mild and suitable for a wide range of alcohols, avoiding the need to prepare the Dess–Martin reagent. The related iodine(III) reagent iodosobenzene (PhI=O), combined with KBr, oxidizes alcohols in water to give high yields of ketones from secondary alcohols or carboxylic acids from primary alcohols.[42]

$$\text{(6.43)}$$

The stable, commercially available nitroxyl radical 2,2,6,6-tetramethylpiperidin-1-oxyl (TEMPO) **51** is an excellent catalyst, in conjunction with a co-oxidant, for the oxidation of alcohols.[43] The most popular co-oxidant is buffered sodium hypochlorite (NaOCl). Oxidation of the nitroxyl radical gives the oxoammonium ion **52**, which acts as the oxidant for the alcohol to form the carbonyl product. Primary alcohols are oxidized faster than secondary and it is often possible to obtain high chemoselectivity for the former. For example, oxidation of the triol **53** gave the aldehyde **54**, with no oxidation of the secondary alcohols (6.44). The use of TEMPO is particularly convenient for the oxidation of primary alcohols in carbohydrates, avoiding the need for protection of the secondary alcohols.

$$\text{(6.44)}$$

[41] J. D. More and N. S. Finney, *Org. Lett.*, **4** (2002), 3001.
[42] M. Tohma, S. Takizawa, T. Maegawa and Y. Kita, *Angew. Chem. Int. Ed.*, **39** (2000), 1306.
[43] A. E. J. de Nooy, A. C. Besemer and H. van Bekkum, *Synthesis* (1996), 1153; W. Adam, C. R. Saha-Möller and P. A. Ganeshpure, *Chem. Rev.*, **101** (2001), 3499.

The Oppenauer oxidation with aluminium alkoxides provides an alternative method for the oxidation of secondary (and less commonly primary) alcohols.[44] The reaction is the reverse of the Meerwein–Pondorff–Verley reduction (see Section 7.3). Typically aluminium triisopropoxide (or aluminium tri-tert-butoxide) is used, which serves to form the aluminium alkoxide of the alcohol. This is then oxidized through a cyclic transition state at the expense of acetone (or cyclohexanone or other carbonyl compound). By use of excess acetone, the equilibrium is forced to the right (6.45).

$$
\begin{array}{c}
R \\
\rangle\!-\!OH \\
R
\end{array}
\xrightleftharpoons{Al(O^iPr)_3}
\cdots
\rightleftharpoons
\begin{array}{c}
R \\
\rangle\!=\!O \\
R
\end{array}
+
\cdots
\qquad (6.45)
$$

The Oppenauer oxidation has been used widely for the oxidation of steroids, particularly for the conversion of allylic secondary hydroxyl groups to α,β-unsaturated ketones. β,γ-Double bonds generally migrate into conjugation with the carbonyl group under the conditions of the reaction (6.46). One drawback of the method is that the rate of oxidation is rather slow and therefore the mixture is normally heated in a solvent such as toluene, although more-active catalysts that are effective at room temperature have been developed.[45]

$$
\xrightarrow[\substack{\text{cyclohexanone} \\ \text{PhMe, heat}}]{Al(O^iPr)_3}
\qquad (6.46)
$$

83%

6.2.6 *Oxidation to carboxylic acids or esters*

Many methods for the oxidation of alcohols to aldehydes or ketones have been described above. Using a primary alcohol substrate, care must be taken to avoid oxidation to the carboxylic acid if the aldehyde is the desired product. Aldehydes are readily oxidized by a number of reagents, including chromic acid, permanganate salts, silver oxide or even by molecular oxygen. If the carboxylic acid is desired,

[44] C. Djerassi, *Org. Reactions*, **6** (1951), 207; C. F. de Graauw, J. A. Peters, H. van Bekkum and J. Huskens, *Synthesis* (1994), 1007.

[45] K. Krohn, B. Knauer, J. Küpke, D. Seebach, A. K. Beck and M. Hayakawa, *Synthesis* (1996), 1341; T. Ooi, H. Otsuka, T. Miura, H. Ichikawa and K. Maruoka, *Org. Lett.*, **4** (2002), 2669.

then it is common practice to prepare the aldehyde first and then to oxidize the aldehyde in a separate step. For example, Swern oxidation of a primary alcohol can be followed by oxidation with sodium chlorite ($NaClO_2$) to provide the required carboxylic acid (6.47).[46] Hydrogen peroxide is added to scavenge the hypochlorite (HOCl) formed during the oxidation reaction (thereby giving HCl, O_2 and H_2O).

$$R \diagup OH \xrightarrow[\text{then Et}_3\text{N}]{\substack{(COCl)_2 \\ \text{DMSO}}} R{-}CHO \xrightarrow[\substack{H_2O_2,\ MeCN \\ H_2O,\ NaH_2PO_4}]{NaClO_2} R{-}CO_2H \qquad (6.47)$$

The direct oxidation of primary alcohols to carboxylic acids is sometimes possible using the methods described in this section (especially with metal-based oxidants). Clearly, it would be advantageous to have a general, one-pot process for the direct formation of carboxylic acids. However, it is often found that by-products such as carboxylic esters, formed by subsequent reaction of the starting alcohol, are produced in addition to the desired carboxylic acid. Despite this, some useful procedures have been developed and these are likely to gain increasing popularity. Excess pyridinium dichromate (PDC) in a moist polar solvent such as DMF can successfully oxidize primary alcohols to carboxylic acids. Alternatively, oxidation of primary alcohols can be achieved with 2,2,6,6-tetramethylpiperidin-1-oxyl (TEMPO) **51**, in conjunction with the co-oxidant sodium chlorite ($NaClO_2$) and sodium hypochlorite (NaOCl) as a catalyst.[47] No epimerization of labile centres occurs and high yields of the carboxylic acids are produced directly from the primary alcohol. Thus, the carboxylic acid **55** was formed without loss of enantiopurity using these conditions (6.48). High yields can also be obtained with periodic acid (H_5IO_6) as the stoichiometric oxidant in conjunction with CrO_3 or $RuCl_3$ as the catalyst, or using IBX (**45**) in DMSO followed by addition of *N*-hydroxysuccinimide.[48]

$$\text{(6.48)}$$

92%   **55**

Oxidation adjacent to the oxygen atom of an ether, to give a carboxylic ester or lactone, is possible using a powerful oxidant.[49] Typically chromium trioxide or

[46] E. Dalcanale and F. Montanari, *J. Org. Chem.*, **51** (1986), 567.
[47] M. Zhao, J. Li, E. Mano, Z. Song, D. M. Tschaen, E. J. J. Grabowski and P. J. Reider, *J. Org. Chem.*, **64** (1999), 2564.
[48] R. Mazitschek, M. Mülbaier and A. Giannis, *Angew. Chem. Int. Ed.*, **41** (2002), 4059.
[49] C. A. Godfrey, in *Comprehensive Organic Synthesis*, ed. B. M. Trost and I. Fleming, vol. 7 (Oxford: Pergamon Press, 1991), p. 235.

ruthenium tetroxide (RuO_4) are used. The latter can be generated *in situ* by oxidation of a ruthenium(II) or (III) complex with NaOCl or $NaIO_4$.[50] The oxidation of cyclic ethers, such as substituted tetrahydrofurans, using RuO_4 provides a useful method for the formation of lactones. For example, in a synthesis of a portion of the macrolide tedanolide, the lactone **57** was prepared by oxidation of the tetrahydrofuran **56** (6.49). The methodology can also be applied to acyclic ethers. For example, oxidation of dibutyl ether with $[RuCl_2(DMSO)_4]$ as the catalyst (0.25 mol%) and NaOCl as the stoichiometric oxidant gave butyl butanoate in high yield (95%).[51] Ruthenium tetroxide can also be used for the oxidation of alkynes to 1,2-diketones and of aromatic rings to carboxylic acids (R—Ar to R—CO_2H). Phenylcyclohexane, for example, gave cyclohexanecarboxylic acid in 94% yield.

$$\text{(6.49)}$$

56   70%   **57**

6.3 Oxidation of ketones

Powerful oxidants (for example chromic acid or permanganate ion) can bring about, under vigorous conditions, the oxidation of ketones, although this usually leads to cleavage of the carbon chain adjacent to the carbonyl group (with formation of carboxylic acids) and is rarely used in synthesis. More important are controlled methods of oxidation leading to α,β-unsaturated ketones, α-hydroxy-ketones or lactones (without rupture of the molecule).

6.3.1 α,β-Unsaturated ketones

Conversion of ketones into α,β-unsaturated ketones has been effected by bromination–dehydrobromination, although a better method involves α-phenyl-seleno ketones as intermediates. These are normally obtained by reaction of the enolate of the ketone with a phenylselenyl halide or diphenyl diselenide at low temperature. Oxidation with hydrogen peroxide, sodium periodate or other oxidant gives the selenoxide which immediately undergoes *syn* β-elimination to form the α,β-unsaturated ketone. The process is tolerant of many functional groups, such as

[50] P. H. J. Carlsen, T. Katsuki, V. S. Martin and K. B. Sharpless, *J. Org. Chem.*, **46** (1981), 3936.
[51] L. Gonsalvi, I. W. C. E. Arends and R. A. Sheldon, *Chem. Commun.* (2002), 202.

alcohols, esters and alkenes. For example, propiophenone is converted (89% yield) into phenyl vinyl ketone, an alkene that is difficult to obtain by other means due to its ready polymerization and susceptibility to nucleophilic attack (6.50). In a similar way, 4-acetoxycyclohexanone gives 4-acetoxycyclohexenone. The procedure can also be used to make α,β-unsaturated esters and lactones from the saturated precursors (6.51).

$$
\text{(scheme 6.50): } \underset{}{\text{Et—C(O)—Ph}} \xrightarrow[\text{ii, PhSeBr}]{\text{i, LDA, THF, }-78\,^\circ\text{C}} \underset{\text{SePh}}{\text{CH(SePh)—C(O)—Ph}} \xrightarrow[\text{MeOH, H}_2\text{O}]{\text{NaIO}_4} \xrightarrow[\text{89\%}]{15\text{--}25\,^\circ\text{C}} \text{CH}_2=\text{CH—C(O)—Ph} \qquad (6.50)
$$

$$
\text{(scheme 6.51): } \text{diene—CO}_2\text{Me} \xrightarrow[\substack{\text{ii, PhSeBr}\\\text{iii, NaIO}_4}]{\text{i, LDA, THF, }-78\,^\circ\text{C}} \text{triene—CO}_2\text{Me} \qquad (6.51)
$$

80%

The sequence provides a method for converting α,β-unsaturated ketones into β-alkyl derivatives by alkylation with an organocuprate (see Section 1.2.1) and reaction of the intermediate copper enolate with phenylselenyl bromide, followed by oxidative elimination (6.52).

$$
\text{(scheme 6.52): } \text{CH}_3\text{CH}=\text{CH—C(O)—Ph} \xrightarrow[\text{ii, PhSeBr}]{\text{i, Me}_2\text{CuLi}} \underset{\text{SePh}}{\text{R—C(O)—Ph}} \xrightarrow[\text{85\%}]{\text{H}_2\text{O}_2} \text{R'—CH=C—C(O)—Ph} \qquad (6.52)
$$

The reaction of enolates with phenylselenyl halides is very fast, even at $-78\,^\circ\text{C}$, and the kinetically generated enolates react without rearrangement to the more stable isomer. Unsymmetrical ketones may therefore be converted into one or other of the two alternative α,β-unsaturated ketones. 2-Methylcyclohexanone, for example, gave 2-methyl-2-cyclohexenone or 6-methyl-2-cyclohexenone selectively by way of the corresponding thermodynamic or kinetic enolate respectively.

The same transformations can be effected by reaction of the trialkylsilyl enol ether of the aldehydes or ketones with palladium acetate.[52] For example, treatment of the trimethylsilyl enol ether of cyclooctanone with 10 mol% Pd(OAc)$_2$ in DMSO under one atmosphere of oxygen at 25 °C for 12 h gave cyclooctenone (82% yield).

Recently, it has been found that hypervalent iodine reagents promote the conversion of ketones to α,β-unsaturated ketones. For example, treatment of octyl phenyl

[52] Y. Ito, T. Hirao and T. Saegusa, *J. Org. Chem.*, **43** (1978), 1011; R. C. Larock, T. R. Hightower, G. A. Kraus, P. Hahn and D. Zheng, *Tetrahedron Lett.*, **36** (1995), 2423.

ketone with IBX (2-iodoxybenzoic acid, **45**) or with iodic acid (HIO_3) in DMSO on warming gave the corresponding α,β-unsaturated ketone (6.53).[53]

88%

$$(6.53)$$

6.3.2 α-Hydroxy-ketones

Oxidation of ketones at the α-carbon atom to give α-hydroxy-ketones is a synthetically useful transformation. This can be carried out using one of a number of oxidants, such as molecular oxygen in the presence of a strong base (e.g. potassium tert-butoxide), followed by reduction of the resulting hydroperoxide using zinc in acetic acid or triethyl phosphite.[54] This method has the disadvantage that cleavage products are often formed, and where the hydroxyperoxide bears an α-hydrogen atom a 1,2-diketone is likely to be produced (by base-catalysed elimination), resulting in poor yields of the α-hydroxy-ketone.

Alternative procedures that avoid these difficulties use the molybdenum peroxide $MoO_5 \cdot$pyridine$\cdot$hexamethylphosphoramide complex (MoOPH),[55] or oxaziridines such as **59**.[56] These reagents react readily with enolates at low temperature to form, after work-up, the desired α-hydroxy-ketone, normally in good yield and without contamination by oxidative cleavage products. Ketones, esters and lactones with an enolizable methylene or methine group are all readily converted into α-hydroxy compounds by this methodology. For example, 2-phenyl-cyclohexanone gave exclusively *trans*-2-hydroxy-6-phenyl-cyclohexanone (6.54), and the ketone **58** gave the α-hydroxy-ketone **60** (6.55). The electron-withdrawing *N*-sulfonyl group activates the oxaziridine to nucleophilic attack at the oxygen atom of the strained three-membered ring.

70%

$$(6.54)$$

[53] K. C. Nicolaou, T. Montagnon and P. S. Baran, *Angew. Chem. Int. Ed.*, **41** (2002), 1386.
[54] J. N. Gardner, F. E. Carlon and O. Gnoj, *J. Org. Chem.*, **33** (1968), 3294.
[55] E. Vedejs, D. A. Engler and J. E. Telschow, *J. Org. Chem.*, **43** (1978), 188.
[56] F. A. Davis and B. Chen, *Chem. Rev.*, **92** (1992), 919.

(6.55)

58   **59**   72%   **60**

> For reaction of oxaziridines or oxaziridinium salts with alkenes, see Section 5.2

Hydroxy-ketones have also been obtained very conveniently by epoxidation or dihydroxylation of silyl enol ethers (derived from ketones with either kinetic or thermodynamic control), for example with mCPBA or osmium tetroxide and N-methylmorpholine-N-oxide. Asymmetric dihydroxylation, for example with AD-mix-α or -β (see Section 5.3), can provide highly enantioenriched products (6.56).[57]

(6.56)

94%   99% ee

Alternatively direct oxidation of ketones with a hypervalent iodine reagent followed by hydrolysis of the resulting acetal can give the α-hydroxy-ketone.[58] Pentan-3-one, for example, gave 2-hydroxy-pentan-3-one in good yield (6.57). Direct asymmetric α-oxidation of aldehydes has been reported recently using nitrosobenzene as the electrophilic source of oxygen with proline as a catalyst in the solvent chloroform or DMSO.[59] Addition of proline to the aldehyde gives an intermediate enamine that attacks nitrosobenzene to form the C—O (rather than C—N) bond α-to the aldehyde. Reduction of the aldehyde and hydrogenolysis of the O—N bond provides the 1,2-diol product (6.58).

[57] T. Hashiyama, K. Morikawa and K. B. Sharpless, *J. Org. Chem.*, **57** (1992), 5067.

[58] R. M. Moriarty and K.-C. Hou, *Tetrahedron Lett.*, **25** (1984), 691; R. M. Moriarty, B. A. Berglund and R. Penmasta, *Tetrahedron Lett.*, **33** (1992), 6065.

[59] S. P. Brown, M. P. Brochu, C. J. Sinz and D. W. C. MacMillan, *J. Am. Chem. Soc.*, **125** (2003), 10 808; G. Zhong, *Angew. Chem. Int. Ed.*, **42** (2003), 4247.

$$(6.57)$$

$$(6.58)$$

6.3.3 Baeyer–Villiger oxidation of ketones

On oxidation with peroxy-acids, ketones are converted into esters or lactones.[60] This reaction was discovered in 1899 by Baeyer and Villiger, who found that reaction of a number of cyclic ketones with Caro's acid (permonosulfuric acid) led to the formation of lactones. Better yields are obtained with organic peroxy-acids, such as perbenzoic acid, peracetic acid and trifluoroperacetic acid, although most reactions are effected with *meta*-chloroperbenzoic acid (mCPBA). This reagent is more stable than the other acids and is commercially available. The reaction occurs under mild conditions and is applicable to open chain and cyclic ketones and to aromatic ketones. It provides a route to alcohols from ketones, through hydrolysis of the esters formed, and of hydroxy-acids from cyclic ketones by way of the lactones; lithium aluminium hydride reduction of the lactones gives diols with a defined arrangement of the two hydroxyl groups (6.59).

$$(6.59)$$

[60] G. R. Krow, *Org. Reactions*, **43** (1993), 251; M. Renz and B. Meunier, *Eur. J. Org. Chem.* (1999), 737.

The Baeyer–Villiger reaction is thought to take place by a concerted intramolecular process, involving migration of a group from carbon to electron-deficient oxygen. In the presence of a strong acid there may be addition of peroxy-acid to the protonated ketone, but additional acid is not needed and in its absence addition may take place to the ketone itself. The general mechanism (6.60) is supported by the fact that the reaction is catalysed by acid and is accelerated by electron-releasing groups in the ketone and by electron-withdrawing groups in the acid. In an elegant experiment using ^{18}O-benzophenone, Doering and Dorfman showed that the phenyl benzoate obtained had the same ^{18}O content as the ketone and that the ^{18}O was contained entirely in the carbonyl oxygen atom.[61] The intramolecular concerted nature of the reaction is also supported by the many demonstrations of complete retention of configuration at the migrating carbon atom. For example, optically active (*S*)-3-phenyl-butan-2-one was converted into (*S*)-1-phenylethyl acetate with no loss in enantiopurity.

An unsymmetrical ketone could obviously give rise to two different products in this reaction. Cyclohexyl phenyl ketone, for example, on reaction with peracetic acid gives both cyclohexyl benzoate and phenyl cyclohexanecarboxylate by migration of the cyclohexyl or the phenyl group respectively. It is found that the relative ease of migration of different groups in the reaction is in the following order.

tert-alkyl > cyclohexyl ~ *sec*-alkyl ~ benzyl ~ phenyl > primary alkyl > methyl

That is, in the alkyl series migratory aptitudes are in the series tertiary > secondary > primary; among benzene derivatives migration is facilitated by electron-releasing substituents, and hindered by electron-withdrawing ones. Thus phenyl *para*-nitrophenyl ketone gives only phenyl *p*-nitrobenzoate and *sec*-butyl methyl ketone is converted into *sec*-butyl acetate. The methyl group shows the least tendency to migrate, so that methyl ketones always give acetates in the Baeyer–Villiger reaction. Electronic factors are evidently important and the ease of migration is

[61] W. von E. Doering and E. Dorfman, *J. Am. Chem. Soc.*, **75** (1953), 5595.

related to the ability of the migrating group to accommodate a partial positive charge in the transition state. However, in some cases steric effects may also be involved and the experimental conditions can influence the regioselectivity. This is particularly noticeable in Baeyer–Villiger reactions of bridged bicyclic ketones. For example, while 1-methylnorcamphor **61** gives the expected lactone on oxidation with peracetic acid, the product obtained from camphor itself depends on the conditions and epicamphor **62** gives only the 'abnormal' product (6.61).

$$\text{(6.61)}$$

Baeyer–Villiger oxidation of bridged bicyclic ketones is valuable in synthesis because it provides a method for preparing derivatives of cyclohexane and cyclopentane with control of the stereochemistry of the substituent groups, and several syntheses of natural products have exploited this possibility. Thus, the lactone **64**, important in the synthesis of prostaglandins, was obtained in a sequence the key step of which was the Baeyer–Villiger oxidation of the bridged bicyclic ketone **63** (6.62).

$$\text{(6.62)}$$

When the ketone contains an alkene group, then mixtures of products sometimes result through competing epoxidation (see Section 5.2). The chemoselectivity of the Baeyer–Villiger reaction depends on the relative reactivities of the ketone and the alkene. Strained ketones tend to favour Baeyer–Villiger oxidation. If mixtures are obtained, then the reagent basic hydrogen peroxide or bis(trimethylsilyl)peroxide, $Me_3SiOOSiMe_3$ and a Lewis acid may be used. Bis(trimethylsilyl)peroxide behaves as a masked form of 100% hydrogen peroxide and brings about the Baeyer–Villiger oxidation of ketones without affecting carbon–carbon double bonds. For example, treatment of the unsaturated ketone **65** with mCPBA promoted preferential epoxidation; however, bis(trimethylsilyl)peroxide and boron trifluoride gave the lactone **66** (6.63).

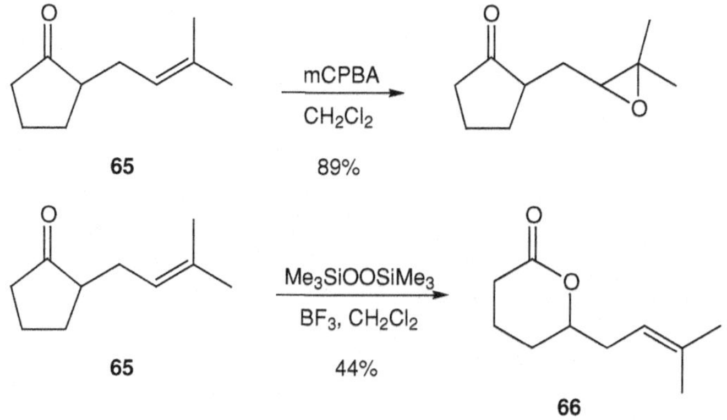

(6.63)

The Baeyer–Villiger reaction can be carried out using isolated enzymes or whole cell systems.[62] Biotransformations of simple cyclic ketones are most effective. For example, 4-methylcyclohexanone is oxidized with high enantioselectivity by using cyclohexanone monooxygenase (6.64).

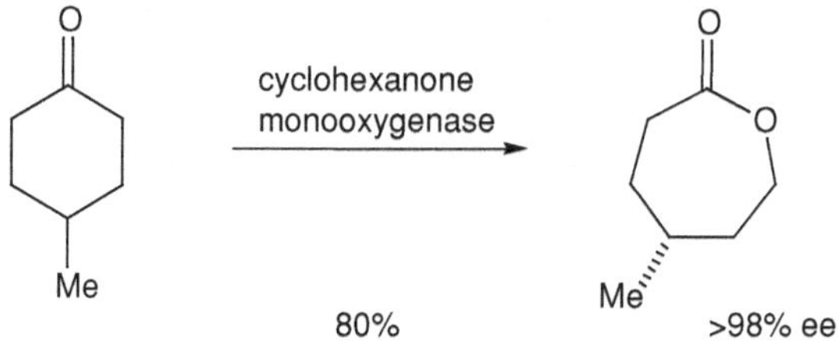

(6.64)

Oxidation of aldehydes with peroxy-acids is not so synthetically useful as oxidation of ketones and generally gives either carboxylic acids or formate esters. However, reaction of *ortho*- and *para*-hydroxy-benzaldehydes or -acetophenones with alkaline hydrogen peroxide (the Dakin reaction) is a useful method for making catechols and quinols.[60] With benzaldehyde itself, only benzoic acid is formed, but *ortho*-hydroxy-benzaldehyde (salicylaldehyde) gives catechol almost quantitatively (6.65) and 3,4-dimethylcatechol was obtained by oxidation of 2-hydroxy-3,4-dimethylacetophenone.

[62] S. M. Roberts and P. W. H. Wan, *J. Mol. Cat. B*, **4** (1998), 111; J. D. Stewart, *Curr. Org. Chem.*, **2** (1998), 195; M. Kayser, G. Chen and J. Stewart, *Synlett* (1999), 153; M. D. Mihovilovic, B. Müller and P. Stanetty, *Eur. J. Org. Chem.* (2002), 3711.

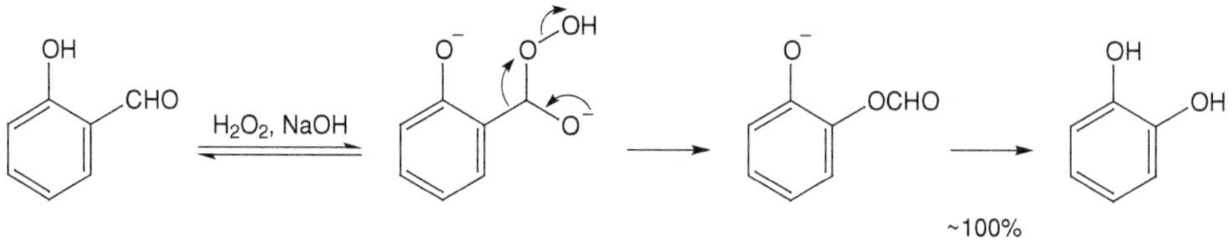

(6.65)

Problems (answers can be found on page 482)

1. Suggest a method for the preparation of naphthaquinone.

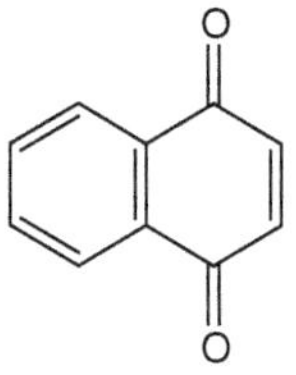

naphthaquinone

2. Draw the structures of the intermediates in the following allylic amination reactions and hence explain the difference in the outcome of these two reactions.

(i)

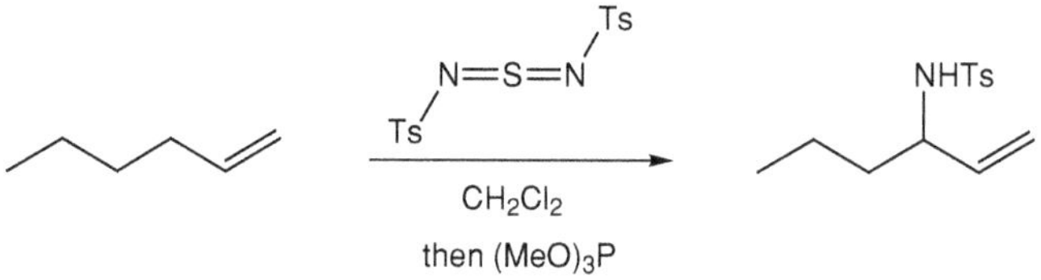

(ii)

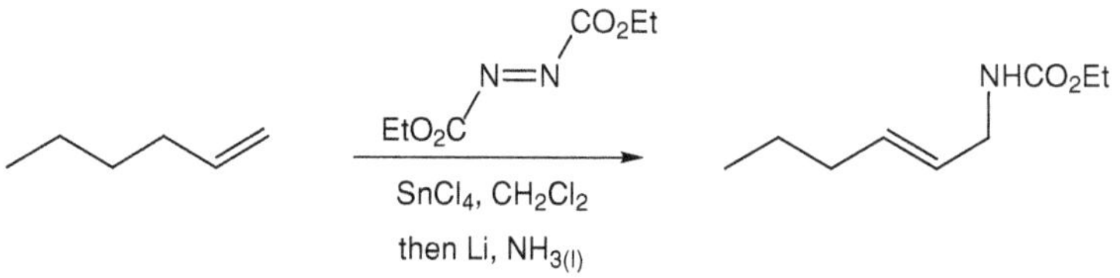

3. Suggest a two-step method for the conversion of the diol **1** to the α,β-unsaturated ketone **2**.

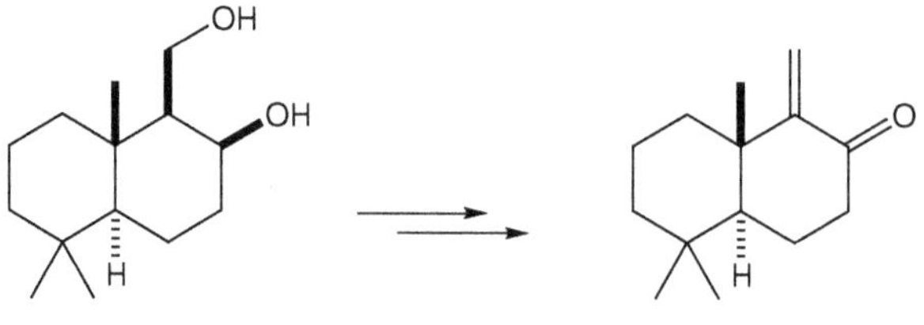

1   2

4. Explain the formation of the products **4** and **5**, formed from the alcohol **3** in the presence or absence of a base.

5. Suggest a reagent for the selective oxidation of the diol **6** to the ketol **7**.

6. Draw the structure of the product **9** from oxidation of the diol **8** with silver carbonate on celite.

7. Suggest reagents for the conversion of the alcohol **10** to the carboxylic acid **11**.

8. Suggest reagents for the conversion of undecanal [$CH_3(CH_2)_9CHO$] to undecenal.

9. Draw the structures of the products **12** and **13** from oxidation of the following ketones:

10. Suggest reagents for the formation of the phenol **15** from the aldehyde **14**.

7

Reduction

There must be few organic syntheses of any complexity that do not involve a reduction at some stage. Reduction is used in the sense of addition of hydrogen to an unsaturated group (such as a carbon–carbon double bond, a carbonyl group or an aromatic ring) or addition of hydrogen with concomitant fission of a bond between two atoms (such as the reduction of a disulfide to a thiol or of an alkyl halide to a hydrocarbon).

Reductions are generally effected either by catalytic hydrogenation or by a reducing agent (such as lithium aluminium hydride). Complete reduction of an unsaturated compound can generally be achieved without undue difficulty, but the aim is often selective reduction of one group in a molecule in the presence of other unsaturated groups. The method of choice in a particular case will often depend on the selectivity required and on the stereochemistry of the desired product.

7.1 Catalytic hydrogenation

Of the many methods available for reduction of organic compounds, catalytic hydrogenation is one of the most convenient.[1] Reduction is carried out easily by simply stirring or shaking the substrate with the catalyst in a suitable solvent (or even without a solvent if the substance being reduced is a liquid) in an atmosphere of hydrogen gas. An apparatus can be used that measures the uptake of hydrogen. At the end of the reaction, the catalyst is filtered off and the product is recovered from the filtrate, often in a high state of purity. The method is easily adapted for work on a micro scale, or on a large, even industrial, scale. In many cases reaction proceeds smoothly at or near room temperature and at atmospheric or slightly elevated

---

[1] S. Siegel, in *Comprehensive Organic Synthesis*, ed. B. M. Trost and I. Fleming, vol. 8 (Oxford: Pergamon Press, 1991), p. 417.

pressure. In other cases, high temperatures (100–200 °C) and pressures (100–300 atmospheres) are necessary, requiring special high-pressure equipment.

Catalytic hydrogenation may result simply in the addition of hydrogen to one or more unsaturated groups in the molecule, or it may be accompanied by fission of a bond between atoms. The latter process is known as hydrogenolysis.

Most of the common unsaturated groups in organic chemistry, such as alkenes, alkynes, carbonyl groups, nitriles, nitro groups and aromatic rings can be reduced catalytically under appropriate conditions, although they are not all reduced with equal ease. Certain groups, notably allylic and benzylic hydroxyl and amino groups and carbon–halogen single bonds readily undergo hydrogenolysis, resulting in cleavage of the bond between the carbon and the heteroatom. Much of the usefulness of the benzyloxycarbonyl protecting group (especially in peptide chemistry) is the result of the ease by which it can be removed by hydrogenolysis over a palladium catalyst (7.1). Hydrogenolysis of the C–O bond gives toluene and an intermediate carbamic acid that loses carbon dioxide to give the deprotected amine product.

$$ (7.1) $$

An alternative procedure that is sometimes advantageous is 'catalytic transfer hydrogenation', in which hydrogen is transferred to the substrate from another organic compound. The reduction is carried out simply by warming the substrate and hydrogen donor (such as isopropanol or a salt of formic acid) together in the presence of a catalyst, usually palladium. Catalytic-transfer hydrogenation can show different selectivity towards functional groups from that shown in catalytic reduction with molecular hydrogen.[2]

7.1.1 The catalyst

Many different catalysts have been used for catalytic hydrogenations; they are mainly finely divided metals, metallic oxides or sulfides. The most commonly used in the laboratory are the platinum metals (platinum, palladium and, increasingly, rhodium and ruthenium) and nickel. The catalysts are not specific and may be used for a variety of different reductions. The most widely used are palladium and platinum catalysts. They are used either as the finely divided metal or, more commonly, supported on a suitable carrier such as activated carbon, alumina or barium sulfate.

[2] R. A. W. Johnstone, A. H. Wilby and I. D. Entwistle, *Chem. Rev.*, **85** (1985), 129.

In general, supported metal catalysts, because they have a larger surface area, are more active than the unsupported metal, but the activity is influenced strongly by the support and by the method of preparation, and this provides a means of preparing catalysts of varying activity. Platinum is often used in the form of its oxide PtO_2 (Adams' catalyst), which is reduced to metallic platinum by hydrogen in the reaction medium. For example, reduction of the dihydropyrrole **1** occurs with good selectivity under these conditions (7.2).

Most platinum metal catalysts (with the exception of Adams' catalyst) are stable and can be kept for many years without appreciable loss of activity, but they can be deactivated by many substances, particularly by compounds of divalent sulfur. Catalytic activity is sometimes increased by addition of small amounts of platinum or palladium salts or mineral acid. The increase in the activity may simply be the result of neutralization of alkaline impurities in the catalyst.

7.1.2 Selectivity of reduction

Many hydrogenations proceed satisfactorily under a wide range of conditions, but where a selective reduction is wanted, conditions may be more critical.

The choice of catalyst for a hydrogenation is governed by the activity and selectivity required. In general, the more active the catalyst the less discriminating it is in its action, and for greatest selectivity reactions should be run with the least active catalyst and under the mildest possible conditions consistent with a reasonable rate of reaction. The rate of a given hydrogenation may be increased by raising the temperature, by increasing the pressure or by an increase in the amount of catalyst used, but all these factors may result in a decrease in selectivity. For example, hydrogenation of ethyl benzoate with copper chromite catalyst under the appropriate conditions leads to benzyl alcohol by reduction of the ester group, while Raney nickel gives ethyl cyclohexanecarboxylate by selective attack on the benzene ring (7.3). At higher temperatures, however, the selective activity of the catalysts is lost and mixtures of the two products and toluene are obtained from both reactions. Raney nickel is a porous, finely divided nickel obtained by treating a powdered nickel–aluminium alloy with sodium hydroxide. Most unsaturated groups can be reduced with Raney nickel, but it is most frequently used for reduction of aromatic rings and hydrogenolysis of sulfur compounds. When freshly prepared, it contains 25–100 ml adsorbed hydrogen per gram of nickel. Raney nickel catalysts are alkaline and may be used only for hydrogenations that are not adversely affected by basic conditions. They are deactivated by acids.

$$PhCH_2OH \xleftarrow[\text{160 °C, 250 atm.}]{\text{H}_2,\ \text{CuCr}_2\text{O}_4} PhCO_2Et \xrightarrow[\text{50 °C, 100 atm.}]{\text{H}_2,\ \text{Raney Ni}} \bigcirc\!\!-CO_2Et \qquad (7.3)$$

Both the rate and, sometimes, the course of a hydrogenation may be influenced by the solvent used. The most common solvents are methanol, ethanol and acetic acid, although other solvents can be used. Many hydrogenations over platinum metal catalysts are favoured by strong acids. For example, reduction of β-nitrostyrene in acetic acid–sulfuric acid is rapid and gives 2-phenyl-ethylamine (90% yield), but in the absence of sulfuric acid reduction is slow and the yield of amine is poor.

Not all functional groups are reduced with equal ease. Table 7.1[3] shows the approximate order of decreasing ease of catalytic hydrogenation of a number of common groups. This order is not invariable and is influenced to some extent by the structure of the compound being reduced and by the catalyst employed. In general, groups near the top of the list can be reduced selectively in the presence of groups near the bottom. For example, reduction of an unsaturated ester or ketone to a saturated ester or ketone is, in most cases, accomplished readily by hydrogenation over palladium or platinum, but selective reduction of the carbonyl group to form an unsaturated alcohol is difficult to achieve by catalytic hydrogenation and is generally effected using a hydride reducing agent (see Section 7.3). Similarly, nitrobenzene is easily converted into aniline, but selective reduction to nitrocyclohexane is not possible.

7.1.3 Hydrogenation of alkenes

Hydrogenation of carbon–carbon double bonds takes place easily and in most cases can be effected under mild conditions. Only a few highly hindered alkenes are resistant to hydrogenation, and even these can generally be reduced under more vigorous conditions. Palladium and platinum are the most-frequently used catalysts. Both are very active and the preference is determined by the nature of other functional groups in the molecule and by the degree of selectivity required (platinum usually

---

[3] H. O. House, *Modern Synthetic Reactions*, (New York: Benjamin, 1965).

Table 7.1. *Approximate order of reactivity of functional groups in catalytic hydrogenation*

Functional group	Reduction product
R—COCl	R—CHO, R—CH$_2$OH
R—NO$_2$	R—NH$_2$
R—C≡C—R	$\begin{array}{c} H \quad\quad H \\ \diagdown \quad\quad \diagup \\ C{=}C \\ \diagup \quad\quad \diagdown \\ R \quad\quad R \end{array}$, RCH$_2CH_2$R
R—CHO	R—CH$_2$OH
R—CH=CH—R	RCH$_2$CH$_2$R
R—CO—R	R—CH(OH)—R, R—CH$_2$—R
C$_6$H$_5$CH$_2$OR	C$_6$H$_5$CH$_3$ + ROH
R—C≡N	R—CH$_2$NH$_2$
Polycyclic aromatic hydrocarbons	Partially reduced products
R—CO$_2$R′	R—CH$_2$OH + R′OH
R—CONHR′	R—CH$_2$NHR′
 ⬡	 ⬡
R—CO$_2^-$Na$^+$	inert

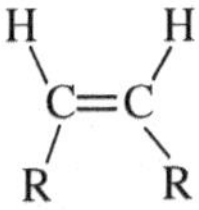

brings about a more exhaustive reduction). For example, the diene **3** is reduced by hydrogenation with palladium on charcoal (7.4). Raney nickel may also be used in certain cases. For example, cinnamyl alcohol is reduced to 3-phenylpropan-1-ol with Raney nickel in ethanol at 20 °C. Hydrogenation of alkenes is often accompanied by concomitant hydrogenolysis of sensitive benzyl ethers, such as *N*-CO$_2$Bn (*N*-Cbz) groups. A few different conditions can be employed to minimize hydrogenolysis, such as the addition of ethylenediamine (en) and THF as solvent, as illustrated in the reduction of the alkene **4** (7.5).[4]

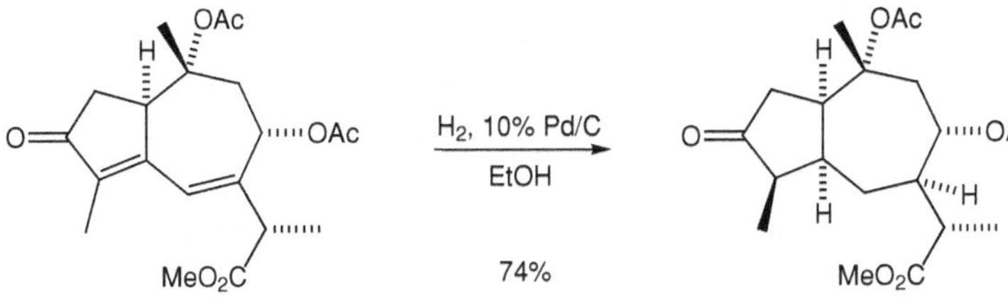

$$\text{H}_2, \text{10\% Pd/C} \atop \text{EtOH}$$

74%

(7.4)

3

---

[4] K. Hattori, H. Sajiki and K. Hirota, *Tetrahedron*, **56** (2000), 8433.

Ph [structure] N N–Cbz   **4**   →   5 atm. H₂, 5% Pd/C(en) / THF   →   Ph [structure] N N–Cbz   93%   (7.5)

en = H₂N ⁀ NH₂

Rhodium and ruthenium catalysts may alternatively be used and sometimes show useful selective properties. Rhodium allows hydrogenation of alkenes without concomitant hydrogenolysis of an oxygen function. For example, hydrogenation of the plant toxin, toxol **5** over rhodium–alumina gave the dihydro compound **6** (7.6); with platinum or palladium catalysts, however, extensive hydrogenolysis took place and a mixture of products was formed.

$$ \text{5} \quad \xrightarrow[\text{EtOH}]{\text{H}_2,\ 5\%\ \text{Rh–Al}_2\text{O}_3} \quad \text{6} \quad (7.6) $$

The ease of reduction of an alkene decreases with the degree of substitution of the double bond, and this sometimes allows selective reduction of one double bond in a molecule which contains several. For example, limonene **7** can be converted into *p*-menthene (by reduction of the terminal alkene) in almost quantitative yield by hydrogenation over platinum oxide if the reaction is stopped after absorption of one molar equivalent of hydrogen. In contrast, the isomeric diene **8**, in which both double bonds are disubstituted, gives only the completely reduced product (7.7).

7   **8**   (7.7)

Selective reduction of carbon–carbon double bonds in compounds containing other unsaturated groups can usually be accomplished, except in the presence of triple bonds, aromatic nitro groups and acyl halides. Palladium is usually the best catalyst. For example, 2-benzylidenecyclopentanone **9** is readily converted into 2-benzylcyclopentanone **10** with hydrogen and palladium in methanol (7.8); with a platinum catalyst, benzylcyclopentanol is formed.[5] Unsaturated nitriles and nitro compounds are also reduced selectively at the double bond with a palladium catalyst.

9	90%	**10**

7.1.4 Stereochemistry and mechanism

Hydrogenation of an unsaturated compound takes place by adsorption of the compound on to the surface of the catalyst, followed by transfer of hydrogen from the catalyst to the side of the molecule that is adsorbed on it. Adsorption onto the catalyst is largely controlled by steric factors, and it is found in general that hydrogenation takes place by *cis* addition of hydrogen atoms to the less-hindered side of the unsaturated centre.

For example, hydrogenation of the *E*-alkene **11** gives the racemic dihydro compound **12** by *cis* addition of hydrogen, while the Z-alkene **13** gives the *meso* isomer **14** (7.9). Hydrogenation of the pinene derivative **15** and of the ketone **17** gave products formed by *cis* addition of hydrogen to the more accessible side of the double bonds (7.10). In these examples the molecule possesses a certain degree of rigidity and it is clear which is the less hindered face of the double bond. With more-flexible molecules, it may be more difficult to decide on which side the molecule will be more easily adsorbed on the catalyst and to predict the steric course of a hydrogenation. In some cases, the affinity of a particular substituent group for the catalyst surface may induce addition of hydrogen from its own side of the molecule. The $-CH_2OH$ group can be effective in this respect. Thus, predominant *cis* addition of hydrogen to the alkene **18** occurs when $R=CH_2OH$, but *trans* addition when $R=CO_2Me$ (7.11).[6]

---

[5] A. P. Phillips and J. Mentha, *J. Am. Chem. Soc.*, **78** (1956), 140.
[6] H. W. Thompson, *J. Org. Chem.*, **36** (1971), 2577.

$$11 \qquad 98\% \qquad (\pm)\text{-}12 \tag{7.9}$$

$$13 \qquad 98\% \qquad meso\text{-}14$$

$$15 \qquad 90\% \qquad 16 \tag{7.10}$$

$$17 \qquad\qquad 83\% \qquad\qquad 17\%$$

	cis	:	trans
18 R = CH$_2$OH	95	:	5
R = CO$_2$Me	15	:	85

$$\tag{7.11}$$

The hydrogenation of substituted cyclic alkenes is anomalous in many cases in that substantial amounts of *trans*-addition products are formed, particularly with palladium catalysts. For example, the alkene **19** on hydrogenation over palladium in acetic acid gives mainly *trans*-decalin (7.12), and the alkene 1,2-dimethylcyclohexene **20** gives variable mixtures of *cis*- and *trans*-1,2-dimethylcyclohexane depending on the conditions (7.13). Similarly, in the hydrogenation of the isomeric dimethylbenzenes (xylenes) over platinum oxide, the *cis*-dimethylcyclohexanes are the main products, but some *trans* isomer is always produced.

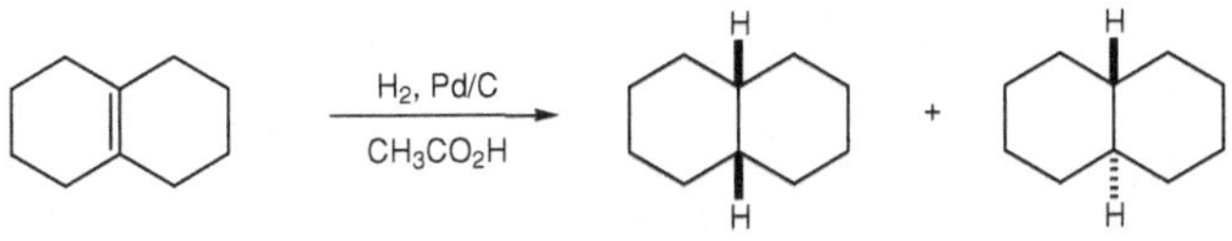

$$(7.12)$$

19   21%   79%

$$(7.13)$$

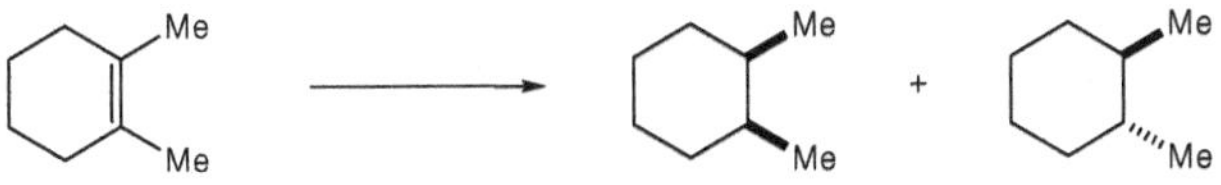

20

	16%	46%
H$_2$, Pd		
CH$_3$CO$_2$H		
H$_2$, PtO$_2$	82%	18%
CH$_3$CO$_2$H		

The reason for the formation of the *trans* products is thought to be because of migration of the double bond in a partially hydrogenated product on the catalyst surface. Although catalytic hydrogenation of alkenes may be accompanied by migration of the double bond, no evidence of migration normally remains on completion of the reduction. Sometimes, however, a tetrasubstituted double bond is formed which resists further reduction.

A satisfactory mechanism for catalytic hydrogenation must explain not only the normal *cis*-addition of hydrogen, but also that alkenes may be isomerized, that *trans*-addition products are formed in some hydrogenations, and the observation that deuteration of an alkene often leads to products containing more or fewer than two atoms of deuterium per molecule. These results can be rationalized on the basis of a mechanism in which transfer of hydrogen atoms from the catalyst to the adsorbed substrate takes place in a stepwise manner. The process is thought to involve equilibria between π-bonded forms **21** and **22** and a half hydrogenated form **23**, which can either take up another atom of hydrogen (to give the reduced product **24**) or revert to starting material or to an isomeric alkene **25** (7.14).

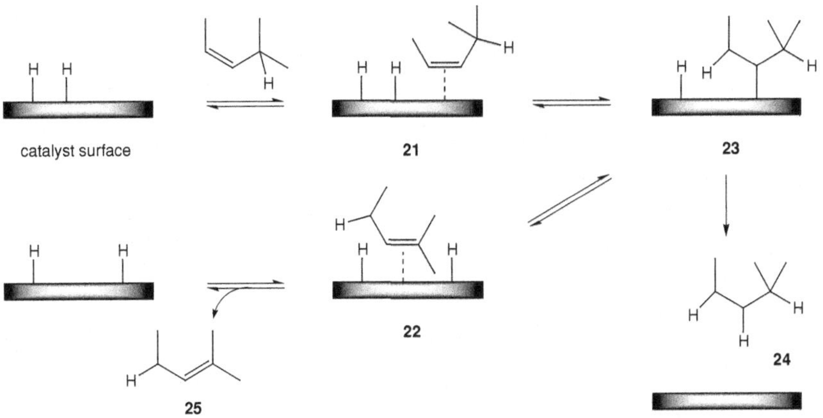

catalyst surface   **21**   **23**

$$(7.14)$$

22

24

25

7.1.5 Hydrogenation of alkynes

Catalytic hydrogenation of alkynes takes place in a stepwise manner, and both the alkene and the alkane can be isolated. Complete reduction of alkynes to the saturated compound is easily accomplished over platinum, palladium or Raney nickel. A complication which sometimes arises, particularly with platinum catalysts, is the hydrogenolysis of hydroxyl groups α- to the alkyne (propargylic hydroxyl groups) (7.15).

$$\text{(7.15)}$$

More useful from a synthetic point of view is the partial hydrogenation of alkynes to Z-alkenes. This reaction can be effected in high yield with a palladium–calcium carbonate catalyst that has been partially deactivated by addition of lead acetate (Lindlar's catalyst) or quinoline.[7] It is aided by the fact that the more electrophilic alkynes are absorbed on the electron-rich catalyst surface more strongly than the corresponding alkenes. An important feature of these reductions is their high stereoselectivity. In most cases the product consists very largely of the thermodynamically less stable Z-alkene and partial catalytic hydrogenation of alkynes provides one of the most convenient routes to Z-1,2-disubstituted alkenes. Thus stearolic acid **26**, on reduction over Lindlar's catalyst, gives oleic acid **27** (7.16). Partial reduction of alkynes with Lindlar's catalyst has been invaluable in the synthesis of many natural products with Z-disubstituted double bonds.

$$CH_3(CH_2)_7 - C \equiv C - (CH_2)_7CO_2H \quad \xrightarrow[\text{Lindlar catalyst}]{H_2} \quad \begin{array}{c} CH_3(CH_2)_7 \quad (CH_2)_7CO_2H \\ C = C \\ H \qquad H \end{array} \quad \text{(7.16)}$$

$$\textbf{26} \hspace{8cm} \textbf{27}$$

7.1.6 Hydrogenation of aromatic compounds

Reduction of aromatic rings by catalytic hydrogenation is more difficult than that of most other functional groups, and selective reduction is not easy. The commonest catalysts are platinum and rhodium, which can be used at ordinary temperatures, and Raney nickel or ruthenium, which require high temperatures and pressures.

Benzene itself can be reduced to cyclohexane with platinum oxide in acetic acid solution. Derivatives of benzene such as benzoic acid, phenol or aniline are

[7] H. Lindlar and R. Dubuis, *Org. Synth.*, Coll. Vol. **5** (1973), 880; J. Rajaram, A. P. S. Narula, H. P. S. Chawla and S. Dev, *Tetrahedron*, **39** (1983), 2315.

reduced more easily. For large scale work, the most convenient method is typically hydrogenation over Raney nickel at 150–200 °C and 100–200 atm. The catalyst rhodium, absorbed on alumina, can be used under mild conditions. Hydrogenation of phenols, followed by oxidation of the resulting cyclohexanols is a convenient method for the preparation of substituted cyclohexanones (7.17).[8] The product **28** was formed by hydrogenation at 55 atm, followed by oxidation with Jones reagent. A small amount (10–15%) of the *trans* compound was also formed.

i, 55 atm. H$_2$, EtOH
5% Rh–Al$_2$O$_3$, AcOH

ii, CrO$_3$, H$_2$SO$_4$
acetone, H$_2$O

71%
(+10–15% *trans*)

28

(7.17)

Reduction of benzene derivatives carrying oxygen or nitrogen functions in benzylic positions is complicated by the easy hydrogenolysis of such groups, particularly over palladium catalysts. Preferential reduction of the benzene ring in these compounds is best achieved with ruthenium or rhodium catalysts, which can be used under mild conditions. For example, mandelic acid is readily converted into the cyclohexyl derivative **29** over rhodium–alumina, whereas with palladium, hydrogenolysis to phenylacetic acid is the main reaction (7.18)

H$_2$, Rh–Al$_2$O$_3$

29

(7.18)

H$_2$, Pd/C

With polycyclic aromatic compounds, it is often possible, by varying the conditions, to obtain either partially or completely reduced products. Naphthalene can be converted into the tetrahydro or decahydro compound over Raney nickel depending on the temperature. With anthracene and phenanthrene, the 9,10-dihydro compounds are obtained by hydrogenation over copper chromite although, in general,

[8] B. B. Snider and Q. Lu, *J. Org. Chem.*, **59** (1994), 8065.

aromatic rings are not reduced with this catalyst. To obtain more-fully hydrogenated compounds, more-active catalysts must be used.

7.1.7 Hydrogenation of aldehydes and ketones

Hydrogenation of the carbonyl group of aldehydes and ketones is easier than that of aromatic rings, but not as easy as that of most carbon–carbon double bonds. Selective hydrogenation of a carbonyl group in the presence of carbon–carbon double bonds is, in most cases, best effected with hydride reducing agents.

For aliphatic aldehydes and ketones, reduction to the alcohol can be carried out under mild conditions over platinum or the more-active forms of Raney nickel. Ruthenium is also an excellent catalyst for reduction of aliphatic aldehydes and can be used to advantage with aqueous solutions. Palladium is not very active for hydrogenation of aliphatic carbonyl compounds, but is effective for the reduction of aromatic aldehydes and ketones; excellent yields of the alcohols can be obtained if the reaction is interrupted after absorption of one mole of hydrogen. Prolonged reaction, particularly at elevated temperatures or in the presence of acid, leads to hydrogenolysis and can therefore be used as a method for the reduction of aromatic ketones to methylene compounds.

Asymmetric reduction of ketones by catalytic homogeneous hydrogenation can be carried out with very high selectivity in many cases and is described below.

7.1.8 Hydrogenation of nitriles, oximes and nitro compounds

Functional groups with multiple bonds to nitrogen are readily reduced by catalytic hydrogenation. Nitriles, oximes, azides and nitro compounds, for example, are all smoothly converted into primary amines. Reduction of nitro compounds takes place easily and is generally faster than reduction of alkenes or carbonyl groups. Raney nickel or any of the platinum metals can be used as the catalyst, and the choice is governed by the nature of other functional groups in the molecule. Thus 2-phenyl-ethylamines, important biologically active molecules and useful for the synthesis of isoquinolines, are conveniently obtained by catalytic reduction of α,β-unsaturated nitro compounds (7.19).

$$\text{Ph}\diagup\!\!\diagup\!\!\diagdown\text{NO}_2 \xrightarrow[\text{EtOH, H}_2\text{SO}_4,\ 25\ ^\circ\text{C}]{\text{H}_2,\ \text{Pd/C}} \text{Ph}\diagup\!\!\diagdown\!\!\diagup\text{NH}_2 \qquad (7.19)$$

Nitriles are reduced with hydrogen and platinum or palladium at room temperature, or with Raney nickel under pressure. Unless precautions are taken, however, large amounts of secondary amines may be formed in a side reaction of the amine with the intermediate imine (7.20).

$$R-C\equiv N \xrightarrow[\text{cat.}]{H_2} \overset{R}{\underset{H}{C}}=NH \xrightarrow[\text{cat.}]{H_2} \overset{R}{\underset{H}{\overset{|}{C}}}-NH_2$$

(7.20)

$$\overset{R}{\underset{H}{C}}=NH \xrightarrow[-NH_3]{H_2N\diagdown R} \overset{R}{\diagdown}=N\diagdown_R \xrightarrow[\text{cat.}]{H_2} \overset{R}{\diagdown}-NH\diagdown_R$$

With the platinum-metal catalysts, this reaction can be suppressed by conducting the hydrogenation in acid solution or in acetic anhydride, which removes the amine from the equilibrium as its salt or as its acetate. For reactions with Raney nickel, where acid cannot be used, secondary amine formation is prevented by addition of ammonia. Hydrogenation of nitriles containing other functional groups may lead to cyclic compounds. For example, indolizidine and quinolizidine derivatives have been obtained in certain cases (7.21).

$$\xrightarrow[\text{EtOH}]{H_2,\ PtO_2} \longrightarrow$$

(7.21)

Reduction of oximes to primary amines takes place under conditions similar to those used for nitriles, with palladium or platinum in acid solution, or with Raney nickel under pressure.

7.1.9 Homogeneous hydrogenation

Catalysts for heterogeneous hydrogenation of the types discussed above, although useful, have some disadvantages. They may show lack of selectivity when more than one unsaturated centre is present, or cause double-bond migration and, in reactions with deuterium, they often bring about allylic interchanges with deuterium. This, in conjunction with double-bond migration, results in unspecific labelling with in many cases introduction of more than two deuterium atoms. The stereochemistry of reduction may not be easy to predict, since it depends on chemisorption and not on reactions between molecules. Some of these difficulties have been overcome by the introduction of soluble catalysts, which allow hydrogenation in homogeneous solution.[9]

A number of soluble-catalyst systems have been used, but the most common are based on rhodium and ruthenium complexes, such as $[(Ph_3P)_3RhCl]$ (Wilkinson's catalyst) and $[(Ph_3P)_3RuClH]$.

[9] A. J. Birch and D. H. Williamson, *Org. Reactions*, **24** (1976), 1; R. E. Harmon, S. K. Gupta and D. J. Brown, *Chem. Rev.*, **73** (1973), 21.

Wilkinson's catalyst is an extremely efficient catalyst for the homogeneous hydrogenation of non-conjugated alkenes and alkynes at ordinary temperature and pressure. Functional groups such as carbonyl, cyano, nitro and chloro are not reduced under these conditions. Mono- and disubstituted double bonds are reduced much more rapidly than tri- or tetrasubstituted ones, permitting the partial hydrogenation of compounds containing different kinds of double bonds. For example, in the reduction of linalool **30**, addition of hydrogen occurred selectively at the vinyl group, giving the dihydro compound **31** in high yield (7.22), and carvone **32** was similarly converted into the ketone **33** (7.23). The selectivity of the catalyst is shown further by the reduction of β-nitrostyrene to 2-phenyl-nitroethane (7.24).

$$H_2, [(Ph_3P)_3RhCl] \quad PhH \qquad (7.22)$$

30   80%   **31**

$$H_2, [(Ph_3P)_3RhCl] \quad PhH \qquad (7.23)$$

32   **33**

$$H_2, [(Ph_3P)_3RhCl] \quad PhH \qquad (7.24)$$

Hydrogenations take place by *cis* addition to the double bond. This has been shown by the catalysed reaction of deuterium with maleic acid to form *meso*-dideuterosuccinic acid, while fumaric acid gave the racemic compound.

An important practical advantage of this catalyst in addition to its selectivity is that deuterium is introduced without scrambling; that is, only two deuterium atoms are added, at the site of the original double bond. Another very valuable feature of this catalyst is that it does not bring about hydrogenolysis, thus allowing the selective hydrogenation of carbon–carbon double bonds without hydrogenolysis of other susceptible groups in the molecule. For example, benzyl cinnamate is converted smoothly into the dihydro compound without attack on the benzyl ester group and allyl phenyl sulfide is reduced to phenyl propyl sulfide (7.25).

$$\text{(7.25)}$$

93%

Wilkinson's catalyst has a strong affinity for carbon monoxide and decarbonylates aldehydes, therefore alkene compounds containing aldehyde groups cannot normally be hydrogenated with this catalyst under the usual conditions. For example, cinnamaldehyde is converted into styrene in 65% yield, and benzoyl chloride gives chlorobenzene in 90% yield.

Addition of hydrogen to Wilkinson's catalyst promotes oxidative addition of hydrogen. Dissociation of a bulky phosphine ligand and co-ordination of the alkene is followed by stepwise stereospecific *cis* transfer of the two hydrogen atoms from the metal to the alkene by way of an intermediate with a carbon–metal bond (7.26). Diffusion of the saturated substrate away from the transfer site allows the released complex to combine with dissolved hydrogen and repeat the catalytic reduction cycle.

$$\text{(7.26)}$$

The ruthenium complex [(Ph$_3$P)$_3$RuClH], formed *in situ* from [(Ph$_3$P)$_3$RuCl$_2$] and molecular hydrogen in the presence of a base (such as Et$_3$N), is an even more-efficient catalyst, which is specific for the hydrogenation of monosubstituted alkenes RCH=CH$_2$. Rates of reduction for other types of alkenes are slower by a factor of at least 2×10^3. Thus 1-heptene was rapidly converted into heptane but 3-heptene was unaffected. Some isomerization of alkenes is observed with this catalyst but the rate is slow compared with the rate of hydrogenation. The catalyst allows the conversion of disubstituted alkynes into Z-alkenes. Very similar behaviour is shown by the rhodium complex [(Ph$_3$P)$_3$RhH(CO)].

Hydrogenations with [(Ph$_3$P)$_3$RuClH] and [(Ph$_3$P)$_3$RhH(CO)] are two-step processes which proceed by the reversible formation of a metal–alkyl intermediate. The high selectivity for reduction of terminal double bonds is attributed to steric hindrance by the bulky Ph$_3$P groups to the formation of the metal–alkyl intermediate with other types of alkenes.

Another useful catalyst is the iridium complex [Ir(COD)py(PCy$_3$)]PF$_4$ (COD = 1,5-cyclooctadiene; py = pyridine; Cy = cyclohexyl).[10] It reduces tri- and

---

[10] R. H. Crabtree and G. E. Morris, *J. Organomet. Chem.*, **135** (1977), 395.

tetrasubstituted double bonds as well as mono- and disubstituted ones, although not so rapidly, and it appears to be unaffected by sulfur in the molecule. A valuable feature of this catalyst is the high degree of stereocontrol that can be achieved in the hydrogenation of cyclic allylic and homoallylic alcohols. Thus, the allylic alcohol **34** gave the saturated compound **35** almost exclusively (7.27), and the homoallylic alcohols **36** and **38** gave the *cis-* and *trans*-indanols **37** and **39** (7.28). There was no stereoselectivity in the hydrogenation of the corresponding acetates, or under heterogeneous conditions, and in these reactions the hydroxyl substituents exert the same kind of directing effect as in the Simmons–Smith (see Scheme 4.89) and the Sharpless epoxidation reactions (see Scheme 5.46).

$$\text{(7.27)}$$

$$\text{(7.28)}$$

7.1.10 Induced asymmetry via homogeneous hydrogenation

The homogeneous hydrogenation reaction is well-suited to asymmetric induction using chiral ligands on the rhodium or ruthenium metal centre.[11] Chiral phosphines are most common and many have been studied. For the hydrogenation of alkenes, some of the most popular include BINAP **40**, DIPAMP **41**, ChiraPHOS **42**, Me-DuPHOS **43** and DIOP **44**. Notice that the chirality can be at the phosphorus atoms or at the carbon backbone of the ligands. Good substrates for asymmetric hydrogenation are *N*-acyl dehydro-amino acids or their corresponding esters, which are reduced enantioselectively to amino-acid derivatives. For example, *N*-benzoyl

---

[11] W. S. Knowles, *Angew. Chem. Int. Ed.*, **41** (2002), 1998; R. Noyori, *Angew. Chem. Int. Ed.*, **41** (2002), 2008.

phenylalanine can be prepared with essentially complete selectivity for either enantiomer, depending on the choice of enantiomer of the chiral ligand (7.29). Recently monodentate phosphine ligands such as MonoPhos **45** have been shown to be very effective for this transformation, for example giving *N*-acetyl phenylalanine in quantitative yield and with 97% ee.[12]

(*S*)-BINAP **40** **41** **42** **43** **44** (*S*)-MonoPhos **45**

$$\text{Ph} \diagdown \overset{CO_2H}{\diagup} \quad \xrightarrow[\text{(}S,S\text{)-42, THF}]{H_2, \text{[Rh(NBD)]}^+\text{ClO}_4^-} \quad \text{Ph} \diagdown \underset{NHCOPh}{\diagup} CO_2H \qquad (7.29)$$

NHCOPh

99% ee

NBD = norbornadiene

The *N*-acyl dehydro-amino acid is thought to complex to the rhodium metal via its alkene and *N*-acyl carbonyl oxygen atom, leaving co-ordination sites for the ligand and hydrogen, which is then delivered asymmetrically to the alkene. Not all alkenes are good substrates for asymmetric hydrogenation, and a functional group other than the alkene that can co-ordinate to the metal catalyst often helps. Some experimentation with the choice of ligand and conditions is normally required for high selectivity. Ruthenium complexes can sometimes be preferable to those based on rhodium. For example, the painkiller naproxen **46** can be formed with high enantioselectivity by hydrogenation using ruthenium acetate as the catalyst with the chiral ligand (*S*)-BINAP **40** (7.30).[13]

135 atm. H₂, MeOH

0.5 mol% [Ru(OAc)₂], (*S*)-**40**

(7.30)

92% **46** 97% ee

Reduction of ketones to give alcohols can be carried out using asymmetric catalytic hydrogenation. For asymmetric hydrogenation, a chiral ruthenium complex is used together with either hydrogen gas,[14] or (by transfer hydrogenation) isopropanol

[12] M. van den Berg, A. J. Minnaard, R. M. Haak, M. Leeman, E. P. Schudde, A. Meetsma, B. L. Feringa, A. H. M. de Vries, C. E. P. Maljaars, C. E. Willans, D. Hyett, J. A. F. Boogers, H. J. W. Henderickx and J. G. de Vries, *Adv. Synth. Catal.*, **345** (2003), 308; see also M. T. Reetz and G. Mehler, *Angew. Chem. Int. Ed.*, **39** (2000), 3889.

[13] T. Ohta, H. Takaya, M. Kitamura, K. Nagai and R. Noyori, *J. Org. Chem.*, **52** (1987), 3174.

[14] R. Noyori and T. Ohkuma, *Angew. Chem. Int. Ed.*, **40** (2001), 40.

or formic acid as the hydrogen source.[15] For example, reduction of the ketones **47** and **50** occur with almost complete selectivity for one enantiomer using the diamine **48** as the chiral ligand (7.31). The reaction is most effective with unsaturated ketone substrates. Reductions of 1,3-dicarbonyl systems, such as β-keto-esters, also occurs with excellent enantioselectivity using ruthenium acetate and BINAP **40** under hydrogen pressure, to give chiral β-hydroxy-esters.[16] This chemistry provides an alternative to the asymmetric aldol reaction.

$$(7.31)$$

7.2 Reduction by dissolving metals

Chemical methods of reduction are of two main types: those that take place by addition of electrons to the unsaturated compound followed or accompanied by transfer of protons; and (more commonly – see Section 7.3) those that take place by addition of hydride ion followed in a separate step by protonation.

Reductions that follow the first path are generally effected by a metal, the source of the electrons, and a proton donor, which may be water, an alcohol or an acid. They can result in the addition of hydrogen atoms to a multiple bond or in fission of a single bond between, usually, carbon and a heteroatom. In these reactions an electron is transferred from the metal surface (or from the metal in solution) to the organic molecule giving, in the case of addition to a multiple bond, a radical anion, which in many cases is immediately protonated. The resulting radical subsequently takes up another electron from the metal to form an anion until work-up. In the absence of a proton source, dimerization or polymerization of the radical anion

---

[15] M. J. Palmer and M. Wills, *Tetrahedron: Asymmetry*, **10** (1999), 2045; R. Noyori and S. Hashiguchi, *Acc. Chem. Res.*, **30** (1997), 97.

[16] D. J. Ager and S. A. Laneman, *Tetrahedron: Asymmetry*, **8** (1997), 3327.

may take place. In some cases a second electron may be added to the radical anion to form a dianion.

In the reduction of benzophenone with sodium in ether or liquid ammonia, the first product is the resonance-stabilized radical anion **52**, which, in the absence of a proton donor, dimerizes to the pinacol (7.32). In the presence of a proton source, however, protonation leads to the radical **53**, which is subsequently converted into the anion and hence to the alcohol **54**. The presence in these radical anions of an unpaired electron that interacts with the atoms in the conjugated system has been established by measurements of the electron spin resonance spectra. Addition of organolithium species to benzophenone may also occur via the radical anion, as demonstrated by the deep blue colour generated in such reactions.

$$(7.32)$$

The metals commonly employed in these reductions include the alkali metals, calcium, zinc, magnesium, tin and iron. The alkali metals are often used in solution in liquid ammonia or as suspensions in inert solvents such as ether or toluene, frequently with addition of an alcohol or water to act as a proton source. Many reductions are also effected by direct addition of sodium, or particularly, zinc or tin to a solution of the compound being reduced in a hydroxylic solvent, such as ethanol, acetic acid or an aqueous mineral acid.

7.2.1 Reduction of carbonyl compounds

Reduction of ketones to secondary alcohols can be effected by catalytic transfer hydrogenation (Scheme 7.31), by complex hydrides (see Section 7.3) or by

Table 7.2. *Ratio **55**:**56** by reduction of
2-methylcyclohexanone*

Reagent	Ratio **55**:**56**
Na–alcohol	99 : 1
LiAlH$_4$	82 : 18
NaBH$_4$	69 : 31
Al(Oi-Pr)$_3$	42 : 58
catalyst + H$_2$	~30 : 70

sodium and an alcohol.[17] One feature of the sodium–alcohol method is that with cyclic ketones it normally gives rise exclusively or predominantly to the thermo-dynamically more stable alcohol. The ratios of the more-stable (*trans*) product **55** (equatorial substituents) and the *cis* product **56** (7.33), formed by reduction of 2-methylcyclohexanone with different reducing agents, are given in Table 7.2. The ketone 4-tert-butylcyclohexanone similarly gives the more stable *trans*-4-tert-butylcyclohexanol almost exclusively on reduction with lithium and propanol in liquid ammonia.

$$(7.33)$$

The high proportion of the more-stable product (equatorial hydroxyl group) formed in the reduction with a metal–alcohol is thought to arise from the preference for the intermediate radical anion (or other intermediate) to adopt the configuration with the equatorial oxygen atom (7.34).

$$(7.34)$$

Reduction of ketones with dissolving metals or low-valent transition metals in the absence of a proton donor leads to the formation of bimolecular products. The

[17] J. W. Huffman, in *Comprehensive Organic Synthesis*, ed. B. M. Trost and I. Fleming, vol. 8 (Oxford: Pergamon Press, 1991), p. 107.

reductive coupling of two aldehydes or ketones is referred to as the McMurry or pinacol coupling reaction and gives rise to a 1,2-diol (a pinacol) (see Section 2.9).[18] A number of reagents can be used, such as magnesium, magnesium amalgam, aluminium amalgam, low valent titanium or chromium species or samarium(II) iodide. A popular reagent is derived from reduction of titanium(III) chloride with zinc–copper couple. For example, treatment of the dialdehyde **57** under these conditions gave the diol sarcophytol **58** (7.35). This methodology is effective for the synthesis of small- or medium-sized rings and even for intermolecular couplings.

$$\begin{array}{ccc} \text{57} & \text{TiCl}_3\text{•(DME)}_2 \xrightarrow{\quad} \text{Zn–Cu, DME} & \text{58} \\ & 46\% & \end{array} \tag{7.35}$$

Carboxylic esters can be reduced by sodium and alcohols to form primary alcohols. This, the Bouveault–Blanc reaction, is one of the oldest established methods of reduction used in organic chemistry, but has now been largely replaced by reduction with lithium aluminium hydride. When the reaction is carried out in the absence of a proton donor, for example with sodium in xylene or liquid ammonia, dimerization takes place, and this is the basis of the acyloin condensation.[19] Intramolecular reaction gives ring compounds, including not only five- and six-membered rings but also medium and large rings. For example, the diester **59** gave the ten-membered ring α-hydroxy-ketone **60** (7.36) and the diester **61** gave the α-hydroxy-ketone **62**, which was readily converted into oestrone (7.37). Improved yields are often obtained by carrying out the reactions in the presence of Me$_3$SiCl, which is thought to serve principally to remove alkoxide ion from the reaction medium, thereby preventing base-catalysed side reactions.

$$\begin{array}{ccc} \text{59} & \xrightarrow[\text{then H}_3\text{O}^+]{\text{Na}} & \text{60} \\ & \text{xylene, heat} & \\ & 63\% & \end{array} \tag{7.36}$$

[18] J. E. McMurry, *Chem. Rev.*, **89** (1989), 1513; E. J. Corey, R. L. Danheiser and S. Chandrasekaran, *J. Org. Chem.*, **41** (1976), 260.

[19] R. Brettle, in *Comprehensive Organic Synthesis*, ed. B. M. Trost and I. Fleming, vol. 3 (Oxford: Pergamon Press, 1991), p. 613; J. J. Bloomfield, D. C. Owsley and J. M. Nelke, *Org. Reactions*, **23** (1976), 259.

$$(7.37)$$

61   91%   **62**

The Clemmensen reduction of aldehydes and ketones to methyl or methylene groups takes place by heating with zinc and hydrochloric acid. A non-miscible solvent can be used and serves to keep the concentration in the aqueous phase low, and thus prevent bimolecular condensations at the metal surface. The choice of acid is confined to the hydrogen halides, which appear to be the only strong acids whose anions are not reduced with zinc amalgam. The Clemmensen reduction employs rather vigorous conditions and is not suitable for the reduction of polyfunctional molecules, such as 1,3- or 1,4-diketones, or of sensitive compounds. However, it is effective for simple compounds that are stable to acid (7.38). A modification under milder conditions uses zinc dust and HCl dissolved in diethyl ether (ethereal HCl).[20] Other methods for converting C=O to CH_2 are described in Schemes 7.87 and 7.105.

$$(7.38)$$

88%

Reductive cleavage of α-substituted ketones, such as α-halo-, α-hydroxy- and α-acyloxy-ketones to the unsubstituted ketone can be carried out with zinc and acetic acid or dilute mineral acid. The reaction is thought to proceed by transfer of two electrons to the carbonyl group, followed by departure of the leaving group as the anion (7.39). This generates an enolate, which is converted into the ketone by acid. For α-halo-ketones, a different mechanism, with attack by zinc on the halogen, is possible.

$$(7.39)$$

Reductive eliminations of this type proceed most readily if the molecule can adopt a conformation where the bond to the group being displaced is orthogonal to the plane of the carbonyl group. Elimination of the substituent group is

[20] E. Vedejs, *Org. Reactions*, **22** (1975), 401.

then eased by continuous overlap of the developing p-orbital at the α-carbon atom with the carbonyl π-system. For this reason, cyclohexanone derivatives with axial α-substituents are reductively cleaved more readily than their equatorial isomers.

7.2.2 Reduction with metal and ammonia or an amine: conjugated systems

Isolated carbon–carbon double bonds are not normally reduced by dissolving metal reducing agents. Reduction is possible when the double bond is conjugated, because the intermediate anion can be stabilized by electron delocalization. The best reagent is a solution of an alkali metal in liquid ammonia, with or without addition of an alcohol – the so-called Birch reduction conditions. Under these conditions conjugated alkenes, α,β-unsaturated ketones and even aromatic rings can be reduced to dihydro derivatives.

Birch reductions are usually carried out with solutions of lithium or sodium in liquid ammonia. Any added alcohol can act as a proton donor to buffer against the accumulation of the strongly basic amide ion. Solutions of alkali metals in liquid ammonia contain solvated metal cations and electrons (7.40), and part of the usefulness of these reagents arises from the small steric requirement of the electrons. This may allow reactions that are difficult to achieve with other reducing agents and can lead to different stereochemical results. Reductions are usually carried out at the boiling point of ammonia ($-33\,°C$). As the solubility of many organic compounds in liquid ammonia is low at this temperature, co-solvents such as Et_2O, THF or DME can be added to aid solubility.

$$M \xrightarrow{\ NH_3\ (l)\ } M^+ (NH_3) \text{-----} e^- (NH_3) \qquad (7.40)$$

Conjugated dienes are readily reduced to the 1,4-dihydro derivatives with metal–ammonia reagents in the absence of added proton donors. For example, isoprene is reduced to 2-methyl-2-butene by sodium in ammonia, by way of an intermediate radical anion (7.41). The protons required to complete the reduction are supplied by the ammonia.

$$\text{(structure)} \xrightarrow{\ Na,\ NH_3\ (l)\ } \text{(structure)} \qquad (7.41)$$

Reduction of α,β-unsaturated ketones gives the saturated ketone or saturated alcohol, depending on the conditions.[21] Thus, cyperone **57** is converted to the ketone **58** with lithium in ammonia (7.42). Reduction in the presence of ethanol as proton source, however, gave the saturated alcohol **59**. In contrast, hydrogenation reduces both double bonds.

---

[21] D. Caine, *Org. Reactions*, **23** (1976), 1.

$$(7.42)$$

The first step in the reduction of α,β-unsaturated ketones is the formation of the radical anion **60**, which subsequently abstracts a proton from the ammonia or from added alcohol to give **61** (7.43). After addition of another electron, the enolate anion **63** is formed. In the absence of a stronger acid, this enolate remains unprotonated and resists addition of another electron, which would correspond to further reduction. Acidification with ammonium chloride then leads to the saturated ketone product. The reaction of ammonium ions with solvated electrons apparently destroys the reducing system before further reduction of the ketone to the alcohol can take place.

In the presence of 'acids' (e.g. ethanol) sufficiently strong to protonate the enolate anion, however, the ketone is generated in the reducing medium and is reduced further to the saturated alcohol. The formation of enolate anions such as **63** during metal–ammonia reduction of α,β-unsaturated ketones is shown by their ready trapping with electrophiles such as iodomethane.

$$(7.43)$$

Reduction of cyclic α,β-unsaturated ketones in which there are substituents on the β- and γ-carbon atoms could give rise to two stereoisomeric products. In many cases one isomer is formed predominantly, generally the more stable of the two. The guiding principle appears to be that protonation of the intermediate anion takes place orthogonal to the enol double bond (axially in six-membered rings). Thus, reduction of the enone **64** led almost exclusively to the *trans*-decalone **65** through axial protonation (7.44).

(7.44)

64   99% **65**   1%

3D conformation of **65**

7.2.3 Reduction with metal and ammonia: aromatic compounds

One of the most useful synthetic applications of metal–ammonia–alcohol reducing agents is in the reduction of benzene rings to 1,4-dihydro derivatives. The reagents are powerful enough to reduce benzene rings, but specific enough to add only two hydrogen atoms. Benzene itself is reduced with lithium and ethanol in liquid ammonia to 1,4-dihydrobenzene by way of the radical anion (7.45).[22] The presence of an alcohol as a proton donor is necessary in these reactions, for the initial radical anion is an insufficiently strong base to abstract a proton from ammonia. The alcohol also acts to prevent the accumulation of the strongly basic amide ion, which might bring about isomerization of the 1,4-dihydro compound to the conjugated 1,2-dihydro isomer (which would be further reduced to tetrahydrobenzene).

(7.45)

Particularly useful synthetically is the reduction of methoxy- or amino-substituted benzenes to dihydro compounds, which are readily hydrolyzed to cyclo-hexenones. Under mild acid conditions, the first-formed β,γ-unsaturated ketones are obtained, but these are readily isomerized to the conjugated α,β-unsaturated compounds (7.46). This is an excellent method for preparing substituted cyclo-hexenones.

---

[22] P. W. Rabideau and Z. Marcinow, *Org. Reactions*, **42** (1992), 1; L. N. Mander, in *Comprehensive Organic Synthesis*, ed. B. M. Trost and I. Fleming, vol. 8 (Oxford: Pergamon Press, 1991), p. 489.

$$(7.46)$$

　　The Birch reduction of benzenes containing electron-donating substituents takes place to give the 1,4-dihydro compounds in which the two new hydrogen atoms avoid the carbon atoms to which the electron-donating substituents are attached. This selectivity can be rationalized in terms of the relative electron densities of the carbon atoms in the intermediate radical anion. An electron-donating substituent destabilizes an adjacent negative charge and therefore the site of highest electron density (and therefore of protonation) is not α- to such substituents. It follows then, that reduction of benzene rings substituted by electron-withdrawing carbonyl groups gives rise to 1,4-dihydrobenzoic acid derivatives, in which the intermediate anion *is* α- to the anion-stabilizing substituent (7.47). These intermediate carbanions can be alkylated in the α-position to the carbonyl group. For example, 2-heptyl-2-cyclohexenone **68** was obtained from *o*-methoxybenzoic acid by way of the dianion **66** (7.48). Alkylation of **66** followed by acid hydrolysis of the enol ether led to the β,γ-unsaturated ketone **67**, which undergoes decarboxylation to give 2-heptyl-2-cyclohexenone.

$$(7.47)$$

$$(7.48)$$

　　Selective reduction of a benzene ring in the presence of another reducible group is possible if the other group is first protected in some way. Ketones, for example, may be converted to acetals or enol ethers to protect them from reduction. Conversely, reduction of benzene rings takes place only slowly in the absence of a proton donor, and selective reduction of an α,β-unsaturated carbonyl system can be effected.

Selective reduction of less-electron-rich aromatic rings occurs in bicyclic aromatic compounds (7.49).

$$\text{(7.49)}$$

98%

Stereoselective Birch reduction is possible and a number of examples have been reported, particularly for selective alkylation of the intermediate enolate anion.[23] For example, reduction of the chiral benzamide **69** with potassium in ammonia, followed by alkylation with ethyl iodide gave essentially a single diastereomer of the cyclohexadiene **70**, which was used in a synthesis of (+)-apovincamine (7.50).

$$\text{(7.50)}$$

69   100%   **70**   dr >100:1

Heterocyclic aromatic compounds can sometimes be reduced, particularly those which are electron-deficient. For example, reduction of pyridines gives 1,4-dihydropyridines (which are readily hydrolysed to 1,5-dicarbonyl compounds). Partial reduction of five-membered heteroaromatic compounds such as furans and pyrroles is also possible if these have electron-withdrawing substituents to stabilize the intermediate radical anion. For example, reduction of the furan **71** occurred with high selectivity to give the dihydrofuran **72**, used in a synthesis of (+)-nemorensic acid (7.51).[24]

$$\text{(7.51)}$$

71   87%   **72**   dr 30:1

[23] A. G. Schultz, *Chem. Commun.* (1999), 1263.
[24] T. J. Donohoe, J.-B. Guillermin and D. S. Walter, *J. Chem. Soc., Perkin Trans. 1* (2002), 1369.

7.2.4 Reduction with metal and ammonia: alkynes

Isolated carbon–carbon double bonds are not normally reduced by metal–ammonia reducing agents and therefore the partial reduction of alkynes is conveniently effected by these reagents. The procedure is highly selective and none of the saturated product is formed. Furthermore, the reduction is completely stereoselective and the only product from a disubstituted alkyne is the corresponding *E*-alkene (7.52). This method thus complements the formation of Z-alkenes by catalytic hydrogenation (see Section 2.6).

$$ \text{(7.52)} $$

Addition of an electron to the triple bond gives an intermediate radical anion which is protonated and then accepts a second electron to give a vinyl anion. The vinyl anion prefers to adopt the *E*-configuration and therefore leads to the *E*-alkene after protonation. An exception to this general rule is found in the reduction of medium-ring cyclic alkynes, where considerable amounts of the corresponding Z-alkene are often produced. Cyclodecyne, for example, gives a mixture (47:1) of *cis*- and *trans*-cyclodecenes. The strained *trans*-isomer is obviously formed with difficulty in such cyclic compounds and the *cis*-alkene is thought to arise by an alternative mechanism involving the corresponding allene, formed by isomerization of the alkyne by the accumulating sodamide. For other examples, see Section 2.6.

7.2.5 Reductive cleavage with metal and ammonia

Metal–amine reducing agents and other dissolving metal systems can bring about a variety of reductive cleavage reactions, some of which are useful in synthesis.

Most of these reactions proceed by direct addition of two electrons from the metal to the bond that is broken. The anions produced may be protonated by an acid in the reaction medium, or may survive until work-up (7.51).

$$ \text{A–B} \xrightarrow{2\,e} \text{A}^- \;+\; {\cdot}\text{B}^- \xrightarrow{2\,\text{H}^+} \text{AH} \;+\; \text{BH} \qquad \text{(7.53)} $$

Reductive cleavage is facilitated when the anions are stabilized by resonance or by an electronegative atom. As expected, therefore, bonds between heteroatoms or between a heteroatom and an unsaturated system which can stabilize a negative charge by resonance, are particularly easily cleaved. Thus allyl and benzyl ethers and esters (and sometimes even allyl or benzyl alcohols) are readily cleaved by metal–amine systems (or by catalytic hydrogenation). This type of reaction has been used widely for the removal of unsaturated protecting groups for hydroxyl, thiol and amino groups. For example, in a synthesis of the naturally occurring peptide glutathione **72**, the *N*-benzyloxycarbonyl and the *S*-benzyl groups in the protected compound **71** were removed by cleavage with sodium in liquid ammonia to give glutathione (7.54).

(7.54)

The cleavage of carbon–oxygen bonds from alkenyl or aryl phosphates can be accomplished under reductive conditions with a low valent metal. As vinyl phosphates can be formed readily from ketones, this procedure provides a method to convert a ketone to an alkene. For example, the alkenyl phosphate **74** was prepared by trapping the enolate formed on reduction of the enone **73** and was converted into the alkene **75** (7.55).[25] The chemistry therefore provides a method to prepare structurally specific alkenes. Low-valent titanium (prepared for example by reduction of titanium(III) chloride with potassium metal) is a convenient alternative to lithium or sodium in liquid ammonia or an amine for the reductive cleavage of alkenyl or aryl phosphates.

(7.55)

Sodium or lithium metal in liquid ammonia have found use for the reductive cleavage of carbon–sulfur bonds in sulfides, sulfoxides and sulfones. For example, phenylthio groups can be removed readily using these conditions (R−SPh to R−H). Reduction of sulfonamides to amines is also effective and the use of sodium in liquid ammonia is one of the best ways to cleave a tosyl group from an amine (R$_2$N−Ts to R$_2$N−H, Ts = SO$_2$C$_6$H$_4$Me). Reductive cleavage of three-membered rings such as epoxides or activated cyclopropanes can be achieved by using lithium and ammonia.

[25] L. C. Garver and E. E. van Tamelen, *J. Am. Chem. Soc.*, **104** (1982), 867.

7.3 Reduction by hydride-transfer reagents

Reactions that proceed by transfer of hydride ions are widespread in organic chemistry, and they are important also in biological systems. Reductions involving the reduced forms of coenzymes I and II, for example, are known to proceed by transfer of hydride ion from a 1,4-dihydropyridine system to the substrate. In the laboratory, the most useful reagents of this type in synthesis are aluminium isopropoxide and various metal hydride reducing agents.

7.3.1 Aluminium alkoxides

The reduction of carbonyl compounds to alcohols with aluminium isopropoxide has long been known under the name of the Meerwein–Pondorff–Verley reduction.[26] The reaction is easily effected by heating the components together in solution in isopropanol. An equilibrium is set up and the product is obtained by using an excess of the reagent or by distilling off the acetone as it is formed. The reaction is thought to proceed by transfer of hydride ion from the isopropoxide to the carbonyl compound through a six-membered cyclic transition state (7.56).

Aldehydes and ketones are reduced to primary and secondary alcohols respectively, often in high yield. The reaction owes its usefulness to the fact that carbon–carbon double bonds and many other unsaturated groups are unaffected, thus allowing selective reduction of carbonyl groups. For example, cinnamaldehyde is converted into cinnamyl alcohol, *o*-nitrobenzaldehyde gives *o*-nitrobenzyl alcohol and phenacyl bromide gives the alcohol **76** (7.57).

Reductions of a similar type can be brought about by other metallic alkoxides, but aluminium alkoxide is particularly effective because it is soluble in both alcohols and hydrocarbons and, being a weak base, it shows little tendency to bring about wasteful condensation reactions of the carbonyl compounds. Reduction of

[26] R. M. Kellogg, in *Comprehensive Organic Synthesis*, ed. B. M. Trost and I. Fleming, vol. 8 (Oxford: Pergamon Press, 1991), p. 79; C. F. de Graauw, J. A. Peters, H. van Bekkum and J. Huskens, *Synthesis* (1994), 1007.

aldehydes by a similar mechanism can occur by using lithium diisopropylamide (LDA). Therefore it is normally advisable to avoid the use of LDA for attempted enolization of aldehydes.

7.3.2 Lithium aluminium hydride and sodium borohydride

A number of metal hydrides have been employed as reducing agents in organic chemistry, but the most commonly used are lithium aluminium hydride and sodium borohydride, both of which are commercially available. These reagents are nucleophilic and as such they normally attack polarized multiple bonds such as C=O or C≡N by transfer of hydride ion to the more-positive atom. They do not usually reduce isolated carbon–carbon double or triple bonds.

With both reagents all four hydrogen atoms may be used for reduction, being transferred in a stepwise manner (7.58). For reductions with lithium aluminium hydride, each successive transfer of hydride ion takes place more slowly than the one before, and this has been exploited for the preparation of modified reagents that are less reactive and more selective than lithium aluminium hydride itself (for example, by replacement of two or three of the hydrogen atoms of the anion by alkoxy groups).

(7.58)

Lithium aluminium hydride is a more powerful reducing agent than sodium borohydride and reduces most of the commonly encountered organic functional groups (see Table 7.3). It reacts readily with water and other compounds that contain active hydrogen atoms and must be used under anhydrous conditions in a non-hydroxylic solvent; diethyl ether and THF are commonly employed. Lithium aluminium hydride has found widespread use for the reduction of carbonyl compounds. Aldehydes, ketones, esters, carboxylic acids and lactones can all be reduced smoothly to the corresponding alcohols under mild conditions. Carboxylic amides are converted into amines or aldehydes, depending on the conditions and on the

Table 7.3. *Common functional groups reduced by lithium aluminium hydride*

Functional group	Reduction product
RCHO	RCH$_2$OH
R$_2$C=O	RCH(OH)R
RCO$_2$R′	RCH$_2$OH + R′OH
RCO$_2$H	RCH$_2$OH
RCONHR′	RCH$_2$NHR′
RCONR′$_2$	RCH$_2$NR′$_2$ or RCH(OH)NR′$_2$ ($\rightarrow$ RCHO + R′$_2$NH)
RC≡N	RCH$_2$NH$_2$ or RCH=NH ($\rightarrow$ RCHO)
RCH=NOH	RCH$_2$NH$_2$
RNO$_2$	RNH$_2$
ArNO$_2$	ArNHNHAr or ArN=NAr
RCH$_2$Br	RCH$_3$
RCH$_2$OSO$_2$Ar	RCH$_3$

type of *N*–substitution. Some examples of the use of lithium aluminium hydride are given in Scheme 7.59.

(7.59)

An exception to the general rule that carbon–carbon double bonds are not attacked by hydride reducing agents is found in the reduction of β-aryl-α,β-unsaturated carbonyl compounds with lithium aluminium hydride, where the carbon–carbon double bond is often reduced as well as the carbonyl group (7.60). Even in these cases, however, selective reduction of the carbonyl group can generally be achieved by working at low temperatures or by using sodium borohydride, alane (AlH$_3$, formed from lithium aluminium hydride and aluminium chloride) or most commonly DIBAL-H (diisobutylaluminium hydride). This type of reduction of the double bond of allylic alcohols is thought to proceed through a cyclic organoaluminium intermediate **77**, for it is found experimentally that only one of the two hydrogen atoms added to the double bond is derived from the hydride, and acidification with a deuterated solvent leads to the deuterated alcohol **78** (7.61). A similar type of aluminium compound is thought to be involved in the reduction of the triple bond of propargylic alcohols (R−C≡C−CH$_2$OH) with lithium aluminium hydride to give *trans* alkenes (see Section 2.6).

(7.60)

(7.61)

77   **78**

Sodium borohydride is less reactive than lithium aluminium hydride and is therefore more discriminating (chemoselective) in its action. It reacts only slowly with water and most alcohols at room temperature and reductions with this reagent are often effected in ethanol solution. At room temperature in ethanol it readily reduces aldehydes and ketones but it does not generally attack esters or amides and it is normally possible to reduce aldehydes and ketones selectively with sodium borohydride in the presence of a variety of other functional groups. For example, ethyl acetoacetate, which contains both an ester and a ketone functional group, on reduction with sodium borohydride gives ethyl 3-hydroxybutanoate, the product from selective reduction of only the keto group (7.62). In contrast, the more reactive lithium aluminium hydride gives 1,3-butanediol, by reduction of both carbonyl groups. To effect selective reduction of the ester, the keto group must be protected as its acetal, and the ester reduced with lithium aluminium hydride. Mild acid hydrolysis then regenerates the ketone to give the keto-alcohol.

$$(7.62)$$

The reducing properties of sodium borohydride are substantially modified in the presence of metal salts, and particularly useful in this respect are lanthanide salts. In the presence of cerium(III) chloride, for example, sodium borohydride reduces α,β-unsaturated ketones with extremely high selectivity, such that 1,2- and almost no 1,4-reduction occurs to give allylic alcohols.[27] This reagent system has therefore found some use for the formation of allylic alcohols from α,β-unsaturated ketones that otherwise lead to reduction of the carbon–carbon double bond as well (7.63). In comparison, sodium borohydride in alcoholic solvent effects conjugate reduction of α,β-unsaturated esters or lactones. Remarkably, sodium borohydride–$CeCl_3$ can discriminate between different ketone and aldehyde groups, effecting the selective reduction of the *less*-reactive carbonyl group. For example, α,β-unsaturated ketones are reduced selectively in the presence of saturated ketones or aldehydes. Ketones can be sometimes be reduced in the presence of an aldehyde (7.64). It is believed that, under the reaction conditions, the more-reactive aldehyde group is protected as the hydrate, which is stabilized by complexation with the cerium ion, and is regenerated during isolation of the product.[27]

no CeCl$_3$	59 : 41	
CeCl$_3$•7H$_2$O	99 : 1	

$$(7.63)$$

[27] A. L. Gemal and J.-L. Luche, *J. Org. Chem.*, **44** (1979), 4187; *J. Am. Chem. Soc.*, **103** (1981), 5454.

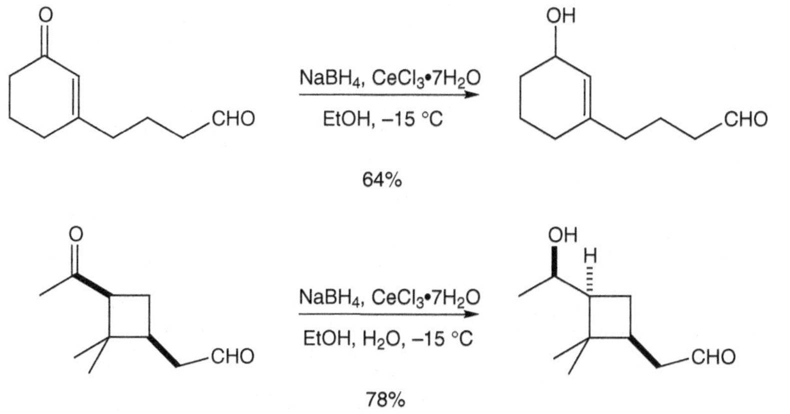

64%

(7.64)

78%

Selective reduction of aldehydes in the presence of ketones can be effected with tetra-*n*-butylammonium triacetoxyborohydride and other reagents. Although lithium aluminium hydride is used most commonly for the reduction of carboxylic esters, sodium borohydride can provide some useful selectivity and its reactivity is enhanced in the presence of metal salts. For example, reduction of carboxylic esters in the presence of carboxylic amides is possible using sodium borohydride and calcium chloride.

The use of solid-supported reagents is gaining in popularity, mostly because of their ease of use and the ability to purify the product by simple filtration rather than column chromatography. For example, borohydride supported on amberlyst was used for the reduction of the ketone **79** to give the alcohol **80**, which was further transformed into the alkaloid epibatidine (7.65).

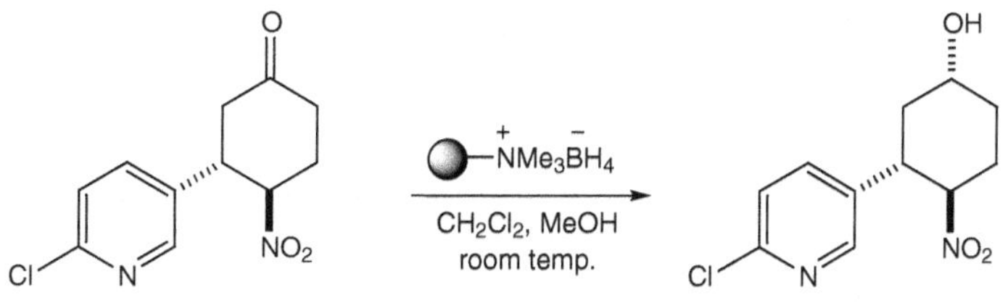

(7.65)

Ordinarily, reduction of the carbonyl group of an unsymmetrical ketone such as ethyl methyl ketone leads to the racemic alcohol. With ketones that contain an asymmetric centre, however, the two diastereomers of the alcohol may not be produced in equal amount. For example, in the reduction of the ketone **81** with lithium aluminium hydride, the *anti* stereoisomer of the alcohol predominates in the product (7.66).

$$ (7.66) $$

81   72 : 28

The main product formed in these reactions can be predicted on the basis of the Felkin–Anh model (see also Cram's rule, see Section 1.1.5). The diastereomer which predominates is that formed by approach of the reagent to the less-hindered side of the carbonyl group when the rotational conformation of the molecule is such that the largest group on the adjacent chiral centre is perpendicular to the carbonyl group. This is best depicted using Newman projections, where S, M and L represent small, medium and large substituents (7.67). Thus, for the reduction of the ketone **81**, the predominant *anti* alcohol arises by attack of the metal hydride on the less hindered side of the carbonyl group in the conformation shown. The selectivity obtained in these reactions increases with the bulk of the reducing agent and some highly stereoselective reductions have been achieved by using complex hydride agents.

$$ (7.67) $$

In cases where there is a polar group on the carbon atom adjacent to the carbonyl group, the Felkin–Anh model may not be followed, because the conformation of the carbonyl compound in the transition state is no longer determined solely by steric factors. In α-hydroxy and α-amino ketones, for example, reaction is thought to proceed through a relatively rigid chelate compound of type **82** (7.68). Reduction of such compounds usually proceeds with a comparatively high degree of stereoselectivity by attack on the chelate from the less-hindered side, but not necessarily in the Cram sense. Thus, reduction of **82** with lithium aluminium hydride leads to the diol **83** (*anti:syn* 80:20). Even higher selectivity can be achieved by using zinc borohydride $Zn(BH_4)_2$ (**83**, *anti:syn* 90:10). When the adjacent carbon atom carries a halogen substituent, the most reactive conformation of the molecule appears to be the one in which the polar halogen atom and the polar carbonyl are *anti* disposed, to minimize dipole–dipole repulsion. The predominant product is then formed by approach of the metal hydride anion to the less-hindered side of this conformation.

$$\text{(7.68)}$$

82   **83** 80 : 20

With ketones that are locked due to a bridged ring system, then it is normally straightforward to predict which is the less-hindered side of the carbonyl group. Reduction of camphor with lithium aluminium hydride, for example, leads mainly to the *exo* alcohol (isoborneol), whereas norcamphor (which lacks the *gem*-dimethyl group on the bridging carbon atom), in which approach of the hydride anion is now easier from the side of the methylene bridge, leads mainly to the *endo* alcohol.

With other less-rigid cyclohexanones, the stereochemical course of the reduction is less easy to predict. In general, a mixture of products is obtained in which, with comparatively unhindered ketones, the more stable equatorial alcohol predominates; with hindered ketones, the axial alcohol is often the main product. Thus, reduction of 4-tert-butylcyclohexanone **84** with lithium aluminium hydride gives predominantly the equatorial trans-4-*tert*-butylcyclohexanol, whereas the hindered 3,3,5-trimethylcyclohexanone **85** gives a mixture containing mainly the axial alcohol **86** (7.69, 7.70). The latter is almost the only product when a more hindered and hence more selective reducing agent such as L-selectride [LiBH(^{s}Bu)$_3$] or lithium hydrido-tri-tert-butoxyaluminate [LiAlH(O^{t}Bu)$_3$] is employed.

$$\text{(7.69)}$$

84   90 : 10

$$\text{(7.70)}$$

85   LiAlH$_4$ 45 : 55 **86**
   or LiBH(^{s}Bu)$_3$ <1 : >99

There have been many attempts to rationalize these stereoselectivities.[28] It is supposed that in general there is a tendency for approach of the reagent to the carbonyl group in an axial direction, leading to the equatorial alcohol. In hindered ketones, however, axial approach may be hampered by steric factors, thereby favouring equatorial approach of the reagent and formation of the axial alcohol.

Alternative reagents can provide high selectivity. Dissolving metal reductions were described in Section 7.2 and the use of lithium in liquid ammonia and an

[28] D. C. Wigfield, *Tetrahedron*, **35** (1979), 449.

alcohol solvent provides the thermodynamically more favourable equatorial alcohol almost exclusively from either ketone **84** or **85**. To obtain the axial alcohol, it is often best to use a highly hindered reducing agent (such as L-selectride, which gives 96% of the axial alcohol product from 4-tert-butylcyclohexanone **84**) or to use catalytic hydrogenation (see Section 7.1) over nickel or rhodium metal. Thus, reduction of 4-tert-butylcyclohexanone **84** with hydrogen and rhodium absorbed on carbon in ethanol solvent gives predominantly (93:7) the axial alcohol product.

Lithium aluminium hydride is a powerful reducing agent and can effect the reduction of a variety of functional groups (see Table 7.3). Treatment of carboxylic amides, nitriles, imines or nitro compounds with lithium aluminium hydride provides amine products. Reductive opening of epoxides proceeds with S_N2 substitution by hydride ion to form a new C−H bond with overall inversion of configuration at the carbon atom attacked. Epoxides can therefore be reduced to alcohols. With unsymmetrical epoxides, reaction takes place at the less-substituted carbon atom to give the more substituted alcohol product. Therefore, the epoxide **87**, for example, gives 1-methylcyclohexanol **88** (7.71). In the presence of a Lewis acid, however, the direction of ring-opening may be reversed. Thus, reduction of the epoxide **87** with sodium cyanoborohydride in the presence of boron trifluoride etherate gives, not 1-methylcyclohexanol **88**, but *cis*-2-methylcyclohexanol **89** by backside attack of hydride on the epoxide–Lewis acid complex. The direction of ring-opening is now dictated by the formation of the more-stable carbocation intermediate.

$$\text{(7.71)}$$

| **88** | | **87** | | **89** |

Primary and secondary alkyl halides are reductively cleaved to the corresponding hydrocarbons with lithium aluminium hydride or lithium triethylborohydride. Tertiary halides react only slowly and give mostly alkenes. Aryl iodides and bromides may also be reduced to the hydrocarbons with lithium aluminium hydride under more vigorous conditions. Sulfonate esters of primary and secondary alcohols are also readily reduced with lithium aluminium hydride and this reaction has been employed in synthesis for the replacement of a hydroxyl group by a hydrogen atom. For example, selective tosylation of the primary alcohol group of the diol **90** to give **91**, followed by treatment with lithium aluminium hydride gave the reduced product **92** (7.72).

$$(7.72)$$

7.3.3 Derivatives of lithium aluminium hydride and sodium borohydride

The high reactivity of lithium aluminium hydride can result in unwanted over-reduction. More-selective reagents can be obtained by modification of lithium aluminium hydride by treatment with alcohols or with aluminium chloride. One such reagent is the sterically bulky lithium tri-t-butoxyaluminium hydride (lithium hydridotri-t-butoxyaluminate), which is readily prepared by action of three equivalents of tert-butyl alcohol on lithium aluminium hydride. Analogous reagents are obtained in the same way from other alcohols and by replacement of only one or two of the hydrogen atoms of the hydride by alkoxy groups, affording a range of reagents of graded activities.[29]

Lithium tri-t-butoxyaluminium hydride is a much milder reducing agent than lithium aluminium hydride. Thus, although aldehydes and ketones are reduced normally to alcohols, carboxylic esters and epoxides react only slowly, and halides, nitriles and nitro groups are not attacked. Aldehydes and ketones can therefore be selectively reduced in the presence of these groups. For example, the aldehyde **93** is reduced selectively to the alcohol **94** (7.73).

$$(7.73)$$

One of the most useful applications of the alkoxy reagents is in the preparation of aldehydes from carboxylic acids by partial reduction of the acid chlorides or dialkylamides. Acid chlorides are readily reduced with lithium aluminium hydride or with sodium borohydride to the corresponding alcohols, but with one equivalent of lithium tri-t-butoxyaluminium hydride, high yields of the aldehyde can be obtained, even in the presence of other functional groups (7.74).

---

[29] H. C. Brown and S. Krishnamurthy, *Tetrahedron*, **35** (1979), 567; J. Málek, *Org. Reactions*, **34** (1985), 1; J. Málek, *Org. Reactions*, **36** (1988), 249.

$$\underset{\text{NC}}{\text{COCl}} \quad \xrightarrow[-78\,°C]{\text{LiAlH(O}^t\text{Bu)}_3} \quad \underset{\text{NC}}{\text{CHO}} \tag{7.74}$$

80%

Reduction of tertiary amides with excess of lithium aluminium hydride gives the corresponding amines in good yield. With the less-active lithium triethoxyaluminium hydride LiAlH(OEt)_3 (the tri-t-butoxy compound is ineffective in this case) reaction stops at the *N,O*-acetal stage and hydrolysis gives the corresponding aldehyde (7.75). Similarly, reduction of a nitrile with lithium aluminium hydride gives a primary amine by way of the imine salt. However, with lithium triethoxyaluminium hydride, reaction stops at the imine stage and hydrolysis gives the aldehyde. The reagent diisobutylaluminium hydride (DIBAL-H) is also effective for reduction of nitriles to aldehydes.

$$\text{CONMe}_2 \quad \xrightarrow[\text{ii, H}_3\text{O}^+]{\text{i, LiAlH(OEt)}_3} \quad \text{CHO} \tag{7.75}$$

85%

7.3.4 Mixed lithium aluminium hydride–aluminium chloride reagents

A useful modification of the properties of lithium aluminium hydride is achieved by addition of aluminium chloride in various proportions. This serves to release mixed chloride–hydrides of aluminium (7.76). The most popular of these is aluminium hydride AlH_3, sometimes referred to as alane.

$$3\ \text{LiAlH}_4 \quad + \quad \text{AlCl}_3 \quad \longrightarrow \quad 3\ \text{LiCl} \quad + \quad 4\ \text{AlH}_3$$

$$\text{LiAlH}_4 \quad + \quad \text{AlCl}_3 \quad \longrightarrow \quad \text{LiCl} \quad + \quad 2\ \text{AlH}_2\text{Cl} \tag{7.76}$$

$$\text{LiAlH}_4 \quad + \quad 3\ \text{AlCl}_3 \quad \longrightarrow \quad \text{LiCl} \quad + \quad 4\ \text{AlHCl}_2$$

The general effect of the addition of aluminium chloride is to lower the reducing power of lithium aluminium hydride and in consequence to produce reagents that are more specific for particular reactions. For example, the carbon–halogen bond is often inert to the mixed hydride reagents. Advantage is taken of this in the reduction of polyfunctional compounds in which retention of the halogen is desired, as in the conversion of ethyl 3-bromobutanoate into 3-bromobutanol (7.77); lithium aluminium hydride alone produces butanol.

$$\underset{\text{Br}}{\text{CO}_2\text{Et}} \quad \xrightarrow[\text{Et}_2\text{O}]{\text{LiAlH}_4-\text{AlCl}_3\ (1:1)} \quad \underset{\text{Br}}{\text{OH}} \tag{7.77}$$

93%

Similarly, nitro groups are not so easily reduced as with lithium aluminium hydride itself, and 4-nitrobenzaldehyde can be converted into 4-nitrobenzyl alcohol in 75% yield. Aldehydes and ketones are reduced to alcohols and there is no advantage in the use of mixed hydrides in these cases (although the stereoselectivity may vary if reduction leads to diastereomeric alcohols). With diaryl ketones and with aryl alkyl ketones, however, the carbonyl group is reduced to the methylene group, and this procedure offers a useful alternative to the Clemmensen or other methods for reduction of this type of ketone.

Reduction with lithium aluminium hydride–aluminium chloride (3:1) provides a good route from α,β-unsaturated carbonyl compounds to unsaturated alcohols (or amines), which are difficult to prepare with lithium aluminium hydride alone because of competing reduction of the carbon–carbon double bond. For example, α,β-unsaturated esters are reduced to allylic alcohols, although diisobutylaluminium hydride (DIBAL-H) is normally the reagent of choice for this transformation. Reduction of carboxylic amides can sometimes be preferable using AlH_3 (7.78).

$$\text{Ph} \diagup \diagdown \text{NMe}_2 \quad \xrightarrow[\text{Et}_2\text{O}]{\text{LiAlH}_4\text{–AlCl}_3 \ (3:1)} \quad \text{Ph} \diagup \diagdown \text{NMe}_2 \qquad (7.78)$$

94%
(0% using $LiAlH_4$)

7.3.5 Diisobutylaluminium hydride (DIBAL-H)

Diisobutylaluminium hydride (DIBAL-H or DIBAL, $^i\text{Bu}_2\text{AlH}$) is a very useful derivative of aluminium hydride and is available commercially as a solution in a variety of solvents. At ordinary temperatures, esters and ketones are reduced to alcohols, nitriles give amines and epoxides are cleaved to alcohols. However, it is particularly useful for the preparation of aldehydes.[30] At low temperatures, esters and lactones are reduced directly to aldehydes (or lactols); nitriles and carboxylic amides give imines which are readily converted into the aldehydes by hydrolysis (7.79–7.81). The lack of further reduction of the aldehyde lies in the relative stability of the intermediate hemiacetal (or imine salt), which hydrolyses to the aldehyde only upon work-up.

$$\xrightarrow[\text{PhMe, }-78\ °\text{C}]{\text{DIBAL-H}} \qquad (7.79)$$

94%

---

[30] N. M. Yoon and Y. S. Gyoung, *J. Org. Chem.*, **50** (1985), 2443; F. Winterfeldt, *Synthesis* (1975), 617.

$$(7.80)$$

85%

$$(7.81)$$

96%

Diisobutylaluminium hydride has found considerable use for the selective 1,2-reduction of α,β-unsaturated carbonyl compounds to allylic alcohols. For example, the ester **95** was reduced to the allylic alcohol **96** (7.82). The reagent has also found use for the reduction of alkynes to *cis*-alkenes (see Section 2.6).

$$(7.82)$$

95   67%   **96**

Boc = $CO_2{}^tBu$

7.3.6 Sodium cyanoborohydride and sodium triacetoxyborohydride

A number of reagents derived from sodium borohydride by replacement of one or more of the hydrogen atoms by other groups allow more-selective reduction than with sodium borohydride itself. Among the most useful are sodium cyanoborohydride and sodium triacetoxyborohydride.

Sodium cyanoborohydride is a weaker and more-selective reducing agent than sodium borohydride because of the electron-withdrawing effect of the cyano group. It has the further advantage that it is stable in acid to pH $= 3$ and can be employed to effect reductions in the presence of functional groups that are sensitive to the more-basic conditions of reduction with sodium borohydride.

Aldehydes and ketones are unaffected by sodium cyanoborohydride in neutral solution, but they are readily reduced to the corresponding alcohol at pH $= 3$–4 by way of the protonated carbonyl group. By previous exchange of the hydrogens of the borohydride for deuterium or tritium, by reaction with D_2O or tritiated water, an efficient and economical route is available for deuteride or tritiide reduction of aldehydes and ketones.

Iminium groups are more easily reduced than carbonyl groups in acid solution, and this has been exploited in a method for reductive amination of aldehydes and ketones by way of the iminium salts formed from the carbonyl compounds and a primary or secondary amine, typically at pH > 5 (7.83);[31] at this pH the carbonyl compounds themselves are unaffected. Some examples of reductive amination using sodium cyanoborohydride or sodium triacetoxyborohydride are given in Schemes 7.84–7.86. Improvement in the yield of the amine product can sometimes be obtained by addition of titanium(IV) salts such as $Ti(O^{i}\text{-}Pr)_4$.[32]

$$RCHO \ + \ HNR'_2 \ \rightleftharpoons \ R\overset{+}{\diagup}NR'_2 \ \xrightarrow{NaBH_3CN} \ R\diagdown NR'_2 \qquad (7.83)$$

(7.84)

83%

(7.85)

93%

98%

Reductive amination with formaldehyde and sodium cyanoborohydride provides a convenient method for methylation of a secondary amine (or dimethylation of a primary amine). An alternative procedure uses formaldehyde together with formic acid (HCOOH) as the source of hydride in what is termed the Eschweiler–Clark reaction. Deprotonation of formic acid provides the formate anion, which delivers hydride to the iminium ion with concomitant formation of carbon dioxide.

Reductive amination of an aldehyde or ketone with ammonia or a primary amine can sometimes be problematic, as the product amine can undergo further reductive amination with the starting carbonyl compound. One method that promotes selective formation of primary amines uses the rhodium(III) complex [RhCp*Cl₂]₂ as a catalyst and ammonium formate (HCOONH₄), which acts as the source of the ammonia and the hydride (7.87).[33]

[31] E. W. Baxter and A. B. Reitz, *Org. Reactions*, **59** (2002), 1; C. F. Lane, *Synthesis* (1975), 135; R. O. Hutchins and M. K. Hutchins, in *Comprehensive Organic Synthesis*, ed. B. M. Trost and I. Fleming, vol. 8 (Oxford: Pergamon Press, 1991), p. 25; A. F. Abdel-Magid, K. G. Carson, B. D. Harris, C. A. Maryanoff and R. D. Shah, *J. Org. Chem.*, **61** (1996), 3849.
[32] R. J. Mattson, K. M. Pham, D. J. Leuck and K. A. Cowen, *J. Org. Chem.*, **55** (1990), 2552.
[33] M. Kitamura, D. Lee, S. Hayashi, S. Tanaka and M. Yoshimura, *J. Org. Chem.*, **67** (2002), 8685.

$$(7.87)$$

95%

(Cp* = pentamethylcyclopentadienyl anion)

Sodium triacetoxyborohydride can reduce aldehydes selectively in the presence of ketones. However, α- and β-hydroxy-ketones are reduced with this reagent. The reduction occurs stereoselectively to give predominantly the *anti* diol product. The reagent tetramethylammonium triacetoxyborohydride $Me_4NBH(OAc)_3$ has shown excellent selectivity for this transformation (7.88).[34] Exchange of one of the acetoxy groups for the alcohol is thought to preceed stereoselective intramolecular transfer of hydride.

$$(7.88)$$

92%   *anti:syn* 95:5

A method for the conversion of carbonyl compounds into the corresponding hydrocarbons involves reduction of the derived toluene-*p*-sulfonyl (tosyl) hydrazones with sodium cyanoborohydride in acidic dimethylformamide (DMF). The reaction is specific for aliphatic carbonyl compounds; aromatic compounds are normally unaffected. The tosyl hydrazone need not be isolated but can be prepared and reduced *in situ*. For example, the ketone **97** was reduced to the alkane **98** (7.89).

$$(7.89)$$

55%

97   **98**

With α,β-unsaturated carbonyl compounds, reduction of the tosyl hydrazone is accompanied by migration of the double bond. Thus, cinnamaldehyde tosylhydrazone gives 3-phenyl-1-propene in 98% yield and the α,β-unsaturated ketone **99** gives the alkene **100** (7.90). The mechanism for this reaction involves reduction of the iminium ion to the tosylhydrazine **101**, elimination of *p*-toluenesulfinic acid

---

[34] D. A. Evans, K. T. Chapman and E. M. Carreira, *J. Am. Chem. Soc.*, **110** (1988), 3560.

and subsequent [1,5]-sigmatropic shift of hydrogen, with loss of nitrogen, to the rearranged alkene.

(7.90)

7.3.7 Borane and derivatives

Borane, BH_3 (which exists as the gaseous dimer diborane B_2H_6) is a powerful reducing agent and attacks a variety of unsaturated groups. It can be prepared by reaction of boron trifluoride etherate with sodium borohydride, but for most purposes the commercially available solutions as complexes with tetrahydrofuran or dimethylsulfide are suitable. The latter, $BH_3 \cdot SMe_2$, has the advantage that it is stable and is soluble in several organic solvents.[35] The use of complexes of borane with amines is becoming increasingly popular and these have lower sensitivity to air and moisture.

Reaction of borane with unsaturated groups takes place readily at room temperature and the products are isolated in high yield after hydrolysis of the intermediate boron compound. The reduction of some common functional groups are given in Table 7.4.[36] Borane reacts rapidly with water, and reactions must be effected under anhydrous conditions and preferably under inert atmosphere since borane may ignite in air.

[35] H. C. Brown, Y. M. Choi and S. Narasimhan, *J. Org. Chem.*, **47** (1982), 3153.
[36] H. C. Brown and S. Krishnamurthy, *Tetrahedron*, **35** (1979), 567; C. F. Lane, *Chem. Rev.*, **76** (1976), 773.

Table 7.4. *Functional groups reduced by borane*

Functional group	Reduction product
RCO_2H	RCH_2OH
$RCH=CHR$	
$RCHO$	RCH_2OH
$R_2C=O$	$RCH(OH)R$
$RC{\equiv}N$	RCH_2NH_2
$RCONR'_2$	$RCH_2NR'_2$
RCO_2COR	RCH_2OH
RCO_2R'	$RCH_2OH + R'OH$ (slow rate of reaction)
$RCOCl$	no reaction
RNO_2	no reaction

Reductions with borane do not simply parallel those with sodium borohydride. This is because sodium borohydride is a nucleophile and reacts by addition of hydride ion to the more-positive end of a polarized multiple bond, whereas borane is a Lewis acid and attacks electron-rich centres. For example, whereas sodium borohydride very rapidly reduces acid chlorides to primary alcohols, the reaction being facilitated by the electron-withdrawing effect of the halogen, borane does not react with acid chlorides under the usual mild conditions. Reduction of carbonyl groups by borane takes place by addition of the electron-deficient borane to the oxygen atom, followed by the irreversible transfer of hydride ion from boron to carbon (7.91). The inertness of acid chlorides can be ascribed to the decreased basic properties of the carbonyl oxygen atom resulting from the electron-withdrawing effect of the halogen atom. For a similar reason esters are reduced only slowly by borane.

$$(7.91)$$

A useful reaction of borane is the reduction of carboxylic acids to alcohols, which occurs very readily and can be achieved selectively in the presence of other functional group, including esters.[37] For example, 4-nitrobenzoic acid is reduced

[37] N. M. Yoon, C. S. Pak, H. C. Brown, S. Krishnamurthy and T. P. Stocky, *J. Org. Chem.*, **38** (1973), 2786.

to 4-nitrobenzyl alcohol in 79% yield, and the carboxylic acid **103** is reduced to the alcohol **104** in the presence of the less reactive ester group (7.92). Reduction of carboxylic acids is believed to proceed by way of a triacyloxyborane, the carbonyl groups of which are reduced rapidly by further reaction with borane.

$$\text{103} \xrightarrow[\text{THF, }-18\ ^\circ\text{C}]{\text{BH}_3\cdot\text{THF}} \text{104} \qquad (7.92)$$

103   88%   **104**

Carboxylic amides and lactams are good functional groups for reduction by borane and this provides a method for the formation of amines. Borane, being electron deficient, reacts fastest with the most-nucleophilic carbonyl group. It is therefore possible to carry out selective reduction of carboxylic amides in the presence of other unsaturated groups such as esters. For example, the amide **105** was reduced to the amine **106** (7.93). In comparison, the reducing agent lithium aluminium hydride reduces both the carboxylic amide and the ester groups of the compound **105**.

$$\text{105} \xrightarrow[\text{THF, room temp.}]{\text{BH}_3\cdot\text{THF}} \text{106} \qquad (7.93)$$

105   67%   **106**

Reduction of aldehydes and ketones is possible with borane. Stereoselective reduction of ketones is possible when a heteroatom is located α- or β- to the carbonyl group. Treatment of a β-hydroxy-ketone with catecholborane results in the selective formation of the *syn* 1,3-diol product (7.94).[38] The borane reacts preferentially with the alcohol to release hydrogen gas and to form the boronic ester **107**. A second equivalent of the borane then effects the reduction to give the *syn* diastereomer. For the preparation of the *anti* diastereomer, triacetoxyborohydride can be used (see Scheme 7.86).

In another example, stereoselective reduction of α-phenylsulfonyl ketones was accomplished to give predominantly the *anti* diastereomer by using borane (as its complex with pyridine) together with titanium tetrachloride, but the *syn* diastereomer by using lithium triethylborohydride (7.95).[39] It is thought that chelation of the carbonyl and sulfonyl oxygen atoms by the titanium(IV) in a non-polar solvent promotes subsequent reduction by borane to give the *anti* product, whereas the non-chelating conditions (Felkin–Anh model) with cerium(III) gives the *syn*

[38] D. A. Evans and A. H. Hoveyda, *J. Org. Chem.*, **55** (1990), 5190.
[39] E. Marcantoni, S. Cingolani, G. Bartoli, M. Bosco and L. Sambri, *J. Org. Chem.*, **63** (1998), 3624.

Reduction

product. Other diastereoselective reductions, such as of α-alkyl-β-keto-esters to give *syn* or *anti* α-alkyl-β-hydroxy esters, have been reported.[40]

(7.94)

82%   107   *syn:anti* 90:10

(7.95)

BH$_3$•pyridine, TiCl$_4$
CH$_2$Cl$_2$, –78 °C   85%   87 : 13

LiBHEt$_3$, CeCl$_3$
THF, –78 °C   85%   <1 : >99

The reduction of epoxides with borane is noteworthy as it gives rise to the *less* substituted alcohol as the major product (7.96), in contrast to reduction with complex hydrides (compare with Scheme 7.71). The reaction is catalysed by small amounts of sodium or lithium borohydride and high yields of the alcohol are obtained. With 1-alkylcycloalkene epoxides, the 2-alkylcycloalkanols produced are entirely *cis*, and this reaction thus complements the hydroboration–oxidation of cycloalkenes described in Section 5.1, which leads to *trans* products. Reaction with borane in the presence of boron trifluoride has also been used for the reduction of epoxides and for the conversion of lactones and some esters into ethers.

(7.96)

82%

Reduction of unsymmetrical ketones generates a new chiral centre. Chiral alcohols are present in a vast number of natural products and biologically active compounds and there are many reports of the enantioselective reduction of ketones.[41] High selectivities have been achieved by using hydride reducing agents such as lithium aluminium hydride in the presence of chiral ligands such as BINOL (2,2′-binaphthol), or chiral organoborane reagents.[42] For example, reduction of the ketone

---

[40] E. Marcantoni, S. Alessandrini, M. Malavolta, G. Bartoli, M. C. Bellucci, L. Sambri and R. Dalpozzo, *J. Org. Chem.*, **64** (1999), 1986.

[41] S. Itsuno, *Org. Reactions*, **52** (1998), 395.

[42] P. Daverio and M. Zanda, *Tetrahedron: Asymmetry*, **12** (2001), 2225; M. Nishizawa and R. Noyori, in *Comprehensive Organic Synthesis*, ed. B. M. Trost and I. Fleming, vol. 8 (Oxford: Pergamon Press, 1991), p. 159; M. M. Midland, *Chem. Rev.*, **89** (1989), 1553.

108 with diisopinocampheylchloroborane, (Ipc)$_2$BCl,[43] gave the chiral alcohol **109** with high enantioselectivity (7.97). The reaction proceeds by co-ordination of the ketone to the borane to give the complex **110**, with the larger group R_L (such as aryl) on the less-hindered side, followed by transfer of hydride predominantly to one face of the carbonyl group.

$$(7.97)$$

A more-efficient method for asymmetric reduction uses the chiral inducing agent as a catalyst. A suitable system involves the oxazaborolidine **111**, which can be prepared from the corresponding amino-alcohol and methylboronic acid, or is available commercially in either enantiomeric form (7.98).[44] Addition of borane to this reagent provides the active reducing agent **112**. The oxazaborolidine **111** enhances the rate of reduction of ketones with borane (or catecholborane) and can be used as a catalyst in substoichiometric amounts. Very high levels of enantioselectivity in the reduction are obtained with a wide selection of ketones (7.98 and 7.99).

$$(7.98)$$

[43] H. C. Brown, J. Chandrasekharan and P. V. Ramachandran, *J. Am. Chem. Soc.*, **110** (1988), 1539; R. K. Dhar, *Aldrichim. Acta*, **27** (1994), 43; M. Zhao, A. O. King, R. D. Larsen, T. R. Verhoeven and P. J. Reider, *Tetrahedron Lett.*, **38** (1997), 2641.
[44] E. J. Corey and C. J. Helal, *Angew. Chem. Int. Ed.*, **37** (1998), 1986; B. T. Cho, *Aldrichim. Acta*, **35** (2002), 3.

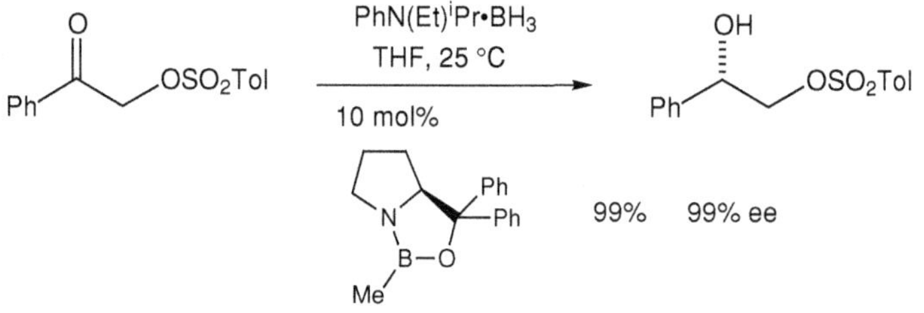

$$\text{(7.99)}$$

111

The mechanism for the reduction involves the oxazaborolidine **111** acting as a Lewis acid for the ketone, with co-ordination of the carbonyl oxygen atom to the boron atom on the less-hindered *exo* face of the bicyclic molecule. This activates the ketone for reduction. The borane is activated by binding to the nitrogen atom, which also occurs on the *exo* face and so is in close proximity to the carbonyl group (7.100). The ketone adopts the least-congested conformation and transfer of hydride then takes place to the carbonyl group. Dissociation of the product **113** to give the boronate $R_2CH-OBH_2$ (which is hydrolysed to the alcohol on work-up) releases the catalyst **111** for further reaction.

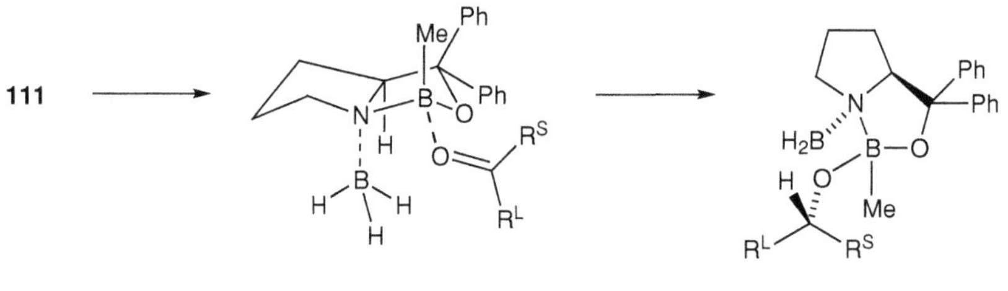

$$111 \longrightarrow \qquad \longrightarrow \qquad \text{(7.100)}$$

113

7.4 Other methods of reduction

7.4.1 Enzyme catalysed

A mild and inexpensive way to reduce aldehydes or ketones uses fermenting Baker's yeast.[45] This is a whole-cell system that contains oxidoreductase enzymes and co-factors that reduce the substrate. The ketonic carbonyl groups of β-keto-esters and cyclic ketones are reduced with high selectivity using Baker's yeast. Typical in this regard is the reduction of ethyl acetoacetate, which gives ethyl 3-hydroxybutyrate as predominantly the (*S*)-stereoisomer (7.101). Similarly, the ketone **114** gave the optically active 3-hydroxyproline derivative **115** (7.102).

---

[45] S. Servi, *Synthesis* (1990), 1; R. Csuk and B. I. Glänzer, *Chem. Rev.*, **91** (1991), 49.

$$ \text{(7.101)} $$

$$ \text{(7.102)} $$

114   85%   95% ee **115**

Interestingly, reduction of ethyl 4-chloroacetoacetate with Baker's yeast gave predominantly the corresponding (*S*)-alcohol (i.e. the opposite configuration from that of the alcohol from ethyl acetoacetate itself) (7.103), but the corresponding octyl ester gave almost entirely the (*R*)-alcohol. The stereochemistry of the reduction depends on the shape of the molecule and it is likely that the yeast contains at least two different oxidoreductase enzymes which produce the two enantiomeric alcohols at different rates.

$$ \text{(7.103)} $$

55% ee

The enantioselectivity on reduction with Baker's yeast is very substrate dependent and although compounds such as β-keto-esters are good substrates, many ketones give low selectivities. As a result, many different microorganisms have been tested for asymmetric reduction of different substituted ketones.[46] As an example, good results have been obtained by using lyophilized whole cells of *Rhodococcus ruber* DSM 44541, as illustrated in the reduction of the ketone **116** (7.104).[47] Aryl ketones and other unfunctionalized ketones can be reduced with high selectivity.

$$ \text{(7.104)} $$

116   67%   99% ee

[46] J. B. Jones, in *Comprehensive Organic Synthesis*, ed. B. M. Trost and I. Fleming, vol. 8 (Oxford: Pergamon Press, 1991), p. 183; E. Santaniello, P. Ferraboschi, P. Grisenti and A. Manzocchi, *Chem. Rev.*, **92** (1992), 1071; K. Nakamura, R. Yamanaka, T. Matsuda and T. Harada, *Tetrahedron: Asymmetry*, **14** (2003), 2659.

[47] W. Stampfer, B. Kosjek, C. Moitzi, W. Kroutil and K. Faber, *Angew. Chem. Int. Ed.*, **41** (2002), 1014.

Reduction with isolated enzymes avoids difficulties associated with diffusion limitations and also avoids the presence of many different enzymes, present in the whole cell, which can cause side reactions or reduced enantioselectivity. The main drawback, however, is the instability of the isolated enzyme and the requirement for added co-factor NAD(H) or NADP(H), which are the oxidized (or reduced) forms of nicotinamide adenine diphosphate or its 2′-phosphate derivative. These co-factors are expensive, but can be used as catalysts in the presence of a co-reductant such as formate ion $HCOO^-$ or an alcohol (e.g. isopropanol or ethanol). The reduction of ketones occurs by transfer of hydride from the C-4 position of the dihydropyridine ring of NADH or NADPH (7.105). Only one of the two hydrogen atoms is transferred and this process occurs within the active site of the enzyme to promote asymmetric reduction.

(7.105)

Many different isolated enzymes can be used and the choice is best made after careful searching of the literature for examples of reduction of related structures. Various dehydrogenase enzymes can promote highly selective reduction of prochiral ketones, sometimes with opposite facial selectivity. The absolute configuration of the product alcohol from reduction of acyclic ketones with many alcohol dehydrogenase enzymes can be predicted by placing the larger group as R′ and the smaller as R in Scheme 7.103. The situation is more complex for cyclic ketones; however, these too can act as suitable substrates. For example, reduction of the symmetrical diketone **117** can be achieved with horse liver alcohol dehydrogenase (HLADH) to give the chiral alcohol **118** (7.106).

(7.106)

7.4.2 Wolff–Kishner reduction

The Wolff–Kishner reduction provides an excellent method for the reduction of the carbonyl group of many aldehydes and ketones to a methyl or methylene group respectively.[48] As originally described the reaction involved preparing the hydrazone or semicarbazone from the carbonyl compound and then heating with sodium ethoxide or other base at 200 °C in a sealed tube. More conveniently, the Wolff–Kishner reduction can be effected by heating a mixture of the carbonyl compound, hydrazine hydrate and sodium or potassium hydroxide in a high-boiling solvent at 180–200 °C for several hours. Use of the high-boiling solvent diethylene glycol promotes removal of excess hydrazine and water after hydrazone formation and reduces the time required for the reduction (Huang–Minlon modification). For example, reduction of the hydrazone of the ketone **119** with sodium in diethylene glycol gave the product **120** (7.107). Alternatively, by using microwave irradiation the reduction can be carried out in minutes in the absence of a solvent (7.108).[49]

The mechanism of the reduction is believed to involve deprotonation of the hydrazone to give the anion **121** (7.109). This is followed by the rate-limiting protonation at the carbon atom and deprotonation of the terminal nitrogen atom to give **122**. Loss of nitrogen and protonation of the carbanion gives the product. In line with this mechanism, the polar aprotic solvent DMSO increases the rate of the reaction, and with potassium tert-butoxide in DMSO reduction can even be carried out at room temperature.

[48] R. O. Hutchins and M. K. Hutchins, in *Comprehensive Organic Synthesis*, ed. B. M. Trost and I. Fleming, vol. 8 (Oxford: Pergamon Press, 1991), p. 327.

[49] S. Gadhwal, M. Baruah and J. S. Sandhu, *Synlett* (1999), 1573.

Reduction of conjugated unsaturated aldehydes or ketones is sometimes accompanied by a shift in the position of the double bond (7.110). In other cases pyrazoline derivatives may be formed; these decompose yielding cyclopropanes isomeric with the expected alkene (7.111).

$$ \text{(7.110)} $$

$$ \text{80\%} $$

$$ \text{(7.111)} $$

With carbonyl compounds carrying a leaving group in the α-position elimination may accompany reduction. Useful in this regard is the reductive opening of α,β-epoxyketones to provide allylic alcohol products. For example, treatment of the ketone **123** with hydrazine and triethylamine gave the allylic alcohol **124** (7.112).[50]

$$ \text{(7.112)} $$

123   71%   **124**

Alternative procedures for deoxygenation of aldehydes and ketones include the Clemmensen reduction (Scheme 7.38) and the reduction of tosylhydrazones, for example with sodium cyanoborohydride (Scheme 7.89). Another method for reduction is desulfurization of the corresponding thio-acetals with Raney nickel in refluxing ethanol. Hydrogenolysis is effected by the hydrogen adsorbed on the nickel during its preparation (7.113). The reduction can be carried out fairly easily, although large amounts of Raney nickel are required, and other unsaturated groups in the compound may also be reduced.

$$ \text{(7.113)} $$

95%   82%

[50] C. Dupuy and J. L. Luche, *Tetrahedron*, **45** (1989), 3437.

7.4.3 Reductions with diimide

It has been known for a long time that isolated carbon–carbon double bonds can be reduced with hydrazine in the presence of oxygen or an oxidizing agent. For example, as early as 1914 it was found that oleic acid is reduced to stearic acid by this method. The actual reducing agent in these reactions is the highly active species diimide, HN=NH, formed *in situ* by oxidation of hydrazine. This compound is a highly selective reducing agent which in many cases offers a useful alternative to catalytic hydrogenation for the reduction of carbon–carbon multiple bonds.[51]

The reagent diimide is prepared by oxidation of hydrazine, by decomposition of tosylhydrazide, or from azodicarboxylic acid (7.114). It is a suitable reducing agent for symmetrical double bonds such as C=C, C≡C and N=N, but unsymmetrical, more polar bonds such as C=N, C≡N, N=O, S=O are not reduced. It is particularly suitable for reduction in the presence of other reactive functional groups, or in cases that are unsuccessful with catalytic hydrogenation. For example, diallyl disulfide is reduced to dipropyl disulfide (7.115). Only the less-conjugated alkene in the substrate **125** is reduced by diimide without affecting the other double bonds or the epoxide ring, whereas catalytic hydrogenation promotes epoxide opening to give the phenol **127** (7.116).

$$NH_2NH_2 \quad \xrightarrow[\text{Cu(II)}]{O_2 \text{ or } H_2O_2} \quad HN{=}NH$$

$$TsNHNH_2 \quad \xrightarrow{\text{heat}} \quad HN{=}NH \qquad (7.114)$$

$$\underset{CO_2K}{\overset{KO_2C}{N{=}N}} \quad \xrightarrow{\text{AcOH}} \quad HN{=}NH$$

$$\xrightarrow[\text{glycol, heat}]{TsNHNH_2} \qquad (7.115)$$

93–100%

$$\xrightarrow[\text{AcOH, DME, 45 °C}]{KO_2CN{=}NCO_2K} \qquad (7.116)$$

125   77%   **126**   **127**

The reactions are highly stereoselective, taking place by *cis*-addition of hydrogen in all cases. Transfer of the hydrogen atoms is thought to take place simultaneously

---

[51] D. J. Pasto and R. T. Taylor, *Org. Reactions*, **40** (1991), 91.

through a cyclic six-membered transition state (7.117). This mechanism explains the stereospecificity of the reaction, and couples the driving force of nitrogen formation with the addition reaction. Concerted *cis*-transfer of hydrogen is a symmetry-allowed process.

$$
\begin{array}{ccccc}
\underset{\substack{H \quad\; H \\ \diagdown\; / \\ N=N}}{\diagup\hspace{-0.3em}=\hspace{-0.3em}\diagdown} & \longrightarrow & \underset{\substack{\; \\ H \quad\quad H \\ \diagdown\,\diagup \\ N\text{--}N}}{} & \longrightarrow & \underset{\substack{H \quad\; H \\ \\ N\equiv N}}{} \quad\quad (7.117)
\end{array}
$$

7.4.4 Reductions with trialkylsilanes

The addition of Si—H to unsaturated substrates is a useful method of reduction in some cases, and is also an important route to complex organosilanes. Addition can be brought about under catalytic or ionic conditions.

Silanes will reduce a variety of functional groups in the presence of transition-metal catalysts. Alkynes undergo *cis*-addition to give vinylsilanes, ketones give silyl ethers of the corresponding secondary alcohols and aromatic imines are readily reduced to amines. A useful reaction is the conversion of acid chlorides or thioesters into aldehydes, which provides an alternative to the Rosenmund reduction (hydrogenolysis of acid chlorides with palladium on barium sulfate) or reduction with complex hydrides (see Scheme 7.74). For example, reduction of octanoyl chloride gave octanal (7.118), and the thioester **128** gave the aldehyde **129** in high yield without racemization or reduction of the ester or carbamate carbonyl groups (7.119).[52] It is possible, however, to reduce carboxylic esters and lactones to the corresponding ethers by using triethylsilane in the presence of $TiCl_4$ and TMSOTf.[53]

$$
\begin{array}{ccc}
\text{COCl} & \xrightarrow[\text{10\% Pd/C}]{\text{Et}_3\text{SiH}} & \text{CHO} \qquad (7.118) \\
& 83\% &
\end{array}
$$

$$
\begin{array}{ccc}
\underset{\text{MeO}_2\text{C}}{\overset{\text{NHBoc}}{\diagup\diagdown\diagup\diagdown}}\text{COSEt} & \xrightarrow[\substack{\text{10\% Pd/C} \\ \text{acetone}}]{\text{Et}_3\text{SiH}} & \underset{\text{MeO}_2\text{C}}{\overset{\text{NHBoc}}{\diagup\diagdown\diagup\diagdown}}\text{CHO} \qquad (7.119) \\
\mathbf{128} & 93\% & \mathbf{129}
\end{array}
$$

The reduction of α,β-unsaturated aldehydes or ketones with trialkylsilanes has proved a valuable reaction for the regioselective preparation of silyl enol ethers. Wilkinson's catalyst [$(Ph_3P)_3RhCl$] is suitable to promote the reaction and subsequent hydrolysis provides the saturated carbonyl compound, as illustrated by the

[52] H. Tokuyama, S. Yokoshima, S.-C. Lin, L. Li and T. Fukuyama, *Synthesis* (2002), 1121.
[53] M. Yato, K. Homma and A. Ishida, *Tetrahedron*, **57** (2001), 5353.

selective reduction of the α,β-unsaturated ketone **130** in the presence of an isolated carbon–carbon double bond (7.120). The intermediate silyl enol ether may be used for further functionalization by reaction with other electrophiles.

(7.120)

An alternative to trialkylsilane reduction of α,β-unsaturated aldehydes or ketones is the use of copper hydride complexes, such as $[Ph_3PCuH]_6$ (Stryker's reagent).[54] This reagent promotes efficient conjugate reduction, and can be used as a catalyst in the presence of stoichiometric reductants such as silyl hydrides (e.g. $PhMe_2SiH$). Copper hydride complexes can be prepared by reduction of various copper(I) salts with mild reducing agents. Recently, asymmetric reduction has been achieved with the chiral ligand (*S*)-*p*-Tol-BINAP, with formation of the chiral ligand-complexed copper hydride from CuCl and the polymeric reducing agent polymethylhydrosiloxane (PMHS) (7.121).[55]

(7.121)

Polymethylhydrosiloxane (PMHS) is easy to handle and is finding increasing popularity as a reducing agent. For example, in the presence of catalytic tris(pentafluorophenyl)borane, $B(C_6F_5)_3$, it is effective for the reduction of ketones (both aromatic and aliphatic) to the corresponding methylene compounds at room temperature.[56]

Ionic hydrogenation with silanes can be accomplished in the presence of an acid or Lewis acid. For example, a combination of triethylsilane and trifluoroacetic acid (TFA) provides a non-catalytic method for hydrogenation of C=C, C=O and C=N double bonds and for hydrogenolysis of some single bonds (such as C–Br or benzylic C–OH). Alkenes can be reduced to saturated hydrocarbons, but only if the double bond is at least trisubstituted, allowing the possibility of selective hydrogenation in a compound containing different types of double bond. A useful

---

[54] W. S. Mahoney, D. M. Brestensky and J. M. Stryker, *J. Am. Chem. Soc.*, **110** (1988), 291; B. H. Lipshutz, W. Chrisman, K. Noson, P. Papa, J. A. Sclafani, R. W. Vivian and J. M. Keith, *Tetrahedron*, **56** (2000), 2779.
[55] Y. Moritani, D. H. Appella, V. Jurkauskas and S. L. Buchwald, *J. Am. Chem. Soc.*, **122** (2000), 6797.
[56] S. Chandrasekhar, Ch. R. Reddy and B. N. Babu, *J. Org. Chem.*, **67** (2002), 9080.

application of this chemistry is for the reduction of aromatic ketones. For example, the ketone **131** gave the reduced compound **132** under these conditions (7.122). This type of reduction is also effective with a Lewis acid such as boron trifluoride.

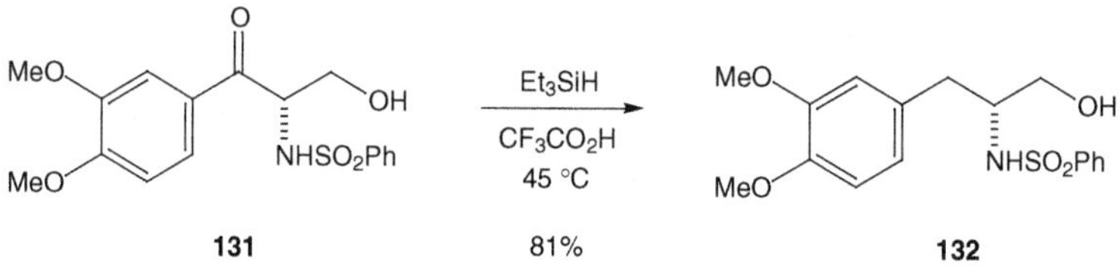

$$\text{(7.122)}$$

131   81%   **132**

Acetals and hemiacetals can be reduced with triethylsilane and an acid or Lewis acid. For example, the benzylidene acetal **133** was treated with triethylsilane and boron trifluoride etherate to give the product **134**, in which the 6-benzyl ether was formed selectively (7.123). In a synthesis of the potent anti-tumor agent phorboxazole B, the hemiacetal **135** was reduced to a single diastereomer of the cyclic ether **136** (7.124).[57] These types of reactions are thought to take place by co-ordination of the Lewis acid to one of the oxygen atoms (or its protonation when using TFA) followed by formation of an oxonium ion, which is reduced by the silane.

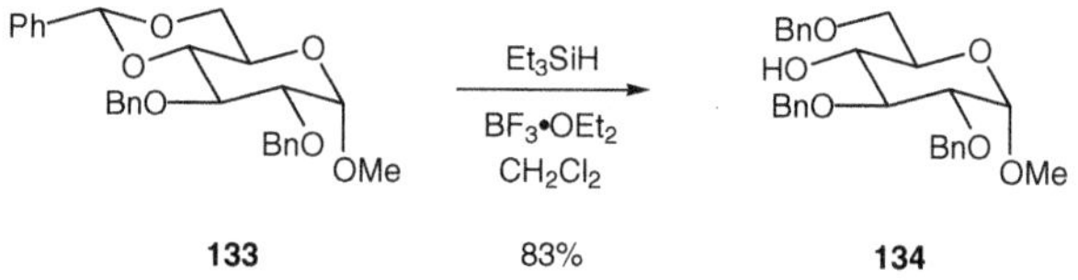

$$\text{(7.123)}$$

133   83%   **134**

$$\text{(7.124)}$$

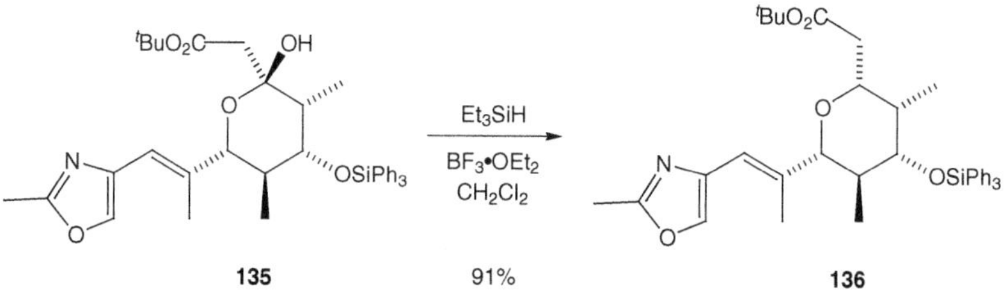

135   91%   **136**

Problems (answers can be found on page 484)

1. Draw the structure of the product from the following reaction.

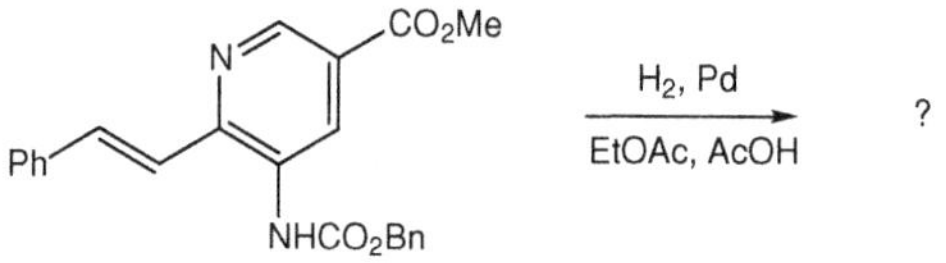

[57] D. A. Evans, D. M. Fitch, T. E. Smith and V. J. Cee, *J. Am. Chem. Soc.*, **122** (2000), 10033.

2. Explain the formation of the product pumiliotoxin C (as its hydrochloride salt) in the following reaction.

pumiliotoxin C

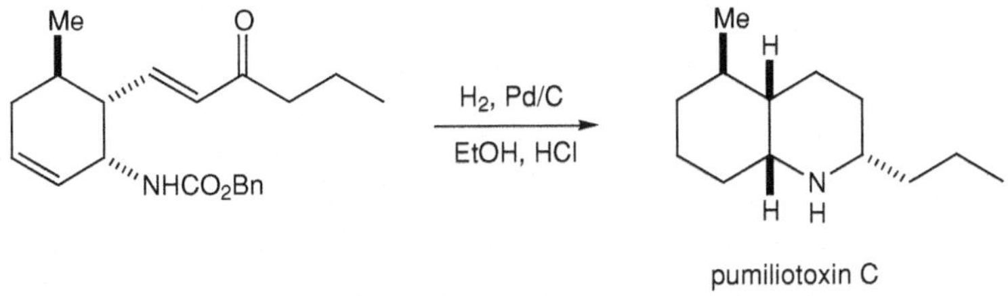

3. Draw the structure of the product from the following reaction.

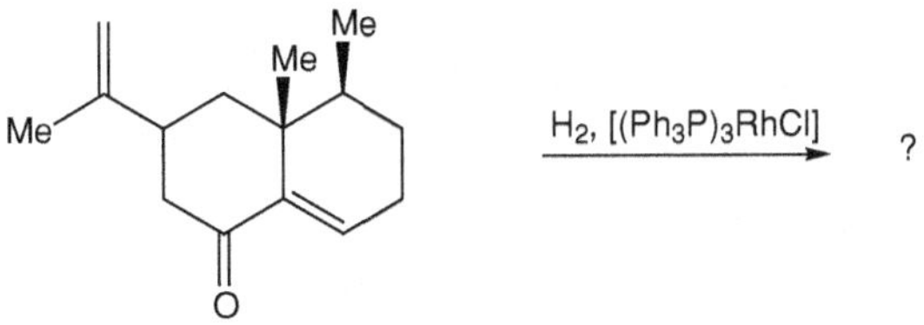

4. Explain the formation of the product from the following reaction.

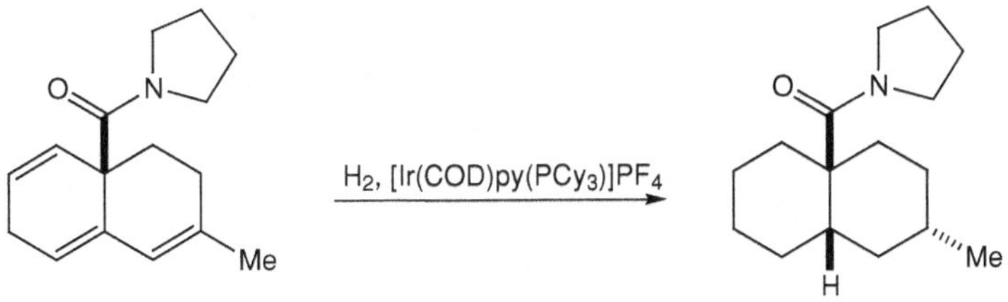

5. Draw the structure of the product from the following reaction.

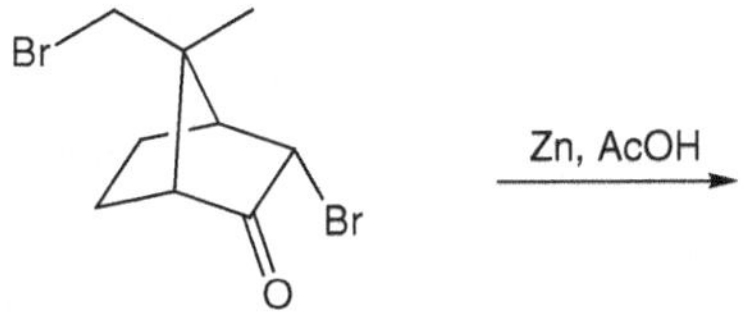

6. Draw the structures of the resonance forms of the intermediate radical anion from addition of an electron to the dienone below. (Hint: compare with structure **60**, Scheme 7.43.)

Addition of a second electron gives a dianion, which leads (after aqueous work-up) exclusively to the stereoisomer shown. Provide an explanation for the stereoselectivity in this reaction.

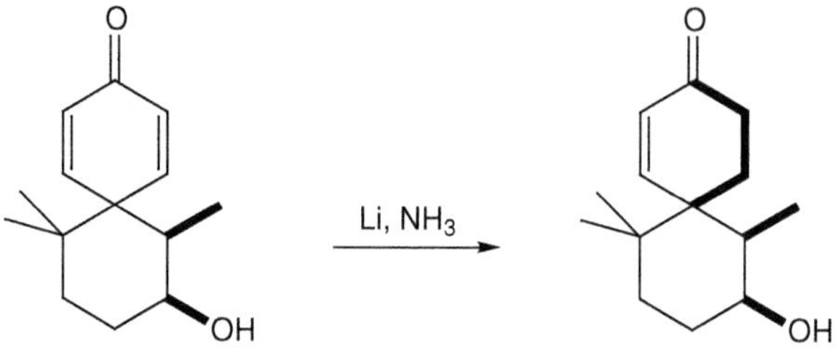

$$\xrightarrow{\text{Li, NH}_3}$$

7. Suggest reagents to effect the transformation shown below.

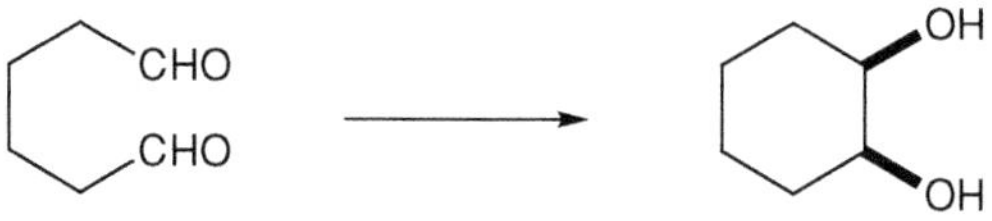

8. Explain the formation of the product from the following reaction.

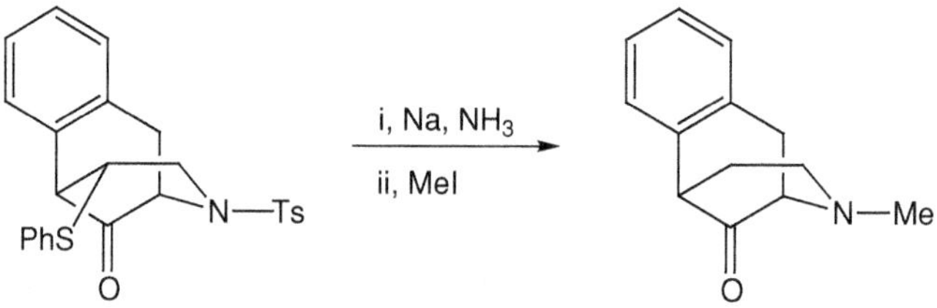

i, Na, NH₃

ii, MeI

9. Draw a mechanism to explain the formation of the primary alcohol from treatment of cyclohexanecarboxaldehyde with LDA.

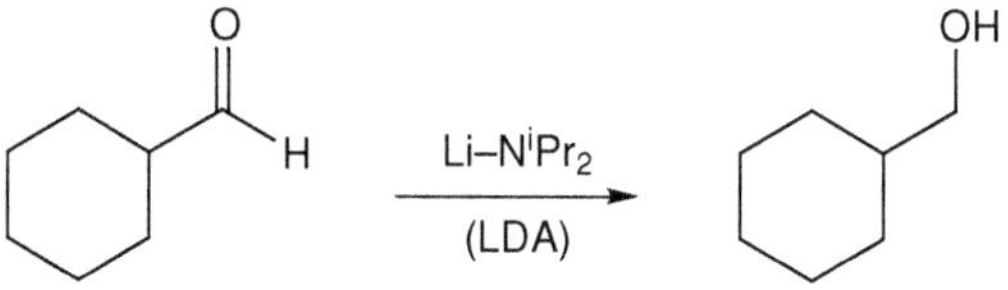

Li–NⁱPr₂

(LDA)

10. Suggest a reagent for the selective reduction of the aldehyde shown below.

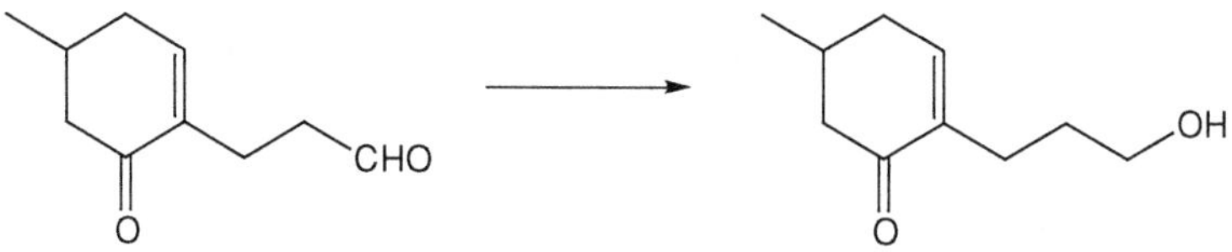

11. Draw the structure of the major enantiomer of the product from reduction of the ketone shown below, used in a synthesis of fredericamycin A.

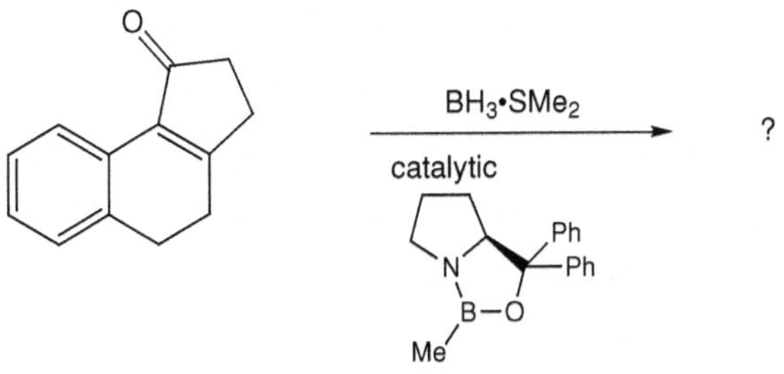

$BH_3 \cdot SMe_2$

catalytic

?

12. Explain the formation of the allylic alcohol product from the reaction shown below.

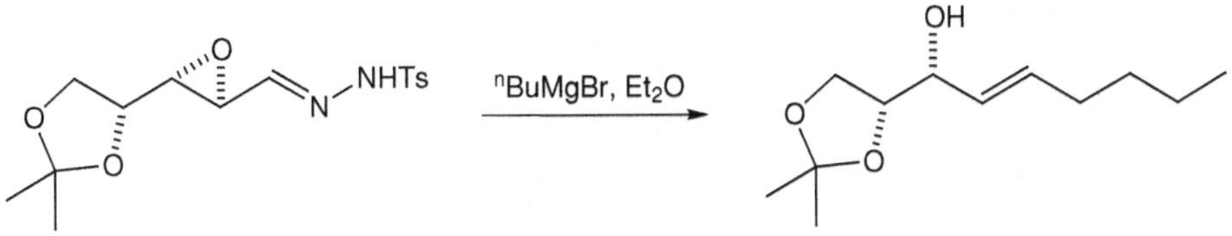

$^nBuMgBr, Et_2O$

Answers to problems

Answers to problems from Chapter 1

1. Methylation of pentan-2-one using a base (to form the enolate) and iodomethane can give either hexan-3-one or 3-methylpentan-2-one. The less-substituted enolate is formed under kinetic conditions (by using a strong base for irreversible proton abstraction) with LDA in THF at low temperature. On addition of iodomethane, this gives hexan-3-one. The more-substituted enolate can be formed under thermodynamic conditions (reversible reaction) with t-BuOK, t-BuOH. However, these conditions tend to result in polyalkylation. More satisfactory results can be obtained by treating the silyl enol ether (formed from pentan-2-one with Me_3SiCl, Et_3N) with methyllithium or benzyltrimethylammonium fluoride (to form the specific enolate) and iodomethane.

2. Compounds (a)–(c) can be formed using the Michael reaction. See (a) H. Feuer, A. Hirschfield and E. D. Bergmann, *Tetrahedron*, **24** (1968), 1187; (b) J. Cason, *Org. Synth. IV* (1963), 630; (c) E. D. Bergmann, D. Ginsburg and R. Pappo, *Org. Reactions*, **10** (1959), 179.

3. The iodide **1** is formed by a Baylis–Hillman reaction (Scheme 1.61). See Z. Han, S. Uehira, H. Shinokubo and K. Oshima, *J. Org. Chem.*, **66** (2001), 7854. Note that the *syn* aldol product is formed from the intermediate Z-titanium enolate.

4. The synthesis of aldol product **2** and cytovaricin is described by D. A. Evans, S. W. Kaldor, T. K. Jones, J. Clardy and T. J. Stout, *J. Am. Chem. Soc.*, **112** (1990), 7001.

   For the preparation of the imide **2**, use Bu_2BOTf and Et_3N (CH_2Cl_2, 0 °C). The boron aldol reaction often benefits from a work-up with H_2O_2, to cleave the boron–oxygen bond and release the alcohol product.

   The stereoselectivity of this reaction can be explained by invoking a Z-enolate and drawing the six-membered, chair-shaped transition state (see Scheme 1.80). The Z-enolate leads, using this transition state, to the *syn* aldol product. The chiral auxiliary directs the approach of the aldehyde to the less hindered face of the enolate. To obtain the other *syn* aldol product, use either the other oxazolidinone auxiliary (**53**) or use the titanium enolate. In the latter case, the stereoselectivity can be rationalized by co-ordination of titanium to the carbonyl oxygen atom of the oxazolidinone group.

5. Deprotonation of the sulfonium salt and addition to the less-hindered face of the ketone gives the intermediate alkoxide, which cyclizes to the epoxide. Ring-opening of the epoxide with chloride (inversion of configuration) gives the chloro-aldehyde **3**. See K. Sato, T. Sekiguchi and S. Akai, *Tetrahedron Lett.*, **42** (2001), 3625.

6. The organolithium compound LiCH$_2$OMOM can be prepared by tin–lithium exchange from the corresponding stannane with butyllithium. The stannane is prepared from formaldehyde by addition of Bu$_3$SnLi followed by *O*-protection. The major product from addition of the 2-phenylpropanal is shown below. Addition takes place with the expected Felkin–Anh selectivity (see Scheme 1.91) to give predominantly the *syn* diastereomer. See W. C. Still, *J. Am. Chem. Soc.*, **100** (1978), 1481; G. A. Molander and A. M. EstevezBraun, *Bull. Soc. Chem. Fr.*, **134** (1997), 275.

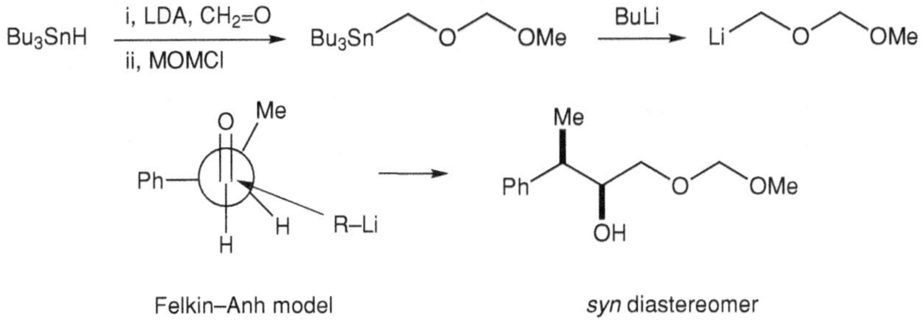

Felkin–Anh model   *syn* diastereomer

7. The alcohol **5** arises from chelation-controlled addition to the ketone **4**. Draw a Newman projection, as shown in **A**, to illustrate this, with attack by the isopropyl group from the less-hindered face. The alcohol **6** is formed by β-hydride delivery from the Grignard reagent. In this case, there must be no O—Mg—N chelation, and the stereochemistry can be explained by using the Felkin–Anh transition state – see the Newman projection **B**. See Schemes 1.125, 1.127 and 1.128.

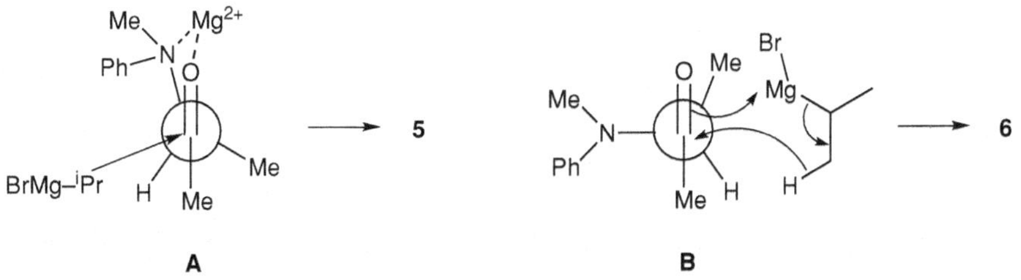

A   B

8. Metallation of 2-chloropyrazine occurs *ortho* to the chlorine atom as this gives the most-stabilized organolithium intermediate **C**. The second lithiation occurs at C-6, to give the 6-iodopyrazine **7**. The regioselectivity of this reaction can be explained if one takes into account the fact that the species in solution is not the alcohol but the lithium alkoxide, which can chelate to the pyrazine nitrogen atom, N-4. This leaves the other pyrazine nitrogen atom (N-1) free and hence results only in the C-6 lithiated intermediate **D**.

Insertion of palladium(0) into the weaker C—I bond (oxidative addition), reaction with the organozinc species (transmetallation – see Scheme 1.201) and reductive elimination gives **8** (Negishi reaction – compare with Scheme 1.140).

See C. Fruit, A. Turck, N. Plé, L. Mojovic and G. Quéguiner, *Tetrahedron*, **57** (2001), 9429.

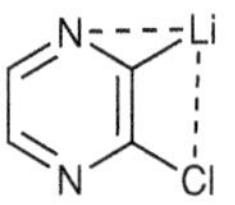

C 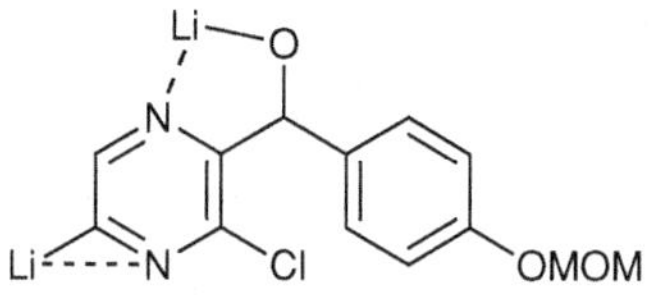   D

9. The zinc metal inserts into the allyl bromide to give an allylzinc bromide that reacts with the ketone. The addition occurs from the less-hindered face to give the alcohol shown below. See P. E. Maligres, M. M. Waters, J. Lee, R. A. Reamer, D. Askin, M. S. Ashwood and M. Cameron, *J. Org. Chem.*, **67** (2002), 1093.

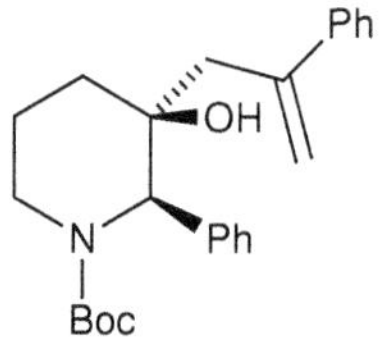

10. Hydroboration (*syn* addition of H and B) followed by transmetallation with diethyl- or (with secondary boranes) diisopropylzinc provides a useful entry to diorganozinc species. Formation of the copper–zinc species allows reaction with activated electrophiles such as allyl bromide. See A. Bourdier, C. Darcel, F. Flachsmann, L. Micouin, M. Oestreich and P. Knochel, *Chem. Eur. J.*, **6** (2000), 2748.

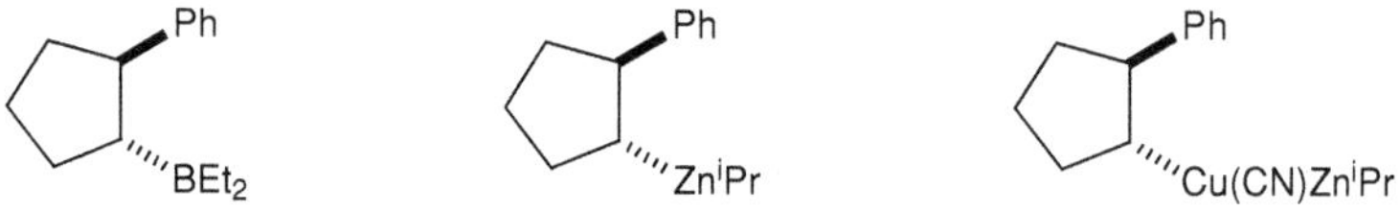

11. Insertion of Ni(0) into the C—Br bond, followed by transmetallation to the alkenyl chromium(III) species and intramolecular addition to the carbonyl group gives the product **11**. Oxidation with pyridinium dichromate (see Section 6.2) gave the natural product *cis*-jasmone. See B. M. Trost and A. B. Pinkerton, *J. Org. Chem.*, **66** (2001), 7714.

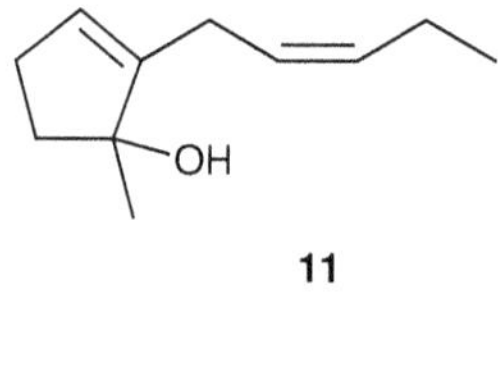

11

12. The cobalt complex **12** can be drawn in either of the forms shown below (**12a** or preferably as **12b**). The Lewis acid TiCl$_4$ promotes cleavage of the C—O bond to give the propargylic cation, stabilized by the cobalt-complexed alkyne (Nicholas reaction, see Schemes 1.194–1.196). Concomitant with this is the formation of a titanium enolate, which traps the cation to give the cyclohexanone product **13**. See D. R. Carbery, S. Reignier, J. W. Myatt, N. D. Miller and J. P. A. Harrity, *Angew. Chem. Int. Ed.*, **41** (2002), 2584.

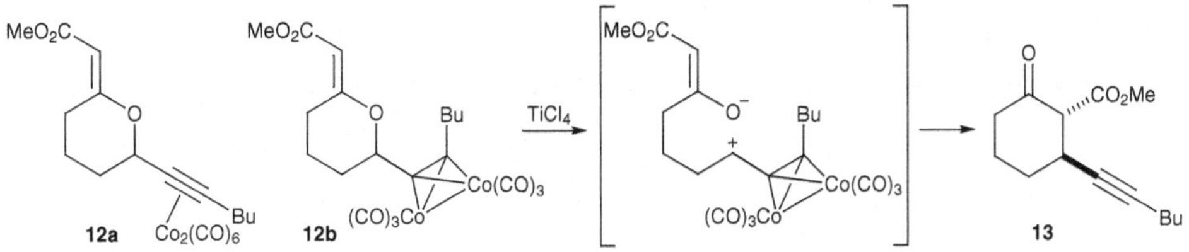

13. These palladium-catalysed reactions involve insertion of the palladium into the carbon–bromine or carbon–triflate bond, reaction with the organometallic species (alkenyl stannane in a Stille reaction, arylboronic acid in a Suzuki reaction, or alkynyl copper species in a Sonogashira reaction) and reductive elimination to give the products shown below.

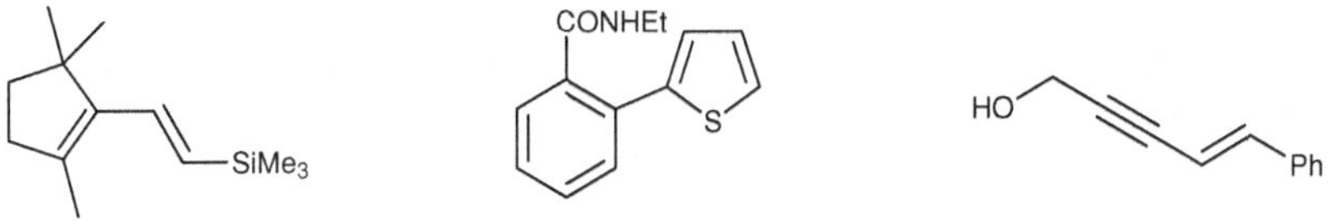

14. The product **14** is formed by initial palladium(0)-catalysed ring-opening of the allylic epoxide to give the π-allyl palladium species. Sodium hydride deprotonates the diester and the resulting enolate attacks the π-allyl palladium species at the less-hindered end. Subsequent insertion of palladium(0) into the C—Br bond and intramolecular Heck reaction forms the five-membered ring. Dehydropalladation (*syn* elimination) releases palladium(0) (to continue the cycle) and gives the product **14**. For a review on palladium-catalysed reactions and domino processes, see G. Poli, G. Giambastiani and A. Heumann, *Tetrahedron*, **56** (2000), 5959.

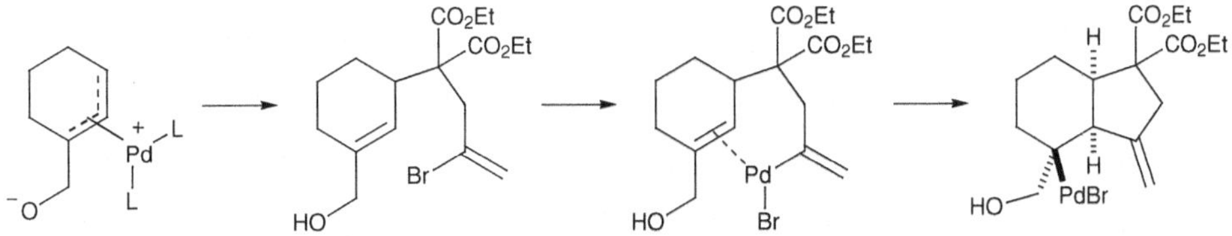

Answers to problems from Chapter 2

1. Amine *N*-oxides eliminate by a concerted, thermal *syn*-elimination process. Hence the diastereomeric amine *N*-oxide *syn*-**1** can eliminate only away from the phenyl group,

whereas the *anti* diastereomer can give a mixture of the two regioisomeric alkenes. Elimination to the more-stable, more-substituted alkene in conjugation with the phenyl group predominates.

2. The Z-α,β-unsaturated ester **2** is formed from the *syn* β-hydroxy selenide by an *anti* elimination. *O*-Mesylation and loss of the leaving group generates the *cis* episelenonium ion and hence the Z-alkene **2** (see Scheme 2.28). The *syn* β-hydroxy selenide is formed by an aldol reaction of the Z-titanium enolate via a chair-shaped transition state (see Section 1.1.3). See S. Nakamura, T. Hayakawa, T. Nishi, Y. Watanabe and T. Toru, *Tetrahedron*, **57** (2001), 6703.

3. Elimination of thionocarbonates is termed the Corey–Winter reaction (see Section 2.5). Attack by the phosphorus reagent on the sulfur atom is followed by fragmentation to give the alkene. See A. Krief, L. Provins and A. Froidbise, *Tetrahedron Lett.*, **43** (2002), 7881.

4. Allylic alcohols can often be prepared by Wittig or Horner–Wadsworth–Emmons reaction followed by DIBAL-H reduction. However, this protocol would give the wrong stereoisomer of the allylic alcohol **3**. A convenient and stereoselective method to prepare trisubstituted alkenes starts from alkynes. Propargyl alcohols are very good substrates and give *E*-allylic alcohols on reduction with LiAlH$_4$ (see Section 2.6, Scheme 2.62). The intermediate alkenyl alane can be quenched with different electrophiles. For the preparation of the allylic alcohol **3**, the aluminium atom needs to be replaced with an aryl substituent. Palladium-catalysed coupling to an aryl iodide provides the answer and this was carried out by addition of dimethyl carbonate to the alkenyl alane, followed by Negishi-type coupling via the alkenyl zinc species (see Section 1.2.4). See M. Havránek and D. Dvořák, *J. Org. Chem.*, **67** (2002), 2125.

5. Alkynes can be converted into Z-alkenes by partial reduction using H$_2$ and Lindlar's catalyst. Alternatively, reduction by hydroboration, followed by addition of acid gives the desired Z-alkene. See Section 2.6. The latter procedure was found to be preferable for the conversion of the alkyne **4** to the alkene **5** using (C$_6$H$_{11}$)$_2$BH followed by MeCO$_2$H; see V. P. Bui, T. Hudlicky, T. V. Hansen and Y. Stenstrom, *Tetrahedron Lett.*, **43** (2002), 2839.

6. Deprotonation of the lactol O—H with a base promotes ring-opening to the alkoxy-aldehyde. The best reagent for formation of the alkene from the aldehyde is a phosphonium ylide, generated from a phosphonium salt. It is possible to use the carboxylic acid (below) and employ two equivalents of base (the first deprotonating the acid, the second forming the phosphonium ylide). A suitable base is NaHMDS [NaN(SiMe$_3$)$_2$] (three equivalents in total) used in toluene. See D. J. Critcher, S. Connolly and M. Wills, *J. Org. Chem.*, **62** (1997), 6638.

7. Deprotonation of the pyrrole N—H and addition of the anion to the alkenyl phosphonium salt gives an intermediate phosphonium ylide. This ylide undergoes intramolecular Wittig reaction to give the pyrrolizine **8**. See E. E. Schweizer and K. K. Light, *J. Am. Chem. Soc.*, **86** (1964), 2963.

8. Nucleophilic addition of the organolithium species to the ketone gives an intermediate adduct which is still a phosphonium ylide. Protonation of this ylide on the carbon atom gives the alkoxy-phosphonium species which cyclizes to the oxaphosphetane. Elimination of triphenylphosphine oxide gives the alkene **9**. See Scheme 2.78 and E. J. Corey, J. Kang and K. Kyler, *Tetrahedron Lett.*, **26** (1985), 555.

9. The alkenyl stannane **10** can be prepared by Swern oxidation (see Section 6.2) of the corresponding allylic alcohol, itself prepared by addition of Bu$_3$SnH to the alkyne (propargyl alcohol or ethyl propiolate then reduction of the ester); see M. E. Jung and L. A. Light, *Tetrahedron Lett.*, **23** (1982), 3851.

   Treatment of the alkenyl stannane **10** with the phosphonate (EtO)$_2$P(O)—CH(Me)CO$_2$Et and *n*-BuLi in THF gives the dienyl stannane with the desired *E,E* geometry. For step ii, conversion of the stannane to the iodide can be achieved by using iodine (in CCl$_4$). Finally, conversion of the dienyl iodide to the diene **11** uses chemistry described in Section 1.2.4: iii, hydroboration of the alkene (9-BBN—H) and iv, Suzuki coupling [PdCl$_2$(dppf), Ph$_3$As, Cs$_2$CO$_3$]. See M. W. Carson, G. Kim, M. F. Hentemann, D. Trauner and S. J. Danishefsky, *Angew. Chem. Int. Ed.*, **40** (2001), 4450.

10. Conversion of the silane **12** to the alkene **13** can be accomplished by using a base such as KHMDS. Basic conditions promote *syn* elimination of β-hydroxysilanes (Peterson elimination, see Scheme 2.86).

   Conversion of the silane **12** to the alkene **14** can be accomplished by using acidic conditions (e.g. H$_2$SO$_4$). Acid-promoted reaction of β-hydroxysilanes occurs by an *anti*-elimination pathway and therefore gives exclusively the other isomeric alkene. See J.-N. Heo, E. B. Holson and W. R. Roush, *Org. Lett.*, **5** (2003), 1697.

11. The base KHMDS deprotonates α- to the sulfone. This carbanion attacks the aldehyde to give an intermediate β-alkoxy-sulfone. Intramolecular attack of the alkoxy anion onto the triazole can then take place. Elimination of SO$_2$ and the alkene then takes place *in situ* to give the required *E*-alkene product. See Schemes 2.94 and 2.95. See A. B. Smith, I. G. Safonov and R. M. Corbett, *J. Am. Chem. Soc.*, **123** (2001), 12 426.

12. Ring-closing metathesis will occur by the lowest-energy transition state to give two six-membered rings, rather than the alternative three- or nine-membered rings. The product (shown below) therefore has a spirocyclic ring structure, present in the natural product histrionicotoxin. See A. S. Edwards, R. A. J. Wybrow, C. Johnstone, H. Adams and J. P. A. Harrity, *Chem. Commun.* (2002), 1542.

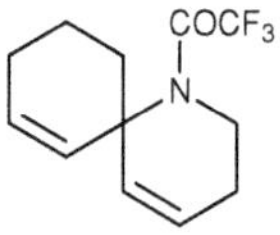

13. The ruthenium catalyst will attack the least-hindered terminal alkene to give an intermediate ruthenium alkylidene species. Ring-closing metathesis takes place onto the alkyne (not the enone as this would generate a seven-membered ring). The new ruthenium alkylidene species then attacks the enone to give the six-membered unsaturated lactone **16** (with simultaneous regeneration of the catalyst). Try drawing the structures of the intermediate metallocycles (compare with Scheme 2.111). See T.-L. Choi and R. H. Grubbs, *Chem. Commun.* (2001), 2648. For a review of enyne metathesis, see C. S. Poulsen and R. Madsen, *Synthesis* (2003), 1.

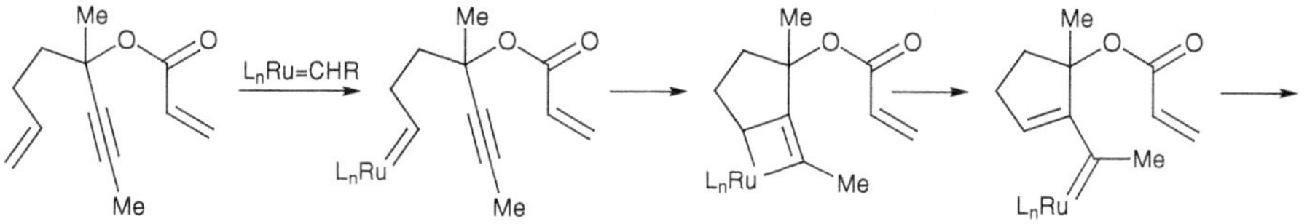

Answers to problems from Chapter 3

1. Intermolecular Diels–Alder cycloaddition reactions usually fail with unactivated dienophiles as the energy of the LUMO of such dienophiles is normally high and therefore the interaction of this molecular orbital with that of the HOMO of the diene is poor (orbitals of similar energy interact more strongly). Filled–unfilled orbital interactions are crucial for successful Diels–Alder reactions, which are typically frontier-orbital controlled. See Section 3.1.1.

2. Four isomers of the cycloadduct **1** are possible. However, there is a preference for two stereoisomers of the expected regioisomer, in which the terminal methylene group of the diene, being more electron-rich (larger coefficient in the HOMO) than the CMe_2 group binds to the β-position of nitroethene (the more electron-deficient end of the dienophile – larger coefficient in the LUMO). Tributyltin hydride effects the denitration reaction under radical conditions (see Section 4.1). See M. Inoue, M. W. Carson, A. J. Frontier and S. J. Danishefsky, *J. Am. Chem. Soc.*, **123** (2001), 1878.

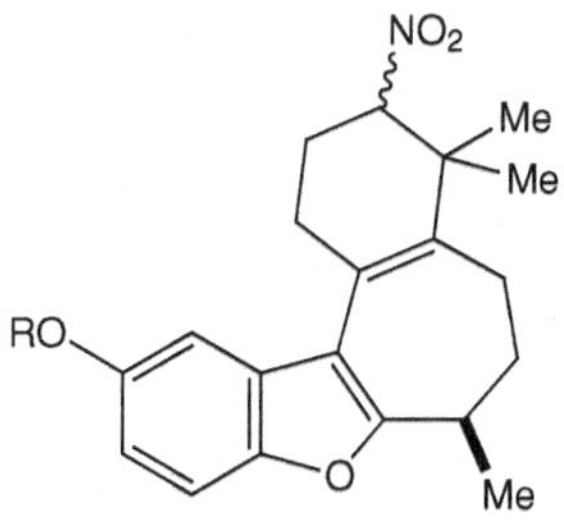

1

3. The cycloadduct **2** is formed by a Diels–Alder reaction of the dienophile methyl 2-ethylacrylate and the diene generated by photochemical activation of the aromatic aldehyde. Irradiation of an *ortho* alkyl aromatic aldehyde (or ketone) gives an intermediate *ortho*-quinodimethane (see Section 3.1.2). Cycloaddition with an activated alkene dienophile then gives the tetrahydronaphthalene product **2**. See K. C. Nicolaou and D. Gray, *Angew. Chem. Int. Ed.*, **40** (2001), 761.

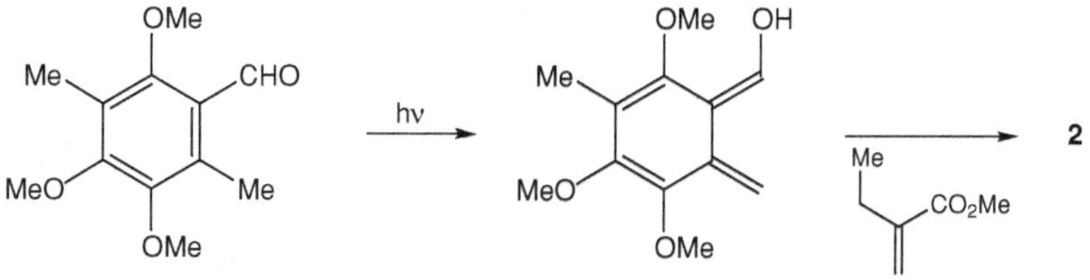

ortho-quinodimethane

4. The diaza-indoline **3** is formed directly on acylation of the amine. Intramolecular cycloaddition of the alkyne onto the tetrazine gives an intermediate bridged bicyclic compound. This intermediate undergoes retro-Diels–Alder reaction with loss of nitrogen to give the product **3**. The product **4** is formed in a similar manner, with intramolecular Diels–Alder cycloaddition to give an intermediate bridged compound that loses methanol and nitrogen to give the tetracycle **4**. See D. L. Boger and S. E. Wolkenberg, *J. Org. Chem.*, **65** (2000), 9120.

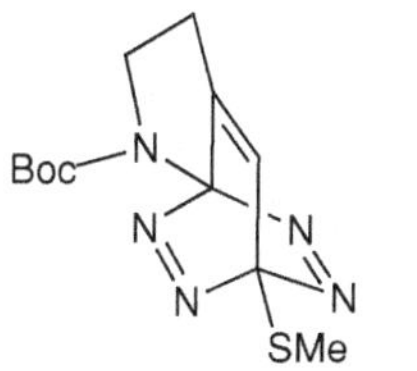

intermediate in formation of **3**

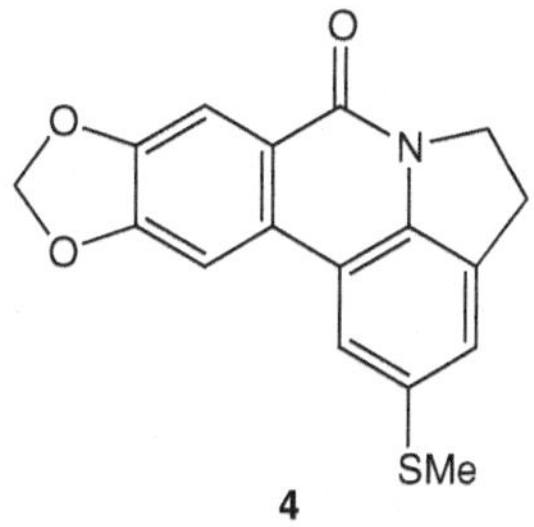

4

5. The cycloadduct **5** is formed by a Diels–Alder reaction of the diene and maleic anhydride. Note the stereoselectivity: the reaction is stereospecific with respect to the dienophile and stereoselective for the *endo* adduct.

   The one-pot procedure will of course be more convenient, but in addition the adduct **5** is formed in higher yield (despite the lower temperature for the cycloaddition reaction). This is probably a result of the presence of the ruthenium complex, which presumably acts as a Lewis acid catalyst. The first step to form the diene is a metathesis reaction (see Section 2.10). See M. Rosillo, L. Casarrubios, G. Domínguez and L. J. Péres-Castells, *Tetrahedron Lett.*, **42** (2001), 7029.

5

6. The zinc-catalysed imino-Diels–Alder reaction is efficient and selective. The major product is the regioisomer in which the imine carbon atom (larger coefficient in the LUMO) interacts with the methylene group of the diene (larger coefficient in the HOMO). The diene approaches from the less hindered face of the imine, as illustrated best by a Newman projection using the Felkin–Anh model – see Section 1.1.5.1, Scheme 1.91). See R. Badorrey, C. Cativiela, M. D. Diaz-de-Villegas and J. A. Galvez, *Tetrahedron*, **55** (1999), 7601. For a review on the imino-Diels–Alder reaction, see P. Buonora, J.-C. Olsen and T. Oh, *Tetrahedron*, **57** (2001), 6099.

6

7. The intermediate **7** is formed by a [4+3] cycloaddition reaction of the 2-oxyallyl cation generated from 1,1,3-trichloroacetone. Cycloaddition occurs onto the electron-rich diene portion of the furan ring. A mixture of regioisomeric dichlorides is formed, but both are reduced with zinc to the same ketone product. See J. C. Lee and J. K. Cha, *J. Am. Chem. Soc.*, **123** (2001), 3243.

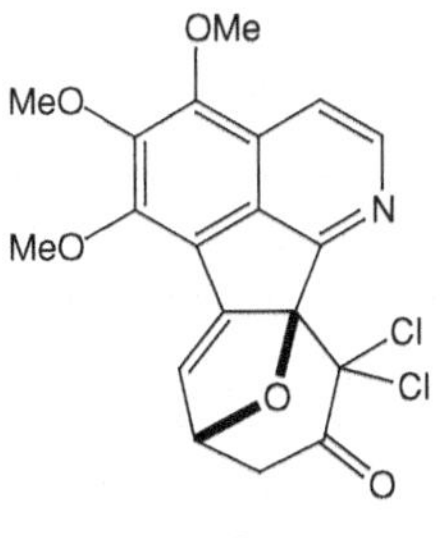

7

8. The isoxazoline **8** is formed by a dipolar cycloaddition reaction between the nitrile oxide formed by oxidation of the oxime. Cycloaddition occurs to give the expected 5-substituted regioisomer of the isoxazoline (see Section 3.4). Cycloaddition also occurs with high stereoselectivity for the Felkin–Anh adduct. See J. W. Bode, N. Fraefel, D. Muri and E. M. Carreira, *Angew. Chem. Int. Ed.*, **40** (2001), 2082.

9. Condensation of the aldehyde **9** and *N*-methyl-glycine gives an intermediate imininum ion which is trapped intramolecularly to give an oxazolidinone. Thermal elimination of CO_2 generates an azomethine ylide that undergoes ready intramolecular dipolar cycloaddition onto the alkene. The *cis*-fused product will have the lower activation energy. See C. J. Lovely and H. Mahmud, *Tetrahedron Lett.*, **40** (1999), 2079.

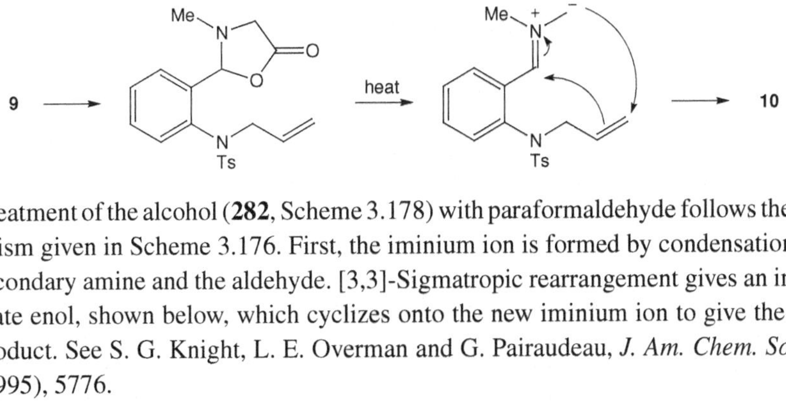

10. Treatment of the alcohol (**282**, Scheme 3.178) with paraformaldehyde follows the mechanism given in Scheme 3.176. First, the iminium ion is formed by condensation of the secondary amine and the aldehyde. [3,3]-Sigmatropic rearrangement gives an intermediate enol, shown below, which cyclizes onto the new iminium ion to give the ketone product. See S. G. Knight, L. E. Overman and G. Pairaudeau, *J. Am. Chem. Soc.*, **117** (1995), 5776.

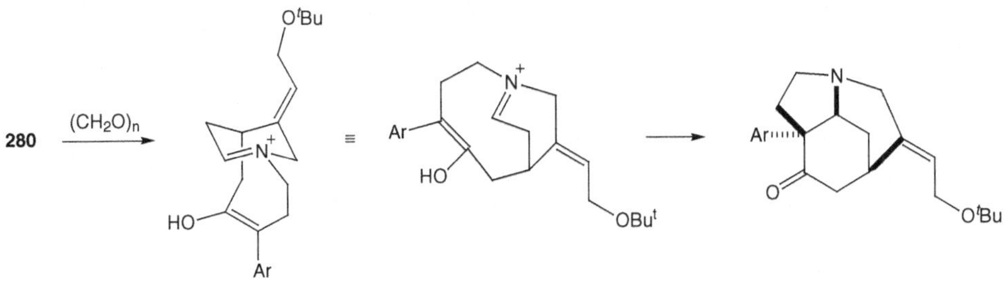

11. Conversion of the alcohol **11** to the ester **12** can be accomplished by using a Claisen [3,3]-sigmatropic rearrangement (see Section 3.6.2). A simple method for the formation of a γ,δ-unsaturated ester is to heat the alcohol with trimethyl orthoacetate MeC(OMe)₃

and an acid catalyst, such as propionic acid, EtCO$_2$H (Johnson–Claisen rearrangement). See G. Stork and S. Raucher, *J. Am. Chem. Soc.*, **98** (1976), 1583.

12. Potassium hydride is a base and abstracts the most acidic proton (on the allylic unit bearing the phenyl group). A [2,3]-sigmatropic rearrangement is followed by an anionic oxy-Cope rearrangement *in situ* to give, after protonation of the resulting enolate, the product **14**. See N. Greeves and W.-M. Lee, *Tetrahedron Lett.*, **38** (1997), 6445.

13. The triene **15** undergoes electrocyclic rearrangement to give a diene intermediate that hydrolyses to the enone **16**. See P. von Zezschwitz, F. Petry and A. de Meijere, *Chem. Eur. J.*, **7** (2001), 4035.

Answers to problems from Chapter 4

1. Decarboxylation of the carboxylic acid **1** can be achieved by conversion to the thiohydroxamic ester and photolysis in the presence of tert-butyl thiol. See Scheme 4.9 and X. Liang, A. Lohse and M. Bols, *J. Org. Chem.*, **65** (2000), 7432.

2. The tributyltin radical adds to the alkenyl epoxide **3** to give an alkoxy radical intermediate. Hydrogen atom abstraction (1,5-H shift) then gives a new radical α- to the benzyloxy group. This radical can cyclize onto the alkene, with concommitant release

of further tributyltin radical, to give the alcohol **4**. See S. Kim, S. Lee and J. S. Koh, *J. Am. Chem. Soc.*, **113** (1991), 5106.

3. Homolysis of the dibenzoyl peroxide gives the benzoyl radical, which is thought to abstract a hydrogen atom from the *N*-hydroxy-phthalimide. The resulting alkoxy radical can abstract a hydrogen atom from the aldehyde to give the acyl radical. Intermolecular addition of the acyl radical to the terminal position of alkene occurs to give the more stable, secondary carbon-centred radical. This species is not capable of abstracting a hydrogen atom from the aldehyde, but can do so from the *N*-hydroxy phthalimide to provide the ketone **5** (and more alkoxy radical to continue the cycle). See S. Tsujimoto, T. Iwahawa, S. Sakaguchi and Y. Ishii, *Chem. Commun.* (2001), 2352.

4. The tin radical abstracts the iodine atom to give the first formed alkenyl radical (not configurationally stable), which cyclizes by a preferential 5-*exo-trig* pathway to give an α-keto radical. Hydrogen atom abstraction gives the ketone **7** but 6-*endo-trig* cyclization gives a tertiary radical (rather than the less stable primary radical that would be formed by 5-*exo-trig* cyclization). Hydrogen atom abstraction gives the ketone **8** but a further radical cyclization (5-*exo-trig*) and subsequent hydrogen atom abstraction gives the ketone **9** (in this case a 6-*endo-trig* cyclization would give a more strained bridged ring system). The final cyclization is more favourable from one of the two diastereomers. See B. P. Haney and D. P. Curran, *J. Org. Chem.*, **65** (2000), 2007.

5. Addition of the phenylthio radical to the alkenyl cyclopropane generates an α-alkoxy radical that can fragment to release benzophenone and an α-keto radical. This radical then adds to the α,β-unsaturated ester to give a new radical that can cyclize onto the allylic sulfide to give the mixture of diastereomeric cyclohexanones **10** (and releasing the phenylthio radical). See K. S. Feldman and A. K. K. Vong, *Tetrahedron Lett.*, **31** (1990), 823.

6. The Wittig alkenylation of the ketone gives a mixture of the two geometrical isomers of the vinyl chloride. Proton abstraction and loss of chloride from both isomers generates the same alkylidene carbene intermediate. The C—H insertion reaction occurs to give the five-membered ring, with insertion into the more reactive methine rather than methyl C—H bond. This process occurs with retention of stereochemistry, providing a single stereoisomer of the product cyclopentene.

alkene product
mixture of *E* and *Z* isomers

alkylidene carbene
intermediate

C–H insertion
product

7. The cyclopropane intermediate results from the rhodium-carbenoid addition to the less-hindered terminal double bond of the diene. This generates the cyclopropane in which both alkenes are *cis* related, which allows the subsequent Cope rearrangement of the divinyl cyclopropane (see Section 3.6.1). The conformation of the boat-shaped transition state for the rearrangement must have the methyl and acetate groups *trans* to one another. See W. R. Cantrell and H. M. L. Davies, *J. Org. Chem.*, **56** (1991), 723.

cyclopropane
intermediate

transition state
leading to cycloheptadiene

8. The lactam product **13** is formed by Wolff rearrangement of the carbene generated from the diazoketone **12** and silver(I), followed by intramolecular trapping of the intermediate ketene by the tosylamine group. See Scheme 4.101 and J. Wang and Y. Hou, *J. Chem. Soc., Perkin Trans. 1* (1998), 1919.

9. Insertion of rhodium into the diazo compound gives an intermediate rhodium carbenoid which is trapped intramolecularly to give the ylide. The regiochemistry of the cycloaddition reaction can be explained by invoking a dipole with a more electron-rich carbon atom α- to the ketone, which interacts with the electron-deficient aldehyde carbon atom. Calculations support this, with the prediction that the α- carbon of the dipole has a larger coefficient in the HOMO and therefore interacts best with the carbon atom of the aldehyde, which will bear a larger coefficient in the LUMO. See A. Padwa, G. E. Fryxell and L. Zhi, *J. Am. Chem. Soc.*, **112** (1990), 3100.

Answers to problems from Chapter 5

1. Regioselectivity: the boron atom adds to the less-hindered, more-electron-rich end of the alkene.

   Stereoselectivity: hydroboration occurs by *syn* addition (four-membered-ring transition state) of the hydrogen and boron atoms. This takes place on the less-hindered face of the alkene, opposite the bulky *gem*-dimethyl group.

   See H. C. Brown, M. C. Desai and P. K. Jadhav, *J. Org. Chem.*, **47** (1982), 5065.

2. Hydroboration of the alkene gives the primary organoborane intermediate. This reacts intramolecularly with the azide to give, after migration of the alkyl group and loss of

nitrogen, the pyrrolidine **3**, R=H. See S. Girard, R. J. Robins, J. Villiéras and J. Lebreton, *Tetrahedron Lett.*, **41** (2000), 9245.

3. Hydroboration and carbonylation of cyclohexene gives an intermediate acylborane, which is reduced by the borohydride reagent. Oxidative work-up gives the aldehyde **4**. See H. C. Brown, J. L. Hubbard and K. Smith, *Synthesis* (1979), 701.

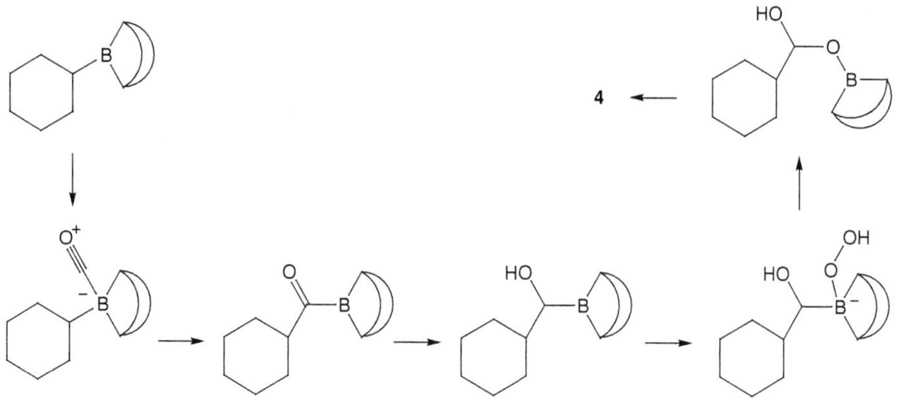

4. For allylic alcohols, use the Sharpless asymmetric epoxidation reaction. Reagents are t-BuOOH, Ti(Oi-Pr)$_4$ and (+)-diethyl (or di-isopropyl) tartrate. Note that the (+)-enantiomer of the ligand is required for the formation of the epoxide **6** (place the alcohol in the *lower right* using the model given in Scheme 5.55).

5. 1,4-Pentadien-3-ol **7** is achiral. However, on complexation with the titanium catalyst and the chiral ligand, the two alkenes become non-equivalent (diastereotopic). The principles of kinetic resolution then come in to effect, with epoxidation of the alkene such that the other alkene substituent does not obstruct the preferred face of oxidation (the lower face as drawn below, with the alcohol in the lower right corner using the (+)-tartrate ligand, model as in Scheme 5.60). Further epoxidation of **8** is slow using the (+)-tartrate ligand. See S. L. Schreiber, T. S. Schreiber and D. B. Smith, *J. Am. Chem. Soc.*, **109** (1987), 1525.

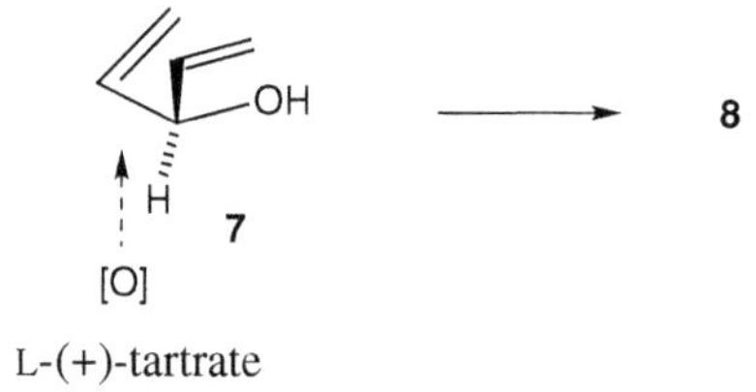

6. Asymmetric epoxidation of *cis*-substituted conjugated alkenes can be achieved efficiently using the Jacobsen–Katsuki conditions (see Section 5.2, Scheme 5.66). For the enantiomer **9**, use the (*S*,*S*)-(salen)Mn(III)Cl catalyst and NaOCl in CH$_2$Cl$_2$ at 4 °C in the presence of an additive such as pyridine *N*-oxide.

7. Dihydroxylation of alkenes can be accomplished conveniently using catalytic OsO$_4$ and the co-oxidant *N*-methylmorpholine *N*-oxide (NMO) in t-BuOH/H$_2$O (Upjohn conditions, see Scheme 5.80). This occurs by *syn* addition of the two hydroxy groups

and therefore, starting from *E*-oct-4-ene, this leads to the diastereomer *syn* **10**. To prepare the diastereomer *anti* **10**, the Prévost reaction may be used (iodine and silver acetate in CCl$_4$), followed by hydrolysis of the diester product (aqueous NaOH) to give the diol.

8. The alkene **11** reacts with electrophiles on the less-hindered (*exo*) face of the double bond. Thus, catalytic osmium tetroxide and NMO (see Scheme 5.80) or KMnO$_4$ provide the *exo-cis*-diol resulting from approach of the osmium from the more accessible face of the molecule. To prepare the isomeric *endo-cis*-diol, the Woodward–Prévost reaction may be used (iodine and silver acetate in the presence of water). In this case, iodine should approach the *exo* face, but subsequent attack on the *exo*-iodonium ion by acetate anion would occur from the opposite (*endo*) face (see Scheme 5.93). Formation and hydrolysis of the cyclic intermediate cation gives the *endo-cis*-diol.

9. Formation of the chiral 1,2-amido-alcohol **12** can be achieved in a single transformation by using the asymmetric amino-hydroxylation reaction (see Section 5.3.3). For the regioisomer **12**, the linker anthraquinone (AQN) rather than the normal phthalazine (PHAL) is required. For the enantiomer **12**, the cinchona alkaloid dihydroquinidine (DHQD) is required. Hence, the reagents and conditions effective for the formation of **12** are:

   BnOCONH$_2$, NaOH, t-BuOCl, [K$_2$OsO$_4$·2H$_2$O], (DHQD)$_2$AQN, *n*-PrOH, H$_2$O.

   See K. C. Nicolaou, S. Natarajan, H. Li, N. F. Jain, R. Hughes, M. E. Solomon, J. M. Ramanjulu, C. N. C. Boddy and M. Takayanagi, *Angew. Chem. Int. Ed.*, **37** (1998), 2708.

10. The structure of the ozonide from the reaction of the allylic alcohol **13** with ozone is shown below. This is the normal product from ozonolysis. Notice that there is no oxidant (such as H$_2$O$_2$) or reductant (such as Me$_2$S) in the formation of the keto-acid **14**. The formation of the product **14** can be explained by fragmentation of the ozonide (see below), which occurs in a similar way to that described for the ozonolysis of α,β-unsaturated carbonyl compounds given in Scheme 5.105. See R. L. Cargill and B. W. Wright, *J. Org. Chem.*, **40** (1975), 120.

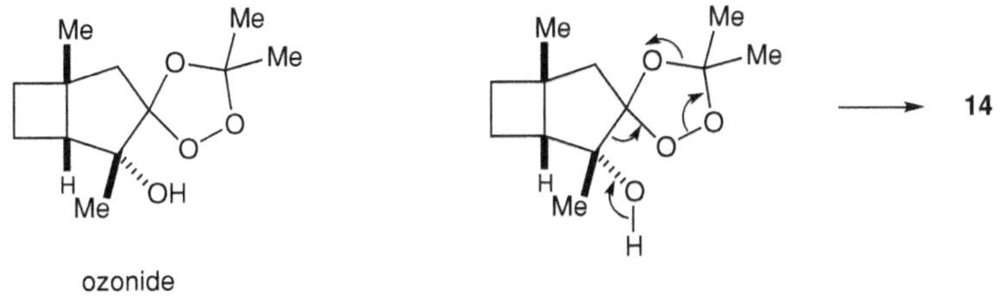

ozonide

11. The alkene **15** can be cleaved by a number of reagents (see Section 5.4). A good method to form carboxylic acids by cleavage of alkenes uses ruthenium tetroxide. Hence, use catalytic RuCl$_3$ with NaIO$_4$ in CCl$_4$, MeCN, H$_2$O (see Scheme 5.109) to obtain the carboxylic acid, from which the ester **16** can be prepared by esterification, for example by using diazomethane in Et$_2$O. See K.-Y. Lee, Y.-H. Kim, M.-S. Park, C.-Y. Oh and W.-H. Ham, *J. Org. Chem.*, **64** (1999), 9450.

12. The cyclopentenone **19** can be prepared by an intramolecular aldol reaction from the diketone **18**. This reaction is best achieved with a base such as KOH in MeOH and heat. The diketone **18** can be prepared by Wacker oxidation of the alkene **17**. Standard conditions for the Wacker oxidation are 10 mol% $PdCl_2$, CuCl, O_2, DMF, H_2O (see Scheme 5.115). The alkene **17** is prepared by allylation of the enamine of cyclohexanone. See J. Tsuji, I. Shimizu and K. Yamamoto, *Tetrahedron Lett.* (1976), 2975.

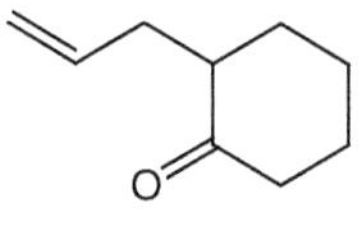

17

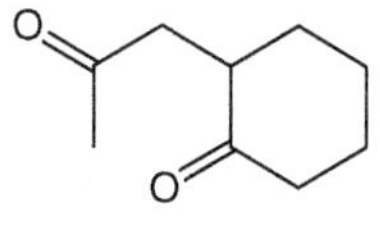

18

Answers to problems from Chapter 6

1. Quinones can be formed by oxidation of phenols (see Section 6.1). To prepare naphthaquinone, use 1-naphthol and an oxidant such as Fremy's salt [$(KSO_3)_2NO$].

2. Reaction (i), of 1-hexene with the sulfur diimido reagent, occurs by an initial ene reaction, followed by a [2,3]-sigmatropic rearrangement. Therefore, formation of the carbon–sulfur bond occurs at the terminal carbon atom and the new carbon–nitrogen bond is then formed at C-3. The trimethyl phosphite cleaves the nitrogen–sulfur bond to give the allylic amine product. Intermediates:

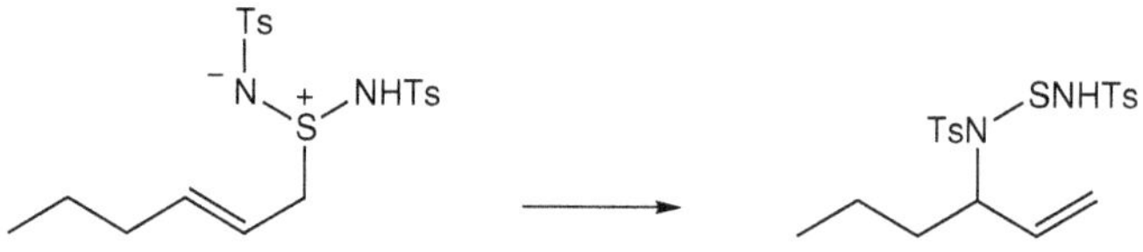

In reaction (ii), 1-hexene reacts with diethyl azodicarboxylate by an ene reaction to give directly the new carbon–nitrogen bond at the terminal carbon atom. The lithium metal in liquid ammonia cleaves the nitrogen–nitrogen bond to give the allylic amine product. For a review on allylic amination, see M. Johannsen and K. A. Jørgensen, *Chem. Rev.*, **98** (1998), 1689. Intermediate:

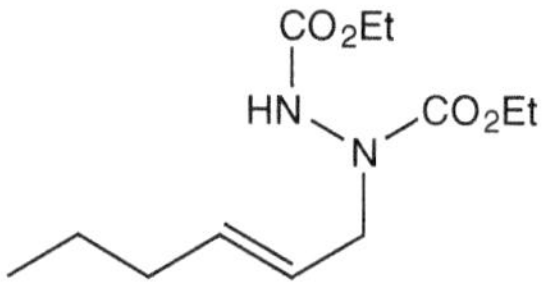

3. The diol **1** must first be protected/activated and this can be achieved by selective acylation (or mesylation) of the less hindered primary alcohol group. For example, treatment with a carboxylic acid, dicyclohexylcarbodiimide (DCC) and 4-dimethylaminopyridine gives the required ester. Oxidation of the secondary alcohol can be achieved

by using one of a number of oxidizing agents (see Section 6.2), such as pyridinium dichromate (PDC). Formation of the ketone was followed by β-elimination on alumina to give the α,β-unsaturated ketone **2**. See H. Toshima, H. Oikawa, T. Toyomasu and T. Sassa, *Tetrahedron*, **56** (2000), 8443.

4. Formation of the aldehyde **4** occurs by the Swern oxidation – see Section 6.2, Scheme 6.28. The base Et$_3$N deprotonates the intermediate alkoxysulfonium salt to promote fragmentation to the aldehyde. Formation of the chloride **5** occurs on warming the alkoxysulfonium salt in the absence of a base, with S$_N$2 displacement of DMSO by chloride ion. See N. Kato, K. Nakanishi and H. Takeshita, *Bull. Chem. Soc. Jpn*, **59** (1986), 1109.

5. Selective oxidation of the benzylic alcohol of the diol **6** to give the ketol **7** is possible by using the oxidant manganese dioxide in a neutral solvent such as acetone at room temperature. Alternatively, silver carbonate on celite heated in a solvent such as acetone or benzene is effective (see Section 6.2).

6. Allylic alcohols are oxidized readily by silver carbonate. Cyclization and further oxidation then occur to give the lactone **9**. See M. Fetizon, M. Golfier and J.-M. Louis, *Tetrahedron*, **31** (1975), 171.

9

7. A good method for the direct conversion of alcohols to carboxylic acid uses 2,2,6,6-tetramethylpiperidin-1-oxyl (TEMPO) **51**, in conjunction with the co-oxidant sodium chlorite (NaClO$_2$) and sodium hypochlorite (NaOCl) as a catalyst. See M. Zhao, J. Li, E. Mano, Z. Song, D. M. Tschaen, E. J. J. Grabowski and P. J. Reider, *J. Org. Chem.*, **64** (1999), 2564.

8. Several possible methods can be used for the conversion of carbonyl compounds to α,β-unsaturated carbonyl compounds. For aldehydes, such as undecanal, it is normally best to prepare the trimethylsilyl enol ether (using Me$_3$SiCl and Et$_3$N in DMF) and then use 10 mol% Pd(OAc)$_2$ in DMSO under an atmosphere of oxygen. See R. C. Larock, T. R. Hightower, G. A. Kraus, P. Hahn and D. Zheng, *Tetrahedron Lett.*, **36** (1995), 2423. Attempts to use LDA (then PhSeBr and oxidative elimination) will result in reduction of the aldehyde.

9. Peroxycarboxylic acids oxidize ketones to esters by the Baeyer–Villiger reaction (see Section 6.3). The more electron-rich substituent migrates to the electron-deficient oxygen

atom. Therefore, the ester **12** results from migration of the *p*-methoxyphenyl group, whereas the ester **13** results from migration of the phenyl (and not the *p*-nitrophenyl group).

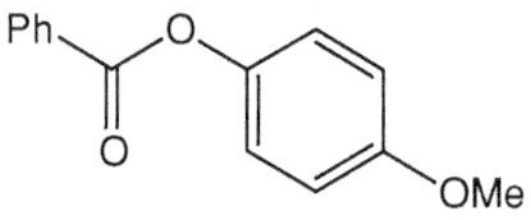

12 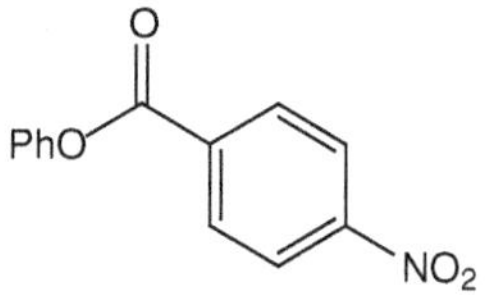   **13**

10. The phenol **15** can be prepared from the aldehyde **14** by the Dakin reaction using H_2O_2 and NaOH (see Section 6.3).

Answers to problems from Chapter 7

1. Heterogeneous hydrogenation, especially with palladium catalysis, is not normally selective and, in addition to hydrogenation of alkenes, hydrogenolysis of benzyl ethers occurs readily (although aromatic heterocycles are not normally reduced under these conditions). Therefore in this case the product is as shown below:

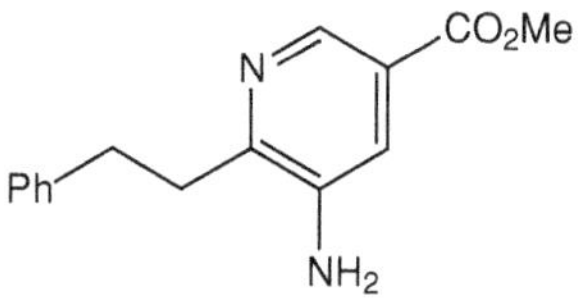

2. Hydrogenation of the two alkenes and hydrogenolysis of the benzyl ether (followed by decarboxylation) gives the intermediate amino-ketone shown below. Subsequent cyclization to the imine/enamine is followed by further reduction on the less-hindered (outer, convex face) to give pumiliotoxin C. See L. E. Overman and P. J. Jessup, *J. Am. Chem. Soc.*, **100** (1978), 5179.

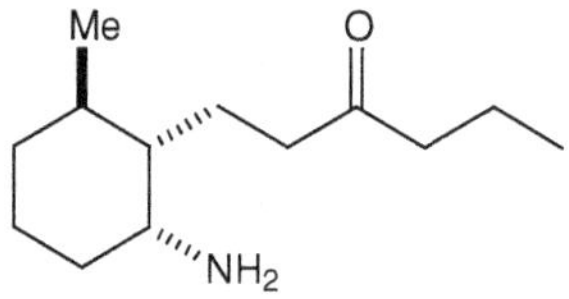

3. Homogeneous hydrogenation with Wilkinson's catalyst is selective for the less-hindered alkene. Therefore in this case the product is as shown below. See M. Brown and L. W. Piszkiewicz, *J. Org. Chem.*, **32** (1967), 2013.

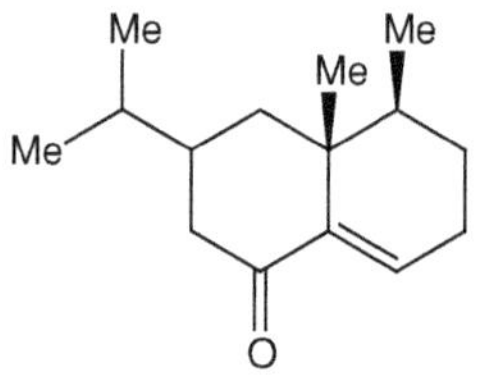

4. Preferential hydrogenation on one side of an alkene can occur either for steric reasons or if a functional group can co-ordinate to the metal and direct the reduction. Alcohols and carbonyl groups are particularly efficient in this regard and hydrogenation of all three alkenes in this substrate occurs from the same side as the carboxylic amide group, thereby leading to the diastereomer shown. See A. G. Schultz and P. J. McCloskey, *J. Org. Chem.*, **50** (1985), 5905.

5. Zinc metal in acid promotes reduction of organic compounds, such as α-substituted ketones (see Section 7.2). Reduction gives the product shown below. The isolated bromine atom remains unaffected. See K. M. Baker and B. R. Davis, *Tetrahedron*, **24** (1968), 1655.

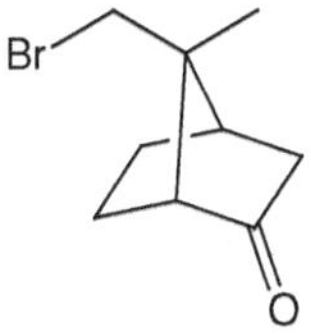

6. The structures of the resonance forms of the intermediate radical anion are shown below. Addition of a second electron gives a dianion, which must be protonated by the intramolecular hydroxyl group, thereby leading exclusively to the stereoisomer shown (with reduction *cis* to the OH group). See C. Iwata, K. Miyashita, Y. Koga, Y. Shinoo, M. Yamada and T. Tanaka, *Chem. Pharm. Bull.*, **31** (1983), 2308.

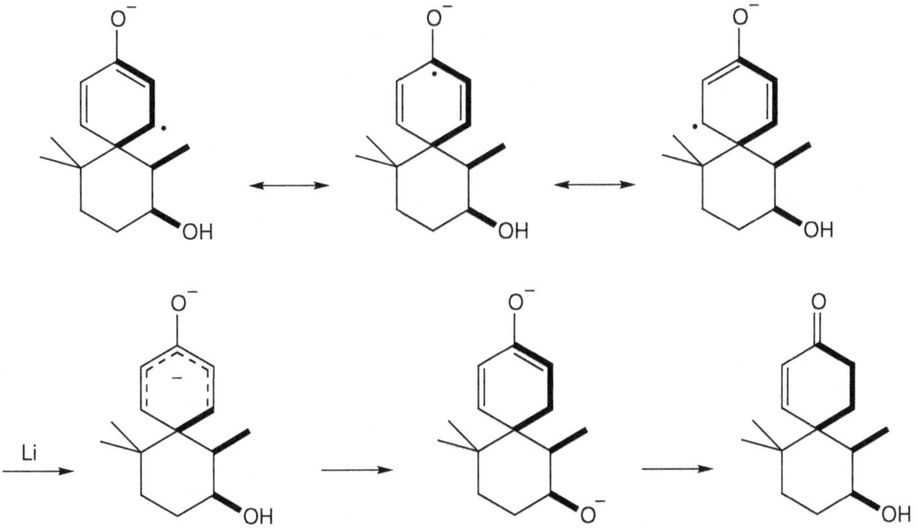

7. The conversion of dicarbonyl compounds to 1,2-diols can be effected by using low valent metals in the absence of a proton source. Cyclohexane-1,2-diol can be formed in high yield by using, for example, $TiCl_3(DME)_2$ and zinc–copper couple (see Sections 2.9 and 7.2). See J. E. McMurry and J. G. Rico, *Tetrahedron Lett.*, **30** (1989), 1169.

8. Dissolving metal conditions are effective for reductive cleavage of C−S and N−S bonds. Hence, removal of both the phenylthio and the tosyl groups occurs. The carbanion

generated from this process is very reactive and picks up a proton from the ammonia solution. However, the less-reactive amide anion ($R_2N^-Na^+$) remains unprotonated and can be alkylated *in situ* by addition of iodomethane. See P. Magnus, J. Lacour, I. Coldham, B. Mugrage and W. B. Bauta, *Tetrahedron*, **51** (1995), 11 087.

9. Hydride transfer from the LDA can take place, similar to the Meerwein–Pondorff–Verley reduction (see Section 7.3). A mechanism is given below.

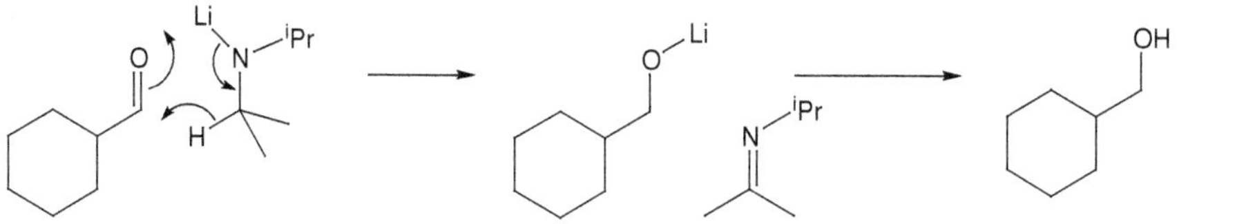

10. A mild reducing agent is required for the selective reduction of aldehydes in the presence of other functional groups that could react, such as ketones. A good reagent is lithium tri-t-butoxyaluminium hydride, $LiAlH(Ot-Bu)_3$, but other reagents such as sodium (or tetra-*n*-butylammonium) triacetoxyborohydride should be effective. See T. Harayama, M. Takatane and Y. Inubushi, *Chem. Pharm. Bull.*, **27** (1979), 726.

11. Asymmetric reduction of the ketone gives the chiral alcohol shown below. The larger (conjugated) side of the ketone sits in the less-hindered position (R_L as described in Section 7.3.7, Scheme 7.100). See Y. Kita, K. Higuchi, Y. Yoshida, K. Iio, S. Kitagaki, K. Ueda, S. Akai and H. Fujioka, *J. Am. Chem. Soc.*, **123** (2001), 3214.

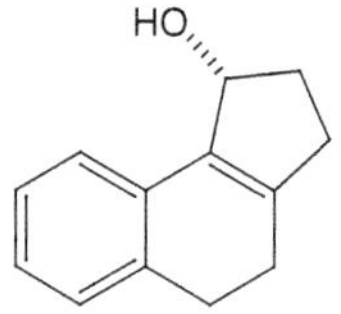

12. Reduction or addition of organometallic reagents to tosyl hydrazones promotes loss of *p*-toluenesulfinic acid and subsequent loss of nitrogen. When a leaving group is present in the α-position, then an alkene is formed (compare with Schemes 7.89 and 7.90). See S. Chandrasekhar, M. Takhi and J. S. Yadav, *Tetrahedron Lett.*, **36** (1995), 307.

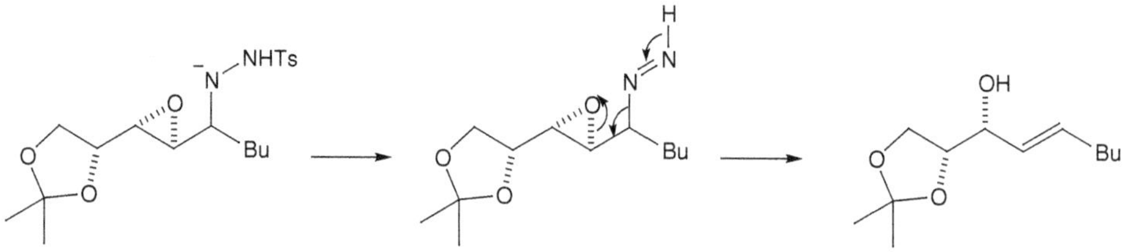

Index

Printed in the United States
By Bookmasters